**CAMBRIDGE MONOGRAPHS ON APPLIED AND COMPUTATIONAL MATHEMATICS**

**Series Editors**
M. ABLOWITZ, S. DAVIS, J. HINCH,
A. ISERLES, J. OCKENDON, P. OLVER

# 20    Greedy Approximation

The *Cambridge Monographs on Applied and Computational Mathematics* series reflects the crucial role of mathematical and computational techniques in contemporary science. The series publishes expositions on all aspects of applicable and numerical mathematics, with an emphasis on new developments in this fast-moving area of research.

State-of-the-art methods and algorithms as well as modern mathematical descriptions of physical and mechanical ideas are presented in a manner suited to graduate research students and professionals alike. Sound pedagogical presentation is a prerequisite. It is intended that books in the series will serve to inform a new generation of researchers.

A complete list of books in the series can be found at
www.cambridge.org/mathematics
Recent titles include the following:

8. Schwarz–Christoffel mapping, *Tobin A. Driscoll & Lloyd N. Trefethen*

9. High-order methods for incompressible fluid flow, *M. O. Deville, P. F. Fischer & E. H. Mund*

10. Practical extrapolation methods, *Avram Sidi*

11. Generalized Riemann problems in computational fluid dynamics, *Matania Ben-Artzi & Joseph Falcovitz*

12. Radial basis functions, *Martin D. Buhmann*

13. Iterative Krylov methods for large linear systems, *Henk van der Vorst*

14. Simulating Hamiltonian dynamics, *Benedict Leimkuhler & Sebastian Reich*

15. Collocation methods for Volterra integral and related functional differential equations, *Hermann Brunner*

16. Topology for computing, *Afra J. Zomorodian*

17. Scattered data approximation, *Holger Wendland*

18. Modern computer arithmetic, *Richard Brent & Paul Zimmermann*

19. Matrix preconditioning techniques and applications, *Ke Chen*

20. Greedy approximation, *Vladimir Temlyakov*

21. Spectral methods for time-dependent problems, *Jan Hesthaven, Sigal Gottlieb & David Gottlieb*

22. The mathematical foundations of mixing, *Rob Sturman, Julio M. Ottino & Stephen Wiggins*

23. Curve and surface reconstruction, *Tamal K. Dey*

24. Learning theory, *Felipe Cucker & Ding Xuan Zhou*

25. Algebraic geometry and statistical learning theory, *Sumio Watanabe*

26. A practical guide to the invariant calculus, *Elizabeth Louise Mansfield*

# Greedy Approximation

VLADIMIR TEMLYAKOV
*University of South Carolina*

# CAMBRIDGE
## UNIVERSITY PRESS

University Printing House, Cambridge CB2 8BS, United Kingdom

One Liberty Plaza, 20th Floor, New York, NY 10006, USA

477 Williamstown Road, Port Melbourne, VIC 3207, Australia

314-321, 3rd Floor, Plot 3, Splendor Forum, Jasola District Centre, New Delhi - 110025, India

79 Anson Road, #06-04/06, Singapore 079906

Cambridge University Press is part of the University of Cambridge.

It furthers the University's mission by disseminating knowledge in the pursuit of education, learning and research at the highest international levels of excellence.

www.cambridge.org
Information on this title: www.cambridge.org/9781107003378

First published 2011

*A catalogue record for this publication is available from the British Library*

*Library of Congress Cataloging in Publication data*
Temlyakov, Vladimir, 1953–
Greedy approximation / Vladimir Temlyakov.
p.   cm. – (Cambridge monographs on applied and computational mathematics ; 20)
ISBN 978-1-107-00337-8 (hardback)
1. Approximation theory.   I. Title.   II. Series.
QA221.T455   2011
518´.5–dc23
2011025053

ISBN  978-1-107-00337-8  Hardback

# Contents

# Preface

From the beginning of time, human beings have been trying to replace complicated with simpler things. From ancient shamans working magic upon clay figures to heal the sick, to Renaissance artists representing God Almighty as a nude painted onto a ceiling, the fundamental problem of representation of the only partially representable continues in contemporary applied mathematics. A generic problem of mathematical and numerical analysis is to represent a given function approximately. It is a classical problem that goes back to the first results on Taylor's and Fourier's expansions of a function.

The first step in solving the representation problem is to choose a representation system. Traditionally, a representation system has natural features such as minimality, orthogonality, simple structure and nice computational characteristics. The most typical representation systems are the trigonometric system $\{e^{ikx}\}$, the algebraic system $\{x^k\}$, the spline system, the wavelet system and their multivariate versions. In general we may speak of a basis $\Psi = \{\psi_k\}_{k=1}^{\infty}$ in a Banach space $X$.

The second step in solving the representation problem is to choose a form of an approximant that is built on the base of the chosen representation system $\Psi$. In a classical way that was used for centuries, an approximant $a_m$ is a polynomial with respect to $\Psi$:

$$a_m := \sum_{k=1}^{m} c_k \psi_k. \tag{1}$$

The complexity of the approximant $a_m$ is characterized by the order $m$ of the polynomial. It is well known in approximation theory that approximation by polynomials is closely related to the smoothness properties of the function being approximated. Approximation of this type is referred to as *linear approximation theory* because, for a fixed $m$, approximants come from a linear subspace spanned by $\psi_1, \ldots, \psi_m$.

It is understood in numerical analysis and approximation theory that in many problems from signal/image processing it is more beneficial to use an $m$-term approximant with respect to $\Psi$ than a polynomial of order $m$. This means that for $f \in X$ we look for an approximant of the form

$$a_m(f) := \sum_{k \in \Lambda(f)} c_k \psi_k, \tag{2}$$

where $\Lambda(f)$ is a set of $m$ indices which is determined by $f$. The complexity of this approximant is characterized by the cardinality $|\Lambda(f)| = m$ of $\Lambda(f)$. Approximation of this type is referred to as *nonlinear approximation theory* because, for a fixed $m$, approximants $a_m(f)$ come from different linear subspaces spanned by $\psi_k, k \in \Lambda(f)$, which depend on $f$. The cardinality $|\Lambda(f)|$ is a fundamental characteristic of $a_m(f)$ called *sparsity* of $a_m(f)$ with respect to $\Psi$. It is now well understood that we need to study nonlinear sparse representations in order to increase significantly our ability to process (compress, denoise, etc.) large data sets. Sparse representations of a function are not only a powerful analytic tool, but also they are utilized in many applications in image/signal processing and numerical computation.

The third step in solving the representation problem is to choose a method of construction of an approximant of desired form. Let us begin with the linear theory. For example, the approximation method that picks the polynomial of degree $m$ with respect to $\Psi$ of best approximation of $f$ in $X$ as an approximant is an optimal method of approximation by polynomials of degree $m$. However, such an obvious optimal method of approximation may not be good from the point of view of practical implementation. Standard methods of approximation that are more practical than the above one are linear methods of approximation, in particular partial sums of the corresponding expansion of $f$ with respect to the basis $\Psi$. Many books (see, for example, DeVore and Lorenz (1993)) discuss this classical topic of approximation theory.

An implementation of the third step in the nonlinear setting is not straightforward. It is clear that an analog of the best polynomial approximant of order $m$ is the best $m$-term approximant. Nonlinearity brings about complications even at this stage. The existence of a best approximant from a finite dimensional subspace is well known; the existence of a best $m$-term approximant is a difficult problem. We discuss this problem in Chapter 1. Next, what are nonlinear analogs of, say, partial sums? We answer this question in Chapter 1.

It turns out that greedy approximants are natural substitutes for the partial sums.

We specify not only a form of an approximant, but also choose a specific method of approximation (for instance, the one that is known to be good in practical implementations). Now, we have a precise mathematical problem of studying the efficiency of our specific method of approximation. We discuss this problem in detail here. It turns out that a convenient and flexible way of measuring the efficiency of a specific approximation method is to prove the corresponding Lebesgue-type inequalities. We would like this method to work for all functions. Therefore, it should converge at least for each $f \in X$. Convergence is a fundamental theoretical problem. In this book we thoroughly discuss the problem of convergence of greedy algorithms.

The fundamental question of nonlinear approximation is how to devise good constructive methods (algorithms) of approximation. This problem has two levels of nonlinearity. The first level of nonlinearity (discussed above) is $m$-term approximation with regard to bases. In this problem one can use the unique function expansion with regard to a given basis to build an approximant. Nonlinearity comes in when we look for $m$-term approximants with terms (i.e. basis elements) that are allowed to depend on a given function. We discuss $m$-term approximation with regard to bases in detail in Chapter 1. On the second level of nonlinearity, we replace a basis by a more general system, which is not necessarily minimal (for example, redundant system, or dictionary). This setting is much more complicated than the first one (bases case); however, there is a solid justification of importance of redundant systems in both theoretical questions and in practical applications (see, for example, Donoho (2001), Huber (1985), Schmidt (1906)). In Chapters 2 and 6 we discuss approximation by linear combinations of elements that are taken from a redundant (overcomplete) system of elements. We briefly discuss the question: Why do we need redundant systems? Answering this question, we first of all mention three classical redundant systems that are used in different areas of mathematics.

Perhaps the first example of $m$-term approximation with regard to redundant dictionary was discussed by Schmidt (1906), who considered the approximation of functions $f(x, y)$ of two variables by bilinear forms $\sum_{i=1}^{m} u_i(x)v_i(y)$ in $L_2([0, 1]^2)$. This problem is closely connected with properties of the integral operator $J_f(g) := \int_0^1 f(x, y)g(y)dy$ with kernel $f(x, y)$.

Another example, which hails from statistics, is the projection pursuit regression problem. In the language of function theory, the problem is to approximate in $L_2(\Omega)$, where $\Omega \subset \mathbb{R}^d$ is a bounded domain, a given function

$f \in L_2(\Omega)$ by a sum of ridge functions, i.e. by $\sum_{j=1}^{m} r_j(\omega_j \cdot x)$, $x, \omega_j \in \mathbb{R}^d$, $j = 1, \ldots, m$, where $r_j$, $j = 1, \ldots, m$, are univariate functions.

The third example is from signal processing. In signal processing the most popular means of approximation are wavelets and the system of Gabor functions $\{g_{a,b}(x-c), g_{a,b}(x) := e^{iax}e^{-bx^2}, \quad a, c \in \mathbb{R}, b \in \mathbb{R}_+\}$. The Gabor system gives more flexibility in constructing an approximant, but it is a redundant (not minimal) system. It also seems natural (see the discussion in Donoho (2001)) to use redundant systems in modeling analyzing elements for the visual system.

Thus, in order to address the contemporary needs of approximation theory and computational mathematics, a very general model of approximation with regard to a redundant system (dictionary) has been considered in many recent papers. As an example of such a model, we choose a Banach space $X$ with elements as target functions and an arbitrary system $\mathcal{D}$ of elements of this space such that the closure of span $\mathcal{D}$ coincides with $X$ as an approximating system. We would like to have an algorithm of constructing $m$-term approximants that adds at each step only one new element from $\mathcal{D}$ and keeps elements of $\mathcal{D}$ obtained at the previous steps. This requirement is an analog of *on-line* computation that is very desirable in practical algorithms. Clearly, we are looking for good algorithms which converge for each target function. It is not obvious that such an algorithm exists in a setting at the above level of generality ($X, \mathcal{D}$ are arbitrary).

The fundamental question is how to construct good methods (algorithms) of approximation. Recent results have established that greedy type algorithms are suitable methods of nonlinear approximation in both $m$-term approximation with regard to bases and $m$-term approximation with regard to redundant systems. It turns out that there is one fundamental principle that allows us to build good algorithms both for arbitrary redundant systems and for very simple well structured bases like the Haar basis. This principle is the use of a greedy step in searching for a new element to be added to a given $m$-term approximant. By a *greedy step*, we mean one which maximizes a certain functional determined by information from the previous steps of the algorithm. We obtain different types of greedy algorithms by varying the above-mentioned functional and also by using different ways of constructing (choosing coefficients of the linear combination) the $m$-term approximant from the already found $m$ elements of the dictionary. In Chapters 2 and 6 we present different greedy type algorithms beginning with a very simple and very natural Pure Greedy Algorithm in a Hilbert space and ending with its rather complicated modifications in a Banach space. Different modifications aim to make the corresponding greedy algorithms more ready for practical implementation. We discuss this issue in detail in Chapters 2 and 6.

It is known that in many numerical problems users are satisfied with a Hilbert space setting and do not consider a more general setting in a Banach space. We now make one remark that justifies our interest in Banach spaces. The first argument is an a-priori argument that the spaces $L_p$ are very natural and should be studied along with the $L_2$ space. The second argument is an a-posteriori argument.

The study of greedy approximation in Banach spaces has revealed that the characteristic of a Banach space $X$ that governs the behavior of greedy approximation is the *modulus of smoothness* $\rho(u)$ of $X$. It is known that the spaces $L_p$, $2 \le p < \infty$, have modulo of smoothness of the same order: $u^2$. Thus, many results that are known for the Hilbert space $L_2$ and proved using some special structure of a Hilbert space can be generalized to Banach spaces $L_p$, $2 \le p < \infty$. The new proofs use only the geometry of the unit sphere of the space expressed in the form $\rho(u) \le \gamma u^2$.

The theory of greedy approximation is developing rapidly, and results are spread over hundreds of papers by different authors. There are several surveys that discuss greedy approximation (see DeVore (1998), Konyagin and Temlyakov (2002), Temlyakov (2003a), Temlyakov (2006b), Temlyakov (2008b), Wojtaszczyk (2002a)). This is the first book on greedy approximation. This book is an extension of Temlyakov (2008b). The book provides a systematic presentation of fundamental results in greedy approximation. It also contains an introduction to two hot topics in numerical mathematics: learning theory and compressed sensing. This book possesses features of both a survey paper and a textbook. The majority of results are given with proofs. However, some important results with technically involved proofs are presented without proof. We included proofs of the most important and typical results; and we tried to include those proofs which demonstrate different ideas and are based on different techniques. In this sense the book has a feature of a survey – it tries to cover broad material. On the other hand, we limit ourselves to a systematic treatment of a specific topic rather than trying to give an overview of all related topics. In this sense the book is close to a textbook. There are many papers on theoretical and computational aspects of greedy approximation, learning theory and compressed sensing. We have chosen to cover the mathematical foundations of greedy approximation, learning theory and compressed sensing.

The book is addressed to researchers working in numerical mathematics, analysis, functional analysis and statistics. It quickly takes the reader from classical results to the frontier of the unknown, but is written at the level of a graduate course and does not require a broad background in order to understand the topics. Graduate students working in different areas of numerical mathematics and analysis may find it useful to learn not only greedy approximation theory, but also theoretical foundations of learning theory and compressed sensing. The author taught three graduate courses, Greedy Approximation, Learning Theory and Compressed Sensing, based on the material of the book, at the University of South Carolina. All three courses were very well accepted by students. The book might be used for designing different graduate courses. It contains a number of important open problems which may assist in uncovering topics for dissertations and research papers.

We use $C$, $C(p, d)$, $C_{p,d}$, *etc.*, to denote various constants, the indices indicating dependence on other parameters. We use the following symbols for brevity. For two non-negative sequences $a = \{a_n\}_{n=1}^{\infty}$ and $b = \{b_n\}_{n=1}^{\infty}$, the relation, or order inequality, $a_n \ll b_n$ means that there is a number $C(a, b)$ such that, for all $n$, we have $a_n \leq C(a, c)b_n$; and the relation $a_n \asymp b_n$ means that $a_n \ll b_n$ and $b_n \ll a_n$. Other notation is defined in the text.

**Acknowledgements** I am grateful to Jessica L. Nelson for help in the preparation of the book.

# 1

## Greedy approximation with regard to bases

### 1.1 Introduction

It is well known that in many problems it is very convenient to represent a function by a series with regard to a given system of functions. For example, in 1807 Fourier suggested representing a $2\pi$-periodic function by its series (known as the Fourier series) with respect to the trigonometric system. A very important feature of the trigonometric system that made it attractive for the representation of periodic functions is orthogonality. For an orthonormal system $\mathcal{B} := \{b_n\}_{n=1}^{\infty}$ of a Hilbert space $H$ with an inner product $\langle \cdot, \cdot \rangle$, one can construct a Fourier series of an element $f$ in the following way:

$$f \sim \sum_{n=1}^{\infty} \langle f, b_n \rangle b_n. \tag{1.1}$$

If the system $\mathcal{B}$ is a basis for $H$, then the series in (1.1) converges to $f$ in $H$ and (1.1) provides a unique representation

$$f = \sum_{n=1}^{\infty} \langle f, b_n \rangle b_n \tag{1.2}$$

of $f$ with respect to $\mathcal{B}$. This representation has nice approximative properties. By Parseval's identity,

$$\|f\|^2 = \sum_{n=1}^{\infty} |\langle f, b_n \rangle|^2, \tag{1.3}$$

we obtain a convenient way to calculate, or estimate, the norm $\|f\|$.

It is known that the partial sums

$$S_m(f, \mathcal{B}) := \sum_{n=1}^{m} \langle f, b_n \rangle b_n \tag{1.4}$$

1

provide the best approximation; that is, defining

$$E_m(f, \mathcal{B}) := \inf_{\{c_n\}} \left\| f - \sum_{n=1}^{m} c_n b_n \right\| \tag{1.5}$$

to be the distance of $f$ from the span$\{b_1, \ldots, b_m\}$, we have

$$\| f - S_m(f, \mathcal{B}) \| = E_m(f, \mathcal{B}). \tag{1.6}$$

Identities (1.3) and (1.6) are fundamental properties of Hilbert spaces and their orthonormal bases. These properties make the theory of approximation in $H$ from the span$\{b_1, \ldots, b_m\}$, or linear approximation theory, simple and convenient.

The situation becomes more complicated when we replace a Hilbert space $H$ by a Banach space $X$. In a Banach space $X$ we consider a Schauder basis $\Psi$ instead of an orthonormal basis $\mathcal{B}$ in $H$. In Section 1.2 we discuss Schauder bases in detail. If $\Psi := \{\psi_n\}_{n=1}^{\infty}$ is a Schauder basis for $X$, then for any $f \in X$ there exists a unique representation

$$f = \sum_{n=1}^{\infty} c_n(f, \Psi) \psi_n$$

that converges in $X$.

Theorem 1.3 from Section 1.2 states that the partial sum operators $S_m$, defined by

$$S_m(f, \Psi) := \sum_{n=1}^{m} c_n(f, \Psi) \psi_n,$$

are uniformly bounded operators from $X$ to $X$. In other words, there exists a constant $B$ such that, for any $f \in X$ and any $m$, we have

$$\| S_m(f, \Psi) \| \le B \| f \|.$$

This inequality implies an analog of (1.6): for any $f \in X$,

$$\| f - S_m(f, \Psi) \| \le (B + 1) E_m(f, \Psi), \tag{1.7}$$

where

$$E_m(f, \Psi) := \inf_{\{c_n\}} \left\| f - \sum_{n=1}^{m} c_n \psi_n \right\|.$$

Inequality (1.7) shows that the $S_m(f, \Psi)$ provides near-best approximation from span$\{\psi_1, \ldots, \psi_m\}$. Thus, if we are satisfied with near-best approximation instead of best approximation, then the linear approximation theory with respect to Schauder bases becomes simple and convenient. The partial sums $S_m(\cdot, \Psi)$ provide near-best approximation for any individual element of $X$.

Motivated by computational issues, researchers became interested in nonlinear approximation with regard to a given system instead of linear approximation. For example, in the case of representation (1.2) in a Hilbert space, one can take an approximant of the form

$$S_\Lambda(f, \mathcal{B}) := \sum_{n \in \Lambda} \langle f, b_n \rangle b_n, \quad |\Lambda| = m,$$

instead of an approximant $S_m(f, \mathcal{B})$ from an $m$-dimensional linear subspace. Then the two approximants $S_m(f, \mathcal{B})$ and $S_\Lambda(f, \mathcal{B})$ have the same sparsity: both are linear combinations of $m$ basis elements. However, we can achieve a better approximation error with $S_\Lambda(f, \mathcal{B})$ than with $S_m(f, \mathcal{B})$ if we choose $\Lambda$ correctly. In the case of a Hilbert space and an orthonormal basis $\mathcal{B}$, an optimal choice $\Lambda_m$ of $\Lambda$ is obvious: $\Lambda_m$ is a set of $m$ indices with the biggest (in absolute value) coefficients $\langle f, b_n \rangle$. Then, by Parseval's identity (1.3), we obtain

$$\|f - S_{\Lambda_m}(f, \mathcal{B})\| \leq \|f - S_m(f, \mathcal{B})\|.$$

Also, it is clear that the $S_{\Lambda_m}(f, \mathcal{B})$ realizes the best $m$-term approximation of $f$ with regard to $\mathcal{B}$,

$$\|f - S_{\Lambda_m}(f, \mathcal{B})\| = \sigma_m(f, \mathcal{B}) := \inf_{\Lambda : |\Lambda| = m} \inf_{\{c_n\}} \|f - \sum_{n \in \Lambda} c_n b_n\|. \qquad (1.8)$$

The approximant $S_{\Lambda_m}(f, \mathcal{B})$ can be obtained as a realization of $m$ iterations of the *greedy approximation step*. For a given $f \in H$ we choose at a *greedy step* an index $n_1$ with the biggest $|\langle f, b_{n_1} \rangle|$. At a greedy approximation step we build a new element $f_1 := f - \langle f, b_{n_1} \rangle b_{n_1}$.

The identity (1.8) shows that the greedy approximation works perfectly in nonlinear approximation in a Hilbert space with regard to orthonormal basis $\mathcal{B}$.

This chapter is devoted to a systematic study of greedy approximation in Banach spaces. In Section 1.2 we discuss the following natural question. Equation (1.8) proves the existence of the best $m$-term approximant in a Hilbert space with respect to an orthonormal basis. Further, we discuss the existence of the best $m$-term approximant in a Banach space with respect to a Schauder basis. That discussion illustrates that the situation regarding existence theorems is much more complex in Banach spaces than in Hilbert spaces. We also give some sufficient conditions on a Schauder basis that guarantee the existence of the best $m$-term approximant. However, the problem is far from being completely solved.

The central issue of this chapter is the question: Which bases are suitable for greedy approximation? Greedy approximation with regard to a Schauder basis is defined in a similar way to the greedy approximation with regard to an orthonormal basis (see above). The greedy algorithm picks the terms with the biggest (in absolute value) coefficients from the expansion

$$f = \sum_{n=1}^{\infty} c_n(f, \Psi)\psi_n \qquad (1.9)$$

and gives a *greedy approximant*

$$G_m(f, \Psi) := S_{\Lambda_m}(f, \Psi) := \sum_{n \in \Lambda_m} c_n(f, \Psi)\psi_n.$$

Here, $\Lambda_m$ is such that $|\Lambda_m| = m$ and

$$\min_{n \in \Lambda_m} |c_n(f, \Psi)| \geq \max_{n \notin \Lambda_m} |c_n(f, \Psi)|.$$

We note that we need some restrictions on the basis $\Psi$ (see Sections 1.3 and 1.4 for a detailed discussion) in order to be able to run the greedy algorithm for each $f \in X$. It is sufficient to assume that $\Psi$ is normalized. We make this assumption for our further discussion in the Introduction. In some later sections we continue to use the normalization assumption; in others, we do not.

An application of the greedy algorithm can also be seen as a rearrangement of the series from (1.9) in a special way: according to the size of coefficients. Let

$$|c_{n_1}(f, \Psi)| \geq |c_{n_2}(f, \Psi)| \geq \dots .$$

Then

$$G_m(f, \Psi) = \sum_{j=1}^{m} c_{n_j}(f, \Psi)\psi_{n_j}.$$

Thus, the greedy approximant $G_m(f, \Psi)$ is a partial sum of the rearranged series

$$\sum_{j=1}^{\infty} c_{n_j}(f, \Psi)\psi_{n_j}. \qquad (1.10)$$

An immediate question arising from (1.10) is: When does this series converge? The theory of convergence of rearranged series is a classical topic in analysis. A series converges *unconditionally* if every rearrangement of this series converges. A basis $\Psi$ of a Banach space $X$ is said to be an *unconditional basis* if, for every $f \in X$, its expansion (1.9) converges unconditionally. For a set of indices $\Lambda$ define

$$S_\Lambda(f, \Psi) := \sum_{n \in \Lambda} c_n(f, \Psi)\psi_n.$$

It is well known that if $\Psi$ is unconditional then there exists a constant $K$ such that, for any $\Lambda$,

$$\|S_\Lambda(f, \Psi)\| \leq K \|f\|. \tag{1.11}$$

This inequality is similar to $\|S_m(f, \Psi)\| \leq B\|f\|$ and implies an analog of inequality (1.7)

$$\|f - S_\Lambda(f, \Psi)\| \leq (K + 1)E_\Lambda(f, \Psi), \tag{1.12}$$

where

$$E_\Lambda(f, \Psi) := \inf_{\{c_n\}} \|f - \sum_{n \in \Lambda} c_n \psi_n\|.$$

Inequality (1.12) indicates that, in the case of an unconditional basis $\Psi$, it is sufficient for finding a near-best $m$-term approximant to optimize only over the sets of indices $\Lambda$. The greedy algorithm $G_m(\cdot, \Psi)$ gives a simple recipe for building $\Lambda_m$: pick the indices with largest coefficients. In Section 1.3 we discuss in detail when the above simple recipe provides a near-best $m$-term approximant. It turns out that the mere assumption that $\Psi$ is unconditional does not guarantee that $G_m(\cdot, \Psi)$ provides a near-best $m$-term approximation. We also discuss a new class of bases (*greedy bases*) that has the property that $G_m(f, \Psi)$ provides a near-best $m$-term approximation for each $f \in X$. We show that the class of greedy bases is a proper subclass of the class of unconditional bases.

It follows from the definition of unconditional basis that any rearrangement of the series in (1.9) converges, and it is known that it converges to $f$. The rearrangement (1.10) is a specific rearrangement of (1.9). Clearly, for an unconditional basis $\Psi$, (1.10) converges to $f$. It turns out that unconditionality of $\Psi$ is not a necessary condition for convergence of (1.10) for each $f \in X$. Bases that have the property of convergence of (1.10) for each $f \in X$ are exactly the *quasi-greedy bases* (see Section 1.4).

Let us summarize our discussion of bases in Banach spaces. Schauder bases are natural for convergence of $S_m(f, \Psi)$ and convenient for linear approximation theory. Other classical bases, namely unconditional bases, are natural for convergence of all rearrangements of expansions. The needs of nonlinear approximation, or, more specifically, the needs of greedy approximation lead us to new concepts of bases: greedy bases and quasi-greedy bases. The relations between these bases are as follows:

$$\{\text{greedy bases}\} \subset \{\text{unconditional bases}\}$$
$$\subset \{\text{quasi-greedy bases}\} \subset \{\text{Schauder bases}\}.$$

All the inclusions $\subset$ are proper inclusions.

In this chapter we provide a justification of the importance of the new classes of bases. With a belief in the importance of greedy bases and quasi-greedy bases, we discuss here the following natural questions: Could we weaken a rule of building $G_m(f, \Psi)$ and still have good approximation and convergence properties? We answer this question in Sections 1.5 and 1.6. What can be said about classical systems, say the Haar system and the trigonometric system, in this regard? We discuss this question in Sections 1.3 and 1.7. How can we build the approximation theory (mostly direct and inverse theorems) for $m$-term approximation with regard to greedy-type bases? Section 1.8 is devoted to this question.

## 1.2 Schauder bases in Banach spaces

Schauder bases in Banach spaces are used to associate a sequence of numbers with an element $f \in X$: these are coefficients of $f$ with respect to a basis. This helps in studying properties of a Banach space $X$. We begin with some classical results on Schauder bases; see, for example, Lindenstrauss and Tzafriri (1977).

**Definition 1.1** A sequence $\Psi := \{\psi_n\}_{n=1}^{\infty}$ in a Banach space $X$ is called a Schauder basis of $X$ (basis of $X$) if, for any $f \in X$, there exists a unique sequence $\{c_n(f)\}_{n=1}^{\infty} := \{c_n(f, \Psi)\}_{n=1}^{\infty}$ such that

$$f = \sum_{n=1}^{\infty} c_n(f)\psi_n.$$

Let

$$S_0(f) := 0, \quad S_m(f) := S_m(f, \Psi) := \sum_{n=1}^{m} c_n(f)\psi_n.$$

For a fixed basis $\Psi$, consider the quantity

$$|||f||| := \sup_m \|S_m(f, \Psi)\|.$$

It is clear that for any $f \in X$ we have

$$\|f\| \leq |||f||| < \infty. \tag{1.13}$$

It is easy to see that $||| \cdot |||$ provides a norm on the linear space $X$. Denote this new normed linear space by $X^s$. The following known proposition is not difficult to prove.

**Proposition 1.2** *The space $X^s$ is a Banach space.*

**Theorem 1.3** *Let $X$ be a Banach space with a Schauder basis $\Psi$. Then the operators $S_m : X \to X$ are bounded linear operators and*

$$\sup_m \|S_m\| < \infty.$$

The proof of this theorem is based on the fundamental theorem of Banach.

**Theorem 1.4** *Let $U$, $V$ be Banach spaces and $T$ be a bounded linear one-to-one operator from $V$ to $U$. Then the inverse operator $T^{-1}$ is a bounded linear operator from $U$ to $V$.*

We specify $U = X$ and $V = X^s$, and let $T$ be the identity map. It follows from (1.13) that $T$ is a bounded operator from $V$ to $U$. Thus, by Theorem 1.4, $T^{-1}$ is also bounded. This means that there exists a constant $C$ such that, for any $f \in X$, we have $|||f||| \leq C\|f\|$. This completes the proof of Theorem 1.3.

The operators $\{S_m\}_{m=1}^\infty$ are called the natural projections associated with a basis $\Psi$. The number $\sup_m \|S_m\|$ is called the basis constant of the basis $\Psi$. A basis whose basis constant is unity is called a *monotone basis*. It is clear that an orthonormal basis in a Hilbert space is a monotone basis. Every Schauder basis $\Psi$ is monotone with respect to the norm $|||f||| := \sup_m \|S_m(f, \Psi)\|$, which was used above. Indeed, we have

$$|||S_m(f)||| = \sup_n \|S_n(S_m(f))\| = \sup_{1 \leq n \leq m} \|S_n(f)\| \leq |||f|||.$$

The above remark means that, for any Schauder basis $\Psi$ of $X$, we can renorm $X$ (take $X^s$) to make the basis $\Psi$ monotone for a new norm.

**Theorem 1.5** *Let $\{x_n\}_{n=1}^\infty$ be a sequence of elements in a Banach space $X$. Then $\{x_n\}_{n=1}^\infty$ is a Schauder basis of $X$ if and only if the following three conditions hold:*

(a)  $x_n \neq 0$ *for all* $n$;

(b)  *there is a constant $K$ such that, for every choice of scalars $\{a_i\}_{i=1}^{\infty}$ and integers $n < m$, we have*

$$\| \sum_{i=1}^{n} a_i x_i \| \leq K \| \sum_{i=1}^{m} a_i x_i \|;$$

(c)  *the closed linear span of $\{x_n\}_{n=1}^{\infty}$ coincides with $X$.*

We note that for a basis $\Psi$ with the basis constant $K$, we have, for any $f \in X$,

$$\| f - S_m(f, \Psi) \| \leq (K + 1) \inf_{\{c_k\}} \| f - \sum_{k=1}^{m} c_k \psi_k \|.$$

Thus, the partial sums $S_m(f, \Psi)$ provide near-best approximation from span$\{\psi_1, \ldots, \psi_m\}$.

Let a Banach space $X$, with a basis $\Psi = \{\psi_k\}_{k=1}^{\infty}$, be given. In order to understand the efficiency of an algorithm providing an $m$-term approximation, we compare its accuracy with the best-possible accuracy when an approximant is a linear combination of $m$ terms from $\Psi$. We define the best $m$-term approximation with regard to $\Psi$ as follows:

$$\sigma_m(f) := \sigma_m(f, \Psi)_X := \inf_{c_k, \Lambda} \| f - \sum_{k \in \Lambda} c_k \psi_k \|_X,$$

where the infimum is taken over coefficients $c_k$ and sets of indices $\Lambda$ with cardinality $|\Lambda| = m$. We note that in the above definition of $\sigma_m(f, \Psi)_X$ the system $\Psi$ may be any system of elements from $X$, not necessarily a basis of $X$.

An immediate natural question is: When does the best $m$-term approximant exist? This question is more difficult than the corresponding question in linear approximation and it has not been studied thoroughly. In what follows, we present some results that may point us in the right direction.

Let us proceed directly to the setting of our approximation problem. Let a subset $A \subset X$ be given. For any $f \in X$, let

$$d(f, A) := d(f, A)_X := \inf_{a \in A} \| f - a \|$$

denote the distance from $f$ to $A$, or, in other words, the best approximation error of $f$ by elements from $A$ in the norm of $X$. To illustrate some appropriate techniques, we prove existence theorems in two settings.

**S1** Let $X = L_p(0, 2\pi)$, $1 \leq p < \infty$, or $X = L_\infty(0, 2\pi) := \mathcal{C}(0, 2\pi)$ be the set of $2\pi$-periodic functions. Consider $A$ to be the set $\Sigma_m$ of all complex trigonometric polynomials or $\Sigma_m(R)$ of all real trigonometric polynomials which have at most $m$ nonzero coefficients:

$$\Sigma_m := \left\{ t : t = \sum_{k \in \Lambda} c_k e^{ikx}, \quad |\Lambda| \le m \right\},$$

$$\Sigma_m(R) := \left\{ t : t = \sum_{k \in \Lambda_1} a_k \cos kx + \sum_{k \in \Lambda_2} b_k \sin kx, \quad |\Lambda_1| + |\Lambda_2| \le m \right\}.$$

We will also use the following notation in this case:

$$\sigma_m(f, \mathcal{T})_X := d(f, \Sigma_m)_X.$$

**S2** Let $X = L_p(0, 1)$, $1 \le p < \infty$, and let $A$ be the set $\Sigma_m^S$ of piecewise constant functions with at most $m - 1$ break-points at $(0, 1)$.

In the setting **S2** we prove here the following existence theorem (see DeVore and Lorenz (1993), p. 363).

**Theorem 1.6** *For any $f \in L_p(0, 1)$, $1 \le p < \infty$, there exists $g \in \Sigma_m^S$ such that*

$$d(f, \Sigma_m^S)_p = \|f - g\|_p.$$

*Proof* Fix the break-points $0 = y_0 \le y_1 \le \cdots \le y_{m-1} \le y_m = 1$, let $y := (y_0, \ldots, y_m)$, and let $S_0(y)$ be the set of piecewise constant functions with break-points $y_1, \ldots, y_{m-1}$. Further, let

$$e_m^y(f)_p := \inf_{a \in S_0(y)} \|f - a\|_p.$$

From the definition of $d(f, \Sigma_m^S)_p$, there exists a sequence $y^i$ such that

$$e_m^{y^i}(f)_p \quad \to \quad d(f, \Sigma_m^S)_p$$

when $i \to \infty$. Considering a subsequence of $\{y^i\}$, if necessary we can assume that $y^i \to y^*$ for some $y^* \in \mathbb{R}^{m+1}$. Now we consider only those indices $j$ for which $y_{j-1}^* \ne y_j^*$. Let $\Lambda$ denote the corresponding set of indices. Take a positive number $\epsilon$ satisfying

$$\epsilon < \min_{j \in \Lambda}(y_j^* - y_{j-1}^*)/3,$$

and consider $i$ such that

$$\|y^* - y^i\|_\infty < \epsilon, \quad \text{where} \quad \|y\|_\infty := \max_k |y_k|. \tag{1.14}$$

By the existence theorem in the case of approximation by elements of a subspace of finite dimension, for each $y^i$ there exists

$$g(f, y^i, c^i) := \sum_{j=1}^{m} c_j^i \chi_{[y_{j-1}^i, y_j^i]},$$

where $\chi_E$ denotes the characteristic function of a set $E$, with the property

$$\|f - g(f, y^i, c^i)\|_p = e_m^{y^i}(f)_p.$$

For $i$ satisfying (1.14) and $j \in \Lambda$ we have $|c_j^i| \le C(f, \epsilon)$, which allows us to assume (passing to a subsequence if necessary) the convergence

$$\lim_{i \to \infty} c_j^i = c_j, \quad j \in \Lambda.$$

Consider

$$g(f, c) := \sum_{j \in \Lambda} c_j \chi_{[y_{j-1}^*, y_j^*]}.$$

Let $U_\epsilon(y^*) := \cup_{j \in \Lambda} (y_j^* - \epsilon, y_j^* + \epsilon)$ and introduce $G := [0, 1] \setminus U_\epsilon(y^*)$. Then we have

$$\int_G |f - g(f, c)|^p = \lim_{i \to \infty} \int_G |f - g(f, y^i, c^i)|^p \le d(f, \Sigma_m^S)_p^p.$$

Making $\epsilon \to 0$, we complete the proof. $\qquad \square$

We proceed now to the trigonometric case **S1**. We will give the proof in the general $d$-variable case for $\mathcal{T}^d := \mathcal{T} \times \cdots \times \mathcal{T}$ ($d$ times) because this generality does not introduce any complications. The following theorem was essentially proved in Baishanski (1983). The presented proof is taken from Temlyakov (1998c).

**Theorem 1.7** *Let $1 \le p \le \infty$. For any $f \in L_p(\mathbb{T}^d)$ and any $m \in \mathbb{N}$, there exists a trigonometric polynomial $t_m$ of the form*

$$t_m(x) = \sum_{n=1}^{m} c_n e^{i(k^n, x)} \tag{1.15}$$

*such that*

$$\sigma_m(f, \mathcal{T}^d)_p = \|f - t_m\|_p. \tag{1.16}$$

*Proof* We prove this theorem by induction. Let us use the abbreviated notation $\sigma_m(f)_p := \sigma_m(f, \mathcal{T}^d)_p$.

    **First step** Let $m = 1$. We assume $\sigma_1(f)_p < \|f\|_p$, because in the case $\sigma_1(f)_p = \|f\|_p$ the proof is trivial: we take $t_1 = 0$. We now prove that polynomials of the form $ce^{i(k,x)}$ with large $k$ cannot provide approximation with error close to $\sigma_1(f)_p$. This will allow us to restrict the search for an optimal approximant $c_1 e^{i(k^1,x)}$ to a finite number of $k^1$, which in turn will imply the existence.

We introduce a parameter $N \in \mathbb{N}$, which will be specified later, and consider the following polynomials:

$$\mathcal{K}_N(u) := \sum_{|k|<N} \left(1 - \frac{|k|}{N}\right) e^{iku}, \quad u \in \mathbb{T}, \tag{1.17}$$

and

$$\mathcal{K}_N(x) := \prod_{j=1}^{d} \mathcal{K}_N(x_j), \quad x = (x_1, \ldots, x_d) \in \mathbb{T}^d.$$

The functions $\mathcal{K}_N$ are the Fejér kernels. These polynomials have the property (see Zygmund (1959), chap. 3, sect. 3 and Section 3.3 here)

$$\|\mathcal{K}_N\|_1 = 1, \quad N = 1, 2, \ldots \tag{1.18}$$

Consider the operator

$$(K_N(g))(x) = (2\pi)^{-d} \int_{\mathbb{T}^d} \mathcal{K}_N(x - y) g(y) dy. \tag{1.19}$$

Let

$$e_N(g) := \|g - K_N(g)\|_p. \tag{1.20}$$

It is known that for any $f \in L_p(\mathbb{T}^d)$ we have $e_N \to 0$ as $N \to \infty$. For fixed $N$ take any $k \in \mathbb{Z}^d$ such that $\|k\|_\infty \geq N$. Consider $g(x) = f(x) - ce^{i(k,x)}$ with some $c$. Using (1.17) and (1.18), we get on the one hand

$$\|K_N(f)\|_p = \|K_N(g)\|_p \leq \|g\|_p. \tag{1.21}$$

On the other hand, we have

$$\|K_N(f)\|_p \geq \|f\|_p - \|f - K_N(f)\|_p \geq \|f\|_p - e_N(f). \tag{1.22}$$

Therefore, combining (1.21) and (1.22) we obtain, for all $k$, $\|k\|_\infty \geq N$, and any $c$

$$\|f(x) - ce^{i(k,x)}\|_p \geq \|f\|_p - e_N(f). \tag{1.23}$$

Making $N$ large enough, we obtain

$$\|f\|_p - e_N(f) \geq (\|f\|_p + \sigma_1(f)_p)/2. \tag{1.24}$$

Relations (1.23) and (1.24) imply

$$\sigma_1(f)_p = \inf_{c, \|k\|_\infty < N} \| f(x) - c e^{i(k,x)} \|_p,$$

which completes the proof for $m = 1$, by the existence theorem in the case of approximation by elements of a subspace of finite dimension.

**General step** Assume that Theorem 1.7 has already been proven for $m - 1$. We now prove it for $m$. If $\sigma_m(f)_p = \sigma_{m-1}(f)_p$, we are done, by the induction assumption. Let $\sigma_m(f)_p < \sigma_{m-1}(f)_p$. The idea of the proof in the general step is similar to that in the first step.

Take any $k^1, \ldots, k^m$. Assume $\|k^j\|_\infty \le \|k^m\|_\infty$, $j = 1, \ldots, m - 1$, and $\|k^m\|_\infty > N$. We prove that a polynomial with frequencies $k^1, \ldots, k^m$ does not provide a good approximation. Take any numbers $c_1, \ldots, c_m$, and consider

$$f_{m-1}(x) := f(x) - \sum_{j=1}^{m-1} c_j e^{i(k^j, x)},$$

$$g(x) := f_{m-1}(x) - c_m e^{i(k^m, x)}.$$

Then, replacing $f$ by $f_{m-1}$, we get in the same way as above the estimate

$$\| f(x) - \sum_{j=1}^{m} c_j e^{i(k^j, x)} \|_p \ge \sigma_{m-1}(f)_p - e_N(f). \tag{1.25}$$

We remark here that the analog to (1.22) appears as follows:

$$\| K_N(f_{m-1}) \|_p \ge \sigma_{m-1}(K_N(f))_p$$
$$\ge \sigma_{m-1}(f)_p - \| f - K_N(f) \|_p \ge \sigma_{m-1}(f)_p - e_N(f).$$

Making $N$ large enough, we derive from (1.25) that

$$\sigma_m(f)_p = \inf \left( \inf_{c_j, j=1, \ldots, m} \| f(x) - \sum_{j=1}^{m} c_j e^{i(k^j, x)} \|_p \right),$$

where the infimum is taken over $k^j$ satisfying the restriction $\|k^j\|_\infty \le N$ for all $j = 1, \ldots, m$. In order to complete the proof of Theorem 1.7, it remains to remark that, by the existence theorem in the case of approximation by elements of a subspace of finite dimension, the inside infimum can always be replaced by the minimum, and the outside infimum is taken over a finite set. This completes the proof.                                          □

Concerning the problem of the uniqueness of the best approximant, we only remark that, in the $m$-term nonlinear approximation, we can hardly expect the unicity. Let us consider problem **S1** about the best $m$-term trigonometric approximation in the particular case $X = L_2(0, 2\pi)$. Take

$$f(x) = \sum_{k=1}^{n} e^{ikx}.$$

Clearly, $\sigma_1(f)_2 = (n-1)^{1/2}$ and each $e^{ikx}$, $k = 1, \dots, n$, may serve as a best approximant.

We can prove the following existence theorem (see Temlyakov (2001a)) in a similar way to the proof of Theorem 1.7.

**Theorem 1.8** *Let $\Psi$ be a monotone basis of $X$. Then, for any $f \in X$ and any $m \in \mathbb{N}$, there exist $\Lambda_m$, $|\Lambda_m| \leq m$, and $\{c_i^*, i \in \Lambda_m\}$ such that*

$$\|f - \sum_{i \in \Lambda_m} c_i^* \psi_i\| = \sigma_m(f, \Psi).$$

Here is another existence theorem from Temlyakov (2001a).

**Theorem 1.9** *Let $\Psi$ be a normalized ($\|\psi_k\| = 1$, $k = 1, \dots$) Schauder basis of $X$ with the additional property that $\psi_k$ converges weakly to $0$. Then, for any $f \in X$, and any $m \in \mathbb{N}$, there exist $\Lambda_m$, $|\Lambda_m| \leq m$, and $\{c_i^*, i \in \Lambda_m\}$ such that*

$$\|f - \sum_{i \in \Lambda_m} c_i^* \psi_i\| = \sigma_m(f, \Psi).$$

*Proof* The proof is a development of ideas from Baishanski (1983). In order to sketch the idea of the proof, let us consider first the case $m = 1$. Let

$$\|f - c_{k_n} \psi_{k_n}\| \quad \to \quad \sigma_1(f, \Psi), \quad n \to \infty. \tag{1.26}$$

If

$$\liminf_{n \to \infty} k_n < \infty$$

then there exists $k$ and a sequence $\{a_n\}$ such that

$$\|f - a_n \psi_k\| \quad \to \quad \sigma_1(f, \Psi), \quad n \to \infty. \tag{1.27}$$

Using the fact that $\Psi$ is a Schauder basis, we infer from (1.27) that the sequence $\{a_n\}$ is bounded. Choosing a convergent subsequence of $\{a_n\}$ we construct an $a$ such that

$$\|f - a\psi_k\| = \sigma_1(f, \Psi),$$

which proves the existence in this case. Assume now that

$$\lim_{n \to \infty} k_n = \infty.$$

Let $F_f$ be a norming (peak) functional for $f$: $F_f(f) = \|f\|$, $\|F_f\| = 1$. Then

$$\|f - c_{k_n}\psi_{k_n}\| \geq F_f(f - c_{k_n}\psi_{k_n}) = \|f\| - c_{k_n}F_f(\psi_{k_n}). \qquad (1.28)$$

Relation (1.26) implies boundedness of $\{c_{k_n}\}$, and therefore, by weak convergence to 0 of $\{\psi_k\}$, we get from (1.28) and (1.26) that

$$\sigma_1(f, \Psi) = \|f\|.$$

Thus we can take 0 as a best approximant. Let us now consider the general case of $m$-term approximation. Let

$$f^n := \sum_{j=1}^{m} c^n_{k^n_j}\psi_{k^n_j}, \quad k^n_1 < k^n_2 < \cdots < k^n_m,$$

be such that

$$\|f - f^n\| \quad \rightarrow \quad \sigma_m(f, \Psi).$$

Then we have

$$|c^n_{k^n_j}| \leq M \qquad (1.29)$$

for all $n$, $j$ with some constant $M$. Assume that we have

$$\liminf_{n \to \infty} k^n_j < \infty, \quad \text{for some (possibly none) } j = 1, \ldots, l \leq m;$$

$$\lim_{n \to \infty} k^n_j = \infty, \quad \text{for some (possibly none) } j = l+1, \ldots, m.$$

Then, as in the case of $m = 1$, we find $\Lambda$, $|\Lambda| \leq l$, and a subsequence $\{n_s\}_{s=1}^{\infty}$ such that

$$\sum_{k \in \Lambda} c^{n_s}_k \psi_k \quad \rightarrow \quad \sum_{k \in \Lambda} c_k \psi_k =: y. \qquad (1.30)$$

Consider the norming functional $F_{f-y}$. We have from (1.29), (1.30) and the weak convergence of $\{\psi_k\}$ to 0 that

$$F_{f-y}(f^{n_s} - y) \quad \rightarrow \quad 0 \quad \text{as} \quad s \to \infty.$$

Thus

$$\|f - y\| = F_{f-y}(f - y) = F_{f-y}(f - f^{n_s} + f^{n_s} - y)$$
$$\leq \|f - f^{n_s}\| + |F_{f-y}(f^{n_s} - y)| \quad \rightarrow \quad \sigma_m(f, \Psi)$$

as $s \to \infty$. This implies that

$$\|f - y\| = \sigma_m(f, \Psi),$$

which completes the proof of Theorem 1.9.                    □

The following observation is from Wojtaszczyk (2002b).

**Remark 1.10** It is clear from the proof of Theorem 1.9 that the condition of weak convergence of $\psi_k$ to 0 can be replaced by the condition $y(\psi_k) \to 0$ for every $y \in Y$. Here, $Y \subset X^*$ is such that, for all $f \in X$,

$$\|f\| = \sup_{y \in Y, \|y\| \le 1} |y(f)|.$$

Also, Wojtaszczyk (2002b) contains an example of an unconditional basis $\Psi$ and an element $f$ such that the best $m$-term approximation of $f$ with regard to $\Psi$ does not exist.

## 1.3 Greedy bases

Let a Banach space $X$, with a basis $\Psi = \{\psi_k\}_{k=1}^{\infty}$, be given. We assume that $\|\psi_k\| \ge C > 0, k = 1, 2, \ldots$, and consider the following theoretical greedy algorithm. For a given element $f \in X$ we consider the expansion

$$f = \sum_{k=1}^{\infty} c_k(f, \Psi)\psi_k. \tag{1.31}$$

For an element $f \in X$ we say that a permutation $\rho$ of the positive integers is decreasing if

$$|c_{k_1}(f, \Psi)| \ge |c_{k_2}(f, \Psi)| \ge \ldots, \tag{1.32}$$

where $\rho(j) = k_j, j = 1, 2, \ldots$, and we write $\rho \in D(f)$. If the inequalities are strict in (1.32), then $D(f)$ consists of only one permutation. We define the $m$th greedy approximant of $f$, with regard to the basis $\Psi$ corresponding to a permutation $\rho \in D(f)$, by the formula

$$G_m(f) := G_m(f, \Psi) := G_m(f, \Psi, \rho) := \sum_{j=1}^{m} c_{k_j}(f, \Psi)\psi_{k_j}.$$

We note that there is another natural greedy-type algorithm based on ordering $\|c_k(f, \Psi)\psi_k\|$ instead of ordering absolute values of coefficients. In this case we do not need the restriction $\|\psi_k\| \ge C > 0, k = 1, 2, \ldots$. Let $\Lambda_m(f)$ be a set of indices such that

$$\min_{k \in \Lambda_m(f)} \|c_k(f, \Psi)\psi_k\| \ge \max_{k \notin \Lambda_m(f)} \|c_k(f, \Psi)\psi_k\|.$$

We define $G_m^X(f, \Psi)$ by the formula

$$G_m^X(f, \Psi) := S_{\Lambda_m(f)}(f, \Psi), \quad \text{where} \quad S_E(f) := S_E(f, \Psi) := \sum_{k \in E} c_k(f, \Psi)\psi_k.$$

It is clear that for a normalized basis ($\|\psi_k\| = 1$, $k = 1, 2, \ldots$) the above two greedy algorithms coincide. It is also clear that the above greedy algorithm $G_m^X(\cdot, \Psi)$ can be considered as a greedy algorithm $G_m(\cdot, \Psi')$, with $\Psi' := \{\psi_k / \|\psi_k\|\}_{k=1}^{\infty}$ being a normalized version of the $\Psi$. Thus, we will concentrate on studying the algorithm $G_m(\cdot, \Psi)$. In the above definition of $G_m(\cdot, \Psi)$ we impose an extra condition on a basis $\Psi$: $\inf_k \|\psi_k\| > 0$. This restriction allows us to define $G_m(f, \Psi)$ for all $f \in X$. For the sake of completeness we will also discuss the case

$$\inf_k \|\psi_k\| = 0. \tag{1.33}$$

In this case we define the $G_m(f, \Psi)$ in the same way as above, but only for $f$ of a special form:

$$f = \sum_{k \in Y} c_k(f, \Psi)\psi_k, \quad |Y| < \infty. \tag{1.34}$$

The above algorithm $G_m(\cdot, \Psi)$ is a simple algorithm which describes the theoretical scheme for $m$-term approximation of an element $f$. We call this algorithm the Greedy Algorithm (GA). In order to understand the efficiency of this algorithm, we compare its accuracy with the best-possible accuracy when an approximant is a linear combination of $m$ terms from $\Psi$. We define the best $m$-term approximation with regard to $\Psi$ as follows:

$$\sigma_m(f) := \sigma_m(f, \Psi)_X := \inf_{c_k, \Lambda} \|f - \sum_{k \in \Lambda} c_k \psi_k\|_X,$$

where the infimum is taken over coefficients $c_k$ and sets of indices $\Lambda$ with cardinality $|\Lambda| = m$. The best we can achieve with the algorithm $G_m$ is

$$\|f - G_m(f, \Psi, \rho)\|_X = \sigma_m(f, \Psi)_X,$$

or the slightly weaker

$$\|f - G_m(f, \Psi, \rho)\|_X \le G\sigma_m(f, \Psi)_X, \tag{1.35}$$

for all elements $f \in X$, and with a constant $G = C(X, \Psi)$ independent of $f$ and $m$. It was mentioned in Section 1.1 (see (1.8)) that, when $X = H$ is a Hilbert space and $\mathcal{B}$ is an orthonormal basis, we have

$$\|f - G_m(f, \mathcal{B}, \rho)\|_H = \sigma_m(f, \mathcal{B})_H.$$

Let us begin our discussion with an important class of bases: wavelet-type bases. In the case $X = L_p$, we will write $p$ instead of $L_p$. Let $\mathcal{H} := \{H_k\}_{k=1}^{\infty}$ denote the Haar basis on $[0, 1)$ normalized in $L_2(0, 1)$. We denote by $\mathcal{H}_p := \{H_{k,p}\}_{k=1}^{\infty}$ the Haar basis $\mathcal{H}$ renormalized in $L_p(0, 1)$, which is defined as follows: $H_{1,p} = 1$ on $[0, 1)$ and, for $k = 2^n + l, l = 1, 2, \ldots, 2^n$, $n = 0, 1, \ldots$

$$H_{k,p}(x) = \begin{cases} 2^{n/p}, & x \in [(2l - 2)2^{-n-1}, (2l - 1)2^{-n-1}) \\ -2^{n/p}, & x \in [(2l - 1)2^{-n-1}, 2l2^{-n-1}) \\ 0, & \text{otherwise.} \end{cases}$$

We will use the following definition of the $L_p$-equivalence of bases. We say that $\Psi = \{\psi_k\}_{k=1}^\infty$ is $L_p$-equivalent to $\Phi = \{\phi_k\}_{k=1}^\infty$ if, for any finite set $\Lambda$ and any coefficients $c_k, k \in \Lambda$, we have

$$C_1(p, \Psi, \Phi)\| \sum_{k \in \Lambda} c_k\phi_k \|_p \le \| \sum_{k \in \Lambda} c_k\psi_k \|_p \le C_2(p, \Psi, \Phi)\| \sum_{k \in \Lambda} c_k\phi_k \|_p$$

(1.36)

with two positive constants $C_1(p, \Psi, \Phi), C_2(p, \Psi, \Phi)$ which may depend on $p$, $\Psi$ and $\Phi$. For sufficient conditions on $\Psi$ to be $L_p$-equivalent to $\mathcal{H}$ see DeVore, Konyagin and Temlyakov (1998), Frazier and Jawerth (1990) and Section 1.10. In particular, it is known that all reasonable univariate wavelet-type bases are $L_p$-equivalent to $\mathcal{H}$ for $1 < p < \infty$. We proved the following theorem in Temlyakov (1998a).

**Theorem 1.11** *Let* $1 < p < \infty$ *and let a basis* $\Psi$ *be* $L_p$*-equivalent to the Haar basis* $\mathcal{H}$*. Then, for any* $f \in L_p(0, 1)$*, we have*

$$\|f - G_m^p(f, \Psi)\|_p \le C(p, \Psi)\sigma_m(f, \Psi)_p$$

*with a constant* $C(p, \Psi)$ *independent of* $f$ *and* $m$*.*

By a simple renormalization argument one obtains the following version of Theorem 1.11.

**Theorem 1.11A** *Let* $1 < p < \infty$ *and let a basis* $\Psi$ *be* $L_p$*-equivalent to the Haar basis* $\mathcal{H}_p$*. Then, for any* $f \in L_p(0, 1)$ *and any* $\rho \in D(f)$*, we have*

$$\|f - G_m(f, \Psi, \rho)\|_p \le C(p, \Psi)\sigma_m(f, \Psi)_p$$

*with a constant* $C(p, \Psi)$ *independent of* $f$*,* $\rho$*, and* $m$*.*

We give a proof of Theorem 1.11 at the end of this section. We note that Temlyakov (1998a) also contains a generalization of Theorem 1.11 to the multivariate Haar basis obtained by the multi-resolution analysis procedure. These theorems motivated us to consider the general setting of greedy approximation in Banach spaces. We concentrated on studying bases which satisfy (1.35) for all individual functions. Definitions 1.12–1.14 below are from Konyagin and Temlyakov (1999a).

**Definition 1.12** We call a basis $\Psi$ a *greedy basis* if, for every $f \in X$ (in the case $\inf_k \|\psi_k\| > 0$) and for $f$ of the form (1.34) (in the case $\inf_k \|\psi_k\| = 0$), there exists a permutation $\rho \in D(f)$ such that the inequality

$$\|f - G_m(f, \Psi, \rho)\|_X \le G\sigma_m(f, \Psi)_X$$

holds with a constant independent of $f$ and $m$.

Theorem 1.11A shows that each basis $\Psi$ which is $L_p$-equivalent to the univariate Haar basis $\mathcal{H}_p$ is a greedy basis for $L_p(0, 1)$, $1 < p < \infty$. We note that for a Hilbert space each orthonormal basis is a greedy basis with a constant $G = 1$ (see (1.35)).

We now give the definitions of unconditional and democratic bases.

**Definition 1.13** A basis $\Psi = \{\psi_k\}_{k=1}^{\infty}$ of a Banach space $X$ is said to be unconditional if, for every choice of signs $\theta = \{\theta_k\}_{k=1}^{\infty}$, $\theta_k = 1$ or $-1$, $k = 1, 2, \ldots$, the linear operator $M_\theta$ defined by

$$M_\theta \left( \sum_{k=1}^{\infty} a_k \psi_k \right) = \sum_{k=1}^{\infty} a_k \theta_k \psi_k$$

is a bounded operator from $X$ into $X$.

**Definition 1.14** We say that a basis $\Psi = \{\psi_k\}_{k=1}^{\infty}$ is a democratic basis for $X$ if there exists a constant $D := D(X, \Psi)$ such that, for any two finite sets of indices $P$ and $Q$ with the same cardinality $|P| = |Q|$, we have

$$\left\| \sum_{k \in P} \psi_k \right\| \le D \left\| \sum_{k \in Q} \psi_k \right\|.$$

We proved in Konyagin and Temlyakov (1999a) the following theorem.

**Theorem 1.15** *A basis is greedy if and only if it is unconditional and democratic.*

This theorem gives a characterization of greedy bases. In the following we give a proof of a generalization of Theorem 1.15 (see Theorem 1.18). Further investigations (Cohen, DeVore and Hochmuth (2000), Gribonval and Nielsen (2001b), Kamont and Temlyakov (2004), Kerkyacharian and Picard (2004), Temlyakov (1998b)) showed that the concept of greedy bases is very useful in direct and inverse theorems of nonlinear approximation and also in applications in statistics.

Let us make a remark on bases $\Psi$ that satisfy condition (1.33). In this case the greedy algorithm $G_m(\cdot, \Psi)$ is defined only for $f$ of the form (1.34). However, if $\Psi$ is a greedy basis, then by Theorem 1.15 it is democratic, and therefore satisfies the condition $\inf_k \|\psi_k\| > 0$. Thus, there are no greedy bases satisfying (1.33).

An interesting generalization of $m$-term approximation was considered in Cohen, DeVore and Hochmuth (2000). Let $\Psi = \{\psi_I\}_I$ be a basis indexed by dyadic intervals. Take an $\alpha$ and assign to each index set $\Lambda$ the measure

$$\Phi_\alpha(\Lambda) := \sum_{I \in \Lambda} |I|^\alpha.$$

In the case $\alpha = 0$ we get $\Phi_0(\Lambda) = |\Lambda|$. An analog of best $m$-term approximation is as follows:

$$\inf_{\Lambda : \Phi_\alpha(\Lambda) \leq m} \inf_{c_I, I \in \Lambda} \left\| f - \sum_{I \in \Lambda} c_I \psi_I \right\|_p.$$

A detailed study of this type of approximation (restricted approximation) can be found in Cohen, DeVore and Hochmuth (2000).

We now elaborate on the idea of assigning to each basis element $\psi_k$ a nonnegative weight $w_k$. We discuss weight-greedy bases and prove a criterion for weight-greedy bases similar to the one for greedy bases.

Let $\Psi$ be a basis for $X$. As above, if $\inf_n \|\psi_n\| > 0$ then $c_n(f) \to 0$ as $n \to \infty$, where

$$f = \sum_{n=1}^\infty c_n(f)\psi_n.$$

Then we can rearrange the coefficients $\{c_n(f)\}$ in the decreasing way

$$|c_{n_1}(f)| \geq |c_{n_2}(f)| \geq \dots$$

and define the $m$th greedy approximant as

$$G_m(f, \Psi) := \sum_{k=1}^m c_{n_k}(f)\psi_{n_k}. \tag{1.37}$$

In the case $\inf_n \|\psi_n\| = 0$ we define $G_m(f, \Psi)$ by (1.37) for $f$ of the form

$$f = \sum_{n \in Y} c_n(f)\psi_n, \quad |Y| < \infty. \tag{1.38}$$

Let a weight sequence $w = \{w_n\}_{n=1}^\infty$, $w_n > 0$, be given. For $\Lambda \subset \mathbb{N}$, denote $w(\Lambda) := \sum_{n \in \Lambda} w_n$. For a positive real number $v > 0$ define

$$\sigma_v^w(f, \Psi) := \inf_{\{b_n\}, \Lambda : w(\Lambda) \leq v} \left\| f - \sum_{n \in \Lambda} b_n \psi_n \right\|,$$

where $\Lambda$ are finite.

We present results from Kerkyacharian, Picard and Temlyakov (2006).

**Definition 1.16** We call a basis $\Psi$ a weight-greedy basis ($w$-greedy basis) if for any $f \in X$ in the case $\inf_n \|\psi_n\| > 0$ or for any $f \in X$ of the form (1.38) in the case $\inf_n \|\psi_n\| = 0$, we have

$$\|f - G_m(f, \Psi)\| \leq C_G \sigma^w_{w(\Lambda_m)}(f, \Psi),$$

where $\Lambda_m$ is obtained from the representation

$$G_m(f, \Psi) = \sum_{n \in \Lambda_m} c_n(f)\psi_n, \quad |\Lambda_m| = m.$$

**Definition 1.17** We call a basis $\Psi$ a weight-democratic basis ($w$-democratic basis) if, for any finite $A, B \subset \mathbb{N}$ such that $w(A) \leq w(B)$, we have

$$\|\sum_{n \in A} \psi_n\| \leq C_D \|\sum_{n \in B} \psi_n\|.$$

**Theorem 1.18** *A basis $\Psi$ is a $w$-greedy basis if and only if it is unconditional and $w$-democratic.*

*Proof* I. We first prove the implication

$$\text{unconditional} + w\text{-democratic} \quad \Rightarrow \quad w\text{-greedy}.$$

Let $f$ be any function or a function of the form (1.38) if $\inf_n \|\psi_n\| = 0$. Consider

$$G_m(f, \Psi) = \sum_{n \in Q} c_n(f)\psi_n =: S_Q(f).$$

We take any finite set $P \subset \mathbb{N}$ satisfying $w(P) \leq w(Q)$. Then our assumption $w_n > 0$, $n \in \mathbb{N}$ implies that either $P = Q$ or $Q \setminus P$ is non-empty. As in Section 1.1 let

$$E_P(f, \Psi) := \inf_{\{b_n\}} \|f - \sum_{n \in P} b_n \psi_n\|.$$

Then, by the unconditionality of $\Psi$, we have (see (1.12))

$$\|f - S_P(f)\| \leq (K+1)E_P(f, \Psi). \tag{1.39}$$

This (with $P = Q$) completes the proof in the case $\sigma^w_{w(Q)}(f, \Psi) = E_Q(f, \Psi)$. Suppose that $\sigma^w_{w(Q)}(f, \Psi) < E_Q(f, \Psi)$. Clearly, we may now consider only those $P$ that satisfy the two conditions

$$w(P) \leq w(Q) \quad \text{and} \quad E_P(f, \Psi) < E_Q(f, \Psi).$$

For $P$ satisfying the above conditions we have $Q \setminus P \neq \emptyset$. We estimate

$$\|f - S_Q(f)\| \leq \|f - S_P(f)\| + \|S_P(f) - S_Q(f)\|. \tag{1.40}$$

We have

$$S_P(f) - S_Q(f) = S_{P \setminus Q}(f) - S_{Q \setminus P}(f). \tag{1.41}$$

As for (1.39) we get

$$\|S_{Q \setminus P}(f)\| \le K E_P(f, \Psi). \tag{1.42}$$

It remains to estimate $\|S_{P \setminus Q}(f)\|$. By unconditionality and $w$-democracy in the case of a real Banach space $X$, we have

$$\|S_{P \setminus Q}(f)\| \le 2K \max_{n \in P \setminus Q} |c_n(f)| \|\sum_{n \in P \setminus Q} \psi_n\|$$
$$\le 2K C_D \min_{n \in Q \setminus P} |c_n(f)| \|\sum_{n \in Q \setminus P} \psi_n\| \le C(K) C_D \|S_{Q \setminus P}(f)\|. \tag{1.43}$$

In the case of a complex Banach space $X$ the above inequalities hold with $2K$ replaced by $4K$. Combining (1.39)–(1.43), we complete the proof of part I.

II. We now prove the implication

$$w\text{-greedy} \quad \Rightarrow \quad \text{unconditional} + w\text{-democratic}.$$

IIa. We begin with the following one:

$$w\text{-greedy} \quad \Rightarrow \quad \text{unconditional}.$$

We will prove a slightly stronger statement.

**Lemma 1.19** *Let $\Psi$ be a basis such that, for any $f$ of the form (1.38), we have*

$$\|f - G_m(f, \Psi)\| \le C E_\Lambda(f, \Psi),$$

*where*

$$G_m(f, \Psi) = \sum_{n \in \Lambda} c_n(f) \psi_n.$$

*Then $\Psi$ is unconditional.*

*Proof* It is clear that it is sufficient to prove that there exists a constant $C_0$ such that, for any finite $\Lambda$ and any $f$ of the form (1.38), we have

$$\|S_\Lambda(f)\| \le C_0 \|f\|.$$

Let $f$ and $\Lambda$ be given and $\Lambda \subset [1, M]$. Consider

$$f_M := S_{[1,M]}(f);$$

then $\|f_M\| \leq C_B\|f\|$. We take $b > \max_{1 \leq n \leq M} |c_n(f)|$ and define a new function

$$g := f_M - S_\Lambda(f_M) + b \sum_{n \in \Lambda} \psi_n.$$

Then

$$G_m(g, \Psi) = b \sum_{n \in \Lambda} \psi_n, \quad m := |\Lambda|$$

and

$$E_\Lambda(g, \Psi) \leq \|f_M\|.$$

Thus, by the assumption,

$$\|f_M - S_\Lambda(f_M)\| = \|g - G_m(g, \Psi)\| \leq CE_\Lambda(g, \Psi) \leq C\|f_M\|,$$

and therefore

$$\|S_\Lambda(f)\| = \|S_\Lambda(f_M)\| \leq C_0\|f\|.$$

$\square$

IIb. It remains to prove the implication

$$w\text{-greedy} \quad \Rightarrow \quad w\text{-democratic}.$$

First, let $A, B \subset \mathbb{N}$, $w(A) \leq w(B)$, be such that $A \cap B = \emptyset$. Consider

$$f := \sum_{n \in A} \psi_n + (1 + \epsilon) \sum_{n \in B} \psi_n, \quad \epsilon > 0.$$

Then

$$G_m(f, \Psi) = (1 + \epsilon) \sum_{n \in B} \psi_n, \quad m := |B|,$$

and

$$E_A(f, \Psi) \leq \|\sum_{n \in B} \psi_n\|(1 + \epsilon).$$

Therefore, by the $w$-greedy assumption we get

$$\|\sum_{n \in A} \psi_n\| \leq C(1 + \epsilon)\|\sum_{n \in B} \psi_n\|.$$

Now let $A, B$ be any finite subsets of $\mathbb{N}$ for which $w(A) \leq w(B)$. Then, using the unconditionality of $\Psi$ proved in IIa and the above part of IIb, we obtain

$$\| \sum_{n \in A} \psi_n \| \leq \| \sum_{n \in A \backslash B} \psi_n \| + \| \sum_{n \in A \cap B} \psi_n \|$$

$$\leq C \| \sum_{n \in B \backslash A} \psi_n \| + K \| \sum_{n \in B} \psi_n \| \leq C_1 \| \sum_{n \in B} \psi_n \|.$$

This completes the proof of Theorem 1.18. $\qquad \square$

Theorems 1.15 and 1.18 show that *greedy = unconditional + democratic*. We now show that unconditionality does not imply democracy, and vice versa.

**Unconditionality does not imply democracy** This follows from properties of the multivariate Haar system $\mathcal{H}^2 = \mathcal{H} \times \mathcal{H}$ defined as the tensor product of the univariate Haar systems $\mathcal{H}$ (see (1.44) below). Relation (1.44) shows that $\mathcal{H}^2$ is not greedy in any $L_p$, $1 < p < \infty$, $p \neq 2$. It is known that $\mathcal{H}^2$ is unconditional in $L_p$, $1 < p < \infty$. Therefore, Theorem 1.15 implies that $\mathcal{H}^2$ is not democratic in any $L_p$, $1 < p < \infty$, $p \neq 2$.

**Democracy does not imply unconditionality** Let $X$ be the set of all real sequences $x = (x_1, x_2, \dots)$ such that

$$\|x\|_X := \sup_{N \in \mathbb{N}} | \sum_{n=1}^{N} x_n |$$

is finite. Clearly, $X$ equipped with the norm $\| \cdot \|_X$ is a Banach space. Let $\psi_k \in X$, $k = 1, 2, \dots$, be defined as $(\psi_k)_n = 1$ if $n = k$ and $(\psi_k)_n = 0$ otherwise. By $X_0$ denote the subspace of $X$ generated by the elements $\psi_k$. It is easy to see that $\{\psi_k\}$ is a democratic basis in $X_0$. However, it is not an unconditional basis, since

$$\| \sum_{k=1}^{m} \psi_k \|_X = m,$$

but

$$\| \sum_{k=1}^{m} (-1)^k \psi_k \|_X = 1.$$

We let $\mathcal{H}_p := \{H_{k,p}\}_{k=1}^{\infty}$ be the Haar basis $\mathcal{H}$ renormalized in $L_p([0, 1))$. We define the multivariate Haar basis $\mathcal{H}_p^d$ to be the tensor product of the univariate Haar bases: $\mathcal{H}_p^d := \mathcal{H}_p \times \cdots \times \mathcal{H}_p$;

$$H_{\mathbf{n},p}(x) := H_{n_1,p}(x_1) \cdots H_{n_d,p}(x_d), \quad x = (x_1, \dots, x_d), \quad \mathbf{n} = (n_1, \dots, n_d).$$

Supports of functions $H_{\mathbf{n},p}$ are arbitrary dyadic parallelepipeds (intervals). It is known (see Temlyakov (2002a)) that the tensor product structure of the multivariate wavelet bases makes them universal for approximation of anisotropic smoothness classes with different anisotropy. It is also known that the study of such bases is more difficult than the study of the univariate bases. In many cases we need to develop new techniques and in some cases we encounter new phenomena. For instance, it turns out that the democratic property does not hold for the multivariate Haar basis $\mathcal{H}_p^d$ for $p \neq 2$. The following relation is known for $1 < p < \infty$:

$$\sup_{f \in L_p} \| f - G_m(f, \mathcal{H}_p^d) \|_p / \sigma_m(f, \mathcal{H}_p^d) \asymp (\log m)^{(d-1)|1/2-1/p|}. \qquad (1.44)$$

The lower bound in (1.44) was proved by R. Hochmuth (see the end of this section); the upper bound in (1.44) was proved in the case $d = 2$, $4/3 \leq p \leq 4$, and was conjectured for all $d$, $1 < p < \infty$, in Temlyakov (1998b). The conjecture was proved in Wojtaszczyk (2000).

Let us return to the problem of finding a near-best $m$-term approximant of $f \in X$ with regard to a basis $\Psi$. This problem consists of two subproblems. First, we need to identify a set $\Lambda_m$ of $m$ indices that can be used in achieving a near-best $m$-term approximation of $f$. Second, we need to find the coefficients $\{c_k\}$, $k \in \Lambda_m$, such that the approximant $\sum_{k \in \Lambda_m} c_k \psi_k$ provides a near-best approximation of $f$. It is clear from the properties of an unconditional basis $\Psi$ that, for any $f \in X$ and any $\Lambda$, we have (see (1.12))

$$\| f - \sum_{k \in \Lambda} c_k(f, \Psi) \psi_k \| \leq C \inf_{\{c_k\}} \| f - \sum_{k \in \Lambda} c_k \psi_k \|.$$

Therefore, in the case of an unconditional basis $\Psi$ the second subproblem is easy: we can always choose the expansion coefficients $c_k(f, \Psi)$, $k \in \Lambda_m$. Theorem 1.15 shows that if a basis $\Psi$ is simultaneously unconditional and democratic, then the first subproblem is also easy: it follows from the definition of greedy basis that the algorithm of choosing the $m$ biggest in absolute value coefficients gives the set $\Lambda_m$.

It would be very interesting to understand how we can find $\Lambda_m$ in the case when we only know that $\Psi$ is unconditional. The following special case of the above problem is of great interest: $X = L_p([0, 1]^d)$, $d \geq 2$, $\Psi$ is the multivariate Haar basis $\mathcal{H}_p^d$, $1 < p < \infty$. It is known from Kamont and Temlyakov (2004), Temlyakov (1998b) and Wojtaszczyk (2000) that the function

$$\mu(m, \mathcal{H}_p^d) := \sup_{k \le m} \left( \sup_{\Lambda : |\Lambda| = k} \| \sum_{\mathbf{n} \in \Lambda} H_{\mathbf{n}, p} \|_p / \inf_{\Lambda : |\Lambda| = k} \| \sum_{\mathbf{n} \in \Lambda} H_{\mathbf{n}, p} \|_p \right)$$

plays a very important role in estimating the $m$-term greedy approximation in terms of the best $m$-term approximation. For instance (see Temlyakov (1998b)),

$$\| f - G_m(f, \mathcal{H}_p^d) \|_p \le C(p, d) \mu(m, \mathcal{H}_p^d) \sigma_m(f, \mathcal{H}_p^d)_p, \quad 1 < p < \infty. \tag{1.45}$$

The following theorem gives, in particular, upper bounds for $\mu(m, \mathcal{H}_p^d)$.

**Theorem 1.20** *Let* $1 < p < \infty$. *Then, for any* $\Lambda$, $|\Lambda| = m$, *we have, for* $2 \le p < \infty$,

$$C_{p,d}^1 m^{1/p} \min_{\mathbf{n} \in \Lambda} |c_{\mathbf{n}}| \le \| \sum_{\mathbf{n} \in \Lambda} c_{\mathbf{n}} H_{\mathbf{n}, p} \|_p \le C_{p,d}^2 m^{1/p} (\log m)^{h(p,d)} \max_{\mathbf{n} \in \Lambda} |c_{\mathbf{n}}|$$

*and, for* $1 < p \le 2$,

$$C_{p,d}^3 m^{1/p} (\log m)^{-h(p,d)} \min_{\mathbf{n} \in \Lambda} |c_{\mathbf{n}}| \le \| \sum_{\mathbf{n} \in \Lambda} c_{\mathbf{n}} H_{\mathbf{n}, p} \|_p \le C_{p,d}^4 m^{1/p} \max_{\mathbf{n} \in \Lambda} |c_{\mathbf{n}}|,$$

*where* $h(p, d) := (d - 1)|1/2 - 1/p|$.

Theorem 1.20 for $d = 1$, $1 < p < \infty$, was proved in Temlyakov (1998a), and for $d = 2$, $4/3 \le p \le 4$, it was proved in Temlyakov (1998b). Theorem 1.20 in the general case was proved in Wojtaszczyk (2000). It is known (Temlyakov (2002c)) that the extra log factors in Theorem 1.20 are sharp.

Let $\Psi$ be a normalized basis for $L_p([0, 1))$. For the space $L_p([0, 1)^d)$ we define $\Psi^d := \Psi \times \cdots \times \Psi$ ($d$ times) and $\psi_{\mathbf{n}}(x) := \psi_{n_1}(x_1) \cdots \psi_{n_d}(x_d)$, $x = (x_1, \ldots, x_d)$, $\mathbf{n} = (n_1, \ldots, n_d)$. In Kerkyacharian, Picard and Temlyakov (2006) we proved the following theorem using a proof whose structure is similar to that in Wojtaszczyk (2000).

**Theorem 1.21** *Let* $1 < p < \infty$ *and let* $\Psi$ *be a greedy basis for* $L_p([0, 1))$. *Then, for any* $\Lambda$, $|\Lambda| = m$, *we have, for* $2 \le p < \infty$,

$$C_{p,d}^5 m^{1/p} \min_{\mathbf{n} \in \Lambda} |c_{\mathbf{n}}| \le \| \sum_{\mathbf{n} \in \Lambda} c_{\mathbf{n}} \psi_{\mathbf{n}} \|_p \le C_{p,d}^6 m^{1/p} (\log m)^{h(p,d)} \max_{\mathbf{n} \in \Lambda} |c_{\mathbf{n}}|$$

*and, for* $1 < p \le 2$,

$$C_{p,d}^7 m^{1/p} (\log m)^{-h(p,d)} \min_{\mathbf{n} \in \Lambda} |c_{\mathbf{n}}| \le \| \sum_{\mathbf{n} \in \Lambda} c_{\mathbf{n}} \psi_{\mathbf{n}} \|_p \le C_{p,d}^8 m^{1/p} \max_{\mathbf{n} \in \Lambda} |c_{\mathbf{n}}|,$$

*where* $h(p, d) := (d - 1)|1/2 - 1/p|$.

Inequality (1.45) was extended in Wojtaszczyk (2000) to a normalized uncon-
ditional basis $\Psi$ for $X$ instead of $\mathcal{H}_p^d$ for $L_p([0, 1)^d)$. Therefore, as a corollary
of Theorem 1.21 we obtain the following inequality for a greedy basis $\Psi$ (for
$L_p([0, 1))$, $1 < p < \infty$):

$$\|f - G_m(f, \Psi^d)\|_p \le C(\Psi, d, p)(\log m)^{h(p,d)}\sigma_m(f, \Psi^d)_p. \qquad (1.46)$$

*Proof of Theorem 1.11.* It will be convenient for us to index elements of bases
by dyadic intervals: $\psi_1 =: \psi_{[0,1]}$ and

$$\psi_{2^n+l} =: \psi_I, \quad I = [(l-1)2^{-n}, l2^{-n}) \quad l = 1, \ldots, 2^n, \quad n = 0, 1, \ldots$$

Then the Haar functions $H_I$ are indexed by their intervals of support. The first
and the second Haar functions are indexed by $[0, 1]$ and $[0, 1)$, respectively.

Let us take a parameter $0 < t \le 1$ and consider the following greedy-
type algorithm $G^{p,t}$ with regard to the Haar system. For the Haar basis $\mathcal{H}$ we
define

$$c_I(f) := \langle f, H_I \rangle = \int_0^1 f(x)H_I(x)dx.$$

Let $\Lambda_m(t)$ denote any set of $m$ dyadic intervals such that

$$\min_{I \in \Lambda_m(t)} \|c_I(f)H_I\|_p \ge t \max_{J \notin \Lambda_m(t)} \|c_J(f)H_J\|_p, \qquad (1.47)$$

and define

$$G_m^{p,t}(f) := G_m^{p,t}(f, \mathcal{H}) := \sum_{I \in \Lambda_m(t)} c_I(f)H_I. \qquad (1.48)$$

For a given function $f \in L_p$ we define

$$g(f) := \sum_I c_I(f, \Psi)H_I. \qquad (1.49)$$

It is clear that for a basis $\Psi$ $L_p$-equivalent to $\mathcal{H}$ we have $g(f) \in L_p$ and

$$\sigma_m(g(f), \mathcal{H})_p \le C_1(p)^{-1}\sigma_m(f, \Psi)_p; \qquad (1.50)$$

here and later on we use brief notation $C_i(p) := C_i(p, \Psi, \mathcal{H})$, $i = 1, 2$, for
the constants from (1.36). Let

$$G_m^p(f, \Psi) = \sum_{I \in \Lambda_m} c_I(f, \Psi)\psi_I.$$

Next, for any two intervals $I \in \Lambda_m$, $J \notin \Lambda_m$, by definition of $\Lambda_m$ we have

$$\|c_I(f, \Psi)\psi_I\|_p \ge \|c_J(f, \Psi)\psi_J\|.$$

Using (1.36), we get

$$\|c_I(g(f))H_I\|_p = \|c_I(f, \Psi)H_I\|_p \geq C_2(p)^{-1}\|c_I(f, \Psi)\psi_I\|_p$$
$$\geq C_2(p)^{-1}\|c_J(f, \Psi)\psi_J\|_p \geq C_1(p)C_2(p)^{-1}\|c_J(g(f))H_J\|_p. \tag{1.51}$$

This inequality implies that, for any $m$, we can find a set $\Lambda_m(t)$, where $t = C_1(p)C_2(p)^{-1}$, such that $\Lambda_m(t) = \Lambda_m$, and therefore

$$\|f - G_m^p(f, \Psi)\|_p \leq C_2(p)\|g(f) - G_m^{p,t}(g(f))\|_p. \tag{1.52}$$

Relations (1.50) and (1.52) show that Theorem 1.11 follows from Theorem 1.22 below. $\qquad\square$

**Theorem 1.22** *Let* $1 < p < \infty$ *and* $0 < t \leq 1$. *Then, for any* $g \in L_p$, *we have*

$$\|g - G_m^{p,t}(g, \mathcal{H})\|_p \leq C(p, t)\sigma_m(g, \mathcal{H})_p.$$

*Proof* The Littlewood–Paley theorem, for the Haar system, gives, for $1 < p < \infty$,

$$C_3(p)\Big\| \Big(\sum_I |c_I(g)H_I|^2\Big)^{1/2} \Big\|_p \leq \|g\|_p \leq C_4(p)\Big\| \Big(\sum_I |c_I(g)H_I|^2\Big)^{1/2} \Big\|_p. \tag{1.53}$$

We first formulate two simple corollaries of (1.53):

$$\|g\|_p \leq C_5(p)\Big(\sum_I \|c_I(g)H_I\|_p^p\Big)^{1/p}, \qquad 1 < p \leq 2, \tag{1.54}$$

$$\|g\|_p \leq C_6(p)\Big(\sum_I \|c_I(g)H_I\|_p^2\Big)^{1/2}, \qquad 2 \leq p < \infty. \tag{1.55}$$

Analogs of these inequalities for the trigonometric system are known (see, for instance, Temlyakov (1989a), p. 37). The same proof yields (1.54) and (1.55).

The dual inequalities to (1.54) and (1.55) are

$$\|g\|_p \geq C_7(p)\Big(\sum_I \|c_I(g)H_I\|_p^2\Big)^{1/2}, \qquad 1 < p \leq 2, \tag{1.56}$$

$$\|g\|_p \geq C_8(p)\Big(\sum_I \|c_I(g)H_I\|_p^p\Big)^{1/p}, \qquad 2 \leq p < \infty. \tag{1.57}$$

We proceed to the proof of Theorem 1.22. Let $T_m$ be an $m$-term Haar polynomial of best $m$-term approximation to $g$ in $L_p$ (for the existence see Section 1.2):

$$T_m = \sum_{I \in \Lambda} a_I H_I, \quad |\Lambda| = m.$$

For any finite set $Q$ of dyadic intervals, we denote by $S_Q$ the projector

$$S_Q(f) := \sum_{I \in Q} c_I(f) H_I.$$

Relation (1.53) implies

$$\|g - S_\Lambda(g)\|_p = \|g - T_m - S_\Lambda(g - T_m)\|_p \le \|Id - S_\Lambda\|_{p \to p} \sigma_m(g, \mathcal{H})_p$$

$$\le C_4(p) C_3(p)^{-1} \sigma_m(g, \mathcal{H})_p, \tag{1.58}$$

where $Id$ denotes the identity operator. Further, we have

$$G_m^{p,t}(g) = S_{\Lambda_m(t)}(g)$$

and

$$\|g - G_m^{p,t}(g)\|_p \le \|g - S_\Lambda(g)\|_p + \|S_\Lambda(g) - S_{\Lambda_m(t)}(g)\|_p. \tag{1.59}$$

The first term on the right-hand side of (1.59) has been estimated in (1.58). We now estimate the second term. We represent it in the form

$$S_\Lambda(g) - S_{\Lambda_m(t)}(g) = S_{\Lambda \setminus \Lambda_m(t)}(g) - S_{\Lambda_m(t) \setminus \Lambda}(g)$$

and note that, as in (1.58), we get

$$\|S_{\Lambda_m(t) \setminus \Lambda}(g)\|_p \le C_9(p) \sigma_m(g, \mathcal{H})_p. \tag{1.60}$$

The key point of the proof of Theorem 1.22 is the inequality

$$\|S_{\Lambda \setminus \Lambda_m(t)}(g)\|_p \le C(p, t) \|S_{\Lambda_m(t) \setminus \Lambda}(g)\|_p, \tag{1.61}$$

which will be derived from the following two lemmas.

**Lemma 1.23** *Consider*

$$f = \sum_{I \in Q} c_I H_I, \quad |Q| = N.$$

*Let $1 \le p < \infty$. Assume*

$$\|c_I H_I\|_p \le 1, \quad I \in Q. \tag{1.62}$$

*Then*

$$\|f\|_p \le C_{10}(p) N^{1/p}.$$

**Lemma 1.24** *Consider*

$$f = \sum_{I \in Q} c_I H_I, \qquad |Q| = N.$$

*Let* $1 < p \le \infty$. *Assume*

$$\|c_I H_I\|_p \ge 1, \qquad I \in Q.$$

*Then*

$$\|f\|_p \ge C_{11}(p) N^{1/p}.$$

*Proof of Lemma 1.23.* We note that in the case $1 < p \le 2$ the statement of Lemma 1.23 follows from (1.54). We will give a proof of this lemma for all $1 \le p < \infty$. We have

$$\|c_I H_I\|_p = |c_I||I|^{1/p-1/2}.$$

Assumption (1.62) implies

$$|c_I| \le |I|^{1/2-1/p}.$$

Next, we have

$$\|f\|_p \le \|\sum_{I \in Q} |c_I H_I|\|_p \le \|\sum_{I \in Q} |I|^{-1/p} \chi_I(x)\|_p, \qquad (1.63)$$

where $\chi_I(x)$ is the characteristic function of the interval $I$.

In order to proceed further we need a lemma.

**Lemma 1.25** *Let* $n_1 < n_2 < \cdots < n_s$ *be integers, and let* $E_j \subset [0, 1]$ *be measurable sets,* $j = 1, \ldots, s$. *Then, for any* $0 < q < \infty$, *we have*

$$\int_0^1 \left( \sum_{j=1}^s 2^{n_j/q} \chi_{E_j}(x) \right)^q dx \le C_{12}(q) \sum_{j=1}^s 2^{n_j} |E_j|.$$

*Proof* Let

$$F(x) := \sum_{j=1}^s 2^{n_j/q} \chi_{E_j}(x),$$

and estimate it on the sets

$$E_l^- := E_l \setminus \cup_{k=l+1}^s E_k, \quad l = 1, \ldots, s-1; \quad E_s^- := E_s.$$

We have, for $x \in E_l^-$,

$$F(x) \le \sum_{j=1}^l 2^{n_j/q} \le C(q) 2^{n_l/q}.$$

Therefore,

$$\int_0^1 F(x)^q \, dx \le C(q)^q \sum_{l=1}^s 2^{n_l} |E_l^-| \le C(q)^q \sum_{l=1}^s 2^{n_l} |E_l|,$$

which proves the lemma. □

We return to the proof of Lemma 1.23. Denote by $n_1 < n_2 < \cdots < n_s$ all integers such that there is $I \in Q$ with $|I| = 2^{-n_j}$. Introduce the sets

$$E_j := \cup_{I \in Q; |I| = 2^{-n_j}} I.$$

Then the number $N$ of elements in $Q$ can be written in the form

$$N = \sum_{j=1}^s |E_j| 2^{n_j}. \tag{1.64}$$

Using this notation, the right-hand side of (1.63) can be rewritten as

$$Y := \left( \int_0^1 \left( \sum_{j=1}^s 2^{n_j/p} \chi_{E_j}(x) \right)^p dx \right)^{1/p}.$$

Applying Lemma 1.25 with $q = p$, we get

$$\|f\|_p \le Y \le C_{13}(p) \left( \sum_{j=1}^s |E_j| 2^{n_j} \right)^{1/p} = C_{13}(p) N^{1/p}.$$

In the last step we used (1.64). Lemma 1.23 is now proved. □

*Proof of Lemma 1.24.* We derive Lemma 1.24 from Lemma 1.23. Define

$$u := \sum_{I \in Q} \bar{c}_I |c_I|^{-1} |I|^{1/p - 1/2} H_I,$$

where the bar denotes the complex conjugate number. Then for $p' = p/(p-1)$ we have

$$\|\bar{c}_I |c_I|^{-1} |I|^{1/p-1/2} H_I\|_{p'} = 1,$$

and, by Lemma 1.23,

$$\|u\|_{p'} \le C_{10}(p) N^{1/p'}. \tag{1.65}$$

Consider $\langle f, u \rangle$. We have, on the one hand,

$$\langle f, u \rangle = \sum_{I \in Q} |c_I| |I|^{1/p - 1/2} = \sum_{I \in Q} \|c_I H_I\|_p \ge N, \tag{1.66}$$

and, on the other hand,

$$\langle f, u \rangle \leq \|f\|_p \|u\|_{p'}. \tag{1.67}$$

Combining (1.65)–(1.67), we obtain the statement of Lemma 1.24. $\square$

We now complete the proof of Theorem 1.22. It remained to prove inequality (1.61). Let

$$A := \max_{I \in \Lambda \backslash \Lambda_m(t)} \|c_I(g) H_I\|_p,$$

and let

$$B := \min_{I \in \Lambda_m(t) \backslash \Lambda} \|c_I(g) H_I\|_p.$$

Then, by definition of $\Lambda_m(t)$, we have

$$B \geq tA. \tag{1.68}$$

Using Lemma 1.23, we get

$$\|S_{\Lambda \backslash \Lambda_m(t)}(g)\|_p \leq AC_{10}(p)|\Lambda \backslash \Lambda_m(t)|^{1/p} \leq t^{-1} B C_{10}(p)|\Lambda \backslash \Lambda_m(t)|^{1/p}. \tag{1.69}$$

Using Lemma 1.24, we deduce

$$\|S_{\Lambda_m(t) \backslash \Lambda}(g)\|_p \geq B C_{11}(p)|\Lambda_m(t) \backslash \Lambda|^{1/p}. \tag{1.70}$$

Taking into account that $|\Lambda_m(t) \backslash \Lambda| = |\Lambda \backslash \Lambda_m(t)|$, we obtain relation (1.61) from (1.69) and (1.70).

The proof of Theorem 1.22 is now complete. $\square$

We now discuss the multivariate analog of Theorem 1.11. There are several natural generalizations of the Haar system to the $d$-dimensional case. We first describe that one for which the statement of Theorem 1.11 and its proof coincide with the one-dimensional version. First of all, we include in the system of functions the constant function

$$H_{[0,1]^d}(x) = 1, \quad x \in [0, 1)^d.$$

Next, we define $2^d - 1$ functions with support $[0, 1)^d$. Take any combination of intervals $Q_1, \ldots, Q_d$, where $Q_i = [0, 1]$ or $Q_i = [0, 1)$ with at least one $Q_j = [0, 1)$, and define for $Q = Q_1 \times \cdots \times Q_d$, $x = (x_1, \ldots, x_d)$,

$$H_Q(x) := \prod_{i=1}^{d} H_{Q_i}(x_i).$$

We shall also denote these functions by $H^k_{[0,1)^d}(x)$, $k = 1, \ldots, 2^d - 1$. We define the basis of the Haar functions with supports on dyadic cubes of the form

$$J = [(j_1 - 1)2^{-n}, j_1 2^{-n}) \times \cdots \times [(j_d - 1)2^{-n}, j_d 2^{-n}), \qquad (1.71)$$
$$j_i = 1, \ldots, 2^n; \quad n = 0, 1, \ldots$$

For each dyadic cube of the form (1.71), we define $2^d - 1$ basis functions

$$H^k_J(x) := 2^{nd/2} H^k_{[0,1)^d}(2^n(x - (j_1 - 1, \ldots, j_d - 1)2^{-n})), \quad k = 1, \ldots, 2^d - 1.$$

We can also use another enumeration of these functions. Let $H^k_{[0,1)^d}(x) = H_Q(x)$ with

$$Q = Q_1 \times \cdots \times Q_d, \quad Q_i = [0, 1), \quad i \in E,$$
$$Q_i = [0, 1], \quad i \in \{1, d\} \setminus E, \quad E \neq \emptyset.$$

Consider a dyadic interval $I$ of the form

$$I = I_1 \times \cdots \times I_d, \quad I_i = [(j_i - 1)2^{-n}, j_i 2^{-n}), \quad i \in E, \qquad (1.72)$$
$$I_i = [(j_i - 1)2^{-n}, j_i 2^{-n}], \quad i \in \{1, d\} \setminus E, \quad E \neq \emptyset,$$

and define $H_I(x) := H^k_J(x)$. Denoting the set of dyadic intervals $D$ as the set of all dyadic cubes of the form (1.72) amended by the cube $[0, 1]^d$, and denoting by $\mathcal{H}$ the corresponding basis $\{H_I\}_{I \in D}$, we get the multivariate Haar system.

**Remark 1.26** Theorem 1.11 holds for the multivariate Haar system $\mathcal{H}$ with the constant $C(p)$ allowed to depend also on $d$.

In this section we have studied approximation in $L_p([0, 1])$ and have made a remark about approximation in $L_p([0, 1]^d)$. We can treat the approximation in $L_p(\mathbb{R}^d)$ in the same way.

**Remark 1.27** Theorem 1.11 holds for approximation in $L_p(\mathbb{R}^d)$.

Let us now discuss another multivariate Haar basis $\mathcal{H}^d := \mathcal{H} \times \cdots \times \mathcal{H}$, which is obtained from the univariate one by the tensor product. We have already discussed above (see (1.44)) some results about greedy approximation with regard to $\mathcal{H}^d$. We now give a proof of the lower bound in (1.44).

For a set $\Lambda$ of indices, we define

$$g_{\Lambda, p} := \sum_{I \in \Lambda} |I|^{1/2 - 1/p} H_I.$$

For each $n \in \mathbb{N}$, we define two sets $A$ and $B$ of dyadic intervals $I$ as follows:

$$A := \{I : \quad |I| = 2^{-n}\};$$

$$B := \{I : \quad I \notin A, \forall I \neq I' \quad \text{we have} \quad I \cap I' = \emptyset; \quad |B| = |A|\}.$$

Let $2 \leq p < \infty$ be given. Let $m := |A|$ and consider

$$f = g_{A,p} + 2g_{B,p}.$$

Then we have, on the one hand,

$$G_m(f, \mathcal{H}_p^d) = G_m^p(f, \mathcal{H}^d) = 2g_{B,p}$$

and

$$\|f - G_m^p(f, \mathcal{H}^d)\|_p = \|g_{A,p}\|_p \gg m^{1/p}(\log m)^{(1/2-1/p)(d-1)}. \tag{1.73}$$

On the other hand, we have

$$\sigma_m(f, \mathcal{H}^d)_p \leq \|2g_{B,p}\|_p \ll m^{1/p}. \tag{1.74}$$

Relations (1.73) and (1.74) imply the required lower estimate in the case $2 \leq p < \infty$. The remaining case $1 < p \leq 2$ can be handled in the same way considering the function $f = 2g_{A,p} + g_{B,p}$.

## 1.4 Quasi-greedy and almost greedy bases

In Section 1.3 we imposed the condition

$$\inf_k \|\psi_k\| > 0, \tag{1.75}$$

on a basis $\Psi$, to define $G_m(f, \Psi)$ for all $f \in X$. We noticed that in the greedy basis case this condition is always satisfied. In this section we assume that (1.75) is satisfied.

Let us discuss the question of weakening the requirement that a basis be a greedy basis. We begin with a concept of quasi-greedy basis that was introduced in Konyagin and Temlyakov (1999a).

**Definition 1.28** We call a basis $\Psi$ a quasi-greedy basis if, for every $f \in X$ and every permutation $\rho \in D(f)$, we have

$$\|G_m(f, \Psi, \rho)\|_X \leq C\|f\|_X \tag{1.76}$$

with a constant $C$ independent of $f$, $m$ and $\rho$.

It is clear that (1.76) is weaker then (1.35). Wojtaszczyk (2000) proved the following theorem.

**Theorem 1.29** *A basis* $\Psi$ *is quasi-greedy if and only if, for any* $f \in X$ *and any* $\rho \in D(f)$, *we have*

$$\|f - G_m(f, \Psi, \rho)\| \to 0 \quad as \quad m \to \infty. \tag{1.77}$$

Theorem 1.29 allows us to use (1.77) as an equivalent definition of a quasi-greedy basis. We give one more equivalent definition of a quasi-greedy basis.

**Definition 1.30** We say that a basis $\Psi$ is quasi-greedy if there exists a constant $C_Q$ such that, for any $f \in X$ and any finite set of indices $\Lambda$ having the property

$$\min_{k \in \Lambda} |c_k(f)| \geq \max_{k \notin \Lambda} |c_k(f)|, \tag{1.78}$$

we have

$$\|S_\Lambda(f, \Psi)\| \leq C_Q \|f\|. \tag{1.79}$$

It is clear that for elements $f$ with the unique decreasing rearrangement of coefficients ($\#D(f) = 1$), inequalities (1.76) and (1.79) are equivalent. By slightly modifying the coefficients and using the continuity argument, we deduce that (1.76) and (1.79) are equivalent for general $f$.

We now continue a discussion from Section 1.3 of relations between the following concepts: greedy basis, unconditional basis, democratic basis and quasi-greedy basis. Theorem 1.15 states that *greedy = unconditional + democratic*. It is clear from the definition of quasi-greedy basis that an unconditional basis is always a quasi-greedy basis. We now give an example from Konyagin and Temlyakov (1999a) of a basis that is quasi-greedy and democratic (even superdemocratic) that is not an unconditional basis.

It is clear that an unconditional and democratic basis $\Psi$ satisfies the following inequality:

$$\left\| \sum_{k \in P} \theta_k \psi_k \right\| \leq D_S \left\| \sum_{k \in Q} \epsilon_k \psi_k \right\| \tag{1.80}$$

for any two finite sets $P$ and $Q$, $|P| = |Q|$, and any choices of signs $\theta_k = \pm 1$, $k \in P$, and $\epsilon_k = \pm 1$, $k \in Q$.

**Definition 1.31** We say that a basis $\Psi$ is a superdemocratic basis if it satisfies (1.80).

Theorem 1.15 implies that a greedy basis is a superdemocratic one. Now we will construct an example of a superdemocratic quasi-greedy basis which is not an unconditional basis and, therefore, by Theorem 1.15, is not a greedy basis.

Let $X$ be the set of all real sequences $x = (x_1, x_2, \dots) \in \ell_2$, $\|x\|_2 := \|x\|_{\ell_2}$, such that

$$\|x\|_+ := \sup_{N \in \mathbb{N}} \left| \sum_{n=1}^{N} x_n/\sqrt{n} \right|$$

is finite. Clearly, $X$ equipped with the norm

$$\|\cdot\| = \max(\|\cdot\|_2, \|\cdot\|_+)$$

is a Banach space. Let $\psi_k \in X$, $k = 1, 2, \dots$, be defined as $(\psi_k)_n = 1$ if $n = k$ and $(\psi_k)_n = 0$ otherwise. By $X_0$ denote the subspace of $X$ generated by the elements $\psi_k$. It is easy to see that $\Psi = \{\psi_k\}$ is a democratic basis in $X_0$. Moreover, it is superdemocratic: for any $k_1, \dots, k_m$, and for any choice of signs,

$$\sqrt{m} \leq \left\| \sum_{j=1}^{m} \pm \psi_{k_j} \right\| < 2\sqrt{m}. \tag{1.81}$$

Indeed, we have

$$\left\| \sum_{j=1}^{m} \pm \psi_{k_j} \right\|_2 = \sqrt{m},$$

$$\left\| \sum_{j=1}^{m} \pm \psi_{k_j} \right\|_+ \leq \sum_{j=1}^{m} 1/\sqrt{j} < 2\sqrt{m},$$

and (1.81) follows. However, $\Psi$ is not an unconditional basis since, for $m \geq 2$,

$$\left\| \sum_{k=1}^{m} \psi_k/\sqrt{k} \right\| \geq \sum_{k=1}^{m} 1/k \asymp \log m,$$

but

$$\left\| \sum_{k=1}^{m} (-1)^k \psi_k/\sqrt{k} \right\| \asymp \sqrt{\log m}.$$

We now prove that the basis $\Psi$ constructed above is a quasi-greedy basis. Assume $\|f\| = 1$. Then, by definition of $\|\cdot\|$ we have

$$\sum_{k=1}^{\infty} |c_k(f)|^2 \leq 1, \tag{1.82}$$

and, for any $M$,

$$\left| \sum_{k=1}^{M} c_k(f)k^{-1/2} \right| \leq 1. \tag{1.83}$$

It is clear that for any $\Lambda$ we have

$$\|S_\Lambda(f, \Psi)\|_2 \leq \|f\|_2 \leq 1. \tag{1.84}$$

We now estimate $\|S_\Lambda(f, \Psi)\|_+$. Let $\Lambda$ be any finite set of indices satisfying (1.78), and let

$$\alpha := \min_{k \in \Lambda} |c_k(f)|.$$

If $\alpha = 0$, then $S_\Lambda(f, \Psi) = f$ and (1.79) holds. Therefore consider $\alpha > 0$, and for any $N$ let

$$\Lambda^+(N) := \{k \in \Lambda : k > N\}, \qquad \Lambda^-(N) := \{k \in \Lambda : k \leq N\}.$$

By Hölder's inequality we have, for any $N$,

$$\sum_{k \in \Lambda^+(N)} |c_k(f)| k^{-1/2} \leq \left( \sum_{k \in \Lambda^+(N)} |c_k(f)|^{3/2} \right)^{2/3} \left( \sum_{k > N} k^{-3/2} \right)^{1/3}$$

$$\ll N^{-1/6} \left( \sum_{k \in \Lambda^+(N)} |c_k(f)|^{3/2} (|c_k(f)|/\alpha)^{1/2} \right)^{2/3}$$

$$\ll (\alpha^2 N)^{-1/6}. \tag{1.85}$$

Choose $N_\alpha := [\alpha^{-2}] + 1$. Then, for any $M \leq N_\alpha$, we have by (1.83) that

$$\left| \sum_{k \in \Lambda^-(M)} c_k(f) k^{-1/2} \right| \leq \left| \sum_{k=1}^{M} c_k(f) k^{-1/2} \right| + \left| \sum_{k \notin \Lambda^-(M), k \leq M} c_k(f) k^{-1/2} \right|$$

$$\leq 1 + \alpha \sum_{k=1}^{M} k^{-1/2} \leq 1 + 2\alpha M^{1/2} \ll 1. \tag{1.86}$$

For $M > N_\alpha$, we get, using (1.85) and (1.86),

$$\left| \sum_{k \in \Lambda^-(M)} c_k(f) k^{-1/2} \right| \leq \left| \sum_{k \in \Lambda^-(N_\alpha)} c_k(f) k^{-1/2} \right| + \sum_{k \in \Lambda^+(N_\alpha)} |c_k(f)| k^{-1/2} \ll 1.$$

Thus

$$\|S_\Lambda(f, \Psi)\|_+ \leq C,$$

which completes the proof.

The above example and Theorem 1.15 show that a quasi-greedy basis is not necessarily a greedy basis. Further results on quasi-greedy bases can be found in Dilworth *et al.* (2003) and Wojtaszczyk (2000).

The above discussion shows that a quasi-greedy basis is not necessarily an unconditional basis. However, quasi-greedy bases have some properties that are close to those of unconditional bases. We formulate two of them (see, for example, Konyagin and Temlyakov (2002)).

**Lemma 1.32** *Let* $\Psi$ *be a quasi-greedy basis. Then, for any two finite sets of indices* $A \subseteq B$ *and coefficients* $0 < t \leq |a_j| \leq 1$, $j \in B$, *we have*

$$\|\sum_{j \in A} a_j \psi_j\| \leq C(X, \Psi, t) \|\sum_{j \in B} a_j \psi_j\|.$$

It will be convenient to define the quasi-greedy constant $K$ to be the least constant such that

$$\|G_m(f)\| \leq K \|f\| \quad \text{and} \quad \|f - G_m(f)\| \leq K \|f\|, \quad f \in X.$$

**Lemma 1.33** *Suppose* $\Psi$ *is a quasi-greedy basis with a quasi-greedy constant* $K$. *Then, for any real numbers* $a_j$ *and any finite set of indices* $P$, *we have*

$$(4K^2)^{-1} \min_{j \in P} |a_j| \|\sum_{j \in P} \psi_j\| \leq \|\sum_{j \in P} a_j \psi_j\| \leq 2K \max_{j \in P} |a_j| \|\sum_{j \in P} \psi_j\|.$$

We note that the $m$th greedy approximant $G_m(x, \Psi)$ changes if we renormalize the basis $\Psi$ (replace it by a basis $\{\lambda_n \psi_n\}$). This gives us more flexibility in adjusting a given basis $\Psi$ for greedy approximation. Let us make one observation from Konyagin and Temlyakov (2003a) along these lines.

**Proposition 1.34** *Let* $\Psi = \{\psi_n\}_{n=1}^{\infty}$ *be a normalized basis for a Banach space* $X$. *Then the basis* $\{e_n\}_{n=1}^{\infty}$, $e_n := 2^n \psi_n$, $n = 1, 2, \ldots$, *is a quasi-greedy basis in* $X$.

We proceed to an intermediate concept of *almost greedy basis*. This concept was introduced and studied in Dilworth *et al.* (2003). Let

$$f = \sum_{k=1}^{\infty} c_k(f) \psi_k.$$

We define the following expansional best $m$-term approximation of $f$:

$$\tilde{\sigma}_m(f) := \tilde{\sigma}_m(f, \Psi) := \inf_{\Lambda, |\Lambda|=m} \|f - \sum_{k \in \Lambda} c_k(f) \psi_k\|.$$

It is clear that

$$\sigma_m(f, \Psi) \leq \tilde{\sigma}_m(f, \Psi).$$

It is also clear that for an unconditional basis $\Psi$ we have

$$\tilde{\sigma}_m(f, \Psi) \le C(X, \Psi)\sigma_m(f, \Psi).$$

**Definition 1.35** We call a basis $\Psi$ an almost greedy basis if, for every $f \in X$, there exists a permutation $\rho \in D(f)$ such that we have the inequality

$$\|f - G_m(f, \Psi, \rho)\|_X \le C\tilde{\sigma}_m(f, \Psi)_X \qquad (1.87)$$

with a constant independent of $f$ and $m$.

The following proposition follows from the proof of theorem 3.3 of Dilworth *et al.* (2003) (see Theorem 1.37 below).

**Proposition 1.36** *If $\Psi$ is an almost greedy basis then (1.87) holds for any permutation $\rho \in D(f)$.*

The following characterization of almost greedy bases was obtained in Dilworth *et al.* (2003).

**Theorem 1.37** *Suppose $\Psi$ is a basis of a Banach space. The following are equivalent:*

  *(A) $\Psi$ is almost greedy;*
  *(B) $\Psi$ is quasi-greedy and democratic;*
  *(C) for any (respectively, every) $\lambda > 1$ there is a constant $C = C_\lambda$ such that*

$$\|f - G_{[\lambda m]}(f, \Psi)\| \le C_\lambda \sigma_m(f, \Psi).$$

In order to give the reader an idea of relations between $\tilde{\sigma}$ and $\sigma$, we present an estimate for $\tilde{\sigma}_n(f, \Psi)$ in terms of $\sigma_m(f, \Psi)$ for a quasi-greedy basis $\Psi$. For a basis $\Psi$ we define the *fundamental function*

$$\varphi(m) := \sup_{A:|A| \le m} \left\| \sum_{k \in A} \psi_k \right\|.$$

We also need the following function:

$$\phi(m) := \inf_{A:|A| = m} \left\| \sum_{k \in A} \psi_k \right\|.$$

The following inequality was obtained in Dilworth *et al.* (2003).

**Theorem 1.38** *Let $\Psi$ be a quasi-greedy basis. Then, for any $m$ and $r$ there exists a set $E$, $|E| \le m + r$, such that*

$$\|f - S_E(f, \Psi)\| \le C \left( 1 + \frac{\varphi(m)}{\phi(r+1)} \right) \sigma_m(f, \Psi).$$

In Section 1.3, in addition to bases $\Psi$ satisfying (1.75), we discussed a more general case that included bases satisfying (1.33). In the latter case we defined the greedy algorithm $G_m(f, \Psi)$ for functions $f$ of the form (1.34). We gave a definition of a greedy basis in the general case, which included those bases satisfying (1.33). However, the characterization of greedy bases given by Theorem 1.15 excluded bases satisfying (1.33). We note that a similar attempt to include bases $\Psi$ satisfying (1.33) into the consideration of quasi-greedy bases does not work. Indeed, let $\Psi$ be a normalized unconditional basis and consider a renormalized basis $\Psi' := \{\psi'_k := k^{-3}\psi_k\}$. Clearly, $\Psi'$ is also an unconditional basis, and therefore inequality (1.76) is satisfied for any $f$ of the form (1.34). However, for the function

$$f := \sum_{k=1}^{\infty} k^{-2}\psi_k = \sum_{k=1}^{\infty} k\psi'_k$$

we cannot apply the algorithm $G_m(\cdot, \Psi')$ because the expansion coefficients are not bounded.

## 1.5 Weak Greedy Algorithms with respect to bases

The greedy approximant $G_m(f, \Psi)$ considered in Sections 1.3 and 1.4 was defined as the sum

$$\sum_{j=1}^{m} c_{k_j}(f, \Psi)\psi_{k_j}$$

of the expansion terms with the $m$ biggest coefficients in absolute value (see (1.32)). In this section we discuss a more flexible way to construct a greedy approximant. The rule for choosing the expansion terms for approximation will be weaker than in the greedy algorithm $G_m(\cdot, \Psi)$. We are motivated by Theorem 1.22. Instead of taking $m$ terms with the biggest coefficients we now take $m$ terms with near-biggest coefficients. We proceed to a formal definition of the Weak Greedy Algorithm with regard to a basis $\Psi$. We assume here that $\Psi$ satisfies (1.75).

Let $t \in (0, 1]$ be a fixed parameter. For a given basis $\Psi$ and a given $f \in X$, let $\Lambda_m(t)$ be any set of $m$ indices such that

$$\min_{k \in \Lambda_m(t)} |c_k(f, \Psi)| \ge t \max_{k \notin \Lambda_m(t)} |c_k(f, \Psi)| \tag{1.88}$$

and define

$$G_m^t(f) := G_m^t(f, \Psi) := \sum_{k \in \Lambda_m(t)} c_k(f, \Psi)\psi_k.$$

We call it the Weak Greedy Algorithm (WGA) with the weakness sequence $\{t\}$ (the weakness parameter $t$). We note that the WGA with regard to a basis was introduced in the very first paper (see Temlyakov (1998a)) on greedy bases. It is clear that $G_m^1(f, \Psi) = G_m(f, \Psi)$. It is also clear that, in the case $t < 1$, we have more flexibility in building a weak greedy approximant $G_m^t(f, \Psi)$ than in building $G_m(f, \Psi)$: it is one advantage of a weak greedy approximant $G_m^t(f, \Psi)$. The question is: How much does this flexibility affect the efficiency of the algorithm? Surprisingly, it turns out that the effect is minimal: it is only reflected in a multiplicative constant (see below).

We begin our discussion with the case when $\Psi$ is a greedy basis. It was proved in Temlyakov (1998a) (see Theorem 1.22) that, when $X = L_p$, $1 < p < \infty$, and $\Psi$ is the Haar system $\mathcal{H}_p$ normalized in $L_p$, we have

$$\|f - G_m^t(f, \mathcal{H}_p)\|_{L_p} \le C(p, t)\sigma_m(f, \mathcal{H}_p)_{L_p}, \qquad (1.89)$$

for any $f \in L_p$. It was noted in Konyagin and Temlyakov (2002) that the proof of (1.89) from Temlyakov (1998a) works for any greedy basis, not merely the Haar system $\mathcal{H}_p$. Thus, we have the following result.

**Theorem 1.39** *For any greedy basis $\Psi$ of a Banach space $X$, and any $t \in (0, 1]$, we have*

$$\|f - G_m^t(f, \Psi)\|_X \le C(\Psi, t)\sigma_m(f, \Psi)_X, \qquad (1.90)$$

*for each $f \in X$.*

We now consider the Weak Greedy Algorithm with regard to a quasi-greedy basis $\Psi$. It was proved in Konyagin and Temlyakov (2002) that the weak greedy approximant has properties similar to those of the greedy approximant.

**Theorem 1.40** *Let $\Psi$ be a quasi-greedy basis. Then, for a fixed $t \in (0, 1]$ and any $m$, we have, for any $f \in X$,*

$$\|G_m^t(f, \Psi)\| \le C(t)\|f\|. \qquad (1.91)$$

The following theorem from Konyagin and Temlyakov (2002) is essentially due to Wojtaszczyk (2000).

**Theorem 1.41** *Let $\Psi$ be a quasi-greedy basis for a Banach space $X$. Then, for any fixed $t \in (0, 1]$, we have, for each $f \in X$, that*

$$G_m^t(f, \Psi) \to f \quad as \quad m \to \infty.$$

Let us now proceed to an almost greedy basis $\Psi$. The following result was established in Konyagin and Temlyakov (2002).

**Theorem 1.42** *Let* $\Psi$ *be an almost greedy basis. Then, for* $t \in (0, 1]$, *we have, for any* $m$,

$$\|f - G_m^t(f, \Psi)\| \le C(t)\tilde{\sigma}_m(f, \Psi). \tag{1.92}$$

*Proof* We drop $\Psi$ from the notation for the sake of brevity. Take any $\epsilon > 0$ and find $P$, $|P| = m$, such that

$$\|f - S_P(f)\| \le \tilde{\sigma}_m(f) + \epsilon.$$

Let $Q := \Lambda_m(t)$ with $\Lambda_m(t)$ from the definition of $G_m^t(f)$. Then

$$\|f - G_m^t(f)\| \le \|f - S_P(f)\| + \|S_P(f) - S_Q(f)\|. \tag{1.93}$$

We have

$$S_P(f) - S_Q(f) = S_{P \setminus Q}(f) - S_{Q \setminus P}(f). \tag{1.94}$$

Let us first estimate $\|S_{Q \setminus P}(f)\|$. Denote $f_1 := f - S_P(f)$. Then

$$S_{Q \setminus P}(f) = S_{Q \setminus P}(f_1).$$

Next,

$$\min_{k \in Q \setminus P} |c_k(f_1)| = \min_{k \in Q \setminus P} |c_k(f)| \ge \min_{k \in Q} |c_k(f)|$$

$$\ge t \max_{k \notin Q} |c_k(f)| \ge t \max_{k \notin Q} |c_k(f_1)| = t \max_{k \notin Q \setminus P} |c_k(f_1)|.$$

Thus $Q \setminus P = \Lambda_n(t)$ for $f_1$ with $n := |Q \setminus P|$. Theorem 1.37 guarantees that an almost greedy basis is a quasi-greedy basis. Therefore, by Theorem 1.40 we have

$$\|S_{Q \setminus P}(f)\| \le C_1(t)\|f_1\|. \tag{1.95}$$

We now estimate $\|S_{P \setminus Q}(f)\|$. From the definition of $Q$ we easily derive

$$at \le b, \quad \text{where} \quad a := \max_{k \in P \setminus Q} |c_k(f)|, \quad b := \min_{k \in Q \setminus P} |c_k(f)|. \tag{1.96}$$

By Lemma 1.33 (see lemma 2.1 of Dilworth *et al.* (2003)),

$$\|S_{P \setminus Q}(f)\| \le 2Ka\|\sum_{k \in P \setminus Q} \psi_k\| \tag{1.97}$$

and (see lemma 2.2 from Dilworth *et al.* (2003))

$$\|S_{Q \setminus P}(f)\| \ge (4K^2)^{-1}b\|\sum_{k \in Q \setminus P} \psi_k\|. \tag{1.98}$$

By Theorem 1.37 an almost greedy basis is a democratic basis. Thus we obtain

$$\| \sum_{k \in P \setminus Q} \psi_k \| \le D \| \sum_{k \in Q \setminus P} \psi_k \|. \tag{1.99}$$

Combining (1.93)–(1.99) we obtain (1.92). Theorem 1.42 is proved.        □

We now discuss the stability of the greedy-type property of a basis. Let $0 < a \le \lambda_k \le b < \infty$, $k = 1, 2, \ldots$, and, for a basis $\Psi = \{\psi_k\}$, consider $\Psi^\lambda := \{\lambda_k \psi_k\}$. The following theorem is from Konyagin and Temlyakov (2002). We note that the case for quasi-greedy bases was proved in Wojtaszczyk (2000).

**Theorem 1.43** *Let a basis $\Psi$ have one of the following properties:*

*(1) greedy;*
*(2) almost greedy;*
*(3) quasi-greedy.*

*Then the basis $\Psi^\lambda$ has the same property.*

*Proof* Let $f \in X$ and

$$f = \sum_k c_k(f) \psi_k = \sum_k c_k(f) \lambda_k^{-1} \lambda_k \psi_k.$$

Consider

$$G_m(f, \Psi^\lambda) = \sum_{k \in \Lambda_m} (c_k(f) \lambda_k^{-1}) \lambda_k \psi_k.$$

Then, using $\lambda_k \in [a, b]$ and the definition of the $G_m(f, \Psi^\lambda)$, we obtain

$$\min_{k \in \Lambda_m} |c_k(f)| \ge a \min_{k \in \Lambda_m} |c_k(f)| \lambda_k^{-1} \ge a \max_{k \notin \Lambda_m} |c_k(f)| \lambda_k^{-1} \ge \frac{a}{b} \max_{k \notin \Lambda_m} |c_k(f)|.$$

Therefore, the set $\Lambda_m$ can be interpreted as a $\Lambda_m(t)$ with $t = a/b$ with regard to the basis $\Psi$. It remains to apply the corresponding results for $G_m^t(f, \Psi)$: (1.90) in case (1); (1.91) in case (3); and (1.92) in case (2). This completes the proof of Theorem 1.43.        □

Kamont and Temlyakov (2004) studied the following modification of the above weak-type greedy algorithm as a way to weaken further restriction (1.88). We call this modification the Weak Greedy Algorithm (WGA) with a weakness sequence $\tau = \{t_k\}$. Let a weakness sequence $\tau := \{t_k\}_{k=1}^\infty$, $t_k \in [0, 1]$, $k = 1, \ldots$, be given. We define the WGA by induction. We take an element $f \in X$, and at the first step we let

$$\Lambda_1(\tau) := \{n_1\}; \qquad G_1^\tau(f, \Psi) := c_{n_1} \psi_{n_1},$$

with any $n_1$ satisfying

$$|c_{n_1}| \geq t_1 \max_n |c_n|,$$

where we write $c_n := c_n(f, \Psi)$ for brevity. Assume we have already defined

$$G^{\tau}_{m-1}(f, \Psi) := G^{X,\tau}_{m-1}(f, \Psi) := \sum_{n \in \Lambda_{m-1}(\tau)} c_n \psi_n.$$

Then, at the $m$th step we define

$$\Lambda_m(\tau) := \Lambda_{m-1}(\tau) \cup \{n_m\}, \qquad G^{\tau}_m(f, \Psi) := G^{X,\tau}_m(f, \Psi) := \sum_{n \in \Lambda_m(\tau)} c_n \psi_n,$$

with any $n_m \notin \Lambda_{m-1}(\tau)$ satisfying

$$|c_{n_m}| \geq t_m \max_{n \notin \Lambda_{m-1}(\tau)} |c_n|.$$

Thus, for an $f \in X$ the WGA builds a rearrangement of a subsequence of the expansion (1.31). If $\Psi$ is an unconditional basis then we always have the limit $G^{\tau}_m(f, \Psi) \to f^*$. It is clear that in this case $f^* = f$ if and only if the sequence $\{n_k\}_{k=1}^{\infty}$ contains indices of all non-zero $c_n(f, \Psi)$. We say that the WGA corresponding to $\Psi$ and $\tau$ converges if, for any realization $G^{\tau}_m(f, \Psi)$, we have

$$\|f - G^{\tau}_m(f, \Psi)\| \to 0 \quad \text{as} \quad m \to \infty$$

for all $f \in X$.

We formulate here only one theorem from Kamont and Temlyakov (2004).

**Theorem 1.44** *Let* $2 \leq p < \infty$, $d \geq 1$, *and let* $\Psi$ *be a normalized unconditional basis in* $L_p([0,1]^d)$. *Let* $\tau = \{t_n, n \geq 1\}$ *be a weakness sequence. Then the WGA corresponding to* $\Psi$ *and* $\tau$ *converges if and only if* $\tau \notin \ell_p$.

## 1.6 Thresholding and minimal systems

In this section we briefly discuss some further generalizations. Here, we assume that $X$ is a quasi-Banach space and replace a basis by a complete minimal system. In addition, we consider the Weak Thresholding Algorithm and prove that its convergence is equivalent to convergence of the Weak Greedy Algorithm (see Proposition 1.47). Thresholding algorithms are very useful in statistics (see, for example, Donoho and Johnstone (1994)).

Let $X$ be a quasi-Banach space (real or complex) with the quasi-norm $\|\cdot\|$ such that for all $x, y \in X$ we have $\|x+y\| \leq \alpha(\|x\| + \|y\|)$ and $\|tx\| = |t| \|x\|$. It is well known (see Kalton, Beck and Roberts (1984), lemma 1.1) that there is a $p$, $0 < p \leq 1$, such that

$$\left\| \sum_n x_n \right\| \le 4^{1/p} \left( \sum_n \|x_n\|^p \right)^{1/p}. \tag{1.100}$$

Let $\{e_n\} \subset X$ be a complete minimal system in $X$ with the conjugate (dual) system $\{e_n^*\} \subset X^*$ $(e_n^*(e_n) = 1, e_n^*(e_k) = 0, k \ne n)$. We assume that $\sup_n \|e_n^*\| < \infty$. This implies that for each $x \in X$ we have

$$\lim_{n \to \infty} e_n^*(x) = 0. \tag{1.101}$$

Any element $x \in X$ has a formal expansion

$$x \sim \sum_n e_n^*(x) e_n, \tag{1.102}$$

and various types of convergence of the series (1.102) can be studied. In this section we deal with greedy-type approximations with regard to the system $\{e_n\}$. We note that in this section we use the notations $x$ and $\{e_n\}$ for an element and for a system, respectively, different from the notation $f$ and $\Psi$ used in the previous sections, to emphasize that we are now in a more general setting. It will be convenient for us to define a unique "greedy ordering" in this section. For any $x \in X$ we define the greedy ordering for $x$ as the map $\rho : \mathbb{N} \to \mathbb{N}$ for which $\{j : e_j^*(x) \ne 0\} \subset \rho(\mathbb{N})$, and such that, if $j < k$, then either $|e_{\rho(j)}^*(x)| > |e_{\rho(k)}^*(x)|$ or $|e_{\rho(j)}^*(x)| = |e_{\rho(k)}^*(x)|$ and $\rho(j) < \rho(k)$. The $m$th greedy approximation is given by

$$G_m(x) := G_m(x, \{e_n\}) := \sum_{j=1}^m e_{\rho(j)}^*(x) e_{\rho(j)}.$$

The system $\{e_n\}$ is a quasi-greedy system (Konyagin and Temlyakov (1999a)) if there exists a constant $C$ such that $\|G_m(x)\| \le C \|x\|$ for all $x \in X$ and $m \in \mathbb{N}$. Wojtaszczyk (2000) proved that these are precisely the systems for which $\lim_{m \to \infty} G_m(x) = x$ for all $x$. If, as in Section 1.4, a quasi-greedy system $\{e_n\}$ is a basis, then we say that $\{e_n\}$ is a quasi-greedy basis. As mentioned above, it is clear that any unconditional basis is a quasi-greedy basis. We note that there are conditional quasi-greedy bases $\{e_n\}$ in some Banach spaces. Hence, for such a basis $\{e_n\}$ there exists a permutation of $\{e_n\}$ which forms a quasi-greedy system but not a basis. This remark justifies the study of the class of quasi-greedy systems rather than the class of quasi-greedy bases.

Greedy approximations are close to thresholding approximations (sometimes they are called *thresholding greedy approximations*). Thresholding approximations are defined by

$$T_\epsilon(x) := \sum_{j : |e_j^*(x)| \ge \epsilon} e_j^*(x) e_j, \quad \epsilon > 0.$$

Clearly, for any $\epsilon > 0$ there exists an $m$ such that $T_\epsilon(x) = G_m(x)$. Therefore, if $\{e_n\}$ is a quasi-greedy system then

$$\forall x \in X \quad \lim_{\epsilon \to 0} T_\epsilon(x) = x. \tag{1.103}$$

Conversely, following the Remark from Wojtaszczyk (2000), pp. 296–297, it is easy to show that condition (1.103) implies that $\{e_n\}$ is a quasi-greedy system.

As in Section 1.5, one can define the Weak Thresholding Approximation. Fix $t \in (0, 1)$. For $\epsilon > 0$ let

$$D_{t,\epsilon}(x) := \{j : t\epsilon \leq |e_j^*(x)| < \epsilon\}.$$

The Weak Thresholding Approximations are defined as all possible sums

$$T_{\epsilon,D}(x) = \sum_{j : |e_j^*(x)| \geq \epsilon} e_j^*(x)e_j + \sum_{j \in D} e_j^*(x)e_j,$$

where $D \subseteq D_{t,\epsilon}(x)$. We say that the Weak Thresholding Algorithm converges for $x \in X$, and write $x \in WT\{e_n\}(t)$ if, for any $D(\epsilon) \subseteq D_{t,\epsilon}$,

$$\lim_{\epsilon \to 0} T_{\epsilon,D(\epsilon)}(x) = x.$$

It is clear that the above relation is equivalent to

$$\lim_{\epsilon \to 0} \sup_{D \subseteq D_{t,\epsilon}(x)} \|x - T_{\epsilon,D}(x)\| = 0.$$

We proved in Konyagin and Temlyakov (2003a) (see Theorem 1.45 below) that the set $WT\{e_n\}(t)$ does not depend on $t \in (0, 1)$. Therefore, we can drop $t$ from the notation: $WT\{e_n\} = WT\{e_n\}(t)$.

It turns out that the Weak Thresholding Algorithm has more regularity than the Thresholding Algorithm: we will see that the set $WT\{e_n\}$ is linear. On the other hand, by "weakening" the Thresholding Algorithm (making convergence stronger), we do not narrow the convergence set too much. It is known that for many natural classes of sets $Y \subseteq X$ the convergence of $T_\epsilon(x)$ to $x$ for all $x \in Y$ is equivalent to the condition $Y \subseteq WT\{e_n\}$. In particular, it can be derived from Wojtaszczyk (2000), prop. 3, that the above two conditions are equivalent for $Y = X$.

We suppose that $X$ and $\{e_n\}$ satisfy the conditions stated at the beginning of this section. The following two theorems were proved in Konyagin and Temlyakov (2003a).

**Theorem 1.45** *Let $t, t' \in (0, 1)$, $x \in X$. Then the following conditions are equivalent:*

*(1)* $\lim_{\epsilon \to 0} \sup_{D \subseteq D_{t,\epsilon}(x)} \|T_{\epsilon,D}(x) - x\| = 0;$

*(2)* $\lim_{\epsilon \to 0} T_{\epsilon}(x) = x$ *and*

$$\lim_{\epsilon \to 0} \sup_{D \subseteq D_{t,\epsilon}(x)} \| \sum_{j \in D} e_j^*(x) e_j \| = 0; \tag{1.104}$$

*(3)* $\lim_{\epsilon \to 0} T_{\epsilon}(x) = x$ *and*

$$\lim_{\epsilon \to 0} \sup_{|a_j| \leq 1(j \in D_{t,\epsilon}(x))} \| \sum_{j \in D_{t,\epsilon}(x)} a_j e_j^*(x) e_j \| = 0; \tag{1.105}$$

*(4)* $\lim_{\epsilon \to 0} T_{\epsilon}(x) = x$ *and*

$$\lim_{\epsilon \to 0} \sup_{|b_j| < \epsilon(j:|e_j^*(x)| \geq \epsilon)} \| \sum_{j:|e_j^*(x)| \geq \epsilon} b_j e_j \| = 0; \tag{1.106}$$

*(5)* $\lim_{\epsilon \to 0} \sup_{D \subseteq D_{t',\epsilon}(x)} \|T_{\epsilon,D}(x) - x\| = 0.$

So, the set $WT\{e_n\}(t)$ defined above is indeed independent of $t \in (0, 1)$.

**Theorem 1.46** *The set $WT\{e_n\}$ is linear.*

Let us discuss relations between the Weak Thresholding Algorithm $T_{\epsilon,D}(x)$ and the Weak Greedy Algorithm $G_m^t(x)$. We define $G_m^t(x)$ with regard to a minimal system $\{e_n\}$ in the same way as it was defined for a basis $\Psi$. For a given system $\{e_n\}$ and $t \in (0, 1]$, we denote for $x \in X$ and $m \in \mathbb{N}$ by $W_m(t)$ any set of $m$ indices such that

$$\min_{j \in W_m(t)} |e_j^*(x)| \geq t \max_{j \notin W_m(t)} |e_j^*(x)|, \tag{1.107}$$

and define

$$G_m^t(x) := G_m^t(x, \{e_n\}) := S_{W_m(t)}(x) := \sum_{j \in W_m(t)} e_j^*(x) e_j.$$

It is clear that for any $t \in (0, 1]$ and any $D \subseteq D_{t,\epsilon}(x)$ there exist $m$ and $W_m(t)$ satisfying (1.107) such that

$$T_{\epsilon,D}(x) = S_{W_m(t)}(x).$$

Thus the convergence $G_m^t(x) \to x$ as $m \to \infty$ implies the convergence $T_{\epsilon,D}(x) \to x$ as $\epsilon \to 0$ for any $t \in (0, 1]$. We will now prove (see Konyagin and Temlyakov (2003a), prop. 2.2) that for $t \in (0, 1)$ the inverse is also true.

**Proposition 1.47** *Let $t \in (0, 1)$ and $x \in X$. Then the following two conditions are equivalent:*

$$\lim_{\epsilon \to 0} \sup_{D \subseteq D_{t,\epsilon}(x)} \|T_{\epsilon,D}(x) - x\| = 0, \tag{1.108}$$

$$\lim_{m \to \infty} \|G_m^t(x) - x\| = 0, \tag{1.109}$$

*for any realization $G_m^t(x)$.*

*Proof* The implication (1.109) $\Rightarrow$ (1.108) is simple and follows from the remark preceding Proposition 1.47. We prove that (1.108) $\Rightarrow$ (1.109). Let

$$\epsilon_m := \max_{j \notin W_m(t)} |e_j^*(x)|.$$

Clearly $\epsilon_m \to 0$ as $m \to \infty$. We have

$$G_m^t(x) = T_{2\epsilon_m}(x) + \sum_{j \in D_m} e_j^*(x)e_j \tag{1.110}$$

with $D_m$ having the following property: for any $j \in D_m$

$$t\epsilon_m \le |e_j^*(x)| < 2\epsilon_m.$$

Thus, by condition (5) from Theorem 1.45, for $t' = t/2$ we obtain (1.109).
Proposition 1.47 is now proved. $\qquad\qquad\square$

Proposition 1.47 and Theorem 1.45 imply that the convergence set of the Weak Greedy Algorithm $G_m^t(\cdot)$ does not depend on $t \in (0, 1)$ and coincides with $WT\{e_n\}$. By Theorem 1.46 this set is a linear set.

Let us comment on the case $t = 1$ that is not covered by Proposition 1.47. It is clear that $T_\epsilon(x) = G_m(x)$ with some $m$, and therefore $G_m(x) \to x$ as $m \to \infty$ implies $T_\epsilon(x) \to x$ as $\epsilon \to 0$. It is also not difficult to understand that, in general, $T_\epsilon(x) \to x$ as $\epsilon \to 0$ does not imply $G_m(x) \to x$ as $m \to \infty$. This can be done, for instance, by considering the trigonometric system in the space $L_p$, $p \ne 2$, and using the Rudin–Shapiro polynomials (see Temlyakov (1998c)). However, if, for the trigonometric system, we put the Fourier coefficients with equal absolute values in a natural order (say, lexicographic), then, in the case $1 < p < \infty$, by Riesz's theorem we obtain convergence of $G_m(f)$ from convergence of $T_\epsilon(f)$. Results from Konyagin and Skopina (2001) show that the situation is different for $p = 1$. In this case the natural order does not help to derive convergence of $G_m(f)$ from convergence of $T_\epsilon(f)$.

## 1.7 Greedy approximation with respect to the trigonometric system

The first results (see Theorem 1.11) on greedy approximation with regard to bases showed that the Haar basis and other bases similar to it are very well designed for greedy approximation. In this section we discuss another classical system, namely the trigonometric system, from the point of view of greedy approximation. It is well known that the trigonometric system is not an unconditional basis for $L_p$, $p \neq 2$. Therefore, by Theorem 1.15 it is not a greedy basis for $L_p$, $p \neq 2$. In this section we mostly discuss convergence properties of the Weak Greedy Algorithm with regard to the trigonometric system. It is a non-trivial problem. We will demonstrate how it relates to some deep results in harmonic and functional analysis.

Consider a periodic function $f \in L_p(\mathbb{T}^d)$, $1 \leq p \leq \infty$, ($L_\infty(\mathbb{T}^d) = C(\mathbb{T}^d)$), defined on the $d$-dimensional torus $\mathbb{T}^d$. Let a number $m \in \mathbb{N}$ and a number $t \in (0, 1]$ be given, and let $\Lambda_m$ be a set of $k \in \mathbb{Z}^d$ with the following properties:

$$\min_{k \in \Lambda_m} |\hat{f}(k)| \geq t \max_{k \notin \Lambda_m} |\hat{f}(k)|, \quad |\Lambda_m| = m, \tag{1.111}$$

where

$$\hat{f}(k) := (2\pi)^{-d} \int_{\mathbb{T}^d} f(x)e^{-i(k,x)} \, dx$$

is a Fourier coefficient of $f$. We define

$$G_m^t(f) := G_m^t(f, \mathcal{T}^d) := S_{\Lambda_m}(f) := \sum_{k \in \Lambda_m} \hat{f}(k)e^{i(k,x)},$$

and call it an $m$th weak greedy approximant of $f$ with regard to the trigonometric system $\mathcal{T}^d := \{e^{i(k,x)}\}_{k \in \mathbb{Z}^d}$, $\mathcal{T} := \mathcal{T}^1$. We write $G_m(f) = G_m^1(f)$ and call it an $m$th greedy approximant. Clearly, an $m$th weak greedy approximant, and even an $m$th greedy approximant, may not be unique. In this section we do not impose any extra restrictions on $\Lambda_m$ in addition to (1.111). Thus, theorems formulated in the following hold for any choice of $\Lambda_m$ satisfying (1.111), or, in other words, for any realization $G_m^t(f)$ of the weak greedy approximation.

We will discuss in detail only results concerning convergence of the WGA with regard to the trigonometric system. Körner (1996), answering a question raised by L. Carleson and R. R. Coifman, constructed a function from $L_2(\mathbb{T})$ and then, in Körner (1999), a continuous function such that $\{G_m(f, \mathcal{T})\}$ diverges almost everywhere. It has been proved in Temlyakov (1998c) for $p \neq 2$ and in Cordoba and Fernandez (1998) for $p < 2$, that there exists an $f \in L_p(\mathbb{T})$ such that $\{G_m(f, \mathcal{T})\}$ does not converge in $L_p$. It was remarked in Temlyakov (2003a) that the method from Temlyakov (1998c) gives a little more.

(1) There exists a continuous function $f$ such that $\{G_m(f, T)\}$ does not converge in $L_p(\mathbb{T})$ for any $p > 2$.
(2) There exists a function $f$ that belongs to any $L_p(\mathbb{T})$, $p < 2$, such that $\{G_m(f, T)\}$ does not converge in measure.

Thus the above negative results show that the condition $f \in L_p(\mathbb{T}^d)$, $p \neq 2$, does not guarantee convergence of $\{G_m(f, T)\}$ in the $L_p$ norm. The main goal of this section is to discuss results in the following setting: find an additional (to $f \in L_p$) condition on $f$ to guarantee that $\|f - G_m(f, T)\|_p \to 0$ as $m \to \infty$. In Konyagin and Temlyakov (2003b) we proved the following theorem.

**Theorem 1.48** *Let* $f \in L_p(\mathbb{T}^d)$, $2 < p \leq \infty$, *and let* $q > p' := p/(p-1)$. *Assume that* $f$ *satisfies the condition*

$$\sum_{|k|>n} |\hat{f}(k)|^q = o(n^{d(1-q/p')}),$$

*where* $|k| := \max_{1 \leq j \leq d} |k_j|$. *Then we have*

$$\lim_{m \to \infty} \|f - G_m^t(f, T)\|_p = 0.$$

It was proved in Konyagin and Temlyakov (2003b) that Theorem 1.48 is sharp.

**Proposition 1.49** *For each* $2 < p \leq \infty$ *there exists* $f \in L_p(\mathbb{T}^d)$ *such that*

$$|\hat{f}(k)| = O(|k|^{-d(1-1/p)}),$$

*and the sequence* $\{G_m(f)\}$ *diverges in* $L_p$.

Let us make some comments. For a given set $\Lambda$ denote

$$E_\Lambda(f)_p := \inf_{c_k, k \in \Lambda} \|f - \sum_{k \in \Lambda} c_k e^{i(k,x)}\|_p, \quad S_\Lambda(f) := \sum_{k \in \Lambda} \hat{f}(k) e^{i(k,x)}.$$

Define a special domain

$$Q(n) := \{k : |k| \leq n^{1/d}\}.$$

**Remark 1.50** Theorem 1.48 implies that if $f \in L_p$, $2 < p \leq \infty$, and

$$E_{Q(n)}(f)_2 = o(n^{1/p-1/2}),$$

then $G_m^t(f) \to f$ in $L_p$.

**Remark 1.51** The proof of Proposition 1.49 (see Konyagin and Temlyakov (2003b)) implies that there is an $f \in L_p(\mathbb{T}^d)$ such that

$$E_{Q(n)}(f)_\infty = O(n^{1/p-1/2})$$

and $\{G_m(f)\}$ diverges in $L_p$, $2 < p \le \infty$.

We note that Remark 1.50 can also be obtained from some general inequalities for $\|f - G_m^t(f)\|_p$. As in the above general definition of best $m$-term approximation, we define the best $m$-term approximation with regard to $\mathcal{T}^d$:

$$\sigma_m(f)_p := \sigma_m(f, \mathcal{T}^d)_p := \inf_{k^j \in \mathbb{Z}^d, c_j} \left\| f - \sum_{j=1}^m c_j e^{i(k^j, x)} \right\|_p.$$

The following inequality was proved in Temlyakov (1998c) for $t = 1$ and in Konyagin and Temlyakov (2003b) for general $t$.

**Theorem 1.52** *For each $f \in L_p(\mathbb{T}^d)$ and any $0 < t \le 1$ we have*

$$\|f - G_m^t(f)\|_p \le (1 + (2 + 1/t)m^{h(p)})\sigma_m(f)_p, \quad 1 \le p \le \infty, \quad (1.112)$$

*where $h(p) := |1/2 - 1/p|$.*

It was proved in Temlyakov (1998c) that the inequality (1.112) is sharp: there is a positive absolute constant $C$ such that, for each $m$ and $1 \le p \le \infty$, there exists a function $f \ne 0$ with the property

$$\|G_m(f)\|_p \ge Cm^{h(p)}\|f\|_p. \quad (1.113)$$

The above inequality (1.113) shows that the trigonometric system is not a quasi-greedy basis for $L_p$, $p \ne 2$. We formulate one more inequality from Konyagin and Temlyakov (2003b).

**Theorem 1.53** *Let $2 \le p \le \infty$. Then, for any $f \in L_p(\mathbb{T}^d)$ and any $Q$, $|Q| \le m$, we have*

$$\|f - G_m^t(f)\|_p \le \|f - S_Q(f)\|_p + (3 + 1/t)(2m)^{h(p)} E_Q(f)_2.$$

We present some results from Konyagin and Temlyakov (2003b) that are formulated in terms of the Fourier coefficients. For $f \in L_1(\mathbb{T}^d)$ let $\{\hat{f}(k(l))\}_{l=1}^\infty$ denote the decreasing rearrangement of $\{\hat{f}(k)\}_{k \in \mathbb{Z}^d}$, i.e.

$$|\hat{f}(k(1))| \ge |\hat{f}(k(2))| \ge \cdots$$

Let $a_n(f) := |\hat{f}(k(n))|$.

**Theorem 1.54** *Let* $2 < p < \infty$ *and let a decreasing sequence* $\{A_n\}_{n=1}^{\infty}$ *satisfy the condition*

$$A_n = o(n^{1/p-1}) \quad as \quad n \to \infty.$$

*Then, for any* $f \in L_p(\mathbb{T}^d)$ *with the property* $a_n(f) \leq A_n$, $n = 1, 2, \ldots$, *we have*

$$\lim_{m \to \infty} \| f - G_m^t(f) \|_p = 0.$$

We also proved in Konyagin and Temlyakov (2003b) that, for any decreasing sequence $\{A_n\}$ satisfying

$$\limsup_{n \to \infty} A_n n^{1-1/p} > 0,$$

there exists a function $f \in L_p$ such that $a_n(f) \leq A_n$, $n = 1, \ldots$, whose sequence of greedy approximants $\{G_m(f)\}$ is divergent in $L_p$.

In Konyagin and Temlyakov (2003b) we proved a necessary and sufficient condition on the majorant $\{A_n\}$ to guarantee, under the assumption that $f$ is a continuous function, uniform convergence of greedy approximants to a function $f$.

**Theorem 1.55** *Let a decreasing sequence* $\{A_n\}_{n=1}^{\infty}$ *satisfy the condition* $(\mathcal{A}_{\infty})$:

$$\sum_{M < n \leq e^M} A_n = o(1) \quad as \quad M \to \infty.$$

*Then, for any* $f \in C(\mathbb{T})$ *with the property* $a_n(f) \leq A_n$, $n = 1, 2, \ldots$, *we have*

$$\lim_{m \to \infty} \| f - G_m^t(f, \mathcal{T}) \|_{\infty} = 0.$$

The condition $(\mathcal{A}_{\infty})$ is very close to the convergence of the series $\sum_n A_n$; if the condition $(\mathcal{A}_{\infty})$ holds then we have

$$\sum_{n=1}^{N} A_n = o(\log_*(N)), \quad as \quad N \to \infty,$$

where a function $\log_*(u)$ is defined to be bounded for $u \leq 0$ and to satisfy $\log_*(u) = \log_*(\log u) + 1$ for $u > 0$. The function $\log_*(u)$ grows slower than any iterated logarithmic function.

The condition $(\mathcal{A}_{\infty})$ in Theorem 1.55 is sharp.

**Theorem 1.56** *Assume that a decreasing sequence* $\{A_n\}_{n=1}^{\infty}$ *does not satisfy the condition* $(\mathcal{A}_{\infty})$. *Then there exists a function* $f \in C(\mathbb{T})$ *with the property* $a_n(f) \leq A_n$, $n = 1, 2, \ldots$, *and such that we have*

$$\limsup_{m \to \infty} \| f - G_m(f, \mathcal{T}) \|_\infty > 0$$

*for some realization $G_m(f, \mathcal{T})$.*

In Konyagin and Temlyakov (2005) we concentrated on imposing extra conditions in the following form. We assume that for some sequence $\{M(m)\}$, $M(m) > m$, we have

$$\| G_{M(m)}(f) - G_m(f) \|_p \to 0, \quad \text{as} \quad m \to \infty.$$

When $p$ is an even number, or $p = \infty$, we found in Konyagin and Temlyakov (2005) necessary and sufficient conditions on the growth of the sequence $\{M(m)\}$ to provide convergence $\| f - G_m(f) \|_p \to 0$ as $m \to \infty$. We proved the following theorem in Konyagin and Temlyakov (2005).

**Theorem 1.57** *Let $p = 2q$, $q \in \mathbb{N}$, be an even integer, $\delta > 0$. Assume that $f \in L_p(\mathbb{T})$ and that there exists a sequence of positive integers $M(m) > m^{1+\delta}$ such that*

$$\| G_m(f) - G_{M(m)}(f) \|_p \to 0, \quad \text{as} \quad m \to \infty.$$

*Then we have*

$$\| G_m(f) - f \|_p \to 0, \quad \text{as} \quad m \to \infty.$$

In Konyagin and Temlyakov (2005) we proved that the condition $M(m) > m^{1+\delta}$ cannot be replaced by the condition $M(m) > m^{1+o(1)}$.

**Theorem 1.58** *For any $p \in (2, \infty)$ there exists a function $f \in L_p(\mathbb{T})$ with an $L_p(\mathbb{T})$-divergent sequence $\{G_m(f)\}$ of greedy approximations with the following property. For any sequence $\{M(m)\}$ such that $m \le M(m) \le m^{1+o(1)}$, we have*

$$\| G_{M(m)}(f) - G_m(f) \|_p \to 0, \quad \text{as} \quad m \to 0.$$

In Konyagin and Temlyakov (2005) we also considered the case $p = \infty$, and proved necessary and sufficient conditions for convergence of greedy approximations in the uniform norm. For a mapping $\alpha : \mathbb{N} \to \mathbb{N}$ we let $\alpha_k$ denote its $k$-fold iteration: $\alpha_k := \alpha \circ \alpha_{k-1}$.

**Theorem 1.59** *Let $\alpha : \mathbb{N} \to \mathbb{N}$ be strictly increasing. Then the following conditions are equivalent.*

  *(a) For some $k \in \mathbb{N}$, and for any sufficiently large $m \in \mathbb{N}$, we have the inequality $\alpha_k(m) > e^m$.*

*(b) If $f \in C(\mathbb{T})$ and*

$$\|G_{\alpha(m)}(f) - G_m(f)\|_\infty \to 0, \quad as \quad m \to 0,$$

*then*

$$\|f - G_m(f)\|_\infty \to 0, \quad as \quad m \to 0.$$

In order to illustrate the techniques used in the proofs of the above results we discuss some inequalities that were used in proving Theorems 1.57 and 1.59. The reader will also see from the further discussion a connection to some deep results in harmonic analysis. The general style of these inequalities is as follows. A function that has a sparse representation with regard to the trigonometric system cannot be approximated in $L_p$ by functions with small Fourier coefficients. We begin our discussion with some concepts introduced in Konyagin and Temlyakov (2005) that are useful in proving such inequalities. The following new characteristic of a Banach space $L_p$ plays an important role in such inequalities. We now introduce some more notation. Let $\Lambda$ be a finite subset of $\mathbb{Z}^d$; we let $|\Lambda|$ denote its cardinality and let $\mathcal{T}(\Lambda)$ be the span of $\{e^{i(k,x)}\}_{k\in\Lambda}$. Denote

$$\Sigma_m(\mathcal{T}) := \cup_{\Lambda:|\Lambda|\le m}\mathcal{T}(\Lambda).$$

For $f \in L_p$, $F \in L_{p'}$, $1 \le p \le \infty$, $p' = p/(p-1)$, we write

$$\langle F, f \rangle := \int_{\mathbb{T}^d} F\bar{f}\, d\mu, \quad d\mu := (2\pi)^{-d}\, dx.$$

**Definition 1.60** Let $\Lambda$ be a finite subset of $\mathbb{Z}^d$ and $1 \le p \le \infty$. We call a set $\Lambda' := \Lambda'(p, \gamma)$, $\gamma \in (0, 1]$, a $(p, \gamma)$-dual to $\Lambda$ if, for any $f \in \mathcal{T}(\Lambda)$, there exists $F \in \mathcal{T}(\Lambda')$ such that $\|F\|_{p'} = 1$ and $\langle F, f \rangle \ge \gamma\|f\|_p$.

Let $D(\Lambda, p, \gamma)$ denote the set of all $(p, \gamma)$-dual sets $\Lambda'$. The following function is important for us:

$$v(m, p, \gamma) := \sup_{\Lambda:|\Lambda|=m} \inf_{\Lambda'\in D(\Lambda,p,\gamma)} |\Lambda'|.$$

We note that in the particular case $p = 2q$, $q \in \mathbb{N}$, we have

$$v(m, p, 1) \le m^{p-1}. \tag{1.114}$$

This follows immediately from the form of the norming functional $F$ for $f \in L_p$:

$$F = f^q(\bar{f})^{q-1}\|f\|_p^{1-p}. \tag{1.115}$$

In Konyagin and Temlyakov (2005) we used the quantity $v(m, p, \gamma)$ in greedy approximation. We first prove a lemma.

**Lemma 1.61** *Let* $2 \le p \le \infty$. *For any* $h \in \Sigma_m(\mathcal{T})$ *and any* $g \in L_p$, *we have*

$$\|h + g\|_p \ge \gamma \|h\|_p - v(m, p, \gamma)^{1-1/p} \|\{\hat{g}(k)\}\|_{\ell_\infty}.$$

*Proof* Let $h \in \mathcal{T}(\Lambda)$ with $|\Lambda| = m$ and let $\Lambda' \in D(\Lambda, p, \gamma)$. Then, using Definition 1.60 we find $F(h, \gamma) \in \mathcal{T}(\Lambda')$ such that

$$\|F(h, \gamma)\|_{p'} = 1 \quad \text{and} \quad \langle F(h, \gamma), h \rangle \ge \gamma \|h\|_p.$$

We have

$$\langle F(h, \gamma), h \rangle = \langle F(h, \gamma), h + g \rangle - \langle F(h, \gamma), g \rangle \le \|h + g\|_p + |\langle F(h, \gamma), g \rangle|.$$

Next,

$$|\langle F(h, \gamma), g \rangle| \le \|\{\hat{F}(h, \gamma)(k)\}\|_{\ell_1} \|\{\hat{g}(k)\}\|_{\ell_\infty}.$$

Using $F(h, \gamma) \in \mathcal{T}(\Lambda')$ and the Hausdorf–Young theorem (Zygmund (1959), chap. 12, sect. 2), we obtain

$$\|\{\hat{F}(h, \gamma)(k)\}\|_{\ell_1} \le |\Lambda'|^{1-1/p} \|\{\hat{F}(h, \gamma)(k)\}\|_{\ell_p}$$
$$\le |\Lambda'|^{1-1/p} \|F(h, \gamma)\|_{p'} = |\Lambda'|^{1-1/p}.$$

We now combine the above inequalities and use the definition of $v(m, p, \gamma)$.
□

**Definition 1.62** Let $X$ be a finite dimensional subspace of $L_p$, $1 \le p \le \infty$. We call a subspace $Y \subset L_{p'}$ a $(p, \gamma)$-dual to $X$, $\gamma \in (0, 1]$, if, for any $f \in X$, there exists $F \in Y$ such that $\|F\|_{p'} = 1$ and $\langle F, f \rangle \ge \gamma \|f\|_p$.

As above, let $D(X, p, \gamma)$ denote the set of all $(p, \gamma)$-dual subspaces $Y$. Consider the following function:

$$w(m, p, \gamma) := \sup_{X:\dim X=m} \inf_{Y \in D(X,p,\gamma)} \dim Y.$$

We begin our discussion with a particular case: $p = 2q$, $q \in \mathbb{N}$. Let $X$ be given and let $e_1, \ldots, e_m$ form a basis of $X$. Using the Hölder inequality for $n$ functions $f_1, \ldots, f_n \in L_n$, we have

$$\int |f_1 \cdots f_n| d\mu \le \|f_1\|_n \cdots \|f_n\|_n.$$

Setting $f_i = |e_j|^{p'}$, $n = p - 1$, we deduce that any function of the form

$$\prod_{i=1}^m |e_i|^{k_i}, \quad k_i \in \mathbb{N}, \quad \sum_{i=1}^m k_i = p - 1,$$

belongs to $L_{p'}$. It now follows from (1.115) that

$$w(m, p, 1) \le m^{p-1}, \quad p = 2q, \quad q \in \mathbb{N}. \tag{1.116}$$

There is a general theory of uniform approximation property (UAP) which provides some estimates for $w(m, p, \gamma)$ and $v(m, p, \gamma)$. We give some definitions from this theory. For a given subspace $X$ of $L_p$, $\dim X = m$, and a constant $K > 1$, let $k_p(X, K)$ be the smallest $k$ such that there is an operator $I_X : L_p \to L_p$, with $I_X(f) = f$ for $f \in X$, $\|I_X\|_{L_p \to L_p} \le K$, and $\mathrm{rank}\, I_X \le k$. Define

$$k_p(m, K) := \sup_{X : \dim X = m} k_p(X, K),$$

and let us discuss how $k_p(m, K)$ can be used in estimating $w(m, p, \gamma)$. Consider the dual operator $I_X^*$ to $I_X$. Then $\|I_X^*\|_{L_{p'} \to L_{p'}} \le K$ and $\mathrm{rank}\, I_X^* \le k_p(m, K)$. Let $f \in X$, $\dim X = m$, and let $F_f$ be the norming functional for $f$. Define

$$F := I_X^*(F_f)/\|I_X^*(F_f)\|_{p'}.$$

Then, for any $f \in X$,

$$\langle f, I_X^*(F_f) \rangle = \langle I_X(f), F_f \rangle = \langle f, F_f \rangle = \|f\|_p$$

and

$$\|I_X^*(F_f)\|_{p'} \le K$$

imply

$$\langle f, F \rangle \ge K^{-1} \|f\|_p.$$

Therefore

$$w(m, p, K^{-1}) \le k_p(m, K). \tag{1.117}$$

We note that the behavior of functions $w(m, p, \gamma)$ and $k_p(m, K)$ may be very different. Bourgain (1992) proved that, for any $p \in (1, \infty)$, $p \ne 2$, the function $k_p(m, K)$ grows faster than any polynomial in $m$. The estimate (1.116) shows that, in the particular case $p = 2q$, $q \in \mathbb{N}$, the growth of $w(m, p, \gamma)$ is at most polynomial. This means that we cannot expect to obtain accurate estimates for $w(m, p, K^{-1})$ using inequality (1.117). We give one more application of the UAP in the style of Lemma 1.61.

**Lemma 1.63** *Let* $2 \le p \le \infty$. *For any* $h \in \Sigma_m(\mathcal{T})$ *and any* $g \in L_p$ *we have*

$$\|h + g\|_p \ge K^{-1} \|h\|_p - k_p(m, K)^{1/2} \|g\|_2; \tag{1.118}$$

$$\|h + g\|_p \ge K^{-2} \|h\|_p - k_p(m, K) \|\{\hat{g}(k)\}\|_{\ell_\infty}. \tag{1.119}$$

*Proof* Let $h \in \mathcal{T}(\Lambda)$, $|\Lambda| = m$. Take $X = \mathcal{T}(\Lambda)$ and consider the operator $I_X$ provided by the UAP. Let $\psi_1, \ldots, \psi_M$ form an orthonormal basis for the range $Y$ of the operator $I_X$. Then $M \leq k_p(m, K)$. Let

$$I_X(e^{i(k,x)}) = \sum_{j=1}^{M} c_j^k \psi_j.$$

Then the property $\|I_X\|_{L_p \to L_p} \leq K$ implies

$$\left( \sum_{j=1}^{M} |c_j^k|^2 \right)^{1/2} = \|I_X(e^{i(k,x)})\|_2 \leq \|I_X(e^{i(k,x)})\|_p \leq K.$$

Consider, along with the operator $I_X$, the new operator

$$A := (2\pi)^{-d} \int_{\mathbb{T}^d} T_t I_X T_{-t} \, dt,$$

where $T_t$ is the shift operator: $T_t(f) = f(\cdot + t)$. Then

$$A(e^{i(k,x)}) = \sum_{j=1}^{M} c_j^k (2\pi)^{-d} \int_{\mathbb{T}^d} e^{-i(k,t)} \psi_j(x+t) dt = \left( \sum_{j=1}^{M} c_j^k \hat{\psi}_j(k) \right) e^{i(k,x)}.$$

Let

$$\lambda_k := \sum_{j=1}^{M} c_j^k \hat{\psi}_j(k).$$

We have

$$\sum_k |\lambda_k|^2 \leq \sum_k \left( \sum_{j=1}^{M} |c_j^k|^2 \right) \left( \sum_{j=1}^{M} |\hat{\psi}_j(k)|^2 \right) \leq K^2 M.$$

Also, $\lambda_k = 1$ for $k \in \Lambda$. For the operator $A$ we have

$$\|A\|_{L_p \to L_p} \leq K \quad \text{and} \quad \|A\|_{L_2 \to L_\infty} \leq K M^{1/2}.$$

Therefore

$$\|A(h+g)\|_p \leq K \|h+g\|_p$$

and

$$\|A(h+g)\|_p \geq \|h\|_p - K M^{1/2} \|g\|_2.$$

This proves the first inequality.

Consider the operator $B := A^2$. Then

$$B(h) = h, \quad h \in \mathcal{T}(\Lambda), \quad \|B\|_{L_p \to L_p} \leq K^2,$$

and

$$\|B(f)\|_\infty \leq K^2 M \|\{\hat{f}(k)\}\|_{\ell_\infty}.$$

Now, on the one hand,

$$\|B(h + g)\|_p \leq K^2 \|h + g\|_p,$$

and, on the other hand,

$$\|B(h + g)\|_p = \|h + B(g)\|_p \geq \|h\|_p - K^2 M \|\{\hat{g}(k)\}\|_{\ell_\infty}.$$

This proves inequality (1.119). □

**Theorem 1.64** *For any $h \in \Sigma_m(\mathcal{T})$ and any $g \in L_\infty$ we have*

$$\|h + g\|_\infty \geq K^{-1} \|h\|_\infty - e^{C(K)m/2} \|g\|_2;$$
$$\|h + g\|_\infty \geq K^{-2} \|h\|_\infty - e^{C(K)m} \|\{\hat{g}(k)\}\|_{\ell_\infty}.$$

*Proof* This theorem is a direct corollary of Lemma 1.63 and the known estimate

$$k_\infty(m, K) \leq e^{C(K)m}$$

(see Figiel, Johnson and Schechtman (1988)). □

As we have already mentioned, $k_p(m, K)$ increases faster than any polynomial. In Konyagin and Temlyakov (2005) we improved inequality (1.118) by using other arguments.

**Lemma 1.65** *Let $2 \leq p \leq \infty$. For any $h \in \Sigma_m(\mathcal{T})$ and any $g \in L_p$ we have*

$$\|h + g\|_p^p \geq 2^{-p-1} \|h\|_p^p - 2m^{p/2} \|h\|_p^{p-2} \|g\|_2^2. \tag{1.120}$$

We mention two inequalities from Konyagin and Temlyakov (2003b) in the style of the inequalities in Lemmas 1.61, 1.63 and 1.65.

**Lemma 1.66** *Let $2 \leq p < \infty$ and $h \in L_p$, $\|h\|_p \neq 0$. Then, for any $g \in L_p$, we have*

$$\|h\|_p \leq \|h + g\|_p + (\|h\|_{2p-2}/\|h\|_p)^{p-1} \|g\|_2.$$

**Lemma 1.67** *Let $h \in \Sigma_m(\mathcal{T})$, $\|h\|_\infty = 1$. Then, for any function $g$ such that $\|g\|_2 \leq \frac{1}{4}(4\pi m)^{-m/2}$, we have*

$$\|h + g\|_\infty \geq 1/4.$$

We proceed to estimate $v(m, p, \gamma)$ and $w(m, p, \gamma)$ for $p \in [2, \infty)$. In the special case of even $p$, we have, by (1.114) and (1.116), that

$$v(m, p, 1) \le m^{p-1}, \quad w(m, p, 1) \le m^{p-1}.$$

The following bound was proved in Konyagin and Temlyakov (2005).

**Lemma 1.68** *Let* $2 \le p < \infty$, *and let* $\alpha := p/2 - [p/2]$. *Then we have*

$$v(m, p, \gamma) \le m^{c(\alpha,\gamma)m^{1/2}+p-1}.$$

## 1.8 Greedy-type bases; direct and inverse theorems

Theorem 1.11 points out the importance of bases $L_p$-equivalent to the Haar basis. We will now discuss necessary and sufficient conditions for $f$ to have a prescribed decay of $\{\sigma_m(f, \Psi)\}$ under the assumption that $\Psi$ is $L_p$-equivalent to the Haar basis $\mathcal{H}$, $1 < p < \infty$. We will express these conditions in terms of coefficients $\{c_n(f)\}$ of the expansion

$$f = \sum_{n=1}^{\infty} c_n(f)\psi_n.$$

The direct theorems of approximation theory provide bounds of approximation error (in our case, $\sigma_m(f, \Psi)$) in terms of smoothness properties of a function $f$. These theorems are also known as Jackson-type inequalities. The inverse theorems of approximation theory (also known as Bernstein-type inequalities) provide some smoothness properties of a function $f$ from the sequence of approximation errors (in our case, $\sigma_m(f, \Psi)$). It is well understood in approximation theory (see DeVore (1998), DeVore and Lorentz (1993) and Petrushev (1988)) how the Jackson-type and Bernstein-type inequalities can be used in order to characterize the corresponding approximation spaces. In our case of interest, when we study best $m$-term approximation with regard to bases that are $L_p$-equivalent to the Haar basis, the theory of Jackson and Bernstein inequalities has been developed in Cohen, DeVore and Hochmuth (2000), where it was used for a description of approximation spaces defined in terms of $\{\sigma_m(f, \Psi)\}$. We want to point out that in the special case of bases that are $L_p$-equivalent to the Haar basis (and also for some more general bases), there exists a simple direct way to describe the approximation spaces defined in terms of $\{\sigma_m(f, \Psi)\}$ (Kamont and Temlyakov (2004), Kerkyacharian and Picard (2004), Temlyakov (1998b)). We present results from Temlyakov (1998b) here. The following lemma from Temlyakov (1998a) (see Lemmas 1.23 and 1.24 above) plays the key role in this consideration.

**Lemma 1.69** *Let a basis* $\Psi$ *be* $L_p$*-equivalent to* $\mathcal{H}_p$, $1 < p < \infty$. *Then, for any finite* $\Lambda$ *and* $a \leq |c_n| \leq b$, $n \in \Lambda$, *we have*

$$C_1(p, \Psi)a(|\Lambda|)^{1/p} \leq \Big\| \sum_{n \in \Lambda} c_n \psi_n \Big\|_p \leq C_2(p, \Psi)b(|\Lambda|)^{1/p}. \qquad (1.121)$$

We note that the results that follow use only the assumption that $\Psi$ is a greedy basis satisfying (1.121). We formulate a general statement and then consider several important particular examples of the rate of decrease of $\{\sigma_m(f, \Psi)_p\}$. We begin by introducing some notation. For a sequence $\mathcal{E} = \{\epsilon_k\}_{k=0}^{\infty}$ of positive numbers monotonically decreasing to zero (we write $\mathcal{E} \in MDP$), we define inductively a sequence $\{N_s\}_{s=0}^{\infty}$ of non-negative integers:

$$N_0 = 0, \quad \text{and} \quad N_s \text{ is the smallest integer satisfying}$$

$$\epsilon_{N_s} < 2^{-s}, \qquad d_s := \max(N_{s+1} - N_s, 1). \qquad (1.122)$$

We are going to consider the following examples of sequences.

**Example 1.70** Take $\epsilon_0 = 1$ and $\epsilon_k = k^{-r}$, $r > 0$, $k = 1, 2, \ldots$ Then

$$N_s \asymp 2^{s/r} \quad \text{and} \quad d_s \asymp 2^{s/r}.$$

**Example 1.71** Fix $0 < b < 1$ and take $\epsilon_k = 2^{-k^b}$, $k = 0, 1, 2, \ldots$ Then

$$N_s = s^{1/b} + O(1) \quad \text{and} \quad d_s \asymp s^{1/b-1}.$$

Let $f \in L_p$. Rearrange the sequence $\|c_n(f)\psi_n\|_p$ in decreasing order,

$$\|c_{n_1}(f)\psi_{n_1}\|_p \geq \|c_{n_2}(f)\psi_{n_2}\|_p \geq \ldots,$$

and define

$$a_k(f, p) := \|c_{n_k}(f)\psi_{n_k}\|_p.$$

We now give some inequalities for $a_k(f, p)$ and $\sigma_m(f, \Psi)_p$. We will use the abbreviations $\sigma_m(f)_p := \sigma_m(f, \Psi)_p$ and $\sigma_0(f)_p := \|f\|_p$.

**Lemma 1.72** *For any two positive integers* $N < M$ *we have*

$$a_M(f, p) \leq C(p, \Psi)\sigma_N(f)_p(M - N)^{-1/p}.$$

*Proof* By Theorem 1.11 we have, for all $m$,

$$\|f - G_m^p(f, \Psi)\|_p \leq C(p, \Psi)\sigma_m(f)_p.$$

Hence, and by the definition of $G_m^p$, we obtain

$$J := \Big\| \sum_{k=N+1}^{M} c_{n_k}(f)\psi_{n_k} \Big\|_p \leq C(p, \Psi)(\sigma_N(f)_p + \sigma_M(f)_p). \qquad (1.123)$$

Next, we have, for $k \in (N, M]$,

$$\|c_{n_k}(f)\psi_{n_k}\|_p \geq \|c_{n_M}(f)\psi_{n_M}\|_p = a_M(f, p),$$

and by Lemma 1.69 we obtain

$$a_M(f, p)(M - N)^{1/p} \leq C(p, \Psi)J. \qquad (1.124)$$

Relations (1.123) and (1.124) imply the conclusion of Lemma 1.72. $\qquad \square$

**Lemma 1.73** *For any sequence* $m_0 < m_1 < m_2 < \ldots$ *of non-negative integers we have*

$$\sigma_{m_s}(f)_p \leq C(p, \Psi) \sum_{l=s}^{\infty} a_{m_l}(f, p)(m_{l+1} - m_l)^{1/p}.$$

*Proof* We have

$$\sigma_{m_s}(f)_p \leq \Big\| \sum_{k > m_s} c_{n_k}(f)\psi_{n_k} \Big\|_p \leq \sum_{l=s}^{\infty} \Big\| \sum_{k \in (m_l, m_{l+1}]} c_{n_k}(f)\psi_{n_k} \Big\|_p.$$

Hence, using Lemma 1.69,

$$\sigma_{m_s}(f)_p \leq C(p, \Psi) \sum_{l=s}^{\infty} a_{m_l}(f, p)(m_{l+1} - m_l)^{1/p}$$

as required. $\qquad \square$

**Theorem 1.74** *Assume that a given sequence* $\mathcal{E} \in MDP$ *satisfies the conditions*

$$\epsilon_{N_s} \geq C_1 2^{-s}, \qquad d_{s+1} \leq C_2 d_s, \qquad s = 0, 1, 2, \ldots$$

*Then we have the equivalence*

$$\sigma_n(f)_p \ll \epsilon_n \quad \Longleftrightarrow \quad a_{N_s}(f, p) \ll 2^{-s} d_s^{-1/p}.$$

*Proof* We first prove $\Rightarrow$. If $N_{s+1} > N_s$, then we use Lemma 1.72 with $M = N_{s+1}$ and $N = N_s$,

$$a_{N_{s+1}}(f, p) \leq C(p, \Psi)\sigma_{N_s}(f)_p d_s^{-1/p} \leq C(p, \Psi)2^{-s-1}(d_{s+1}/C_2)^{-1/p},$$

which implies the statement of Theorem 1.74 in this case. Let $N_{s+1} = N_s = \cdots = N_{s-j} > N_{s-j-1}$. The assumption $\epsilon_{N_s} \geq C_1 2^{-s}$ combined with the definition of $N_s$: $\epsilon_{N_s} < 2^{-s}$ implies that $j \leq C_3$. Then, from the above case we get

$$a_{N_{s-j}}(f, p) \ll 2^{-s+j}(d_{s-j})^{-1/p},$$

and therefore

$$a_{N_{s+1}}(f, p) \ll 2^{-s-1}(d_{s+1})^{-1/p}.$$

The implication $\Rightarrow$ has been proved.

We now prove the inverse statement $\Leftarrow$. Using Lemma 1.73, we get

$$\sigma_{N_s}(f)_p \ll \sum_{l=s}^{\infty} a_{N_l}(f, p)(N_{l+1} - N_l)^{1/p} \ll \sum_{l=s}^{\infty} 2^{-l} \ll 2^{-s} \ll \epsilon_{N_s}$$

and, for $n \in [N_s, N_{s+1})$,

$$\sigma_n(f)_p \leq \sigma_{N_s}(f)_p \ll \epsilon_{N_s} \ll 2^{-s} \ll \epsilon_{N_{s+1}} \leq \epsilon_n.$$

$\square$

**Corollary 1.75** *Theorem 1.74 applied to Examples 1.70 and 1.71 gives the following relations:*

$$\sigma_m(f)_p \ll (m + 1)^{-r} \iff a_n(f, p) \ll n^{-r-1/p}, \qquad (1.125)$$

$$\sigma_m(f)_p \ll 2^{-m^b} \iff a_n(f, p) \ll 2^{-n^b} n^{(1-1/b)/p}. \qquad (1.126)$$

**Remark 1.76** Making use of Lemmas 1.72 and 1.73 we can prove a version of Corollary 1.75 with the $\ll$ sign replaced by $\asymp$.

Theorem 1.74 and Corollary 1.75 are in the spirit of classical Jackson–Bernstein direct and inverse theorems in linear approximation theory, where conditions of the form

$$E_n(f)_p \ll \epsilon_n \quad \text{or} \quad \|E_n(f)_p/\epsilon_n\|_{\ell_\infty} < \infty \qquad (1.127)$$

are imposed on the corresponding sequences of approximating characteristics. It is well known (see DeVore (1998)) that, in studying many questions of approximation theory, it is convenient to consider, along with the restriction (1.127), its following generalization:

$$\|E_n(f)_p/\epsilon_n\|_{\ell_q} < \infty. \qquad (1.128)$$

Lemmas 1.72 and 1.73 are also useful in handling this more general case. For instance, in the particular case of Example 1.70 we have the following statement.

**Theorem 1.77** *Let* $1 < p < \infty$ *and* $0 < q < \infty$. *Then, for any positive* $r$ *we have the equivalence relation*

$$\sum_m \sigma_m(f)_p^q m^{rq-1} < \infty \iff \sum_n a_n(f, p)^q n^{rq-1+q/p} < \infty.$$

**Remark 1.78** The condition

$$\sum_n a_n(f, p)^q n^{rq-1+q/p} < \infty$$

with $q = \beta := (r + 1/p)^{-1}$ takes the following very simple form:

$$\sum_n a_n(f, p)^\beta = \sum_n \|c_n(f)\psi_n\|_p^\beta < \infty. \tag{1.129}$$

In the case $\Psi = \mathcal{H}_p$, condition (1.129) is equivalent to $f$ being in Besov space $B_\beta^r(L_\beta)$.

**Corollary 1.79** *Theorem 1.77 implies the following relation:*

$$\sum_m \sigma_m(f, \mathcal{H})_p^\beta m^{r\beta-1} < \infty \quad \Longleftrightarrow \quad f \in B_\beta^r(L_\beta),$$

*where $\beta := (r + 1/p)^{-1}$.*

The statement similar to Corollary 1.79 for free-knot spline approximation was proved in Petrushev (1988). Corollary 1.79 and further results in this direction can be found in DeVore and Popov (1988) and DeVore, Jawerth and Popov (1992). We want to remark here that conditions in terms of $a_n(f, p)$ are convenient in applications. For instance, (1.125) can be rewritten using the idea of thresholding. For a given $f \in L_p$ denote

$$T(\epsilon) := \#\{k : a_k(f, p) \geq \epsilon\}.$$

Then (1.125) is equivalent to

$$\sigma_m(f)_p \ll (m + 1)^{-r} \quad \Longleftrightarrow \quad T(\epsilon) \ll \epsilon^{-(r+1/p)^{-1}}.$$

For further results in this direction see Cohen, DeVore and Hochmuth (2000), DeVore (1998) and Oswald (2001).

The above direct and inverse Theorem 1.77 that holds for greedy bases satisfying (1.121) was extended in Dilworth *et al.* (2003) and Kerkyacharian and Picard (2004) to the case of quasi-greedy bases satisfying (1.121). Kerkyacharian and Picard (2004) say that a basis $\Psi$ of a Banach space $X$ has the $p$-Temlyakov property if there exists $0 < C < \infty$ such that, for any finite set of indices $\Lambda$, we have

$$C^{-1}(\min_{n\in\Lambda} |c_n|)|\Lambda|^{1/p} \leq \|\sum_{n\in\Lambda} c_n \psi_n\|_X \leq C(\max_{n\in\Lambda} |c_n|)|\Lambda|^{1/p}. \tag{1.130}$$

Now let

$$f = \sum_{k=1}^\infty c_k(f)\psi_k$$

and

$$|c_{k_1}| \geq |c_{k_2}| \geq \ldots$$

be a decreasing reordering of the coefficients. The following result is from Kerkyacharian and Picard (2004).

**Theorem 1.80** *Let $\Psi$ be a quasi-greedy basis.*

*(1) If $\Psi$ has the p-Temlyakov property (1.130), then, for any $0 < r < \infty$, $0 < q < \infty$, we have*

$$\sum_m \sigma_m(f, \Psi)_X^q m^{rq-1} < \infty \quad \Longleftrightarrow \quad \sum_n |c_{k_n}(f)|^q n^{rq-1+q/p} < \infty.$$

$$(1.131)$$

*(2) If (1.131) holds with some $r > 0$, then $\Psi$ has the p-Temlyakov property (1.130).*

We note that property (1.130) implies that $\Psi$ is democratic. Therefore, by Theorem 1.37, a quasi-greedy basis satisfying (1.130) is an almost greedy basis. The basis $\mathcal{H}_p^d$ is not a democratic basis for $L_p$, $p \neq 2$, $d > 1$. So, we cannot apply the above results in this case. Some direct and inverse theorems for $\mathcal{H}_p^d$ are obtained in Kamont and Temlyakov (2004).

## 1.9 Some further results

We begin our discussion with the case of $X = L_p$, $p = 1$ or $p = \infty$, and $\Psi = \mathcal{H}_p^d$. It turns out that the efficiency of greedy algorithms $G_m(\cdot, \mathcal{H}_p^d)$, $p = 1, \infty$, drops dramatically compared with the case $1 < p < \infty$. We formulate a result from Temlyakov (1998b).

**Theorem 1.81** *Let $p = 1$ or $p = \infty$. Then we have, for each $f \in L_p$,*

$$\|f - G_m(f, \mathcal{H}_p^d)\|_p \leq (3m + 1)\sigma_m(f, \mathcal{H}^d)_p.$$

*The extra factor $(3m + 1)$ cannot be replaced by a factor $c(m)$ such that $c(m)/m \to 0$ as $m \to \infty$.*

This particular result indicates that there are problems with greedy approximation in $L_1$ and in $\mathcal{C}$ with regard to the Haar basis. We note that, as is proved in Oswald (2001), the extra factor $3m + 1$ is the best-possible extra factor in Theorem 1.81. The greedy-type bases have nice properties and they are important in nonlinear $m$-term approximation. Therefore, one of the new directions of research in functional analysis and in approximation theory is to understand which Banach spaces may have such bases. Another direction is to understand in which Banach spaces some classical bases are of greedy type. Some results in this direction can be derived immediately from known results on Banach spaces that have unconditional bases, and from the characterization Theorem 1.15. For instance, it is well known that the spaces $L_1$ and $C$ do not have unconditional bases. Therefore, Theorem 1.15 implies that there are no greedy bases in $L_1$ and in $C$.

It was proved in Dilworth, Kutzarova and Wojtaszczyk (2002) that the Haar basis $\mathcal{H}_1$ is not a quasi-greedy basis for $L_1$. We saw in Section 1.6 that the use of the Weak Greedy Algorithm has some advantages over the Greedy Algorithm. Theorem 1.46 states that the convergence set $WT\{e_n\}$ of the WGA is linear for any $t \in (0, 1)$, while the convergence set may not be linear for the Greedy Algorithm. Recently, Gogyan (2009) proved that, for any $t \in (0, 1)$ and for any $f \in L_1(0, 1)$, there exists a realization of the WGA with respect to the Haar basis that converges to $f$ in $L_1$.

It was proved in Dilworth, Kutzarova and Wojtaszczyk (2002) that there exists an increasing sequence of integers $\{n_j\}$ such that the lacunary Haar system $\{H^1_{2^{n_j}+l}; l = 1, \ldots, 2^{n_j}, j = 1, 2, \ldots\}$ is a quasi-greedy basis for its linear span in $L_1$. Gogyan (2005) proved that the above property holds if either $\{n_j\}$ is a sequence of all even numbers or $\{n_j\}$ is a sequence of all odd numbers. We also note that the space $L_1(0, 1)$ has a quasi-greedy basis (Dilworth, Kalton and Kutzarova (2003)). The reader can find further results on existence (and nonexistence) of quasi-greedy and almost greedy bases in Dilworth, Kalton and Kutzarova (2003). In particular, it is proved therein that the $C[0, 1]$ does not have quasi-greedy bases.

We pointed out in Section 1.7 that the trigonometric system is not a quasi-greedy basis for $L_p$, $p \neq 2$. The question of when (and for which weights $w$) the trigonometric system forms a quasi-greedy basis for a weighted space $L_p(w)$ was studied in Nielsen (2009). The author proved that this can happen only for $p = 2$ and, whenever the system forms a quasi-greedy basis, the basis must be a Riesz basis.

Theorem 1.11A shows that, in the case when a basis $\Psi$ is $L_p$-equivalent to the Haar basis $\mathcal{H}_p$, $1 < p < \infty$, the Greedy Algorithm $G_m(f, \Psi)$ provides near-best approximation for each individual function $f \in L_p$. For a function class $F \subset X$ denote

$$\sigma_m(F, \Psi)_X := \sup_{f \in F} \sigma_m(F, \Psi)_X,$$

$$G_m(F, \Psi)_X := \sup_{f \in F} \|f - G_m(f, \Psi)\|_X.$$

Obviously, if $G_m(\cdot, \Psi)$ provides near-best approximation for each individual function, then it provides near-best approximation for each function class $F$:

$$G_m(F, \Psi)_X \leq C\sigma_m(F, \Psi)_X.$$

In Section 1.7 we pointed out that the trigonometric system is not a quasi-greedy basis for $L_p$, $p \neq 2$ (see (1.113)). Thus, the trigonometric system is not a greedy basis for $L_p$, $p \neq 2$, and for some functions $f \in L_p$, $p \neq 2$, the $G_m(f, \mathcal{T})$ does not provide near-best approximation. However, it was proved in Temlyakov (1998c) that, in many cases, the algorithm $G_m(\cdot, \mathcal{T})$ is optimal for a given class of functions. The reader can find further results on $\sigma_m(F, \mathcal{T}^d)_p$ and $G_m(F, \mathcal{T}^d)_p$ for different classes $F$ in DeVore and Temlyakov (1995) and Temlyakov (1998c, 2000a, 2002a).

Consideration of approximation in a function class leads to the concept of the *optimal (best) basis* for a given class. The first results for best-basis approximation were given by Kashin (1985), who showed that, for any orthonormal basis $\Psi$ and any $0 < \alpha \leq 1$, we have

$$\sigma_m(\text{Lip } \alpha, \Psi)_{L_2} \geq cm^{-\alpha}, \tag{1.132}$$

where the constant $c$ depends only on $\alpha$. It follows from this that any of the standard wavelet or Fourier bases are best for the Lipschitz classes, when the approximation is carried out in $L_2$ and the competition is held over all orthonormal bases. The estimate (1.132) rests on some fundamental estimates for the best-basis approximation of finite-dimensional hypercubes using orthonormal bases.

Kashin (2002) considered a function class that is much thinner than the Lip $\alpha$:

$$\mathcal{X} := \{\chi_{[t,1]}, t \in [0, 1]\},$$

where $\chi_{[a,b]}$ is the characteristic function of $[a, b]$. Kashin (2002) proved the following lower bounds. For any orthonormal system $\Psi \subset L_2([0, 1])$ we have

$$\sigma_m(\mathcal{X}, \Psi)_{L_2} \geq C^{-m}. \tag{1.133}$$

For any complete uniformly bounded ($\|\psi_j\|_\infty \leq M$) orthonormal system $\Psi \subset L_2([0, 1])$ we have

$$\sigma_m(\mathcal{X}, \Psi)_{L_2} \geq C(M)m^{-1/2}. \tag{1.134}$$

In the proof of estimates (1.133) and (1.134) the technique from the theory of general orthogonal series, developed in Kashin (1977b), was used. It is known and easy to see that bounds (1.133) and (1.134) are sharp. The complementary to the (1.133) upper bound follows from approximation by the Haar basis and the complementary to the (1.134) upper bound follows from approximation by the trigonometric system.

The problem of best-basis selection was studied in Coifman and Wickerhauser (1992). Donoho (1993, 1997) also studied the problem of best bases for a function class $F$. He calls a basis $\Psi$ from a collection $\mathbb{B}$ best for $F$ if

$$\sigma_m(F, \Psi)_X = O(m^{-\alpha}), \quad m \to \infty,$$

and no other basis $\Psi'$ from $\mathbb{B}$ satisfies

$$\sigma_n(F, \Psi')_X = O(n^{-\beta}), \quad n \to \infty,$$

for a value of $\beta > \alpha$. Donoho has shown that in some cases it is possible to determine a best basis (in the above sense) for the class $F$ by intrinsic properties of how the class is represented with respect to the basis. In Donoho's analysis (as was the case for Kashin as well) the space $X$ is $L_2$ (or equivalently any Hilbert space), and the competition for a best basis takes place over all complete orthonormal systems (i.e. $\mathbb{B}$ consists of all complete orthonormal bases for $L_2$).

In DeVore, Petrova and Temlyakov (2003) we continued to study the problem of optimal bases selection with regard to natural collections of bases. We worked on the following problem in this direction. We say that a function class $F$ is aligned to the basis $\Psi$ if, whenever $f = \sum a_k \psi_k$ is in $F$, then

$$\sum a_k' \psi_k \in F \quad \text{for any} \quad |a_k'| \le c|a_k|, \quad k = 1, 2, \ldots,$$

where $c > 0$ is a fixed constant. We pointed out in DeVore, Petrova and Temlyakov (2003) that the results from Kashin (1985) and Donoho (1993) imply the following result.

**Theorem 1.82** *Let $\Phi$ be an orthonormal basis for a Hilbert space $H$ and let $F$ be a function class aligned with $\Phi$ such that, for some $\alpha > 0$, $\beta \in \mathbb{R}$, we have*

$$\limsup_{m \to \infty} m^\alpha (\log m)^\beta \sigma_m(F, \Phi) > 0.$$

*Then, for any orthonormal basis $B$ we have*

$$\limsup_{m \to \infty} m^\alpha (\log m)^\beta \sigma_m(F, B) > 0.$$

We have obtained in DeVore, Petrova and Temlyakov (2003) a generalization of this important result in the following way. We replaced the Hilbert space with the Banach space and also widened the search for optimal basis selection from the collection of orthonormal bases to the collection of unconditional bases. Here is the corresponding theorem from DeVore, Petrova and Temlyakov (2003).

**Theorem 1.83** *Let* $\Psi$ *be a normalized unconditional basis for X with the property*

$$\left\| \sum_{j \in A} \psi_j \right\|_X \asymp (|A|)^{\mu},$$

*for some* $\mu > 0$. *Assume that the function class F is aligned with* $\Psi$, *and that, for some* $\alpha > 0$, $\beta \in \mathbb{R}$, *we have*

$$\limsup_{m \to \infty} m^{\alpha} (\log m)^{\beta} \sigma_m(F, \Psi) > 0.$$

*Then, for any unconditional basis B we have*

$$\limsup_{m \to \infty} m^{\alpha} (\log m)^{\alpha+\beta} \sigma_m(F, B) > 0. \tag{1.135}$$

Theorem 1.83 is weaker than Theorem 1.82 in the sense that we have an extra factor $(\log m)^{\alpha}$ in (1.135). Recently, Bednorz (2008) proved Theorem 1.83 with (1.135) replaced by (1.136):

$$\limsup_{m \to \infty} m^{\alpha} (\log m)^{\beta} \sigma_m(F, B) > 0. \tag{1.136}$$

The following nonlinear analogs of the Kolmogorov widths and the orthowidths (see, for example, Temlyakov (1989a)) were considered in Temlyakov (2000a, 2002a, 2003a). Let a function class $F$ and a Banach space $X$ be given. Assume that, on the basis of some additional information, we know that our basis for $m$-term approximation should satisfy some structural properties; for instance, it has to be orthogonal. Let $\mathbb{B}$ be a collection of bases satisfying a given property.

**I** Define an analog of the Kolmogorov width

$$\sigma_m(F, \mathbb{B})_X := \inf_{\Psi \in \mathbb{B}} \sup_{f \in F} \sigma_m(f, \Psi)_X.$$

**II** Define an analog of the orthowidth

$$\gamma_m(F, \mathbb{B})_X := \inf_{\Psi \in \mathbb{B}} \sup_{f \in F} \| f - G_m(f, \Psi) \|_X.$$

In the papers cited above some results were obtained when $\mathbb{B} = \mathbb{O}$, the set of orthonormal bases, and $F$ is either a multivariate smoothness class of an anisotropic Sobolev–Nikol'skii kind, or a class of functions with bounded mixed derivative.

We conclude this section with a recent result from Wojtaszczyk (2006). Theorem 1.11A says that the univariate Haar basis $\mathcal{H}_p$ is a greedy basis for $L_p := L_p([0, 1]), 1 < p < \infty$. The spaces $L_p$ are examples of rearrangement-invariant spaces. Let us recall that a rearrangement-invariant space of functions defined on $[0, 1]$ is a Banach space $X$ with norm $\| \cdot \|$ whose elements are measurable (in the sense of Lebesgue) functions defined on $[0, 1]$ satisfying the following conditions.

(1) If $f \in X$ and $g$ is a measurable function such that $|g(x)| \le |f(x)|$ a.e., then $g \in X$ and $\|g\| \le \|f\|$.

(2) If $f \in X$ and $g$ has the same distribution as $f$, i.e. for all $\lambda$

$$\text{measure}(\{x \in [0, 1] : f(x) \le \lambda\}) = \text{measure}(\{x \in [0, 1] : g(x) \le \lambda\}),$$

then $g \in X$ and $\|g\| = \|f\|$.

The following result was proved in Wojtaszczyk (2006).

**Theorem 1.84** *Let $X$ be a rearrangement-invariant space on $[0, 1]$. If the Haar system normalized in $X$ is a greedy basis for $X$, then $X = L_p([0, 1])$ with some $1 < p < \infty$.*

It is a very interesting result that singles out the $L_p$-spaces with $1 < p < \infty$ from the collection of rearrangement-invariant spaces. Theorem 1.84 emphasizes the importance of the $L_p$ spaces in the theory of greedy approximation.

## 1.10 Systems $L_p$-equivalent to the Haar basis

In the preceding sections of this chapter we have presented elements of a general theory of greedy-type bases. In this section we concentrate on the construction of greedy bases and related bases that are useful in the approximation of functions in the $L_p$ norm. Theorem 1.11 indicates the importance of bases that are $L_p$-equivalent to the Haar basis $\mathcal{H}$. It says that such bases (normalized in $L_p$) are greedy bases for $L_p$, $1 < p < \infty$. Theorem 1.11 addresses the case of $L_p([0, 1])$. The same proof works for $L_p(\mathbb{R})$. In this section we will give some sufficient conditions on a system of functions in order to be $L_p$-equivalent to the Haar basis. It is more convenient to give these conditions in the case of $L_p(\mathbb{R})$. These results are part of the general Littlewood–Paley theory. We begin this section by introducing various forms of the Littlewood–Paley theory for systems of functions. From the univariate wavelet $\psi$, we can construct efficient bases for $L_2(\mathbb{R})$ and other function spaces by dilation and shifts (see, for example, DeVore (1998)). For example, the functions

$$\psi_{j,k} := 2^{k/2} \psi(2^k \cdot -j), \quad j, k \in \mathbb{Z},$$

form a stable basis (orthogonal basis in the case of an orthogonal wavelet $\psi$) for $L_2(\mathbb{R})$.

It is convenient to use a different indexing for the functions $\psi_{j,k}$. Let $D :=$ $D(\mathbb{R})$ denote the set of dyadic intervals. Each such interval $I$ is of the form $I = [j2^{-k}, (j+1)2^{-k}]$. We define

$$\psi_I := \psi_{j,k}, \quad I = [j2^{-k}, (j+1)2^{-k}]. \tag{1.137}$$

Thus the basis $\{\psi_{j,k}\}_{j,k \in \mathbb{Z}}$ is the same as $\{\psi_I\}_{I \in D(\mathbb{R})}$.

We consider in this section systems of functions $\{\eta(I, \cdot)\}_{I \in D}$ defined on $\mathbb{R}$. If $1 < p < \infty$, we say that a family of real-valued functions $\eta(I, \cdot)$, $I \in D$, satisfies the *strong Littlewood–Paley property* for $p$ if, for any finite sequence $(c_I)$ of real numbers, we have

$$\left\| \sum_{I \in D} c_I \eta(I, \cdot) \right\|_p \asymp \left\| \left( \sum_{I \in D} [c_I \eta(I, \cdot)]^2 \right)^{1/2} \right\|_p \tag{1.138}$$

with constants of equivalency depending at most on $p$. Here and later we use the notation $A \asymp B$ to mean that there are two constants $C_1, C_2 > 0$ such that

$$C_1 A \leq B \leq C_2 A.$$

We shall indicate what the constants depend on (in the case of (1.138) they may depend on $p$).

Here is a useful remark concerning (1.138). From the validity of (1.138) for finite sequences, we can deduce its validity for infinite sequences by a limiting argument. For example, if $(c_I)_{I \in D}$ is an infinite sequence for which the sum on the left-hand side of (1.138) converges in $L_p(\mathbb{R})$ with respect to some ordering of the $I \in D$, then the right-hand side of (1.138) will converge with respect to the same ordering and the right-hand side of (1.138) will be less than a multiple of the left. Likewise, we can reverse the roles of the left- and right-hand sides. Similar remarks hold for other statements like (1.138).

The term *strong Littlewood–Paley inequality* is used to differentiate (1.138) from other possible forms of Littlewood–Paley inequalities. For example, the Littlewood–Paley inequalities for the complex exponentials take a different form (see Zygmund (1959), chap. XV). Another form of interest in our considerations is the following:

$$\left\| \sum_{I \in D} c_I \eta(I, \cdot) \right\|_p \asymp \left\| \left( \sum_{I \in D} [c_I \chi_I]^2 \right)^{1/2} \right\|_p. \tag{1.139}$$

We use the notation $\chi$ for the characteristic function of $[0, 1]$ and $\chi_I$ for its $L_2(\mathbb{R})$-normalized, shifted dilates given by (1.137) (with $\psi = \chi$).

The two forms (1.138) and (1.139) are equivalent under very mild conditions on the functions $\eta(I, \cdot)$. To see this, we shall use the Hardy–Littlewood maximal operator, which is defined for a locally integrable function $g$ on $\mathbb{R}$ by

$$Mg(x) := \sup_{J \ni x} \frac{1}{|J|} \int_J |g(y)| \, dy,$$

with the supremum taken over all intervals $J$ that contain $x$. It is well known that $M$ is a bounded operator on $L_p(\mathbb{R})$ for all $1 < p \leq \infty$. The Fefferman–Stein inequality (Fefferman and Stein (1972)) bounds the mapping $M$ on sequences of functions. We only need the following special case of this inequality, which says that for any functions $\eta(I, \cdot)$ and constants $c_I$, $I \in D$, we have, for $1 < p \leq \infty$,

$$\left\| \left( \sum_{I \in D} (c_I M \eta(I, \cdot))^2 \right)^{1/2} \right\|_p \leq A \left\| \left( \sum_{I \in D} (c_I \eta(I, \cdot))^2 \right)^{1/2} \right\|_p, \qquad (1.140)$$

with an absolute constant $A$.

Consider now as an example the equivalence of (1.138). If the functions $\eta(I, \cdot)$, $I \in D$, satisfy

$$|\eta(I, x)| \leq C M \chi_I(x), \quad \chi_I(x) \leq C M \eta(I, x), \quad \text{for almost all} \quad x \in \mathbb{R},$$

$$(1.141)$$

then, using (1.140), we see that (1.138) holds if and only if (1.139) holds. The first inequality in (1.141) is a decay condition on $\eta(I, \cdot)$. For example, if $\eta(I, \cdot)$ is given by the normalized, shifted dilates of the function $\psi$, $\eta(I, \cdot) = \psi_I$, then the first inequality in (1.141) holds whenever

$$|\psi(x)| \leq C[\max(1, |x|)]^{-\lambda}, \quad \text{for almost all} \quad x \in \mathbb{R},$$

with $\lambda \geq 1$. The second condition in (1.141) is extremely mild. For example, it is always satisfied when the family $\eta(I, \cdot)$ is generated by the shifted dilates of a non-zero function $\psi$.

Suppose that we have two families $\eta(I, \cdot), \mu(I, \cdot), I \in D(\mathbb{R})$. We shall use the notation $\{\eta(I, \cdot)\}_{I \in D} \prec \{\mu(I, \cdot)\}_{I \in D}$, if there is a constant $C > 0$ such that

$$\left\| \sum_{I \in D} c_I \eta(I, \cdot) \right\|_p \leq C \left\| \sum_{I \in D} c_I \mu(I, \cdot) \right\|_p \qquad (1.142)$$

holds for all finite sequences $(c_I)_{I \in D}$ with $C$ independent of the sequence. If $\{\eta(I, \cdot)\}_{I \in D} \prec \{\mu(I, \cdot)\}_{I \in D}$ and $\{\mu(I, \cdot)\}_{I \in D} \prec \{\eta(I, \cdot)\}_{I \in D}$, then we write $\{\eta(I, \cdot)\}_{I \in D} \approx \{\mu(I, \cdot)\}_{I \in D}$ and say that these systems are $L_p$-equivalent.

Given two families $\eta(I, \cdot), \mu(I, \cdot), I \in D(\mathbb{R})$, we define the operator $T$ which maps $\mu(I, \cdot)$ into $\eta(I, \cdot)$ for all $I \in D$, and we extend $T$ to finite linear combinations of the $\mu(I, \cdot)$ by linearity. Then (1.142) holds if and only if $T$ is a bounded operator with respect to the $L_p$ norm, and $\{\mu(I, \cdot)\}_{I \in D} \prec \{\eta(I, \cdot)\}_{I \in D}$ holds if and only if $T$ has a bounded inverse with respect to the $L_p$ norm.

The strong Littlewood–Paley inequalities (1.139) are the same as the $L_p$-equivalence $\{\eta(I, \cdot)\} \approx \{H_I\}$. We begin with a presentation of sufficient conditions in order that $\{\eta(I, \cdot)\} \prec \{H_I\}$. Let $\xi_I, I \in D$, denote the center of the dyadic interval $I$. We shall assume in this section that $\eta(I, \cdot), I \in D$, is a family of univariate functions that satisfy the following assumptions.

**A1** There is an $\epsilon > 0$ and a constant $C_1$ such that, for all $t \in \mathbb{R}$ and all $J \in D$, we have

$$|\eta(J, \xi_J + t|J|)| \le C_1 |J|^{-1/2}(1 + |t|)^{-1-\epsilon}.$$

**A2** There is an $\epsilon > 0$, a constant $C_2$ and a partition of $[-1/2, 1/2]$ into intervals $J_1, \ldots, J_m$ that are dyadic with respect to $[-1/2, 1/2]$ such that, for any $J \in D$, any $j \in \mathbb{Z}$ and any $t_1, t_2$ in the interior of the same interval $J_k, k = 1, \ldots, m$, we have

$$|\eta(J, \xi_J + j|J| + t_1|J|) - \eta(J, \xi_J + j|J| + t_2|J|)|$$
$$\le C_2 |J|^{-1/2}(1 + |j|)^{-1-\epsilon}|t_2 - t_1|^\epsilon.$$

**A3** For any $J \in D$, we have

$$\int_{\mathbb{R}} \eta(J, x) dx = 0.$$

When $\eta(J, \cdot) = \psi_J$ for a function $\psi$, it is enough to check these assumptions for $J = [0, 1]$, i.e. for the function $\psi$ alone. They follow for all other dyadic intervals $J$ by dilation and translation.

Condition **A1** is a standard decay assumption and **A3** is the zero moment condition. Condition **A2** requires that the functions $\eta(I, \cdot)$ be piecewise in Lip $\epsilon$.

Let $T$ be the linear operator which satisfies

$$T\left(\sum_{I \in D} c_I H_I\right) = \sum_{I \in D} c_I \eta(I, \cdot) \tag{1.143}$$

for each finite linear combination $\sum_{I \in D} c_I H_I$ of the $H_I$. We wish to show that

$$\left\|T\left(\sum_{I \in D} c_I H_I\right)\right\|_p \le C \left\|\sum_{I \in D} c_I H_I\right\|_p$$

for each such sum. From this it would follow that $T$ extends (by continuity) to a bounded operator on all of $L_p(\mathbb{R})$ and therefore $\{\eta(I, \cdot)\} \prec \{H_I\}$.

We can expand $\eta(J, \cdot)$ into its Haar decomposition. Let

$$\lambda(I, J) := \int_{\mathbb{R}} \eta(J, x) H_I(x) dx, \qquad (1.144)$$

so that

$$\eta(J, \cdot) = \sum_{I \in D} \lambda(I, J) H_I.$$

It follows that

$$T\left(\sum_{J \in D} c_J H_J\right) = \sum_{I \in D} \sum_{J \in D} \lambda(I, J) c_J H_I. \qquad (1.145)$$

Thus the mapping $T$ is tied to the bi-infinite matrix $\Lambda := (\lambda(I, J))_{I, J \in D}$ which maps the sequence $c := (c_J)$ into the sequence

$$(c'_I) := \Lambda c.$$

One approach to proving Littlewood–Paley inequalities is to show that the matrix $\Lambda$ decays sufficiently quickly away from the diagonal (see Frazier and Jawerth (1990), sect. 3). Following Frazier and Jawerth (1990), we say that a matrix $A = (a(I, J))_{I, J \in D}$ is *almost diagonal* if, for some $\epsilon > 0$, we have

$$|a(I, J)| \leq C\omega(I, J), \qquad (1.146)$$

with

$$\omega(I, J) := \left(1 + \frac{|\xi_I - \xi_J|}{\max(|I|, |J|)}\right)^{-1-\epsilon} \left(\min\left(\frac{|I|}{|J|}, \frac{|J|}{|I|}\right)\right)^{(1+\epsilon)/2}. \qquad (1.147)$$

In DeVore, Konyagin and Temlyakov (1998) we used the following special case of Frazier and Jawerth (1990), theorem 3.3, concerning almost-diagonal operators.

**Theorem 1.85** *If* $(a(I, J))_{I, J \in D}$ *is an almost-diagonal matrix, then the operator* $A$ *defined by*

$$A\left(\sum_{J \in D} c_J H_J\right) := \sum_{I \in D} \sum_{J \in D} a(I, J) c_J H_I \qquad (1.148)$$

*is bounded on* $L_p(\mathbb{R})$ *for each* $1 < p < \infty$.

In DeVore, Konyagin and Temlyakov (1998) we proved the following theorems.

**Theorem 1.86** *If $\eta(I, \cdot)$, $I \in D$, satisfy assumptions* **A1–A3**, *then the operator T defined by (1.143) is bounded from $L_p(\mathbb{R})$ into itself for each $1 < p < \infty$.*

**Corollary 1.87** *If $\eta(I, \cdot)$, $I \in \mathcal{D}$, satisfy assumptions* **A1–A3**, *then $\{\eta(I, \cdot\}_{I \in D}$ $\prec \{H_I\}_{I \in D}$.*

We can use a duality argument to provide sufficient conditions that the operator $T$ of (1.143) is boundedly invertible. For this, we assume that $\eta(I, \cdot)$, $I \in D$, is a family of functions for which there is a dual family $\eta^*(I, \cdot)$, $I \in D$, that satisfies

$$\langle \eta(I, \cdot), \eta^*(J, \cdot) \rangle = \delta(I, J), \quad I, J \in D.$$

**Theorem 1.88** *If the functions $\eta^*(I, \cdot)$, $I \in D$, satisfy assumptions* **A1–A3**, *then $\{H_I\}_{I \in D} \prec \{\eta(I, \cdot\}_{I \in D}$.*

**Theorem 1.89** *If the systems of functions $\{\eta(I, \cdot)\}_{I \in D}$, $\{\eta^*(I, \cdot)\}_{I \in D}$, satisfy assumptions* **A1–A3**, *then the system $\{\eta(I, \cdot)\}_{I \in D}$ is $L_p$-equivalent to the Haar system $\{H_I\}_{I \in D}$ for $1 < p < \infty$.*

It is known from different results (see DeVore (1998), DeVore, Jawerth and Popov (1992) and Temlyakov (2003a)) that wavelets are well designed for non-linear approximation. We present here one general result in this direction. We fix $p \in (1, \infty)$ and consider in $L_p([0, 1]^d)$ a basis $\Psi := \{\psi_I\}_{I \in D}$ indexed by dyadic intervals $I$ of $[0, 1]^d$, $I = I_1 \times \cdots \times I_d$, where $I_j$ is a dyadic interval of $[0, 1]$, $j = 1, \ldots, d$, which satisfies certain properties. Set $L_p := L_p(\Omega)$ with a normalized Lebesgue measure on $\Omega$, $|\Omega| = 1$. First of all we assume that, for all $1 < q, p < \infty$ and $I \in D$, where $D := D([0, 1]^d)$ is the set of all dyadic intervals of $[0, 1]^d$, we have

$$\|\psi_I\|_p \asymp \|\psi_I\|_q |I|^{1/p - 1/q}, \tag{1.149}$$

with constants independent of $I$. This property can be easily checked for a given basis.

Next, assume that for any $s = (s_1, \ldots, s_d) \in \mathbb{Z}^d$, $s_j \geq 0$, $j = 1, \ldots, d$, and any $\{c_I\}$, we have, for $1 < p < \infty$,

$$\left\| \sum_{I \in D_s} c_I \psi_I \right\|_p^p \asymp \sum_{I \in D_s} \|c_I \psi_I\|_p^p, \tag{1.150}$$

where

$$D_s := \{I = I_1 \times \cdots \times I_d \in D \quad : \quad |I_j| = 2^{-s_j}, \quad j = 1, \ldots, d\}.$$

This assumption allows us to estimate the $L_p$ norm of a dyadic block in terms of Fourier coefficients.

The third assumption is that $\Psi$ is a basis satisfying the following version (weak form) of the Littlewood–Paley inequality. Let $1 < p < \infty$ and assume $f \in L_p$ has the expansion

$$f = \sum_I f_I \psi_I.$$

We assume that

$$\lim_{\min_j \mu_j \to \infty} \left\| f - \sum_{s_j \le \mu_j, j=1,\ldots,d} \sum_{I \in D_s} f_I \psi_I \right\|_p = 0 \qquad (1.151)$$

and

$$\|f\|_p \asymp \left\| \left( \sum_s \left| \sum_{I \in D_s} f_I \psi_I \right|^2 \right)^{1/2} \right\|_p. \qquad (1.152)$$

Let $\mu \in \mathbb{Z}^d$, $\mu_j \ge 0$, $j = 1, \ldots, d$. Denote by $\Psi(\mu)$ the subspace of polynomials of the form

$$\psi = \sum_{s_j \le \mu_j, j=1,\ldots,d} \sum_{I \in D_s} c_I \psi_I.$$

We now define a function class. Let $R = (R_1, \ldots, R_d)$, $R_j > 0$, $j = 1, \ldots, d$, and

$$g(R) := \left( \sum_{j=1}^d R_j^{-1} \right)^{-1}.$$

For any natural number $l$, define

$$\Psi(R, l) := \Psi(\mu), \qquad \mu_j = [g(R)l/R_j], \quad j = 1, \ldots, d.$$

We define the class $H_q^R(\Psi)$ as the set of functions $f \in L_q$ representable in the form

$$f = \sum_{l=1}^\infty t_l, \quad t_l \in \Psi(R, l), \quad \|t_l\|_q \le 2^{-g(R)l}.$$

We proved the following theorem in Temlyakov (2002a).

**Theorem 1.90** *Let $1 < q, p < \infty$ and $g(R) > (1/q - 1/p)_+$. Then, for $\Psi$ satisfying (1.149)–(1.152) we have*

$$\sup_{f \in H_q^R(\Psi)} \| f - G_m^p(f, \Psi) \|_p \ll m^{-g(R)}.$$

In the periodic case the following basis $U^d := U \times \cdots \times U$ can be taken in place of $\Psi$ in Theorem 1.90. We define the system $U := \{U_I\}$ in the univariate case. Denote

$$U_n^+(x) := \sum_{k=0}^{2^n-1} e^{ikx} = \frac{e^{i2^n x} - 1}{e^{ix} - 1}, \quad n = 0, 1, 2, \ldots;$$

$$U_{n,k}^+(x) := e^{i2^n x} U_n^+(x - 2\pi k 2^{-n}), \quad k = 0, 1, \ldots, 2^n - 1;$$

$$U_{n,k}^-(x) := e^{-i2^n x} U_n^+(-x + 2\pi k 2^{-n}), \quad k = 0, 1, \ldots, 2^n - 1.$$

We normalize the system of functions $\{U_{n,k}^+, U_{n,k}^-\}$ in $L_2$ and enumerate it by dyadic intervals. We write

$$U_I(x) := 2^{-n/2} U_{n,k}^+(x) \quad \text{with} \quad I = [(k + 1/2)2^{-n}, (k + 1)2^{-n}),$$

$$U_I(x) := 2^{-n/2} U_{n,k}^-(x) \quad \text{with} \quad I = [k2^{-n}, (k + 1/2)2^{-n})$$

and

$$U_{[0,1)}(x) := 1.$$

Wojtaszczyk (1997) proved that $U$ is an unconditional basis of $L_p$, $1 < p < \infty$. It is well known that $H_q^R(U^d)$ is equivalent to the standard anisotropic multivariate periodic Hölder–Nikol'skii classes $NH_p^R$. We define these classes in the following way (see Nikol'skii (1975)). The class $NH_p^R$, $R = (R_1, \ldots, R_d)$ and $1 \le p \le \infty$, is the set of periodic functions $f \in L_p([0, 2\pi]^d)$ such that, for each $l_j = [R_j] + 1$, $j = 1, \ldots, d$, the following relations hold:

$$\| f \|_p \le 1, \qquad \| \Delta_t^{l_j, j} f \|_p \le |t|^{R_j}, \quad j = 1, \ldots, d, \tag{1.153}$$

where $\Delta_t^{l,j}$ is the $l$th difference with step $t$ in the variable $x_j$. For $d = 1$, $NH_p^R$ coincides with the standard Hölder class $H_p^R$. Theorem 1.90 gives the following result.

**Theorem 1.91** *Let $1 < q, p < \infty$; then for $R$ such that $g(R) > (1/q - 1/p)_+$, we have*

$$\sup_{f \in NH_q^R} \| f - G_m^p(f, U^d) \|_p \ll m^{-g(R)}.$$

We also proved in Temlyakov (2002a) that the basis $U^d$ is an optimal orthonormal basis for approximation of classes $NH_q^R$ in $L_p$:

$$\sigma_m(NH_q^R, \mathbb{O})_p \asymp \sigma_m(NH_q^R, U^d)_p \asymp m^{-g(R)} \tag{1.154}$$

for $1 < q < \infty$, $2 \le p < \infty$, $g(R) > (1/q - 1/p)_+$. Here $\mathbb{O}$ is a collection of orthonormal bases. It is important to note that Theorem 1.91 guarantees that the estimate in (1.154) can be realized by the greedy algorithm $G_m^p(\cdot, U^d)$ with regard to $U^d$. Another important feature of (1.154) is that the basis $U^d$ is optimal (in the sense of order) for each class $NH_q^R$ independently of $R = (R_1, \ldots, R_d)$ and $q$. This property is known as universality for a collection of classes (in the above case, the collection $\{NH_q^R\}$). Further discussion of this important issue can be found in Temlyakov (2002a, 2003a).

## 1.11 Open problems

We formulate here some open problems related to the results discussed in this chapter.

**1.1.** Characterize Schauder bases with the following property: for any $f \in X$ and any $m \in \mathbb{N}$ there exists the best $m$-term approximant of $f$ with respect to a given basis.

**1.2.** Solve Problem 1.1 with Schauder bases replaced by one of the following bases: greedy, unconditional, almost greedy, quasi-greedy.

**1.3.** Find greedy-type algorithms realizing near-best approximation for individual functions in the $L_p([0, 1]^d)$, $1 < p < \infty$, $d \ge 2$, with regard to the multivariate Haar basis $\mathcal{H}_p^d$.

**1.4.** Find the right order of the function $v(m, p, \gamma)$ (see Section 1.7) as a function of $m$.

**1.5.** Find the right order of the function $w(m, p, \gamma)$ (see Section 1.7) as a function of $m$.

# 2

# Greedy approximation with respect to dictionaries: Hilbert spaces

## 2.1 Introduction

In this chapter we discuss greedy approximation with regard to redundant systems. Greedy approximation is a special form of nonlinear approximation. The basic idea behind nonlinear approximation is that the elements used in the approximation do not come from a fixed linear space but are allowed to depend on the function being approximated. The standard problem in this regard is the problem of $m$-term approximation, where one fixes a basis and aims to approximate a target function $f$ by a linear combination of $m$ terms of the basis. We discussed this problem in detail in Chapter 1. When the basis is a wavelet basis or a basis of other waveforms, then this type of approximation is the starting point for compression algorithms. An important feature of approximation using a basis $\Psi := \{\psi_k\}_{k=1}^{\infty}$ of a Banach space $X$ is that each function $f \in X$ has a unique representation

$$f = \sum_{k=1}^{\infty} c_k(f)\psi_k, \tag{2.1}$$

and we can identify $f$ with the set of its coefficients $\{c_k(f)\}_{k=1}^{\infty}$. The problem of $m$-term approximation with regard to a basis has been studied thoroughly and rather complete results have been established (see Chapter 1). In particular, it was established that the greedy-type algorithm which forms a sum of $m$ terms with the largest $\|c_k(f)\psi_k\|_X$ out of expansion (2.1) realizes in many cases near-best $m$-term approximation for function classes (DeVore, Jawerth and Popov (1992)) and even for individual functions (see Chapter 1).

Recently, there has emerged another more complicated form of nonlinear approximation, which we call highly nonlinear approximation. It takes many forms but has the basic ingredient that a basis is replaced by a larger system of functions that is usually redundant. We call such systems dictionaries. On

77

the one hand, redundancy offers much promise for greater efficiency in terms of the approximation rate, but on the other hand it gives rise to highly non-trivial theoretical and practical problems. The problem of characterizing the approximation rate for a given function or function class is now much more substantial and results are quite fragmentary. However, such results are very important for understanding what this new type of approximation offers. Perhaps the first example of this type was considered by Schmidt (1906), who studied the approximation of functions $f(x, y)$ of two variables by bilinear forms,

$$\sum_{i=1}^{m} u_i(x)v_i(y),$$

in $L_2([0, 1]^2)$. This problem is closely connected with properties of the integral operator

$$J_f(g) := \int_0^1 f(x, y)g(y)dy$$

with kernel $f(x, y)$. Schmidt (1906) gave an expansion (known as the Schmidt expansion):

$$f(x, y) = \sum_{j=1}^{\infty} s_j(J_f)\phi_j(x)\psi_j(y),$$

where $\{s_j(J_f)\}$ is a non-increasing sequence of singular numbers of $J_f$, i.e. $s_j(J_f) := \lambda_j(J_f^*J_f)^{1/2}$, where $\{\lambda_j(A)\}$ is a sequence of eigenvalues of an operator $A$, and $J_f^*$ is the adjoint operator to $J_f$. The two sequences $\{\phi_j(x)\}$ and $\{\psi_j(y)\}$ form orthonormal sequences of eigenfunctions of the operators $J_f J_f^*$ and $J_f^*J_f$, respectively. He also proved that

$$\left\| f(x, y) - \sum_{j=1}^{m} s_j(J_f)\phi_j(x)\psi_j(y) \right\|_{L_2}$$

$$= \inf_{u_j, v_j \in L_2, \; j=1,\dots,m} \left\| f(x, y) - \sum_{j=1}^{m} u_j(x)v_j(y) \right\|_{L_2}.$$

It was understood later that the above best bilinear approximation can be realized by the following greedy algorithm. Assume that $c_j$, $u_j(x)$, $v_j(y)$, $\|u_j\|_{L_2} = \|v_j\|_{L_2} = 1$ and $j = 1, \dots, m - 1$ have been constructed after $m - 1$ steps of the algorithm. At the $m$th step we choose $c_m$, $u_m(x)$, $v_m(y)$, $\|u_m\|_{L_2} = \|v_m\|_{L_2} = 1$, to minimize

$$\|f(x, y) - \sum_{j=1}^{m} c_j u_j(x) v_j(y)\|_{L_2}.$$

We call this type of algorithm the Pure Greedy Algorithm (PGA) (see the general definition below).

Another problem of this type which is well known in statistics is the projection pursuit regression problem, mentioned in the Preface. The problem is to approximate in $L_2$ a given function $f \in L_2$ by a sum of ridge functions, i.e. by

$$\sum_{j=1}^{m} r_j(\omega_j \cdot x), \quad x, \omega_j \in \mathbb{R}^d, \quad j = 1, \ldots, m,$$

where $r_j$, $j = 1, \ldots, m$, are univariate functions. The following greedy-type algorithm (projection pursuit) was proposed in Friedman and Stuetzle (1981) to solve this problem. Assume functions $r_1, \ldots, r_{m-1}$ and vectors $\omega_1, \ldots, \omega_{m-1}$ have been determined after $m-1$ steps of the algorithm. Choose at the $m$th step a unit vector $\omega_m$ and a function $r_m$ to minimize the error

$$\|f(x) - \sum_{j=1}^{m} r_j(\omega_j \cdot x)\|_{L_2}.$$

This is one more example of the Pure Greedy Algorithm. The Pure Greedy Algorithm and some other versions of greedy-type algorithms have recently been intensively studied: see Barron (1993), Davis, Mallat and Avellaneda (1997), DeVore and Temlyakov (1996, 1997), Donahue *et al.* (1997), Dubinin (1997), Huber (1985), Jones (1987, 1992), Konyagin and Temlyakov (1999b), Livshitz (2006, 2007, 2009), Livshitz and Temlyakov (2001, 2003), Temlyakov (1999, 2000b, 2002b, 2003b). There are several survey papers that discuss greedy approximation with regard to redundant systems: see DeVore (1998) and Temlyakov (2003a, 2006a). In this chapter we discuss along with the PGA some of its modifications which are more suitable for implementation. This new type of greedy algorithms will be termed Weak Greedy Algorithms.

In order to orient the reader we recall some notations and definitions from the theory of greedy algorithms. Let $H$ be a real Hilbert space with an inner product $\langle \cdot, \cdot \rangle$ and the norm $\|x\| := \langle x, x \rangle^{1/2}$. We say a set $\mathcal{D}$ of functions (elements) from $H$ is a dictionary if each $g \in \mathcal{D}$ has norm one ($\|g\| = 1$) and the closure of span $\mathcal{D}$ is equal to $H$. Sometimes it will be convenient for us also to consider the symmetrized dictionary $\mathcal{D}^{\pm} := \{\pm g, g \in \mathcal{D}\}$. In DeVore and Temlyakov (1996) we studied the following two greedy algorithms. If $f \in H$, we let $g = g(f) \in \mathcal{D}$ be the element from $\mathcal{D}$ which maximizes $|\langle f, g \rangle|$ (we make an additional assumption that a maximizer exists) and define

$$G(f) := G(f, \mathcal{D}) := \langle f, g \rangle g \qquad (2.2)$$

and

$$R(f) := R(f, \mathcal{D}) := f - G(f).$$

**Pure Greedy Algorithm (PGA)** We define $f_0 := R_0(f) := R_0(f, \mathcal{D}) := f$ and $G_0(f) := G_0(f, \mathcal{D}) := 0$. Then, for each $m \geq 1$, we inductively define

$$G_m(f) : = G_m(f, \mathcal{D}) := G_{m-1}(f) + G(R_{m-1}(f)),$$

$$f_m := R_m(f) : = R_m(f, \mathcal{D}) := f - G_m(f) = R(R_{m-1}(f)).$$

We note that the Pure Greedy Algorithm is known under the name Matching Pursuit in signal processing (see, for example, Mallat and Zhang (1993)).

If $H_0$ is a finite-dimensional subspace of $H$, we let $P_{H_0}$ be the orthogonal projector from $H$ onto $H_0$. That is, $P_{H_0}(f)$ is the best approximation to $f$ from $H_0$.

**Orthogonal Greedy Algorithm (OGA)** We define $f_0^o := R_0^o(f) := R_0^o(f, \mathcal{D}) := f$ and $G_0^o(f) := G_0^o(f, \mathcal{D}) := 0$. Then, for each $m \geq 1$, we inductively define

$$H_m := H_m(f) := \mathrm{span}\{g(R_0^o(f)), \ldots, g(R_{m-1}^o(f))\},$$

$$G_m^o(f) := G_m^o(f, \mathcal{D}) := P_{H_m}(f),$$

$$f_m^o := R_m^o(f) := R_m^o(f, \mathcal{D}) := f - G_m^o(f).$$

We remark that for each $f$ we have

$$\|f_m^o\| \leq \|f_{m-1}^o - G_1(f_{m-1}^o, \mathcal{D})\|. \qquad (2.3)$$

In Section 1.5 we realized that the Weak Greedy Algorithms with regard to bases work as well as the corresponding Greedy Algorithms. In this chapter we study similar modifications of the Pure Greedy Algorithm (PGA) and the Orthogonal Greedy Algorithm (OGA), which we call, respectively, the Weak Greedy Algorithm (WGA) and the Weak Orthogonal Greedy Algorithm (WOGA). We now give the corresponding definitions from Temlyakov (2000b). Let a sequence $\tau = \{t_k\}_{k=1}^{\infty}, 0 \leq t_k \leq 1$, be given.

**Weak Greedy Algorithm (WGA)** We define $f_0^{\tau} := f$. Then, for each $m \geq 1$ we have the following inductive definition.

(1) $\varphi_m^{\tau} \in \mathcal{D}$ is any element satisfying

$$|\langle f_{m-1}^{\tau}, \varphi_m^{\tau} \rangle| \geq t_m \sup_{g \in \mathcal{D}} |\langle f_{m-1}^{\tau}, g \rangle|.$$

(2)

$$f_m^\tau := f_{m-1}^\tau - \langle f_{m-1}^\tau, \varphi_m^\tau \rangle \varphi_m^\tau.$$

(3)

$$G_m^\tau(f, \mathcal{D}) := \sum_{j=1}^m \langle f_{j-1}^\tau, \varphi_j^\tau \rangle \varphi_j^\tau.$$

We note that, for a particular case $t_k = t$, $k = 1, 2, \ldots$, this algorithm was considered in Jones (1987). Thus, the WGA is a generalization of the PGA, making it easier to construct an element $\varphi_m^\tau$ at the $m$th greedy step. We point out that the WGA contains, in addition to the first (greedy) step, the second step (see (2) and (3) in the above definition) where we update an approximant by adding an orthogonal projection of the residual $f_{m-1}^\tau$ onto $\varphi_m^\tau$. Therefore, the WGA provides for each $f \in H$ an expansion into a series (greedy expansion)

$$f \sim \sum_{j=1}^\infty c_j(f)\varphi_j^\tau, \quad c_j(f) := \langle f_{j-1}^\tau, \varphi_j^\tau \rangle.$$

In general, it is not an orthogonal expansion, but it has some similar properties. The coefficients $c_j(f)$ of an expansion are obtained by the Fourier formulas with $f$ replaced by the residuals $f_{j-1}^\tau$. It is easy to see that

$$\|f_m^\tau\|^2 = \|f_{m-1}^\tau\|^2 - |c_m(f)|^2.$$

We prove convergence of greedy expansion (see, for example, Theorem 2.4 below), and therefore, from the above equality, we get for this expansion an analog of the Parseval formula for orthogonal expansions:

$$\|f\|^2 = \sum_{j=1}^\infty |c_j(f)|^2.$$

**Weak Orthogonal Greedy Algorithm (WOGA)** We define $f_0^{o,\tau} := f$ and $f_1^{o,\tau} := f_1^\tau$; $\varphi_1^{o,\tau} := \varphi_1^\tau$, where $f_1^\tau$, $\varphi_1^\tau$ are given in the above definition of the WGA. Then, for each $m \geq 2$, we have the following inductive definition.

(1) $\varphi_m^{o,\tau} \in \mathcal{D}$ is any element satisfying

$$|\langle f_{m-1}^{o,\tau}, \varphi_m^{o,\tau} \rangle| \geq t_m \sup_{g \in \mathcal{D}} |\langle f_{m-1}^{o,\tau}, g \rangle|.$$

(2)

$$G_m^{o,\tau}(f, \mathcal{D}) := P_{H_m^\tau}(f), \quad \text{where} \quad H_m^\tau := \text{span}(\varphi_1^{o,\tau}, \ldots, \varphi_m^{o,\tau}).$$

(3)

$$f_m^{o,\tau} := f - G_m^{o,\tau}(f, \mathcal{D}).$$

It is clear that $G_m^\tau$ and $G_m^{o,\tau}$ in the case $t_k = 1, k = 1, 2, \ldots,$ coincide with the PGA $G_m$ and the OGA $G_m^o$, respectively. It is also clear that the WGA and the WOGA are more ready for implementaion than the PGA and the OGA. The WOGA has the same greedy step as the WGA and differs in the construction of a linear combination of $\varphi_1, \ldots, \varphi_m$. In the WOGA we do our best to construct an approximant out of $H_m := \text{span}(\varphi_1, \ldots, \varphi_m)$: we take an orthogonal projection onto $H_m$. Clearly, in this way we lose a property of the WGA to build an expansion into a series in the case of the WOGA. However, this modification pays off in the sense of improving the convergence rate of approximation. To see this, compare Theorems 2.18 and 2.19.

There is one more greedy-type algorithm that works well for functions from the convex hull of $\mathcal{D}^\pm$, where $\mathcal{D}^\pm := \{\pm g, \quad g \in \mathcal{D}\}$.

For a general dictionary $\mathcal{D}$ we define the class of functions

$$\mathcal{A}_1^o(\mathcal{D}, M) := \left\{ f \in H : f = \sum_{k \in \Lambda} c_k w_k, \quad w_k \in \mathcal{D}, \ \#\Lambda < \infty, \quad \sum_{k \in \Lambda} |c_k| \leq M \right\},$$

and we define $\mathcal{A}_1(\mathcal{D}, M)$ to be the closure (in $H$) of $\mathcal{A}_1^o(\mathcal{D}, M)$. Furthermore, we define $\mathcal{A}_1(\mathcal{D})$ to be the union of the classes $\mathcal{A}_1(\mathcal{D}, M)$ over all $M > 0$. For $f \in \mathcal{A}_1(\mathcal{D})$, we define the norm

$$|f|_{\mathcal{A}_1(\mathcal{D})}$$

to be the smallest $M$ such that $f \in \mathcal{A}_1(\mathcal{D}, M)$. For $M = 1$ we denote $A_1(\mathcal{D}) := \mathcal{A}_1(\mathcal{D}, 1)$. In a similar way we define the classes $\mathcal{A}_\beta(\mathcal{D}, M)$ and the quantity $|f|_{\mathcal{A}_\beta(\mathcal{D})}$, $0 < \beta < 1$, replacing $\sum_{k \in \Lambda} |c_k| \leq M$ by $\sum_{k \in \Lambda} |c_k|^\beta \leq M^\beta$.

We proceed to discuss the relaxed type of greedy algorithms. We begin with the simplest one.

**Relaxed Greedy Algorithm (RGA)** Let $f_o^r := R_0^r(f) := R_0^r(f, \mathcal{D}) := f$ and $G_0^r(f) := G_0^r(f, \mathcal{D}) := 0$. For $m = 1$, we define $G_1^r(f) := G_1^r(f, \mathcal{D}) := G_1(f)$ and $f_1^r := R_1^r(f) := R_1^r(f, \mathcal{D}) := R_1(f)$. For a function $h \in H$, let $g = g(h)$ denote the function from $\mathcal{D}^\pm$ which maximizes $\langle h, g \rangle$ (we assume the existence of such an element). Then, for each $m \geq 2$, we inductively define

$$G_m^r(f) := G_m^r(f, \mathcal{D}) := (1 - \frac{1}{m}) G_{m-1}^r(f) + \frac{1}{m} g(R_{m-1}^r(f)),$$

$$f_m^r := R_m^r(f) := R_m^r(f, \mathcal{D}) := f - G_m^r(f).$$

There are several modifications of the Relaxed Greedy Algorithm (see, for example, Barron (1993) and DeVore and Temlyakov (1996)). Before giving the definition of the Weak Relaxed Greedy Algorithm (WRGA), we make one remark which helps to motivate the corresponding definition. Assume $G_{m-1} \in A_1(\mathcal{D})$ is an approximant to $f \in A_1(\mathcal{D})$ obtained at the $(m-1)$th step. The major idea of relaxation in greedy algorithms is to look for an approximant at the $m$th step of the form $G_m := (1-a)G_{m-1} + ag$, $g \in \mathcal{D}^{\pm}$, $0 \le a \le 1$. This form guarantees that $G_m \in A_1(\mathcal{D})$. Thus we are looking for co-convex approximants. The best we can do at the $m$th step is to achieve

$$\delta_m := \inf_{g \in \mathcal{D}^{\pm}, 0 \le a \le 1} \| f - ((1-a)G_{m-1} + ag) \|.$$

Denote $f_n := f - G_n$, $n = 1, \ldots, m$. It is clear that for a given $g \in \mathcal{D}^{\pm}$ we have

$$\inf_a \| f_{m-1} - a(g - G_{m-1}) \|^2 = \| f_{m-1} \|^2 - \langle f_{m-1}, g - G_{m-1} \rangle^2 \| g - G_{m-1} \|^{-2},$$

and this infimum is attained for

$$a(g) = \langle f_{m-1}, g - G_{m-1} \rangle \| g - G_{m-1} \|^{-2}.$$

Next, it is not difficult to derive from the definition of $A_1(\mathcal{D})$ and from our assumption on the existence of a maximizer that, for any $h \in H$ and $u \in A_1(\mathcal{D})$, there exists $g \in \mathcal{D}^{\pm}$ such that

$$\langle h, g \rangle \ge \langle h, u \rangle. \tag{2.4}$$

Taking $h = f_{m-1}$ and $u = f$, we get from (2.4) that there exists $g_m \in \mathcal{D}^{\pm}$ such that

$$\langle f_{m-1}, g_m - G_{m-1} \rangle \ge \langle f_{m-1}, f - G_{m-1} \rangle = \| f_{m-1} \|^2. \tag{2.5}$$

This implies in particular that we get for $g_m$

$$\| g_m - G_{m-1} \| \ge \| f_{m-1} \| \tag{2.6}$$

and $0 \le a(g_m) \le 1$. Thus,

$$\delta_m^2 \le \| f_{m-1} \|^2 - \frac{1}{4} \sup_{g \in \mathcal{D}^{\pm}} \langle f_{m-1}, g - G_{m-1} \rangle^2.$$

We now give the definition of the WRGA for $f \in A_1(\mathcal{D})$.

**Weak Relaxed Greedy Algorithm (WRGA)** We define $f_0 := f$ and $G_0 := 0$. Then, for each $m \ge 1$, we have the following inductive definition.

(1) $\varphi_m \in \mathcal{D}^{\pm}$ is any element satisfying

$$\langle f_{m-1}, \varphi_m - G_{m-1} \rangle \ge t_m \| f_{m-1} \|^2. \tag{2.7}$$

(2)

$$G_m := G_m(f, \mathcal{D}) := (1 - \beta_m)G_{m-1} + \beta_m \varphi_m,$$

$$\beta_m := t_m \left(1 + \sum_{k=1}^{m} t_k^2\right)^{-1} \quad \text{for} \quad m \geq 1.$$

(3)

$$f_m := f - G_m.$$

## 2.2 Convergence

We begin this section with convergence of the Weak Orthogonal Greedy Algorithm (WOGA). The following theorem was proved in Temlyakov (2000b).

**Theorem 2.1** *Assume*

$$\sum_{k=1}^{\infty} t_k^2 = \infty. \tag{2.8}$$

*Then, for any dictionary $\mathcal{D}$ and any $f \in H$, we have for the WOGA*

$$\lim_{m \to \infty} \|f_m^{o,\tau}\| = 0. \tag{2.9}$$

**Remark 2.2** It is easy to see that if $\mathcal{D} = \mathcal{B}$, an orthonormal basis, the assumption (2.8) is also necessary for convergence (2.9) for all $f$.

*Proof of Theorem 2.1* Let $f \in H$ and let $\varphi_1^{o,\tau}, \varphi_2^{o,\tau}, \ldots$ be as given in the definition of the WOGA. Let

$$H_n := H_n^\tau = \text{span}(\varphi_1^{o,\tau}, \ldots, \varphi_n^{o,\tau}).$$

It is clear that $H_n \subseteq H_{n+1}$, and therefore $\{P_{H_n}(f)\}$ converges to some function $v$. The following Lemma 2.3 says that $v = f$ and completes the proof of Theorem 2.1. $\square$

**Lemma 2.3** *Assume that (2.8) is satisfied. Then, if $\{f_m^\tau\}_{m=1}^{\infty}$ or $\{f_m^{o,\tau}\}_{m=1}^{\infty}$ converges, it converges to zero.*

*Proof of Lemma 2.3* We prove this lemma by contradiction. Let us consider first the case of $\{f_m^\tau\}_{m=1}^{\infty}$. Assume $f_m^\tau \to u \neq 0$ as $m \to \infty$. It is clear that

$$\sup_{g \in \mathcal{D}} |\langle u, g \rangle| \geq 2\delta$$

with some $\delta > 0$. Therefore, there exists $N$ such that, for all $m \geq N$, we have

$$\sup_{g \in \mathcal{D}} |\langle f_m^\tau, g \rangle| \geq \delta.$$

From the definition of the WGA we get, for all $m > N$,

$$\|f_m^\tau\|^2 = \|f_{m-1}^\tau\|^2 - |\langle f_{m-1}^\tau, \varphi_m^\tau \rangle|^2 \leq \|f_N^\tau\|^2 - \delta^2 \sum_{k=N+1}^m t_k^2,$$

which contradicts (2.8).

We now proceed to the case $\{f_m^{o,\tau}\}_{m=1}^\infty$. Assume $f_m^{o,\tau} \to u \neq 0$ as $m \to \infty$. Then, as in the above proof, there exist $\delta > 0$ and $N$ such that, for all $m \geq N$, we have

$$\sup_{g \in \mathcal{D}} |\langle f_m^{o,\tau}, g \rangle| \geq \delta.$$

Next, as in (2.3) we have

$$\|f_m^{o,\tau}\|^2 \leq \|f_{m-1}^{o,\tau}\|^2 - t_m^2 (\sup_{g \in \mathcal{D}} |\langle f_{m-1}^{o,\tau}, g \rangle|)^2 \leq \|f_N^{o,\tau}\|^2 - \delta^2 \sum_{k=N+1}^m t_k^2,$$

which contradicts the divergence of $\sum_k t_k^2$. □

Theorem 2.1 and Remark 2.2 show that (2.8) is a necessary and sufficient condition on the weakness sequence $\tau = \{t_k\}$ in order that the WOGA converges for each $f$ and all $\mathcal{D}$. Condition (2.8) can be rewritten as $\tau \notin \ell_2$. It turns out that the convergence of the PGA is more delicate. We now proceed to the corresponding results. The following theorem gives a criterion of convergence in a special case of monotone weakness sequences $\{t_k\}$. Sufficiency was proved in Temlyakov (2000b) and necessity in Livshitz and Temlyakov (2001).

**Theorem 2.4** *In the class of monotone sequences $\tau = \{t_k\}_{k=1}^\infty$, $1 \geq t_1 \geq t_2 \geq \cdots \geq 0$, the condition*

$$\sum_{k=1}^\infty \frac{t_k}{k} = \infty \tag{2.10}$$

*is necessary and sufficient for convergence of the Weak Greedy Algorithm for each $f$ and all Hilbert spaces $H$ and dictionaries $\mathcal{D}$.*

**Remark 2.5** We note that the sufficiency part of Theorem 2.4 (see Temlyakov (2000b)) does not need the monotonicity of $\tau$.

*Proof of sufficiency condition in Theorem 2.4* This proof (see Temlyakov (2000b)) is a refinement of the original proof of Jones (1987). The following lemma, Lemma 2.6, combined with Lemma 2.3, implies sufficiency in Theorem 2.4. □

**Lemma 2.6** *Assume (2.10) is satisfied. Then* $\{f_m^\tau\}_{m=1}^\infty$ *converges.*

*Proof of Lemma 2.6* It is easy to derive from the definition of the WGA the following two relations:

$$f_m^\tau = f - \sum_{j=1}^m \langle f_{j-1}^\tau, \varphi_j^\tau \rangle \varphi_j^\tau, \qquad (2.11)$$

$$\|f_m^\tau\|^2 = \|f\|^2 - \sum_{j=1}^m |\langle f_{j-1}^\tau, \varphi_j^\tau \rangle|^2. \qquad (2.12)$$

Let $a_j := |\langle f_{j-1}^\tau, \varphi_j^\tau \rangle|$. We get from (2.12) that

$$\sum_{j=1}^\infty a_j^2 \leq \|f\|^2. \qquad (2.13)$$

We take any two indices $n < m$ and consider

$$\|f_n^\tau - f_m^\tau\|^2 = \|f_n^\tau\|^2 - \|f_m^\tau\|^2 - 2\langle f_n^\tau - f_m^\tau, f_m^\tau \rangle.$$

Let

$$\theta_{n,m}^\tau := |\langle f_n^\tau - f_m^\tau, f_m^\tau \rangle|.$$

Using (2.11) and the definition of the WGA, we obtain, for all $n < m$ and all $m$ such that $t_{m+1} \neq 0$,

$$\theta_{n,m}^\tau \leq \sum_{j=n+1}^m |\langle f_{j-1}^\tau, \varphi_j^\tau \rangle| |\langle f_m^\tau, \varphi_j^\tau \rangle| \leq \frac{a_{m+1}}{t_{m+1}} \sum_{j=1}^{m+1} a_j. \qquad (2.14)$$

□

We now need a property of the $\ell_2$-sequences.

**Lemma 2.7** *Assume* $y_j \geq 0$, $j = 1, 2, \ldots$, *and*

$$\sum_{k=1}^\infty \frac{t_k}{k} = \infty, \qquad \sum_{j=1}^\infty y_j^2 < \infty.$$

*Then*

$$\lim_{n \to \infty} \frac{y_n}{t_n} \sum_{j=1}^{n} y_j = 0.$$

*Proof* Let $P(\tau) := \{n \in \mathbb{N} : t_n \neq 0\}$. Consider a series

$$\sum_{n \in P(\tau)} \frac{t_n}{n} \frac{y_n}{t_n} \sum_{j=1}^{n} y_j. \tag{2.15}$$

We shall prove that this series converges. It is clear that convergence of this series, together with the assumption $\sum_{k=1}^{\infty} t_k / k = \infty$, imply the statement of Lemma 2.7.

We use the following known fact. If $\{y_j\}_{j=1}^{\infty} \in \ell_2$ then $\{n^{-1} \sum_{j=1}^{n} y_j\}_{n=1}^{\infty} \in \ell_2$ (see Zygmund (1959), chap. 1, sect. 9). By the Cauchy inequality, we have

$$\sum_{n \in P(\tau)} \frac{t_n}{n} \frac{y_n}{t_n} \sum_{j=1}^{n} y_j \leq \left( \sum_{n=1}^{\infty} y_n^2 \right)^{1/2} \left( \sum_{n=1}^{\infty} \left( n^{-1} \sum_{j=1}^{n} y_j \right)^2 \right)^{1/2} < \infty.$$

This completes the proof of Lemma 2.7. $\qquad\square$

The relation (2.14) and Lemma 2.7 imply that

$$\lim_{m \to \infty} \max_{n < m} \theta_{n,m}^{\tau} = 0.$$

It remains to use the following simple lemma.

**Lemma 2.8** *In a Banach space $X$, let a sequence $\{x_n\}_{n=1}^{\infty}$ be given such that, for any $k, l$, we have*

$$\|x_k - x_l\|^2 = y_k - y_l + \vartheta_{k,l},$$

*where $\{y_n\}_{n=1}^{\infty}$ is a convergent sequence of real numbers and the real sequence $\vartheta_{k,l}$ satisfies the property*

$$\lim_{l \to \infty} \max_{k < l} |\vartheta_{k,l}| = 0.$$

*Then $\{x_n\}_{n=1}^{\infty}$ converges.*

The necessary condition in Theorem 2.4 was proved in Livshitz and Temlyakov (2001). We do not present it here.

Theorem 2.4 solves the problem of convergence of the WGA in the case of monotone weakness sequences. We now consider the case of general weakness sequences. In Theorem 2.4 we reduced the proof of convergence of the WGA with weakness sequence $\tau$ to some properties of $\ell_2$-sequences with regard to $\tau$. The sufficiency part of Theorem 2.4 was derived from the following two statements.

**Proposition 2.9** *Let $\tau$ be such that, for any $\{a_j\}_{j=1}^\infty \in \ell_2$, $a_j \geq 0$, $j = 1, 2, \ldots$, we have*

$$\liminf_{n \to \infty} a_n \sum_{j=1}^n a_j / t_n = 0.$$

*Then, for any $H$, $\mathcal{D}$ and $f \in H$, we have*

$$\lim_{m \to \infty} \|f_m^\tau\| = 0.$$

**Proposition 2.10** *If $\tau$ satisfies condition (2.10) then $\tau$ satisfies the assumption of Proposition 2.9.*

We proved in Temlyakov (2002b) a criterion on $\tau$ for convergence of the WGA. Let us introduce some notation. We define by $\mathcal{V}$ the class of sequences $x = \{x_k\}_{k=1}^\infty$, $x_k \geq 0$, $k = 1, 2, \ldots$, with the following property: there exists a sequence $0 = q_0 < q_1 < \cdots$ that may depend on $x$ such that

$$\sum_{s=1}^\infty \frac{2^s}{\Delta q_s} < \infty \tag{2.16}$$

and

$$\sum_{s=1}^\infty 2^{-s} \sum_{k=1}^{q_s} x_k^2 < \infty, \tag{2.17}$$

where $\Delta q_s := q_s - q_{s-1}$.

**Remark 2.11** It is clear from this definition that if $x \in \mathcal{V}$ and for some $N$ and $c$ we have $0 \leq y_k \leq c x_k$, $k \geq N$, then $y \in \mathcal{V}$.

**Theorem 2.12** *The condition $\tau \notin \mathcal{V}$ is necessary and sufficient for convergence of all realizations of the Weak Greedy Algorithm with weakness sequence $\tau$ for each $f$ and all Hilbert spaces $H$ and dictionaries $\mathcal{D}$.*

The proof of the sufficiency part of Theorem 2.12 is a refinement of the corresponding proof of Theorem 2.4. The study of the behavior of sequences $a_n \sum_{j=1}^n a_j$ for $\{a_j\}_{j=1}^\infty \in \ell_2$, $a_j \geq 0$, $j = 1, 2, \ldots$, plays an important role in both proofs. It turns out that the class $\mathcal{V}$ appears naturally in the study of the above-mentioned sequences. We proved the following theorem in Temlyakov (2002b).

**Theorem 2.13** *The following two conditions are equivalent:*

$$\tau \notin \mathcal{V}, \tag{2.18}$$

$$\forall \{a_j\}_{j=1}^{\infty} \in \ell_2, \quad a_j \geq 0, \quad \liminf_{n \to \infty} a_n \sum_{j=1}^{n} a_j / t_n = 0. \tag{2.19}$$

Theorem 2.12 solves the problem of convergence of the WGA in a very general situation. The sufficiency part of Theorem 2.12 guarantees that, whenever $\tau \notin \mathcal{V}$, the WGA converges for each $f$ and all $\mathcal{D}$. The necessity part of Theorem 2.12 states that, if $\tau \in \mathcal{V}$, then there exist an element $f$ and a dictionary $\mathcal{D}$ such that some realization $G_m^\tau(f, \mathcal{D})$ of the WGA does not converge to $f$. However, Theorem 2.12 leaves open the following interesting and important problem. Let a dictionary $\mathcal{D} \subset H$ be given. Find necessary and sufficient conditions on a weakness sequence $\tau$ in order that $G_m^\tau(f, \mathcal{D}) \to f$ for each $f \in H$. The corresponding open problems for special dictionaries are formulated in Temlyakov (2003a), pp. 78 and 81. They concern the following two classical dictionaries:

$$\Pi_2 := \{u(x)v(y), \quad u, v \in L_2([0, 1]), \quad \|u\|_2 = \|v\|_2 = 1\}$$

and

$$\mathcal{R}_2 := \{g(x) = r(\omega \cdot x), \quad \|g\|_2 = 1\},$$

where $r$ is a univariate function and $\omega \cdot x$ is the scalar product of $x$, $\|x\|_{\ell_2} \leq 1$, and a unit vector $\omega \in \mathbb{R}^2$.

## 2.3 Rate of convergence

### 2.3.1 Upper bounds for approximation by general dictionaries

We shall discuss here approximation from a general dictionary $\mathcal{D}$. We begin with a discussion of the approximation properties of the Relaxed Greedy Algorithm. The result we give below in Theorem 2.15 is from DeVore and Temlyakov (1996), and can be found in Jones (1992) in a different form. We begin with the following elementary lemma about numerical sequences.

**Lemma 2.14** *If $A > 0$ and $\{a_n\}_{n=1}^{\infty}$ is a sequence of non-negative numbers satisfying $a_1 \leq A$ and*

$$a_m \leq a_{m-1} - \frac{2}{m} a_{m-1} + \frac{A}{m^2}, \quad m = 2, 3, \ldots, \tag{2.20}$$

*then*

$$a_m \leq \frac{A}{m}. \tag{2.21}$$

*Proof* The proof is by induction. Suppose we have

$$a_{m-1} \le \frac{A}{m-1}$$

for some $m \ge 2$. Then, from our assumption (2.20), we have

$$a_m \le \frac{A}{m-1}\left(1 - \frac{2}{m}\right) + \frac{A}{m^2} = A\left(\frac{1}{m} - \frac{1}{(m-1)m} + \frac{1}{m^2}\right) \le \frac{A}{m}.$$

□

If $f \in \mathcal{A}_1^o(\mathcal{D})$, then $f = \sum_j c_j g_j$, for some $g_j \in \mathcal{D}$ and with $\sum_j |c_j| \le 1$. Since the functions $g_j$ all have norm one, it follows that

$$\|f\| \le \sum_j |c_j| \|g_j\| \le 1.$$

Since the functions $g \in \mathcal{D}$ have norm one, it follows that $G_1^r(f) = G_1(f)$ also has norm at most one. By induction, we find that $\|G_m^r(f)\| \le 1, m \ge 1$.

**Theorem 2.15** *For the Relaxed Greedy Algorithm we have, for each $f \in A_1(\mathcal{D})$, the estimate*

$$\|f - G_m^r(f)\| \le \frac{2}{\sqrt{m}}, \quad m \ge 1. \tag{2.22}$$

*Proof* We use the abbreviation $r_m := G_m^r(f)$ and $g_m := g(R_{m-1}^r(f))$. From the definition of $r_m$, we have

$$\|f - r_m\|^2 = \|f - r_{m-1}\|^2 + \frac{2}{m}\langle f - r_{m-1}, r_{m-1} - g_m\rangle + \frac{1}{m^2}\|r_{m-1} - g_m\|^2. \tag{2.23}$$

The last term on the right-hand side of (2.23) does not exceed $4/m^2$. For the middle term, we have

$$\langle f - r_{m-1}, r_{m-1} - g_m\rangle = \inf_{g \in \mathcal{D}^\pm} \langle f - r_{m-1}, r_{m-1} - g\rangle$$

$$= \inf_{\phi \in A_1(\mathcal{D})} \langle f - r_{m-1}, r_{m-1} - \phi\rangle$$

$$\le \langle f - r_{m-1}, r_{m-1} - f\rangle = -\|f - r_{m-1}\|^2.$$

We substitute this into (2.23) to obtain

$$\|f - r_m\|^2 \le \left(1 - \frac{2}{m}\right)\|f - r_{m-1}\|^2 + \frac{4}{m^2}. \tag{2.24}$$

Thus the theorem follows from Lemma 2.14 with $A = 4$ and $a_m := \|f - r_m\|^2$.

□

We now turn our discussion to the approximation properties of the Pure Greedy Algorithm and the Orthogonal Greedy Algorithm.

We shall need the following simple known lemma (see, for example, DeVore and Temlyakov (1996)).

**Lemma 2.16** *Let* $\{a_m\}_{m=1}^{\infty}$ *be a sequence of non-negative numbers satisfying the inequalities*

$$a_1 \leq A, \quad a_{m+1} \leq a_m(1 - a_m/A), \quad m = 1, 2, \ldots$$

*Then we have, for each m,*

$$a_m \leq A/m.$$

*Proof* The proof is by induction on $m$. For $m = 1$ the statement is true by assumption. We assume $a_m \leq A/m$ and prove that $a_{m+1} \leq A/(m + 1)$. If $a_{m+1} = 0$ this statement is obvious. Assume therefore that $a_{m+1} > 0$. Then we have

$$a_{m+1}^{-1} \geq a_m^{-1}(1 - a_m/A)^{-1} \geq a_m^{-1}(1 + a_m/A) = a_m^{-1} + A^{-1} \geq (m + 1)A^{-1},$$

which implies $a_{m+1} \leq A/(m + 1)$. $\qquad\square$

We now want to estimate the decrease in error provided by one step of the Pure Greedy Algorithm. Let $\mathcal{D}$ be an arbitrary dictionary. If $f \in H$ and

$$\rho(f) := \langle f, g(f) \rangle / \|f\|, \tag{2.25}$$

where as before $g(f) \in \mathcal{D}^{\pm}$ satisfies

$$\langle f, g(f) \rangle = \sup_{g \in \mathcal{D}^{\pm}} \langle f, g \rangle,$$

then

$$R(f)^2 = \|f - G(f)\|^2 = \|f\|^2(1 - \rho(f)^2). \tag{2.26}$$

The larger $\rho(f)$, the better the decrease of the error in the step of the Pure Greedy Algorithm. The following lemma estimates $\rho(f)$ from below.

**Lemma 2.17** *If* $f \in \mathcal{A}_1(\mathcal{D}, M)$, *then*

$$\rho(f) \geq \|f\|/M. \tag{2.27}$$

*Proof* It is sufficient to prove (2.27) for $f \in \mathcal{A}_1^o(\mathcal{D}, M)$ since the general result follows from this by taking limits. We can write $f = \sum c_k g_k$, where this sum has a finite number of terms and $g_k \in \mathcal{D}$ and $\sum |c_k| \leq M$. Hence,

$$\|f\|^2 = \langle f, f \rangle = \langle f, \sum c_k g_k \rangle = \sum c_k \langle f, g_k \rangle \leq M\rho(f)\|f\|$$

and (2.27) follows. $\qquad\square$

The following theorem was proved in DeVore and Temlyakov (1996).

**Theorem 2.18** *Let $\mathcal{D}$ be an arbitrary dictionary in $H$. Then, for each $f \in \mathcal{A}_1(\mathcal{D}, M)$, we have*

$$\|f - G_m(f, \mathcal{D})\| \le Mm^{-1/6}.$$

*Proof* It is enough to prove the theorem for $f \in \mathcal{A}_1(\mathcal{D}, 1)$; the general result then follows by rescaling. We shall use the abbreviated notation $f_m := R_m(f)$ for the residual. Let

$$a_m := \|f_m\|^2 = \|f - G_m(f, \mathcal{D})\|^2, \quad m = 0, 1, \ldots, \quad f_0 := f,$$

and define the sequence $\{b_m\}_{m=0}^{\infty}$ by

$$b_0 := 1, \quad b_{m+1} := b_m + \rho(f_m)\|f_m\|, \quad m = 0, 1, \ldots$$

Since $f_{m+1} := f_m - \rho(f_m)\|f_m\|g(f_m)$, we obtain by induction that

$$f_m \in \mathcal{A}_1(\mathcal{D}, b_m), \quad m = 0, 1, \ldots,$$

and consequently we have the following relations for $m = 0, 1, \ldots$:

$$a_{m+1} = a_m(1 - \rho(f_m)^2), \tag{2.28}$$

$$b_{m+1} = b_m + \rho(f_m)a_m^{1/2}, \tag{2.29}$$

$$\rho(f_m) \ge a_m^{1/2}b_m^{-1}. \tag{2.30}$$

Equations (2.29) and (2.30) yield

$$b_{m+1} = b_m(1 + \rho(f_m)a_m^{1/2}b_m^{-1}) \le b_m(1 + \rho(f_m)^2). \tag{2.31}$$

Combining this inequality with (2.28) we find

$$a_{m+1}b_{m+1} \le a_m b_m(1 - \rho(f_m)^4),$$

which in turn implies, for all $m$,

$$a_m b_m \le a_0 b_0 = \|f\|^2 \le 1. \tag{2.32}$$

Further, using (2.28) and (2.30) we get

$$a_{m+1} = a_m(1 - \rho(f_m)^2) \le a_m(1 - a_m/b_m^2).$$

Since $b_n \le b_{n+1}$, this gives

$$a_{n+1}b_{n+1}^{-2} \le a_n b_n^{-2}(1 - a_n b_n^{-2}).$$

Applying Lemma 2.16 to the sequence $\{a_m b_m^{-2}\}$ we obtain

$$a_m b_m^{-2} \le m^{-1}. \tag{2.33}$$

Relations (2.32) and (2.33) imply

$$a_m^3 = (a_m b_m)^2 a_m b_m^{-2} \leq m^{-1}.$$

In other words,

$$\|f_m\| = a_m^{1/2} \leq m^{-1/6},$$

which proves the theorem. □

The next theorem (DeVore and Temlyakov (1996)) estimates the error in approximation by the Orthogonal Greedy Algorithm.

**Theorem 2.19** *Let $\mathcal{D}$ be an arbitrary dictionary in $H$. Then, for each $f \in \mathcal{A}_1(\mathcal{D}, M)$ we have*

$$\|f - G_m^o(f, \mathcal{D})\| \leq M m^{-1/2}.$$

*Proof* The proof of this theorem is similar to the proof of Theorem 2.18, but is technically even simpler. We can again assume that $M = 1$. We let $f_m^o := R_m^o(f)$ be the residual in the Orthogonal Greedy Algorithm. Then, from the definition of the Orthogonal Greedy Algorithm, we have

$$\|f_{m+1}^o\| \leq \|f_m^o - G_1(f_m^o, \mathcal{D})\|. \tag{2.34}$$

From (2.26) we obtain

$$\|f_{m+1}^o\|^2 \leq \|f_m^o\|^2 (1 - \rho(f_m^o)^2). \tag{2.35}$$

By the definition of the Orthogonal Greedy Algorithm, $G_m^o(f) = P_{H_m} f$ and hence $f_m^o = f - G_m^o(f)$ is orthogonal to $G_m^o(f)$. Using this as in the proof of Lemma 2.17, we obtain

$$\|f_m^o\|^2 = \langle f_m^o, f \rangle \leq \rho(f_m^o) \|f_m^o\|.$$

Hence,

$$\rho(f_m^o) \geq \|f_m^o\|.$$

Using this inequality in (2.35), we find

$$\|f_{m+1}^o\|^2 \leq \|f_m^o\|^2 (1 - \|f_m^o\|^2).$$

In order to complete the proof it remains to apply Lemma 2.16 with $A = 1$ and $a_m = \|f_m^o\|^2$. □

### 2.3.2 Upper estimates for weak-type greedy algorithms

We begin this subsection with an error estimate for the Weak Orthogonal Greedy Algorithm. The following theorem from Temlyakov (2000b) is a generalization of Theorem 2.19.

**Theorem 2.20** *Let $\mathcal{D}$ be an arbitrary dictionary in $H$. Then, for each $f \in \mathcal{A}_1(\mathcal{D}, M)$ we have*

$$\|f - G_m^{o,\tau}(f, \mathcal{D})\| \leq M \left(1 + \sum_{k=1}^{m} t_k^2\right)^{-1/2}.$$

We now turn to the Weak Relaxed Greedy Algorithms. The following theorem from Temlyakov (2000b) shows that the WRGA performs on the $\mathcal{A}_1(\mathcal{D}, M)$ similar to the WOGA. We note that the approximation step of building the $G_m(f, \mathcal{D})$ in the WRGA is simpler than the corresponding step of building the $G_m^{o,\tau}(f, \mathcal{D})$ in the WOGA.

**Theorem 2.21** *Let $\mathcal{D}$ be an arbitrary dictionary in $H$. Then, for each $f \in \mathcal{A}_1(\mathcal{D}, M)$ we have, for the Weak Relaxed Greedy Algorithm,*

$$\|f - G_m(f, \mathcal{D})\| \leq 2M \left(1 + \sum_{k=1}^{m} t_k^2\right)^{-1/2}.$$

We now proceed to the Weak Greedy Algorithm. The construction of an approximant $G_m^\tau(f, \mathcal{D})$ in the WGA is the simplest out of the three types of algorithms (WGA, WOGA, WRGA) discussed here. We pointed out above that the WGA provides for each $f \in H$ an expansion into a series that satisfies an analog of the Parseval formula. The following theorem from Temlyakov (2000b) gives the upper bounds for the residual $\|f_m^\tau\|$ of the WGA that are not as good as in Theorems 2.20 and 2.21 for the WOGA and WRGA, respectively. The next theorem, Theorem 2.23, will show that the bound (2.36) is sharp in a certain sense.

**Theorem 2.22** *Let $\mathcal{D}$ be an arbitrary dictionary in $H$. Assume $\tau := \{t_k\}_{k=1}^{\infty}$ is a non-increasing sequence. Then, for $f \in \mathcal{A}_1(\mathcal{D}, M)$ we have*

$$\|f - G_m^\tau(f, \mathcal{D})\| \leq M \left(1 + \sum_{k=1}^{m} t_k^2\right)^{-t_m/2(2+t_m)}. \tag{2.36}$$

In a particular case $\tau = \{t\}$ $(t_k = t, k = 1, 2, \dots)$, (2.36) gives

$$\|f - G_m^t(f, \mathcal{D})\| \leq M(1 + mt^2)^{-t/(4+2t)}, \quad 0 < t \leq 1. \tag{2.37}$$

This estimate implies the inequality

$$\|f - G_m^t(f, \mathcal{D})\| \leq C_1(t) m^{-at} |f|_{\mathcal{A}_1(\mathcal{D})}, \tag{2.38}$$

with the exponent $at$ approaching zero linearly in $t$. We proved in Livshitz and Temlyakov (2003) that this exponent cannot decrease to zero at a slower rate than linear.

**Theorem 2.23** *There exists an absolute constant $b > 0$ such that, for any $t > 0$, we can find a dictionary $\mathcal{D}_t$ and a function $f_t \in \mathcal{A}_1(\mathcal{D}_t)$ such that, for some realization $G_m^t(f_t, \mathcal{D}_t)$ of the Weak Greedy Algorithm, we have*

$$\liminf_{m \to \infty} \|f_t - G_m^t(f_t, \mathcal{D}_t)\| m^{bt} / |f_t|_{\mathcal{A}_1(\mathcal{D}_t)} > 0. \tag{2.39}$$

**Remark 2.24** The estimate (2.37) implies that for small $t$ the parameter $a$ in (2.38) can be taken close to $1/4$. The proof from Livshitz and Temlyakov (2003) implies that the parameter $b$ in (2.39) can be taken close to $(\ln 2)^{-1}$.

We now discuss some further results on the rate of convergence of the PGA and related results on greedy expansions. Theorem 2.18 states that for a general dictionary $\mathcal{D}$ the Pure Greedy Algorithm provides the estimate

$$\|f - G_m(f, \mathcal{D})\| \leq |f|_{\mathcal{A}_1(\mathcal{D})} m^{-1/6}.$$

The above estimate was improved a little in Konyagin and Temlyakov (1999b) to

$$\|f - G_m(f, \mathcal{D})\| \leq 4 |f|_{\mathcal{A}_1(\mathcal{D})} m^{-11/62}.$$

We now discuss recent progress on the following open problem (see Temlyakov (2003a), p. 65, open prob. 3.1). This problem is a central theoretical problem in greedy approximation in Hilbert spaces.

**Open problem** Find the order of decay of the sequence

$$\gamma(m) := \sup_{f, \mathcal{D}, \{G_m\}} (\|f - G_m(f, \mathcal{D})\| |f|_{\mathcal{A}_1(\mathcal{D})}^{-1}),$$

where the supremum is taken over all dictionaries $\mathcal{D}$, all elements $f \in \mathcal{A}_1(\mathcal{D}) \setminus \{0\}$ and all possible choices of $\{G_m\}$.

Recently, the known upper bounds in approximation by the Pure Greedy Algorithm were improved in Sil'nichenko (2004), who proved the estimate

$$\gamma(m) \leq C m^{-s/2(2+s)},$$

where $s$ is a solution from $[1, 1.5]$ of the equation

$$(1 + x)^{1/(2+x)} \left( \frac{2+x}{1+x} \right) - \frac{1+x}{x} = 0.$$

Numerical calculations of $s$ (see Sil'nichenko (2004)) give

$$\frac{s}{2(2+s)} = 0.182 \cdots > 11/62.$$

The technique used in Sil'nichenko (2004) is a further development of a method from Konyagin and Temlyakov (1999b).

There is also some progress in the lower estimates. The estimate

$$\gamma(m) \geq Cm^{-0.27},$$

with a positive constant $C$, was proved in Livshitz and Temlyakov (2003). For previous lower estimates see Temlyakov (2003a), p. 59. Very recently, Livshitz (2009), using the technique from Livshitz and Temlyakov (2003), proved the following lower estimate:

$$\gamma(m) \geq Cm^{-0.1898}. \tag{2.40}$$

We mentioned above that the PGA and its generalization the Weak Greedy Algorithm (WGA) give, for every element $f \in H$, a convergent expansion in a series with respect to a dictionary $\mathcal{D}$. We discuss a further generalization of the WGA that also provides a convergent expansion. We consider here a generalization of the WGA obtained by introducing to it a tuning parameter $b \in (0, 1]$ (see Temlyakov (2007a)). Let a sequence $\tau = \{t_k\}_{k=1}^{\infty}, 0 \leq t_k \leq 1$, and a parameter $b \in (0, 1]$ be given. We define the Weak Greedy Algorithm with parameter $b$ as follows.

**Weak Greedy Algorithm with parameter $b$ (WGA($b$))** We define $f_0^{\tau,b} := f$. Then, for each $m \geq 1$ we have the following inductive definition.

(1) $\varphi_m^{\tau,b} \in \mathcal{D}$ is any satisfying

$$|\langle f_{m-1}^{\tau,b}, \varphi_m^{\tau,b} \rangle| \geq t_m \sup_{g \in \mathcal{D}} |\langle f_{m-1}^{\tau,b}, g \rangle|.$$

(2)

$$f_m^{\tau,b} := f_{m-1}^{\tau,b} - b\langle f_{m-1}^{\tau,b}, \varphi_m^{\tau,b} \rangle \varphi_m^{\tau,b}.$$

(3)

$$G_m^{\tau,b}(f, \mathcal{D}) := b \sum_{j=1}^{m} \langle f_{j-1}^{\tau,b}, \varphi_j^{\tau,b} \rangle \varphi_j^{\tau,b}.$$

We note that the WGA($b$) can be seen as a realization of the Approximate Greedy Algorithm studied in Galatenko and Livshitz (2003, 2005) and Gribonval and Nielsen (2001a).

We point out that the WGA($b$), like the WGA, contains, in addition to the first (greedy) step, the second step (see (2) and (3) in the above definition), where we update an approximant by adding an orthogonal projection of the residual $f_{m-1}^{\tau,b}$ onto $\varphi_m^{\tau,b}$ multiplied by $b$. The WGA($b$) therefore provides, for each $f \in H$, an expansion into a series (greedy expansion):

$$f \sim \sum_{j=1}^{\infty} c_j(f)\varphi_j^{\tau,b}, \quad c_j(f) := b\langle f_{j-1}^{\tau,b}, \varphi_j^{\tau,b}\rangle.$$

We begin with a convergence result from Temlyakov (2007a).

**Theorem 2.25** *Let* $\tau \notin V$. *Then the WGA($b$) with* $b \in (0, 1]$ *converges for each* $f$ *and all Hilbert spaces* $H$ *and dictionaries* $\mathcal{D}$.

Theorem 2.25 is an extension of the corresponding result for the WGA (see Theorem 2.12).

We proved in Temlyakov (2007a) the following convergence rate of the WGA($b$).

**Theorem 2.26** *Let* $\mathcal{D}$ *be an arbitrary dictionary in* $H$. *Assume* $\tau := \{t_k\}_{k=1}^{\infty}$ *is a non-increasing sequence and* $b \in (0, 1]$. *Then, for* $f \in A_1(\mathcal{D})$ *we have*

$$\|f - G_m^{\tau,b}(f, \mathcal{D})\| \le \left(1 + b(2-b)\sum_{k=1}^{m} t_k^2\right)^{-(2-b)t_m/2(2+(2-b)t_m)}. \quad (2.41)$$

This theorem is an extension of the corresponding result for the WGA (see Theorem 2.22). In the particular case $t_k = 1$, $k = 1, 2, \ldots$, we get the following rate of convergence:

$$\|f - G_m^{1,b}(f, \mathcal{D})\| \le Cm^{-r(b)}, \quad r(b) := \frac{2-b}{2(4-b)}.$$

We note that $r(1) = 1/6$ and $r(b) \to 1/4$ as $b \to 0$. Thus we can offer the following observation. At each step of the Pure Greedy Algorithm we can choose a fixed fraction of the optimal coefficient for that step instead of the optimal coefficient itself. Surprisingly, this leads to better upper estimates than those known for the Pure Greedy Algorithm.

## 2.4 Greedy algorithms for systems that are not dictionaries

In this section we discuss greedy algorithms with regard to a system $\mathcal{G}$ that is not a dictionary. Here, we will discuss a variant of the RGA that is a generalization of the version of the RGA suggested by Barron (1993). Let $H$ be a real Hilbert space and let $\mathcal{G} := \{g\}$ be a system of elements $g \in H$

such that $\|g\| \leq C_0$. Usually, in the theory of greedy algorithms we consider approximation with regard to a dictionary $\mathcal{D}$. One of the properties of a dictionary $\mathcal{D}$ is that the closure of span $\mathcal{D}$ is equal to $H$. In this section we do not assume that the system $\mathcal{G}$ is a dictionary. In particular, we do not assume that the closure of span $\mathcal{G}$ is $H$. This setting is motivated by applications in Learning Theory (see Chapter 4). We present here results from Temlyakov (2006e). Let $\mathcal{G}^{\pm} := \{\pm g, g \in \mathcal{G}\}$ denote the symmetrized system $\mathcal{G}$, and let $\theta > 0$.

**RGA($\theta$) with respect to $\mathcal{G}$** For $f \in H$ we define $f_0 := f$, $G_0 := G_0(f) := 0$. Then, for each $n \geq 1$ we have the following inductive definition.

(1) $\varphi_n \in \mathcal{G}^{\pm}$ is an element satisfying (we assume existence)

$$\langle f_{n-1}, \varphi_n \rangle = \max_{g \in \mathcal{G}^{\pm}} \langle f_{n-1}, g \rangle.$$

(2)

$$G_n := G_n(f) := \left(1 - \frac{\theta}{n+\theta}\right) G_{n-1} + \frac{\theta}{n+\theta} \varphi_n, \quad f_n := f - G_n.$$

Let $A_1(\mathcal{G})$ denote the closure in $H$ of the convex hull of $\mathcal{G}^{\pm}$. Then, for $f \in H$ there exists a unique element $f' \in A_1(\mathcal{G})$ such that

$$d(f, A_1(\mathcal{G}))_H = \|f - f'\| \leq \|f - \phi\|, \quad \phi \in A_1(\mathcal{G}). \tag{2.42}$$

In analysis of the RGA($\theta$) we will use the following simple lemma (see DeVore and Temlyakov (1996) and Lemma 2.14 above for a variant of this lemma). Our analysis is similar to that of DeVore and Temlyakov (1996) and Lee, Bartlett and Williamson (1996).

**Lemma 2.27** *Let a sequence $\{a_n\}_{n=0}^{\infty}$ of non-negative numbers satisfy the following relations (with $\beta > 1$, $B > 0$):*

$$a_n \leq \frac{n}{n+\beta} a_{n-1} + \frac{B}{(n+\beta)^2}, \quad n = 1, 2, \ldots; \quad a_0 \leq \frac{B}{(\beta-1)\beta}.$$

*Then, for all $n$,*

$$a_n \leq \frac{B}{(\beta-1)(n+\beta)}.$$

*Proof* Setting $A := B/(\beta-1)$, we obtain by induction

$$a_n \leq \frac{A}{n-1+\beta} \frac{n}{n+\beta} + \frac{B}{(n+\beta)^2} = \frac{A}{n+\beta} - \frac{A(\beta-1)}{(n+\beta)(n-1+\beta)} + \frac{B}{(n+\beta)^2}.$$

Taking into account the inequality

$$\frac{A(\beta-1)}{(n+\beta)(n-1+\beta)} \geq \frac{A(\beta-1)}{(n+\beta)^2} = \frac{B}{(n+\beta)^2},$$

we complete the proof. $\qquad\square$

**Theorem 2.28** *For $\theta > 1$ there exists a constant $C(\theta)$ such that, for any $f \in H$, we have*

$$\|f_n\|^2 \le d(f, A_1(\mathcal{G}))_H^2 + C(\theta)(\|f\| + C_0)^2 n^{-1}.$$

*Proof* From the definition of $G_n$ and $f_n$ we get, setting $\alpha := \theta/(n + \theta)$,

$$f_n = f - G_n = (1 - \alpha)f_{n-1} + \alpha(f - \varphi_n)$$

and

$$\|f_n\|^2 = (1-\alpha)^2\|f_{n-1}\|^2 + 2\alpha(1-\alpha)\langle f_{n-1}, f - \varphi_n \rangle + \alpha^2\|f - \varphi_n\|^2. \quad (2.43)$$

It is known (and easy to check) that, for any $h \in H$, we have

$$\sup_{g \in \mathcal{G}^{\pm}} \langle h, g \rangle = \sup_{\phi \in A_1(\mathcal{G})} \langle h, \phi \rangle. \quad (2.44)$$

Denote $f'$ as above and set $f^* := f - f'$. Using (2.44) and the definition of $\varphi_n$, we obtain from (2.43)

$$
\begin{aligned}
\|f_n\|^2 &\le (1 - \alpha)^2\|f_{n-1}\|^2 + 2\alpha(1-\alpha)\langle f_{n-1}, f - f' \rangle + \alpha^2\|f - \varphi_n\|^2 \\
&= (1 - \alpha)(\|f_{n-1}\|^2 - \alpha\|f_{n-1}\|^2 + 2\alpha\langle f_{n-1}, f^* \rangle - \alpha\|f^*\|^2) \\
&\quad + \alpha(1 - \alpha)\|f^*\|^2 + \alpha^2\|f - \varphi_n\|^2 \\
&\le (1 - \alpha)\|f_{n-1}\|^2 + \alpha\|f^*\|^2 + \alpha^2\|f - \varphi_n\|^2.
\end{aligned}
$$

This implies

$$\|f_n\|^2 - \|f^*\|^2 \le (1 - \alpha)(\|f_{n-1}\|^2 - \|f^*\|^2) + \alpha^2(\|f\| + C_0)^2.$$

Setting $a_n := \|f_n\|^2 - \|f^*\|^2$, $\beta := \theta$ and applying Lemma 2.27 we complete the proof. $\qquad\square$

**Theorem 2.29** *For $\theta > 1/2$ there exists a constant $C := C(\theta, C_0)$ such that, for any $f \in H$, we have*

$$\|f' - G_n(f)\|^2 \le C/n.$$

*Proof* If $f \in A_1(\mathcal{G})$ then the statement of Theorem 2.29 follows from known results (see Barron (1993) and Theorem 2.15). If $d(f, A_1(\mathcal{G})) > 0$, then the property (2.42) implies that, for any $\phi \in A_1(\mathcal{G})$, we have

$$\langle f^*, \phi - f' \rangle \le 0. \quad (2.45)$$

It follows from the definition of $f_n$ that

$$f_n = \left(1 - \frac{\theta}{n + \theta}\right)f_{n-1} + \frac{\theta}{n + \theta}(f - \varphi_n).$$

We set $f_n' := f_n - f^*$. Then, from the above representation we obtain

$$f_n' = \left(1 - \frac{\theta}{n+\theta}\right) f_{n-1}' + \frac{\theta}{n+\theta}(f' - \varphi_n).$$

We note that $f_n' = f' - G_n(f)$. Let us estimate

$$\|f_n'\|^2 = \|f_{n-1}'\|^2 \left(1 - \frac{\theta}{n+\theta}\right)^2$$

$$+ \frac{2\theta}{n+\theta} \left(1 - \frac{\theta}{n+\theta}\right) \langle f_{n-1}', f' - \varphi_n \rangle + \frac{\theta^2}{(n+\theta)^2} \|f' - \varphi_n\|^2. \quad (2.46)$$

Next,

$$\langle f_{n-1}', f' - \varphi_n \rangle = \langle f_{n-1}' + f^*, f' - \varphi_n \rangle - \langle f^*, f' - \varphi_n \rangle$$

$$= \langle f_{n-1}, f' - \varphi_n \rangle + \langle f^*, \varphi_n - f' \rangle. \quad (2.47)$$

First, we prove that

$$\langle f_{n-1}, f' - \varphi_n \rangle \leq 0. \quad (2.48)$$

It easily follows from $f' \in A_1(\mathcal{G})$ that

$$\langle f_{n-1}, f' \rangle \leq \max_{g \in \mathcal{G}^\pm} \langle f_{n-1}, g \rangle. \quad (2.49)$$

By the definition of $\varphi_n$ we get

$$\max_{g \in \mathcal{G}^\pm} \langle f_{n-1}, g \rangle = \langle f_{n-1}, \varphi_n \rangle. \quad (2.50)$$

Thus, (2.48) follows from (2.49) and (2.50).

Secondly, we note that (2.45) implies

$$\langle f^*, \varphi_n - f' \rangle \leq 0. \quad (2.51)$$

Therefore, by (2.47), (2.48) and (2.51) we obtain

$$\langle f_{n-1}', f' - \varphi_n \rangle \leq 0. \quad (2.52)$$

Substitution of (2.52) into (2.46) gives

$$\|f_n'\|^2 \leq \|f_{n-1}'\|^2 \left(1 - \frac{2\theta}{n+\theta}\right) + \frac{\theta^2}{(n+\theta)^2}(\|f_{n-1}'\|^2 + \|f' - \varphi_n\|^2). \quad (2.53)$$

Using bounds $\|f_{n-1}'\| \leq 2C_0$ and $\|f' - \varphi_n\| \leq 2C_0$, we find

$$\|f_n'\|^2 \leq \|f_{n-1}'\|^2 \left(1 - \frac{2\theta}{n+\theta}\right) + \frac{8C_0^2\theta^2}{(n+\theta)^2}.$$

We note that

$$1 - \frac{2\theta}{n+\theta} < 1 - \frac{2\theta}{n+2\theta}.$$

We now apply Lemma 2.27 with $a_n = \|f'_n\|^2$, $\beta = 2\theta$ and obtain

$$\|f'_n\|^2 \le C(\theta, C_0)/n. \tag{2.54}$$

This completes the proof.                                    □

## 2.5  Greedy approximation with respect to λ-quasi-orthogonal dictionaries

In this section we present some partial progress in the following general problem.

**Problem 2.30** *Let* $0 < r \le 1/2$ *be given. Characterize dictionaries* $\mathcal{D}$ *that possess the property: For any* $f \in H$ *such that*

$$\sigma_m(f, \mathcal{D}) \le m^{-r}, \quad m = 1, 2, \ldots,$$

*we have*

$$\|f - G_m(f, \mathcal{D})\| \le C(r, \mathcal{D})m^{-r}, \quad m = 1, 2, \ldots$$

We impose the restriction $r \le 1/2$ in Problem 2.30 because of the following result (to be discussed in Section 2.7). We will construct in Section 2.7 a dictionary $\mathcal{D} = \{\varphi_k\}_{k=1}^{\infty}$ such that, for the function $f = \varphi_1 + \varphi_2$, we have

$$\|f - G_m(f, \mathcal{D})\| \ge m^{-1/2}, \quad m \ge 4.$$

It is clear that $\sigma_m(f, \mathcal{D}) = 0$ for $m \ge 2$. This example of a dictionary shows that in general we cannot improve on an $m^{-1/2}$ rate of approximation by the Pure Greedy Algorithm, even if we impose extremely tough restrictions on $\sigma_m(f, \mathcal{D})$. We call this phenomenon a saturation property.

We give here a sufficient condition on $\mathcal{D}$ to have the property formulated in Problem 2.30. We consider dictionaries which we call λ-quasi-orthogonal.

**Definition 2.31** We say $\mathcal{D}$ is a λ-quasi-orthogonal dictionary if, for any $n \in \mathbb{N}$ and any $g_i \in \mathcal{D}$, $i = 1, \ldots, n$, there exists a collection $\varphi_j \in \mathcal{D}$, $j = 1, \ldots, M$, $M \le N := \lambda n$, with the properties

$$g_i \in X_M := \operatorname{span}(\varphi_1, \ldots, \varphi_M); \tag{2.55}$$

and, for any $f \in X_M$, we have

$$\max_{1 \le j \le M} |\langle f, \varphi_j \rangle| \ge N^{-1/2}\|f\|. \tag{2.56}$$

**Remark 2.32** It is clear that an orthonormal dictionary is a 1-quasi-orthogonal dictionary.

We shall prove the following theorem and its slight generalization on an asymptotically $\lambda$-quasi-orthogonal dictionary. Examples of asymptotically $\lambda$-quasi-orthogonal dictionaries are also given in this section.

**Theorem 2.33** *Let a given dictionary $\mathcal{D}$ be $\lambda$-quasi-orthogonal and let $0 < r < (2\lambda)^{-1}$ be a real number. Then, for any $f$ such that*

$$\sigma_m(f, \mathcal{D}) \leq m^{-r}, \quad m = 1, 2, \ldots,$$

*we have*

$$\|f - G_m(f, \mathcal{D})\| \leq C(r, \lambda)m^{-r}, \quad m = 1, 2, \ldots$$

We will make a comment about this theorem before proceeding to the proof. In Section 2.3 we formulated an open problem about the rate of convergence of the PGA with respect to a general dictionary $\mathcal{D}$ for elements of $A_1(\mathcal{D})$. Theorem 2.33 allows us to give an example of a collection of dictionaries for which the behavior of the PGA is better than its behavior for a general dictionary. Theorem 2.19 guarantees that

$$\sigma_m(f, \mathcal{D}) \leq m^{-1/2}, \quad f \in A_1(\mathcal{D}).$$

Therefore, for any $\lambda$-quasi-orthogonal dictionary we have, for any $r < (2\lambda)^{-1}$,

$$\|f - G_m(f, \mathcal{D})\| \leq C(r, \lambda)m^{-r}, \quad f \in A_1(\mathcal{D}).$$

For example, if $\lambda < 2.5$ then the rate of convergence of the PGA with respect to the $\lambda$-quasi-orthogonal dictionary is better than $m^{-0.2}$, which, in turn, is better than the known lower bound for the rate of convergence of the PGA for general dictionaries.

We now proceed to the proof of Theorem 2.33. We begin with a numerical lemma.

**Lemma 2.34** *Let three positive numbers $\alpha < \gamma \leq 1$, $A > 1$ be given and let a sequence of positive numbers $1 \geq a_1 \geq a_2 \geq \cdots$ satisfy the following condition: If, for some $\nu \in \mathbb{N}$ we have*

$$a_\nu \geq A\nu^{-\alpha},$$

*then*

$$a_{\nu+1} \leq a_\nu(1 - \gamma/\nu). \tag{2.57}$$

*Then there exists $B = B(A, \alpha, \gamma)$ such that, for all $n = 1, 2, \ldots$, we have*

$$a_n \leq Bn^{-\alpha}.$$

*Proof* We have $a_1 \leq 1 < A$, which implies that the set

$$V := \{v : a_v \geq A v^{-\alpha}\}$$

does not contain $v = 1$. We now prove that for any segment $[n, n+k] \subset V$ we have $k \leq C(\alpha, \gamma)n$. Indeed, let $n \geq 2$ be such that $n - 1 \notin V$, which means

$$a_{n-1} < A(n-1)^{-\alpha}, \tag{2.58}$$

and $[n, n+k] \subset V$, which in turn means

$$a_{n+j} \geq A(n+j)^{-\alpha}, \quad j = 0, 1, \ldots, k. \tag{2.59}$$

Then by condition (2.57) of Lemma 2.34 we get

$$a_{n+k} \leq a_n \prod_{v=n}^{n+k-1} (1 - \gamma/v) \leq a_{n-1} \prod_{v=n}^{n+k-1} (1 - \gamma/v). \tag{2.60}$$

Combining (2.58)–(2.60) we obtain

$$(n+k)^{-\alpha} \leq (n-1)^{-\alpha} \prod_{v=n}^{n+k-1} (1 - \gamma/v). \tag{2.61}$$

Taking logarithms and using the inequalities

$$\ln(1 - x) \leq -x, \quad x \in [0, 1);$$

$$\sum_{v=n}^{m-1} v^{-1} \geq \int_n^m x^{-1} \, dx = \ln(m/n),$$

we get, from (2.61),

$$-\alpha \ln \frac{n+k}{n-1} \leq \sum_{v=n}^{n+k-1} \ln(1 - \gamma/v) \leq - \sum_{v=n}^{n+k-1} \gamma/v \leq -\gamma \ln \frac{n+k}{n}.$$

Hence,

$$(\gamma - \alpha) \ln(n+k) \leq (\gamma - \alpha) \ln n + \alpha \ln \frac{n}{n-1},$$

which implies

$$n + k \leq 2^{\alpha/(\gamma-\alpha)} n$$

and

$$k \leq C(\alpha, \gamma)n.$$

Let us take any $\mu \in \mathbb{N}$. If $\mu \notin V$ we have the desired inequality with $B = A$. Assume $\mu \in V$, and let $[n, n + k]$ be the maximal segment in $V$ containing $\mu$. Then

$$a_\mu \leq a_{n-1} \leq A(n - 1)^{-\alpha} = A\mu^{-\alpha}\left(\frac{n - 1}{\mu}\right)^{-\alpha}. \tag{2.62}$$

Using the inequality $k \leq C(\alpha, \gamma)n$ proved above we get

$$\frac{\mu}{n - 1} \leq \frac{n + k}{n - 1} \leq C_1(\alpha, \gamma). \tag{2.63}$$

Substituting (2.63) into (2.62) we complete the proof of Lemma 2.34 with $B = AC_1(\alpha, \gamma)^\alpha$. □

*Proof of Theorem 2.33* Let $v(r, \lambda)$ be such that for $v > v(r, \lambda)$ we have

$$(\lambda(v + 1))^{-1} \geq (r/2 + 3/(4\lambda))/v.$$

Take two positive numbers $C \geq v(r, \lambda)^r$ and $\kappa$ which will be chosen later.

We consider the sequence $a_v := 1$ for $v < v(r, \lambda)$ and $a_v := \|f_v\|^2$, $v \geq v(r, \lambda)$, where

$$f_v := f - G_v(f, \mathcal{D}).$$

The assumption $\sigma_1(f, \mathcal{D}) \leq 1$ implies

$$a_{v(r,\lambda)} := \|f_{v(r,\lambda)}\|^2 \leq \|f_1\|^2 \leq 1.$$

Let us assume that for some $v$ we have $a_v \geq C^2 v^{-2r}$. We want to prove that for those same $v$ we have

$$a_{v+1} \leq a_v(1 - \gamma/v)$$

with some $\gamma > 2r$. We shall specify the numbers $C$ and $\kappa$ in this proof. The assumptions $C \geq v(r, \lambda)^r$ and $a_v \geq C^2 v^{-2r}$ imply $v \geq v(r, \lambda)$ and $\|f_v\| \geq Cv^{-r}$, or

$$v^{-r} \leq C^{-1}\|f_v\|. \tag{2.64}$$

We know that $f_v$ has the form

$$f_v = f - \sum_{i=1}^{v} c_i\phi_i, \quad \phi_i \in \mathcal{D}, \quad i = 1, \ldots, v.$$

Therefore, by the assumption of Theorem 2.33 we have

$$\sigma_{[(1+\kappa)v]+1}(f_v) \leq \sigma_{[\kappa v]+1}(f) < (\kappa v)^{-r},$$

where $[x]$ denotes the integer part of the number $x$. This inequality implies that there are $l := [(1 + \kappa)\nu] + 1$ elements $g_1, \ldots, g_l \in \mathcal{D}$ such that

$$\left\| f_\nu - \sum_{i=1}^{l} c_i g_i \right\| \leq (\kappa \nu)^{-r}. \tag{2.65}$$

Now we use the assumption that $\mathcal{D}$ is a $\lambda$-quasi-orthogonal dictionary. We find $M \leq N = \lambda l$ elements $\varphi_j \in \mathcal{D}$, $j = 1, \ldots, M$, satisfying properties (2.55) and (2.56). Denote by $u$ an orthogonal projection of $f_\nu$ onto $X_M = \text{span}(\varphi_1, \ldots, \varphi_M)$ and set $v := f_\nu - u$. Property (2.55) and inequality (2.65) imply

$$\|v\| \leq (\kappa \nu)^{-r},$$

and, therefore, by (2.64) we have

$$\|u\|^2 = \|f_\nu\|^2 - \|v\|^2 \geq \|f_\nu\|^2 (1 - (C\kappa^r)^{-2}).$$

Making use of property (2.56) we get

$$\sup_{g \in \mathcal{D}} |\langle f_\nu, g \rangle| \geq \max_{1 \leq j \leq M} |\langle f_\nu, \varphi_j \rangle| = \max_{1 \leq j \leq M} |\langle u, \varphi_j \rangle| \geq N^{-1/2} \|u\|.$$

Hence,

$$\|f_{\nu+1}\|^2 \leq \|f_\nu\|^2 - \frac{\|u\|^2}{N} \leq \|f_\nu\|^2 (1 - (1 - (C\kappa^r)^{-2})(\lambda([(1+\kappa)\nu]+1))^{-1}).$$

It is clear that taking a small enough $\kappa > 0$ and a sufficiently large $C$ we can make for $\nu \geq \nu(r, \lambda)$

$$\nu(1 - (C\kappa^r)^{-2})(\lambda([(1+\kappa)\nu]+1))^{-1} \geq \gamma > 2r.$$

With the $C$ as chosen we get a sequence $\{a_\nu\}_{\nu=1}^{\infty}$ satisfying the hypotheses of Lemma 2.34 with $A = C^2$, $\alpha = 2r$, $\gamma > \alpha$. Applying Lemma 2.34 we obtain

$$\|f_n\| = a_n^{1/2} \leq C(r, \lambda) n^{-r}, \quad n = 1, 2, \ldots,$$

which completes the proof of Theorem 2.33. $\qquad\square$

The above proof of Theorem 2.33 gives a slightly more general result, with a $\lambda$-quasi-orthogonal dictionary replaced by an asymptotically $\lambda$-quasi-orthogonal dictionary. We formulate the corresponding definition and statements.

**Definition 2.35** We say $\mathcal{D}$ is an asymptotically $\lambda$-quasi-orthogonal dictionary if for any $n \in \mathbb{N}$ and any $g_i \in \mathcal{D}$, $i = 1, \ldots, n$, there exists a collection $\varphi_j \in \mathcal{D}$, $j = 1, \ldots, M$, $M \leq N(n)$, with the following properties:

$$\limsup_{n \to \infty} N(n)/n = \lambda;$$

$$g_i \in X_M := \text{span}(\varphi_1, \ldots, \varphi_M); \qquad (2.66)$$

and for any $f \in X_M$ we have

$$\max_{1 \le j \le M} |\langle f, \varphi_j \rangle| \ge N(n)^{-1/2} \|f\|. \qquad (2.67)$$

**Theorem 2.36** *Let a given dictionary $\mathcal{D}$ be asymptotically $\lambda$-quasi-orthogonal and let $0 < r < (2\lambda)^{-1}$ be a real number. Then, for any $f$ such that*

$$\sigma_m(f, \mathcal{D}) \le m^{-r}, \quad m = 1, 2, \ldots,$$

*we have*

$$\|f - G_m(f, \mathcal{D})\| \le C(r, \lambda, \mathcal{D}) m^{-r}, \quad m = 1, 2, \ldots$$

*Proof* In the proof of this theorem we use the following Lemma 2.37 instead of Lemma 2.34.

**Lemma 2.37** *Let four positive numbers $\alpha < \gamma \le 1$, $A > 1$, $U \in \mathbb{N}$ be given and let a sequence of positive numbers $1 \ge a_1 \ge a_2 \ge \ldots$ satisfy the condition: If for some $v \in \mathbb{N}$, $v \ge U$ we have*

$$a_v \ge A v^{-\alpha},$$

*then*

$$a_{v+1} \le a_v (1 - \gamma/v).$$

*Then there exists $B = B(A, \alpha, \gamma, U)$ such that, for all $n = 1, 2, \ldots$, we have*

$$a_n \le B n^{-\alpha}. \qquad \square$$

We now proceed to a discussion of $\lambda$-quasi-orthogonal dictionaries.

**Proposition 2.38** *Let a system $\{\varphi_1, \ldots, \varphi_M\}$ and its linear span $X_M$ satisfy (2.56). If $M = N$ and $\dim X_M = N$, then $\{\varphi_j\}_{j=1}^N$ is an orthonormal system.*

*Proof* Our proof is by contradiction. The system $\{\varphi_j\}_{j=1}^N$ is normalized and we assume that it is not orthogonal. Consider a system $\{v_j\}_{j=1}^N$ biorthogonal to $\{\varphi_j\}_{j=1}^N$:

$$\langle \varphi_i, v_j \rangle = \delta_{i,j}, \quad 1 \le i, j \le N.$$

Our assumption implies that $\{v_j\}_{j=1}^N$ is also not orthogonal. Consider

$$u_j := v_j / \|v_j\|, \quad j = 1, 2, \ldots N,$$

and form a vector

$$y_t := N^{-1/2} \sum_{i=1}^{N} r_i(t) u_i,$$

where the $r_i(t)$ are the Rademacher functions. Then for all $j = 1, 2, \ldots, N$ and $t \in [0, 1]$ we have

$$|\langle y_t, \varphi_j \rangle| = N^{-1/2} |\langle u_j, \varphi_j \rangle| \leq N^{-1/2}; \qquad (2.68)$$

and

$$\|y_t\|^2 = N^{-1} \sum_{i=1}^{N} \langle u_i, u_i \rangle + N^{-1} \sum_{i \neq j} r_i(t) r_j(t) \langle u_i, u_j \rangle$$

$$= 1 + 2N^{-1} \sum_{1 \leq i < j \leq N} r_i(t) r_j(t) \langle u_i, u_j \rangle.$$

From this we get

$$\int_0^1 \|y_t\|^4 \, dt = 1 + 4N^{-2} \sum_{1 \leq i < j \leq N} |\langle u_i, u_j \rangle|^2 > 1.$$

This inequality implies that for some $t^*$ we have $\|y_{t^*}\| > 1$ and by (2.68) for this $t^*$ we get, for all $1 \leq j \leq N$,

$$|\langle y_{t^*}, \varphi_j \rangle| < N^{-1/2} \|y_{t^*}\|,$$

which contradicts (2.56). $\qquad\qquad\square$

**Definition 2.39** For given $\mu, \gamma \geq 1$, a dictionary $\mathcal{D}$ is called $(\mu, \gamma)$-semistable if, for any $g_i \in \mathcal{D}, i = 1, \ldots, n$, there exist elements $h_j \in \mathcal{D}, j = 1, \ldots, M \leq \mu n$, such that

$$g_i \in \text{span}\{h_1, \ldots, h_M\}$$

and, for any $c_1, \ldots, c_M$, we have

$$\|\sum_{j=1}^{M} c_j h_j\| \geq \gamma^{-1/2} \left( \sum_{j=1}^{M} c_j^2 \right)^{1/2}. \qquad (2.69)$$

**Proposition 2.40** *A $(\mu, \gamma)$-semistable dictionary $\mathcal{D}$ is $\mu\gamma$-quasi-orthogonal.*

*Proof* It is clear from (2.69) that $\{h_1, \ldots, h_M\}$ are linearly independent. Let $\psi_1, \ldots, \psi_M$ be the biorthogonal system to $\{h_1, \ldots, h_M\}$. We shall derive from (2.69) that, for any $a_1, \ldots, a_M$, we have

$$\|\sum_{j=1}^{M} a_j \psi_j\| \leq \gamma^{1/2} \left( \sum_{j=1}^{M} a_j^2 \right)^{1/2}. \qquad (2.70)$$

Indeed, using the representation

$$g = \sum_{j=1}^{M} c_j(g) h_j$$

and (2.69) we get

$$\|\sum_{j=1}^{M} a_j \psi_j\| = \sup_{\|g\| \leq 1} \langle \sum_{j=1}^{M} a_j \psi_j, g \rangle = \sup_{\|g\| \leq 1} \sum_{j=1}^{M} a_j c_j(g)$$

$$\leq \sup_{\|(c_1,\dots,c_M)\| \leq \gamma^{1/2}} \sum_{j=1}^{M} a_j c_j = \gamma^{1/2} \left( \sum_{j=1}^{M} a_j^2 \right)^{1/2}.$$

Take any $f \in \text{span}\{h_1, \dots, h_M\} = \text{span}\{\psi_1, \dots, \psi_M\}$. Let

$$f = \sum_{j=1}^{M} a_j(f) \psi_j.$$

Then

$$\langle f, h_j \rangle = a_j(f).$$

Inequality (2.70) implies

$$\max_{1 \leq j \leq M} |a_j(f)| \geq (\gamma M)^{-1/2} \|f\| \geq (\gamma \mu n)^{-1/2} \|f\|.$$

The proof of Proposition 2.40 is complete.                    □

We now give two concrete examples of asymptotically λ-quasi-orthogonal dictionaries.

**Example 2.41** The dictionary $\chi := \{f = |J|^{-1/2} \chi_J, \quad J \subset [0, 1)\}$, where $\chi_J$ is the characteristic function of an interval $J$, is an asymptotically 2-quasi-orthogonal dictionary.

*Proof* The statement of this example follows from Remark 2.32 and from the known simple Lemma 2.42.                    □

**Lemma 2.42** *For any system of intervals $J_i \subset [0, 1)$, $i = 1, \dots, n$, there exists a system of disjoint intervals $J_i^d \subset [0, 1)$, $i = 1, \dots, 2n + 1$, $[0, 1) = \cup_{i=1}^{2n+1} J_i^d$, such that each $J_i$ can be represented as a union of some $J_j^d$.*

*Proof* Our proof is by induction. Let $n = 1$ and $J_1 = [a, b)$. Take $J_1^d = [0, a)$, $J_2^d = [a, b)$ and $J_3^d = [b, 1)$. Assume now that the statement is true for $n - 1$. Consider $n$ intervals $J_1, \dots, J_{n-1}, J_n$. Let $J_j^d = [a_j, a_{j+1})$, $j = 1, \dots, 2n - 1$ be the disjoint system of intervals corresponding to

$J_1, \ldots, J_{n-1}$ and let $J_n = [a, b)$. Then for at most two intervals $J_k^d$ and $J_l^d$ we have $a \in J_k^d$ and $b \in J_l^d$. If $k = l$ we split $J_k^d$ into three intervals $[a_k, a), [a, b)$ and $[b, a_{k+1})$. If $k \neq l$ we split each $J_k^d$ and $J_l^d$ into two intervals $[a_k, a), [a, a_{k+1})$ and $[a_l, b), [b, a_{l+1})$. In both cases the total number of intervals is $2n + 1$. □

Another corollary of Lemma 2.42 can be formulated as follows.

**Example 2.43** The dictionary $\mathcal{P}(r)$ that consists of functions of the form $f = p\chi_J, \|f\| = 1$, where $p$ is an algebraic polynomial of degree $r - 1$ and $\chi_J$ is the characteristic function of an interval $J$, is asymptotically $2r$-quasi-orthogonal.

Theorems 2.33 and 2.36 work for small smoothness $r < (2\lambda)^{-1}$. It is known (see Section 2.7) that there are dictionaries which have the saturation property for the Pure Greedy Algorithm. Namely, there is a dictionary $\mathcal{D}$ such that

$$\sup_{f \in \Sigma_2(\mathcal{D})} \|f - G_m(f, \mathcal{D})\| / \|f\| \geq Cm^{-1/2},$$

where

$$\Sigma_n(\mathcal{D}) := \left\{ f : f = \sum_{i \in \Lambda} c_i g_i, \ g_i \in \mathcal{D}, \ |\Lambda| = n \right\}.$$

We shall prove that the dictionary $\chi$ from Example 2.41 does not have the saturation property.

**Theorem 2.44** *For any $f \in \Sigma_n(\chi)$ we have*

$$\|f - G_m(f, \chi)\| \leq \left(1 - \frac{1}{2n + 1}\right)^{m/2} \|f\|.$$

*Proof* We prove a variant of Theorem 2.44 for functions of the form

$$f = \sum_{j=1}^{n} c_j g_{I_j}, \quad \cup_{j=1}^{n} I_j = [0, 1), \quad g_J := |J|^{-1/2} \chi_J, \qquad (2.71)$$

where the $I_1, \ldots, I_n$ are disjoint.

**Lemma 2.45** *For any $f$ of the form (2.71) we have*

$$\|f - G_m(f, \chi)\| \leq (1 - 1/n)^{m/2} \|f\|.$$

*Proof* We begin with the following lemma.

**Lemma 2.46** *Let $I^1 = [a, b)$ and $I^2 = [b, d)$ be two adjacent intervals. Assume that a function $f$ is integrable on $I^1$ and equals a constant $c$ on $I^2$. Then we have the following inequality ($g_I := |I|^{-1/2} \chi_I$):*

$$|\langle f, g_J \rangle| \leq \max(|\langle f, g_{I^1} \rangle|, |\langle f, g_{I^1 \cup I^2} \rangle|) \qquad (2.72)$$

*for any $J = [a, y)$, $b \leq y \leq d$. Moreover, if the right-hand side of (2.72) is non-zero, we have a strict inequality in (2.72) for all $b < y < d$.*

*Proof* Denote

$$A := \int_{I^1} f(x)dx.$$

Then we have

$$\langle f, g_J \rangle = |J|^{-1/2} \left( A + \int_b^y c\,dx \right) = (|I^1| + y - b)^{-1/2}(A + c(y - b)),$$

and hence

$$\langle f, g_J \rangle = \frac{P + cy}{(Q + y)^{1/2}}, \quad b \leq y \leq d,$$

where $P = A - cb$ and $Q = |I^1| - b$. Let $z = (Q + y)^{1/2}$. Then

$$\frac{P + cy}{(Q + y)^{1/2}} = \frac{(P + c(z^2 - Q))}{z} = \frac{(P - cQ)}{z} + cz =: F(z).$$

In the case $P - cQ = 0$, $c \neq 0$, or $P - cQ \neq 0$, $c = 0$, the statement is trivial. It remains to consider the case $P - cQ \neq 0$, $c \neq 0$. Assume $P - cQ < 0$, $c > 0$. Then

$$F'(z) = -\frac{P - cQ}{z^2} + c > 0$$

and the statement is true. Assume $P - cQ > 0$, $c > 0$. Then

$$F''(z) = 2\frac{P - cQ}{z^3} > 0, \quad z > 0.$$

It follows that $F(z) > 0$ is a convex function and the statement is also true. $\square$

We use Lemma 2.46 to prove one more lemma.

**Lemma 2.47** *For each function $f$ of the form (2.71), $\max_J |\langle f, g_J \rangle|$ is attained on an interval $J^*$ of the form $J^* = \cup_{j=k}^l I_j$.*

*Proof* The function

$$F(x, y) := (y - x)^{-1/2} \int_x^y f(t)dt, \quad 0 \leq x < y \leq 1;$$

$$F(x, x) = 0, \quad 0 \leq x \leq 1,$$

is continuous on $Y := \{(x, y) : 0 \leq x \leq y \leq 1\}$ for any $f$ of the form (2.71). This implies the existence of $J^*$ such that

$$|\langle f, g_{J^*} \rangle| = \max_J |\langle f, g_J \rangle|. \tag{2.73}$$

Clearly, $|\langle f, g_{J^*} \rangle| > 0$ if $f$ is non-trivial. We complete the proof by contradiction. Assume $J^* = [a, t)$ and, for instance, $t$ is an interior point of $I_s = [b, d)$. Apply Lemma 2.46 with $I^1 = [a, b), I^2 = [b, d), J = J^*$. We get strict inequality, which contradicts (2.73). Hence, $t$ is an endpoint of one of the intervals $I_j$. The same argument proves that $a$ is also an endpoint of one of the intervals $I_j$. This completes the proof of Lemma 2.47. □

Lemma 2.47 implies that, for $f$ of the form (2.71), all the residuals of the PGA $R_j(f)$ are also of the form (2.71). Next, for $f$ of the form (2.71) we have

$$\max_J |\langle f, g_J \rangle| \geq \max_{I_j} |\langle f, g_{I_j} \rangle| \geq n^{-1/2} \|f\|.$$

Consequently,

$$\|R_m(f)\|^2 \leq (1 - 1/n)\|R_{m-1}(f)\|^2 \leq \cdots \leq (1 - 1/n)^m \|f\|^2,$$

which completes the proof of Lemma 2.45. □

The statement of Theorem 2.44 follows from Lemmas 2.45 and 2.42. □

## 2.6 Lebesgue-type inequalities for greedy approximation

### 2.6.1 Introduction

Lebesgue (1909) proved the following inequality: For any $2\pi$-periodic continuous function $f$ we have

$$\|f - S_n(f)\|_\infty \leq \left(4 + \frac{4}{\pi^2} \ln n\right) E_n(f)_\infty, \tag{2.74}$$

where $S_n(f)$ is the $n$th partial sum of the Fourier series of $f$ and $E_n(f)_\infty$ is the error of the best approximation of $f$ by the trigonometric polynomials of order $n$ in the uniform norm $\| \cdot \|_\infty$. The inequality (2.74) relates the error of a particular method ($S_n$) of approximation by the trigonometric polynomials of order $n$ to the best-possible error $E_n(f)_\infty$ of approximation by the trigonometric polynomials of order $n$. By the Lebesgue-type inequality we mean an inequality that provides an upper estimate for the error of a particular method of approximation of $f$ by elements of a special form, say form $\mathcal{A}$, by the best-possible approximation of $f$ by elements of the form $\mathcal{A}$. In the case of approximation with regard to bases (or minimal systems), the Lebesgue-type inequalities are known both in linear and in nonlinear settings (see Chapter 1 and Konyagin and Temlyakov (2002) and Temlyakov (2003a)). It would be very interesting to prove the Lebesgue-type inequalities for redundant systems (dictionaries). However, there are substantial difficulties.

We begin our discussion with the Pure Greedy Algorithm (PGA). It is natural to compare the performance of the PGA with the best $m$-term approximation with regard to a dictionary $\mathcal{D}$. We let $\Sigma_m(\mathcal{D})$ denote the collection of all functions (elements) in $H$ which can be expressed as a linear combination of at most $m$ elements of $\mathcal{D}$. Thus, each function $s \in \Sigma_m(\mathcal{D})$ can be written in the form

$$s = \sum_{g \in \Lambda} c_g g, \quad \Lambda \subset \mathcal{D}, \quad \#\Lambda \le m,$$

where the $c_g$ are real or complex numbers. In some cases, it may be possible to write an element from $\Sigma_m(\mathcal{D})$ in this form in more than one way. The space $\Sigma_m(\mathcal{D})$ is not linear: the sum of two functions from $\Sigma_m(\mathcal{D})$ is generally not in $\Sigma_m(\mathcal{D})$.

For a function $f \in H$ we define its best $m$-term approximation error

$$\sigma_m(f) := \sigma_m(f, \mathcal{D}) := \inf_{s \in \Sigma_m(\mathcal{D})} \|f - s\|.$$

It seems that there is no hope of proving a non-trivial multiplicative Lebesgue-type inequality for the PGA in the case of an arbitrary dictionary $\mathcal{D}$. This pessimism is based on the following result from DeVore and Temlyakov (1996) (see Section 2.7 below).

Let $\mathcal{B} := \{h_k\}_{k=1}^{\infty}$ be an orthonormal basis in a Hilbert space $H$. Consider the following element:

$$g := Ah_1 + Ah_2 + aA \sum_{k \ge 3} (k(k+1))^{-1/2} h_k$$

with

$$A := (33/89)^{1/2} \quad \text{and} \quad a := (23/11)^{1/2}.$$

Then $\|g\| = 1$. We define the dictionary $\mathcal{D} = \mathcal{B} \cup \{g\}$. It has been proved in DeVore and Temlyakov (1996) (see Section 2.7) that, for the function

$$f = h_1 + h_2,$$

we have

$$\|f - G_m(f, \mathcal{D})\| \ge m^{-1/2}, \quad m \ge 4.$$

It is clear that $\sigma_2(f, \mathcal{D}) = 0$.

Therefore, we look for conditions on a dictionary $\mathcal{D}$ that allow us to prove the Lebesgue-type inequalities. The condition $\mathcal{D} = \mathcal{B}$, an orthonormal basis for $H$, guarantees that

$$\|R_m(f, \mathcal{B})\| = \sigma_m(f, \mathcal{B}).$$

This is an ideal situation. The results that we will discuss here concern the case when we replace an orthonormal basis $\mathcal{B}$ by a dictionary that is, in a certain sense, not far from an orthonormal basis.

Let us begin with results from Donoho, Elad and Temlyakov (2007) that are close to the results from Temlyakov (1999) discussed in Section 2.5. We give a definition of a $\lambda$-quasi-orthogonal dictionary with depth $D$. In the case $D = \infty$ this definition coincides with the definition of a $\lambda$-quasi-orthogonal dictionary from Temlyakov (1999) (see Definition 2.31 above).

**Definition 2.48** We say that $\mathcal{D}$ is a $\lambda$-quasi-orthogonal dictionary with depth $D$ if, for any $n \in [1, D]$ and any $g_i \in \mathcal{D}, i = 1, \ldots, n$, there exists a collection $\varphi_j \in \mathcal{D}, j = 1, \ldots, J, J \leq N := \lambda n$, with the properties

$$g_i \in X_J := \text{span}(\varphi_1, \ldots, \varphi_J), \quad i = 1, \ldots, n,$$

and for any $f \in X_J$ we have

$$\max_{1 \leq j \leq J} |\langle f, \varphi_j \rangle| \geq N^{-1/2} \|f\|.$$

It is pointed out in Donoho, Elad and Temlyakov (2007) that the proof of Theorem 1.1 from Temlyakov (1999) (see Theorem 2.33 above) also works in the case $D < \infty$ and gives the following result.

**Theorem 2.49** *Let a given dictionary $\mathcal{D}$ be $\lambda$-quasi-orthogonal with depth $D$, and let $0 < r < (2\lambda)^{-1}$ be a real number. Then, for any $f$ such that*

$$\sigma_m(f, \mathcal{D}) \leq m^{-r}, \quad m = 1, 2, \ldots, D,$$

*we have*

$$\|f_m\| = \|f - G_m(f, \mathcal{D})\| \leq C(r, \lambda) m^{-r}, \quad m \in [1, D/2].$$

In this section we consider dictionaries that have become popular in signal processing. Denote by

$$M(\mathcal{D}) := \sup_{g \neq h; g, h \in \mathcal{D}} |\langle g, h \rangle|$$

the coherence parameter of a dictionary $\mathcal{D}$. For an orthonormal basis $\mathcal{B}$ we have $M(\mathcal{B}) = 0$. It is clear that the smaller the $M(\mathcal{D})$, the more $\mathcal{D}$ resembles an orthonormal basis. However, we should note that in the case $M(\mathcal{D}) > 0$ the $\mathcal{D}$ can be a redundant dictionary. We showed in Donoho, Elad and Temlyakov (2007) (see Proposition 2.59 below) that a dictionary with coherence $M := M(\mathcal{D})$ is a $(1 + 4\delta)$-quasi-orthogonal dictionary with depth $\delta/M$, for any $\delta \in (0, 1/7]$. Therefore, Theorem 2.49 applies to $M$-coherent dictionaries.

We now proceed to a discussion of the Orthogonal Greedy Algorithm (OGA). It is clear from the definition of the OGA that at each step we have

$$\|f_m^o\|^2 \le \|f_{m-1}^o\|^2 - |\langle f_{m-1}^o, g(f_{m-1}^o)\rangle|^2.$$

We noted in Donoho, Elad and Temlyakov (2007) that the use of this inequality instead of the equality

$$\|f_m\|^2 = \|f_{m-1}\|^2 - |\langle f_{m-1}, g(f_{m-1})\rangle|^2,$$

which holds for the PGA, allows us to prove an analog of Theorem 2.49 for the OGA. The proof repeats the corresponding proof from Temlyakov (1999) (see the proof of Theorem 2.33 above). We formulate this as a remark.

**Remark 2.50** Theorem 2.49 holds for the OGA instead of the PGA (for $\|f_m^o\|$ instead of $\|f_m\|$).

The first general Lebesgue-type inequality for the OGA for the $M$-coherent dictionary was obtained in Gilbert, Muthukrishnan and Strauss (2003). They proved that

$$\|f_m^o\| \le 8m^{1/2}\sigma_m(f) \quad \text{for} \quad m < 1/(32M).$$

The constants in this inequality were improved in Tropp (2004):

$$\|f_m^o\| \le (1 + 6m)^{1/2}\sigma_m(f) \quad \text{for} \quad m < 1/(3M). \tag{2.75}$$

We proved in Donoho, Elad and Temlyakov (2007) the following inequalities.

**Theorem 2.51** *Let a dictionary $\mathcal{D}$ have the mutual coherence $M = M(\mathcal{D})$. Then for any $S \le 1/(2M)$ we have the following inequalities:*

$$\|f_S^o\|^2 \le 2\|f_k^o\|(\sigma_{S-k}(f_k^o) + 3MS\|f_k^o\|), \quad 0 \le k \le S; \tag{2.76}$$

$$\|f_S\|^2 \le 2\|f\|(\sigma_S(f) + 5MS\|f\|).$$

These inequalities were improved in Temlyakov and Zheltov (2010).

**Theorem 2.52** *Let a dictionary $\mathcal{D}$ have the mutual coherence $M = M(\mathcal{D})$. Then for any $S \le 1/(2M)$ we have the following inequalities:*

$$\|f_S^o\|^2 \le \sigma_{S-k}(f_k^o)^2 + 5MS\|f_k^o\|^2, \quad 0 \le k \le S; \tag{2.77}$$

$$\|f_S\|^2 \le \sigma_S(f)^2 + 7MS\|f\|^2. \tag{2.78}$$

The inequality (2.76) can be used for improving (2.75) for small $m$. The following inequalities were proved in Donoho, Elad and Temlyakov (2007).

**Theorem 2.53** *Let a dictionary $\mathcal{D}$ have the mutual coherence $M = M(\mathcal{D})$. Assume $m \leq 0.05M^{-2/3}$. Then, for $l \geq 1$ satisfying $2^l \leq \log m$, we have*

$$\|f_{m(2^l-1)}^o\| \leq 6m 2^{-l} \sigma_m(f).$$

**Corollary 2.54** *Let a dictionary $\mathcal{D}$ have the mutual coherence $M = M(\mathcal{D})$. Assume $m \leq 0.05M^{-2/3}$. Then we have*

$$\|f_{[m \log m]}^o\| \leq 24\sigma_m(f).$$

It was pointed out in Donoho, Elad and Temlyakov (2007) that the inequality $\|f_{[m \log m]}^o\| \leq 24\sigma_m(f)$ from Corollary 2.54 is almost (up to a $\log m$ factor) perfect Lebesgue inequality. However, we are paying a big price for it in the sense of a strong assumption on $m$. It was mentioned in Donoho, Elad and Temlyakov (2007) that it was not known if the assumption $m \leq 0.05M^{-2/3}$ can be substantially weakened. It was shown in Temlyakov and Zheltov (2010) that the use of Theorem 2.52 instead of Theorem 2.51 allows us to weaken substantially the assumption $m \leq 0.05M^{-2/3}$.

**Theorem 2.55** *Let a dictionary $\mathcal{D}$ have the mutual coherence $M = M(\mathcal{D})$. For any $\delta \in (0, 1/4]$ set $L(\delta) := [1/\delta] + 1$. Assume $m$ is such that $20Mm^{1+\delta}2^{L(\delta)} \leq 1$. Then we have*

$$\|f_{m(2^{L(\delta)+1}-1)}^o\| \leq \sqrt{3}\sigma_m(f).$$

Very recently, Livshitz (2010) improved the above Lebesgue-type inequality by proving that

$$\|f_{2m}^o\| \leq 3\sigma_m(f)$$

for $m \leq (20M)^{-1}$. The proof in Livshitz (2010) is different from the proof of Theorem 2.55 given below; it is much more technically involved.

We now demonstrate the use of the inequality (2.78). The following result from Temlyakov and Zheltov (2010) is a corollary of (2.78).

**Theorem 2.56** *Let a dictionary $\mathcal{D}$ have the mutual coherence $M = M(\mathcal{D})$. For any $r > 0$ and $\delta \in (0, 1]$ set $L(r, \delta) := [2r/\delta] + 1$. Let $f$ be such that*

$$\sigma_m(f) \leq m^{-r}\|f\|, \quad m \leq 2^{-L(r,\delta)}(14M)^{-1/(1+\delta)}.$$

*Then, for all $n$ such that $n \leq (14M)^{-1/(1+\delta)}$, we have*

$$\|f_n\| \leq C(r, \delta)n^{-r}\|f\|.$$

The classical Lebesgue inequality is a multiplicative inequality, where the quality of an approximation method is measured by the growth of an extra factor. It turns out that multiplicative Lebesgue-type inequalities are rather rare in nonlinear approximation. The above discussed example from DeVore and Temlyakov (1996) shows that, even for a simple dictionary that differs from an orthonormal basis by one element, there is no multiplicative Lebesgue-type inequality for the Pure Greedy Algorithm. Theorems 2.51 and 2.52 are not multiplicative Lebesgue-type inequalities – on the right-hand side they have an additive term of the form $MS\|f_k^o\|^2$. Their applications in the proofs of Theorems 2.55 and 2.56 indicate that these inequalities are useful and rather powerful. It seems that the additive Lebesgue-type inequalities are an appropriate tool in nonlinear approximation.

### 2.6.2 Proofs

We will use the following simple known lemma (see, for example, Donoho, Elad and Temlyakov (2007)).

**Lemma 2.57** *Assume a dictionary $\mathcal{D}$ has mutual coherence $M$. Then we have, for any distinct $g_j \in \mathcal{D}$, $j = 1, \ldots, N$, and, for any $a_j$, $j = 1, \ldots, N$, the inequalities*

$$\left(\sum_{j=1}^{N} |a_j|^2\right)(1 - M(N-1)) \le \|\sum_{j=1}^{N} a_j g_j\|_2^2 \le \left(\sum_{j=1}^{N} |a_j|^2\right)(1 + M(N-1)).$$

*Proof* We have

$$\|\sum_{j=1}^{N} a_j g_j\|_2^2 = \sum_{j=1}^{N} |a_j|^2 + \sum_{i \ne j} a_i \bar{a}_j \langle g_i, g_j \rangle.$$

Next,

$$|\sum_{i \ne j} a_i \bar{a}_j \langle g_i, g_j \rangle| \le M \sum_{i \ne j} |a_i a_j| = M\left(\sum_{i,j} |a_i a_j| - \sum_{i=1}^{N} |a_i|^2\right)$$

$$= M\left(\left(\sum_{i=1}^{N} |a_i|\right)^2 - \sum_{i=1}^{N} |a_i|^2\right) \le \left(\sum_{i=1}^{N} |a_i|^2\right)M(N-1). \qquad \square$$

We now proceed to one more technical lemma (see Donoho, Elad and Temlyakov (2007)).

**Lemma 2.58** *Suppose that $g_1, \ldots, g_N$ are such that $\|g_i\| = 1$, $i = 1, \ldots, N$; $|\langle g_i, g_j \rangle| \le M$, $1 \le i \ne j \le N$. Let $H_N := \text{span}(g_1, \ldots, g_N)$. Then, for any $f$, we have*

$$\left(\sum_{i=1}^{N} |\langle f, g_i \rangle|^2\right)^{1/2} \geq \left(\sum_{i=1}^{N} |c_i|^2\right)^{1/2} (1 - M(N-1)),$$

where $\{c_i\}$ are from the representation of the orthogonal projection of $f$ onto $H_N$:

$$P_{H_N}(f) = \sum_{j=1}^{N} c_j g_j.$$

*Proof* We have $\langle f - P_{H_N}(f), g_i \rangle = 0$, $i = 1, \ldots, N$, and therefore

$$|\langle f, g_i \rangle| = |\langle P_{H_N}(f), g_i \rangle| = |\sum_{j=1}^{N} c_j \langle g_j, g_i \rangle| \geq |c_i|(1 + M) - M \sum_{j=1}^{N} |c_j|.$$

Next, denoting $\sigma := (\sum_{j=1}^{N} |c_j|^2)^{1/2}$ and using the inequality $\sum_{j=1}^{N} |c_j| \leq N^{1/2}\sigma$, we get

$$\left(\sum_{i=1}^{N} |\langle f, g_i \rangle|^2\right)^{1/2} \geq \sigma(1 - M(N-1)).$$

$\square$

The following proposition is a direct corollary of Lemmas 2.57 and 2.58.

**Proposition 2.59** *Let $\delta \in (0, 1/7]$. Then any dictionary with mutual coherence $M$ is a $(1 + 4\delta)$-quasi-orthogonal dictionary with depth $\delta/M$.*

*Proof* Let $n \leq \delta/M$. Consider any distinct $g_i \in \mathcal{D}$, $i = 1, \ldots, n$. Following Definition 2.48 we specify $J = n$, $\varphi_j = g_j$, $j = 1, \ldots, n$. For any $f = \sum_{j=1}^{n} a_j g_j$, we have, by Lemma 2.58,

$$\max_{1 \leq j \leq n} |\langle f, g_j \rangle| \geq n^{-1/2} \left(\sum_{j=1}^{n} |\langle f, g_j \rangle|^2\right)^{1/2} \geq n^{-1/2} \left(\sum_{j=1}^{n} |a_j|^2\right)^{1/2} (1 - Mn).$$

Using the assumption $n \leq \delta/M$, we get from this, by Lemma 2.57,

$$\max_{1 \leq j \leq n} |\langle f, g_j \rangle| \geq n^{-1/2} \frac{1 - \delta}{(1 + \delta)^{1/2}} \|f\| \geq (n(1 + 4\delta))^{-1/2} \|f\|.$$

This completes the proof of Proposition 2.59. $\square$

*Proof of Theorem 2.52.* Denote

$$d(f) := \sup_{g \in \mathcal{D}} |\langle f, g \rangle|. \tag{2.79}$$

For simplicity we assume that the maximizer in (2.79) exists. Then

$$\| f_m \|^2 = \| f_{m-1} \|^2 - d(f_{m-1})^2 \quad \text{and} \quad \| f_m^o \|^2 \leq \| f_{m-1}^o \|^2 - d(f_{m-1}^o)^2.$$

We carry out the proof for the OGA and later point out the necessary changes for the PGA. For $k = S$ the inequality (2.77) is obvious because $\sigma_0(f_S^o) = \| f_S^o \|$. Let $k \in [0, S)$ be fixed. Assume $f_k^o \neq 0$. Denote by $g_1, \ldots, g_{S-k} \subset \mathcal{D}$ the elements that have the biggest inner products with $f_k^o$:

$$|\langle f_k^o, g_1 \rangle| \geq |\langle f_k^o, g_2 \rangle| \geq \cdots \geq |\langle f_k^o, g_{S-k} \rangle| \geq \sup_{g \in \mathcal{D}, g \neq g_i, i=1,\ldots,S-k} |\langle f_k^o, g \rangle|.$$

We define a natural number $s$ in the following way. If $\langle f_k^o, g_{S-k} \rangle \neq 0$ then we set $s := S - k$; otherwise $s$ is chosen such that $\langle f_k^o, g_s \rangle \neq 0$ and $\langle f_k^o, g_{s+1} \rangle = 0$. Let $m \in [k, k + s)$ and

$$f_m^o = f - P_{H_m}(f) = f_k^o - P_{H_m}(f_k^o), \quad H_m = \text{span}(\varphi_1, \ldots, \varphi_m), \quad \varphi_j \in \mathcal{D},$$

where $\varphi_j \in \mathcal{D}$ are obtained by realization of the OGA. We note that $\langle f_k^o, \varphi_l \rangle = 0, l \in [1, k]$. Therefore, each $g_i, i \in [1, s]$, is different from all $\varphi_l, l = 1, \ldots, k$. By the counting argument there exists an index $i \in [1, m + 1 - k]$ such that $g_i \neq \varphi_j, j = 1, \ldots, m$. For this $i$ we estimate

$$\langle f_m^o, g_i \rangle = \langle f_k^o, g_i \rangle - \langle P_{H_m}(f_k^o), g_i \rangle. \tag{2.80}$$

Let

$$P_{H_m}(f_k^o) = \sum_{j=1}^{m} c_j \varphi_j.$$

Clearly, $\| P_{H_m}(f_k^o) \| \leq \| f_k^o \|$. Then, by Lemma 2.57,

$$\left( \sum_{j=1}^{m} |c_j|^2 \right)^{1/2} \leq \| f_k^o \| (1 - M(m - 1))^{-1/2}.$$

We continue, to obtain

$$|\langle P_{H_m}(f_k^o), g_i \rangle| \leq M \sum_{j=1}^{m} |c_j| \leq M m^{1/2} \left( \sum_{j=1}^{m} |c_j|^2 \right)^{1/2}$$

$$\leq M S^{1/2} \| f_k^o \| (1 - MS)^{-1/2}. \tag{2.81}$$

Thus we get from (2.80) and (2.81) that

$$d(f_m^o) \geq |\langle f_m^o, g_i \rangle| \geq |\langle f_k^o, g_i \rangle| - M S^{1/2} \| f_k^o \| (1 - MS)^{-1/2}, \quad i \in [1, m+1-k],$$

and, using the inequality $|\langle f_k^o, g_i \rangle| \geq |\langle f_k^o, g_{m+1-k} \rangle|$ that follows from the definition of $\{g_j\}$, we obtain

$$d(f_m^o) \geq |\langle f_k^o, g_{m+1-k} \rangle| - MS^{1/2}\|f_k^o\|(1 - MS)^{-1/2}.$$

Therefore,

$$\left(\sum_{v=k}^{k+s-1} d(f_v^o)^2\right)^{1/2} \geq \left(\sum_{i=1}^{s}(|\langle f_k^o, g_i \rangle| - MS^{1/2}\|f_k^o\|(1 - MS)^{-1/2})^2\right)^{1/2}$$

$$\geq \left(\sum_{i=1}^{s}|\langle f_k^o, g_i \rangle|^2\right)^{1/2} - MS\|f_k^o\|(1 - MS)^{-1/2}.$$

$$(2.82)$$

Next, let

$$\sigma_{S-k}(f_k^o) = \|f_k^o - P_{H(n)}(f_k^o)\|, \quad P_{H(n)}(f_k^o) = \sum_{j=1}^{n} b_j \psi_j, \quad n \leq s,$$

where $\psi_j \in \mathcal{D}$, $j = 1, \ldots, n$, are distinct. Then

$$\|P_{H(n)}(f_k^o)\|^2 = \|f_k^o\|^2 - \sigma_{S-k}(f_k^o)^2$$

and, by Lemma 2.57,

$$\sum_{j=1}^{n}|b_j|^2 \geq (\|f_k^o\|^2 - \sigma_{S-k}(f_k^o)^2)(1 + MS)^{-1}. \qquad (2.83)$$

By Lemma 2.58,

$$\sum_{j=1}^{n}|\langle f_k^o, \psi_j \rangle|^2 \geq \left(\sum_{j=1}^{n}|b_j|^2\right)(1 - MS)^2. \qquad (2.84)$$

We get from (2.83) and (2.84) the following:

$$\sum_{i=1}^{s}|\langle f_k^o, g_i \rangle|^2 \geq \sum_{j=1}^{n}|\langle f_k^o, \psi_j \rangle|^2 \geq (\|f_k^o\|^2 - \sigma_{S-k}(f_k^o)^2)(1+MS)^{-1}(1-MS)^2.$$

Finally, by (2.82) we obtain

$$\left(\sum_{v=k}^{k+s-1} d(f_v^o)^2\right)^{1/2} \geq (\|f_k^o\|^2 - \sigma_{S-k}(f_k^o)^2)^{1/2}$$

$$\times \frac{1 - MS}{(1 + MS)^{1/2}} - MS\|f_k^o\|(1 - MS)^{-1/2}. \qquad (2.85)$$

Let $MS \leq 1/2$. Denote $x := MS$. We use the following simple inequalities:

$$\frac{1-x}{(1+x)^{1/2}} \geq 1 - \frac{3}{2}x, \quad x \leq 1/2,$$

and

$$(1-x)^{-1/2} \leq 1+x, \quad x \in [0, 1/2].$$

Next, we use the following inequality for $0 \leq B \leq A$, $a, b \geq 0$, $a^2 - 2b + 1 \geq 0$:

$$((A^2 - B^2)^{1/2}(1 - ax) - x(1 + bx)A)^2$$
$$\geq (A^2 - B^2)(1 - ax)^2 - 2x(1 + bx)A^2 + x^2(1 + bx)^2 A^2$$
$$\geq A^2 - B^2 - (2a + 2)x A^2. \tag{2.86}$$

Using (2.86) with $A := \|f_k^o\|$, $B := \sigma_{S-k}(f_k^o)$, $a = 3/2$, $b = 1$, we obtain

$$\left( \sum_{v=k}^{k+s-1} d(f_v^o)^2 \right)^{1/2} \geq \|f_k^o\|^2 - \sigma_{S-k}(f_k^o)^2 - 5MS\|f_k^o\|^2.$$

Thus,

$$\|f_S^o\|^2 \leq \sigma_{S-k}(f_k^o)^2 + 5MS\|f_k^o\|^2.$$

This completes the proof of Theorem 2.52 for the OGA.

A few changes adapt the proof for $k = 0$ to the PGA setting. As above, we write

$$f_m = f - G_m(f); \quad G_m(f) = \sum_{j=1}^{m} b_j \psi_j, \quad \psi_j \in \mathcal{D},$$

and estimate $|\langle f_m, g_i \rangle|$ with $i \in [1, m+1]$ such that $g_i \neq \psi_j$, $j = 1, \ldots, m$. Using instead of $\|P_{H_m}(f)\| \leq \|f\|$ the inequality

$$\|G_m(f)\| \leq \|f\| + \|f_m\| \leq 2\|f\|,$$

we obtain the following analog of (2.85):

$$\left( \sum_{v=0}^{s-1} d(f_v)^2 \right)^{1/2} \geq (\|f\|^2 - \sigma_{S-k}(f)^2)^{1/2} \frac{1 - MS}{(1 + MS)^{1/2}} - 2MS\|f\|(1 - MS)^{-1/2}. \tag{2.87}$$

We use the inequality

$$((A^2 - B^2)^{1/2}(1 - ax) - 2x(1 + bx)A)^2 \geq A^2 - B^2 - (2a + 4)x A^2,$$

provided $a^2 - 4b + 4 \geq 0$; this is the case for $a = 3/2$ and $b = 1$. Therefore, for the PGA we get

$$\|f_S\|^2 \leq \sigma_S(f_k)^2 + 7MS\|f\|^2. \tag{2.88}$$

$\square$

*Proof of Theorem 2.55* We now show how one can combine the inequalities from Theorem 2.52 with the inequality (2.75). For a given natural number $m$, consider a sequence $m_l := m(2^l - 1)$, $l = 1, 2, \ldots$. We estimate $\|f^o_{m_l}\|$, $l = 1, 2, \ldots$. For $l = 1$ we have $m_1 = m$, and by (2.75) we get

$$\|f^o_m\|^2 \leq (1 + 6m)\sigma_m(f)^2, \quad m < 1/(3M).$$

By Theorem 2.52 with $S = m_l$, $k = m_{l-1}$, we obtain for $l \geq 2$

$$\|f^o_{m_l}\|^2 \leq \sigma_{m_l - m_{l-1}}(f^o_{m_{l-1}})^2 + 5Mm_l\|f^o_{m_{l-1}}\|^2 \tag{2.89}$$

provided $Mm_l \leq 1/2$. It is easy to see that

$$\sigma_{m_l - m_{l-1}}(f^o_{m_{l-1}}) \leq \sigma_{m_l - 2m_{l-1}}(f) = \sigma_m(f).$$

Let $\delta > 0$ be an arbitrary fixed number. Suppose $m$ and $L$ are such that

$$5Mm2^{L+1} \leq m^{-\delta}/2. \tag{2.90}$$

Then for all $l \leq L + 1$ we have

$$5Mm_l \leq 5Mm2^l = 5Mm2^{L+1}2^{l-L-1} \leq m^{-\delta}2^{l-L-2}.$$

Applying (2.89) recursively from $l = 1$ to $l = L + 1$ we obtain

$$\|f^o_{m_{L+1}}\|^2 \leq \sigma_m(f)^2 \sum_{j=0}^{L} 2^{-j} + (1 + 6m)\sigma_m(f)^2 \prod_{l=2}^{L+1}(5Mm_l)$$

$$\leq 2\sigma_m(f)^2 + (1 + 6m)m^{-\delta L}2^{-L^2/2}\sigma_m(f)^2.$$

For $\delta \leq 1/4$, $L \geq 1/\delta$, we get

$$(1 + 6m)m^{-\delta L}2^{-L^2/2} \leq 1.$$

Therefore, for these $\delta$ and $L$

$$\|f^o_{m_{L+1}}\|^2 \leq 3\sigma_m(f)^2. \tag{2.91}$$

Let us specify $L := L(\delta) := [1/\delta] + 1$ and rewrite condition (2.90) as

$$20Mm^{1+\delta}2^{L(\delta)} \leq 1.$$

The inequality (2.91) gives the required bound in Theorem 2.55. $\qquad\square$

*Proof of Theorem 2.56* Let the sequence $\{m_l\}$ be as in the proof of Theorem 2.55. We do not have an analog of (2.75) for the PGA. We write for $l = 1$ the inequality (2.78) with $S = m$:

$$\|f_m\|^2 \leq \sigma_m(f)^2 + 7Mm\|f\|^2. \tag{2.92}$$

Applying Theorem 2.52 to $f_{m_{l-1}}$ with $S = m_l$ we get the following analog of (2.89):

$$\|f_{m_l}\|^2 \le \sigma_m(f)^2 + 7Mm_l\|f_{m_{l-1}}\|^2.$$

Assuming instead of (2.90) that

$$7Mm2^L \le m^{-\delta}/2,$$

we obtain

$$\|f_{m_L}\|^2 \le 2\sigma_m(f)^2 + \|f\|^2 \prod_{l=1}^{L}(7Mm_l) \le 2\sigma_m(f)^2 + \|f\|^2 m^{-\delta L} 2^{-L^2/2}.$$

(2.93)

Let $r > 0$ be a fixed number. Set $L(r, \delta) := [2r/\delta] + 1$. Let $n \le (14M)^{-1/(1+\delta)}$ and let $m$ be the largest natural number such that $m2^{L(r,\delta)} \le n$. Then by (2.93) we obtain

$$\|f_n\|^2 \le \|f_{m_{L(r,\delta)}}\|^2 \le 2\sigma_m(f)^2 + m^{-2r}\|f\|^2.$$

Using the assumption $\sigma_m(f) \le m^{-r}\|f\|$ we continue:

$$\|f\|^2 \le 3m^{-2r}\|f\|^2 \le C(r, \delta)n^{-2r}\|f\|^2.$$

This completes the proof of Theorem 2.56.                                   □

## 2.7 Saturation property of greedy-type algorithms

### 2.7.1 Saturation of the Pure Greedy Algorithm

In this section we give an example from DeVore and Temlyakov (1996) which shows that replacing a dictionary $\mathcal{B}$ given by an orthogonal basis by a non-orthogonal redundant dictionary $\mathcal{D}$ may damage the efficiency of the Pure Greedy Algorithm. The dictionary $\mathcal{D}$ in our example differs from the dictionary $\mathcal{B}$ by the one addition of the element $g$ for a certain suitably chosen function $g$.

Let $\mathcal{B} := \{h_k\}_{k=1}^{\infty}$ be an orthonormal basis in a Hilbert space $H$. Consider the following element:

$$g := Ah_1 + Ah_2 + aA\sum_{k\ge 3}(k(k+1))^{-1/2}h_k \qquad (2.94)$$

with

$$A := (33/89)^{1/2} \quad \text{and} \quad a := (23/11)^{1/2}.$$

Then $\|g\| = 1$. We define the dictionary $\mathcal{D} := \mathcal{B} \cup \{g\}$.

**Theorem 2.60** *For the function*

$$f := h_1 + h_2,$$

*we have*

$$\|f - G_m(f)\| \geq m^{-1/2}, \quad m \geq 4.$$

*Proof* We shall examine the steps of the Pure Greedy Algorithm applied to the function $f = h_1 + h_2$. We shall use the abbreviated notation $f_m := R_m(f) := f - G_m(f)$ for the residual at step $m$.

**First step** We have

$$\langle f, g \rangle = 2A > 1, \quad |\langle f, h_k \rangle| \leq 1, \quad k = 1, 2, \ldots$$

This implies

$$G_1(f, D) = \langle f, g \rangle g$$

and

$$f_1 = f - \langle f, g \rangle g = (1 - 2A^2)(h_1 + h_2) - 2aA^2 \sum_{k \geq 3}(k(k+1))^{-1/2} h_k.$$

**Second step** We have

$$\langle f_1, g \rangle = 0, \quad \langle f_1, h_k \rangle = (1 - 2A^2), \quad k = 1, 2,$$
$$\langle f_1, h_3 \rangle = -aA^2 3^{-1/2}.$$

Comparing $\langle f_1, h_1 \rangle$ and $|\langle f_1, h_3 \rangle|$, we get

$$|\langle f_1, h_3 \rangle| = (23/89)(33/23)^{1/2} > 23/89 = 1 - 2A^2 = \langle f_1, h_1 \rangle.$$

This implies that the second approximation $G_1(f_1, D)$ is $\langle f_1, h_3 \rangle h_3$ and

$$f_2 = f_1 - \langle f_1, h_3 \rangle h_3 = (1 - 2A^2)(h_1 + h_2) - 2aA^2 \sum_{k \geq 4}(k(k+1))^{-1/2} h_k.$$

**Third step** We have

$$\langle f_2, g \rangle = -\langle f_1, h_3 \rangle \langle h_3, g \rangle = (A/2)(23/89),$$
$$\langle f_2, h_1 \rangle = \langle f_2, h_2 \rangle = 1 - 2A^2 = 23/89,$$
$$\langle f_2, h_4 \rangle = -aA^2 5^{-1/2} = -(23/89)(99/115)^{1/2}.$$

Therefore, the third approximation should be $\langle f_2, h_1 \rangle h_1$ or $\langle f_2, h_2 \rangle h_2$. Let us take the first of these so that

$$f_3 = f_2 - \langle f_2, h_1 \rangle h_1.$$

**Fourth step** It is clear that, for all $k \neq 1$, we have

$$\langle f_3, h_k \rangle = \langle f_2, h_k \rangle.$$

This equality and the calculations from the third step show that it is sufficient to compare $\langle f_3, h_2 \rangle$ and $\langle f_3, g \rangle$ . We have

$$\langle f_3, g \rangle = \langle f_2, g \rangle - \langle f_2, h_1 \rangle \langle h_1, g \rangle = -(23/89)(A/2).$$

This means that

$$f_4 = f_3 - \langle f_3, h_2 \rangle h_2 = -2aA^2 \sum_{k \geq 4} (k(k+1))^{-1/2} h_k. \tag{2.95}$$

$m$**th step** ($m > 4$) We prove by induction that, for all $m \geq 4$, we have

$$f_m = -2aA^2 \sum_{k \geq m} (k(k+1))^{-1/2} h_k. \tag{2.96}$$

For $m = 4$ this relation follows from (2.95). We assume we have proved (2.96) for some $m$ and derive that (2.96) also holds true for $m + 1$. To find $f_{m+1}$, we only have to compare the two inner products: $\langle f_m, h_m \rangle$ and $\langle f_m, g \rangle$. We have

$$|\langle f_m, h_m \rangle| = 2aA^2 (m(m+1))^{-1/2}$$

and

$$|\langle f_m, g \rangle| = 2a^2 A^3 \sum_{k \geq m} (k(k+1))^{-1} = 2a^2 A^3 m^{-1}.$$

Since

$$(|\langle f_m, g \rangle|/|\langle f_m, h_m \rangle|)^2 = (aA)^2 (1 + 1/m) \leq 345/356 < 1,$$

we have that

$$|\langle f_m, g \rangle| < |\langle f_m, h_m \rangle|, \quad m \geq 4.$$

This proves (2.96) with $m$ replaced by $m + 1$. From (2.96), we obtain

$$\|f - G_m(f, D)\| = \|f_m\| = 2aA^2 m^{-1/2} > m^{-1/2}, \quad m \geq 4.$$

$\square$

### 2.7.2 A generalization of the Pure Greedy Algorithm

Results of this subsection are from Temlyakov (1999). In this subsection we consider a generalization of the Pure Greedy Algorithm. We study the $n$-Greedy Algorithm which differs from the Pure Greedy Algorithm in the basic step: instead of finding a single element $g(f) \in \mathcal{D}$ with the largest projection of $f$ on it, we are looking for $n$ elements $g_1(f), \ldots, g_n(f) \in \mathcal{D}$ with the largest projection $G^n(f, \mathcal{D})$ of $f$ onto their span. It is clear that

$$\|f - G^n(f, \mathcal{D})\| \leq \|f - G_n(f, \mathcal{D})\|.$$

However, we construct here an example of a dictionary $\mathcal{D}$ and a non-zero function $f \in \Sigma_{6n}(\mathcal{D})$ such that

$$\|f - G^n_m(f, \mathcal{D})\| \geq C(nm)^{-1/2}\|f\|. \tag{2.97}$$

This relation implies that, like the Pure Greedy Algorithm, the $n$-Greedy Algorithm has a saturation property.

We now give the definition of the $n$-Greedy Algorithm. Take a fixed number $n \in \mathbb{N}$ and define the basic step of the $n$-Greedy Algorithm as follows. Find an $n$-term polynomial

$$p_n(f) := p_n(f, \mathcal{D}) = \sum_{n=1}^{n} c_i g_i, \quad g_i \in \mathcal{D}, \quad i = 1, \ldots, n,$$

such that (we assume its existence)

$$\|f - p_n(f)\| = \sigma_n(f, \mathcal{D}).$$

Denote

$$G^n(f) := G^n(f, \mathcal{D}) := p_n(f), \qquad R^n(f) := R^n(f, \mathcal{D}) := f - p_n(f).$$

$n$-**Greedy Algorithm** We define $R^n_0(f) := R^n_0(f, \mathcal{D}) := f$ and $G^n_0(f) := 0$. Then, for each $m \geq 1$, we inductively define

$$\begin{aligned} G^n_m(f) &:= G^n_m(f, \mathcal{D}) := G^n_{m-1}(f) + G^n(R^n_{m-1}(f)); \\ R^n_m(f) &:= R^n_m(f, \mathcal{D}) := f - G^n_m(f) = R^n(R^n_{m-1}(f)). \end{aligned} \tag{2.98}$$

It is clear that a 1-Greedy Algorithm is a Pure Greedy Algorithm.

We prove in this subsection that the $n$-Greedy Algorithm, like the Pure Greedy Algorithm, has a saturation property.

**Theorem 2.61** *For any orthonormal basis $\{\varphi_k\}_{k=1}^{\infty}$, and for any given natural number $n$, there exists an element $g$ such that, for the dictionary $\mathcal{D} = g \cup \{\varphi_k\}_{k=1}^{\infty}$, there is an element $f \in \Sigma_{6n}(\mathcal{D})$ which has the property: For any $0 < \beta \leq 1$,*

$$\|f - G_m^n(f)\|/|f|_{A_\beta(\mathcal{D})} \geq C(\beta)n^{-1/\beta}(m+2)^{-1/2}.$$

*Proof* Let $n \geq 2$ be given. Define

$$g := An^{-1/2}\sum_{k=1}^{2n}\varphi_k + \frac{1}{3}\sum_{k=3n}^{\infty}(k(k+1))^{-1/2}\varphi_k,$$

with

$$A := \left(\frac{1}{2} - \frac{1}{54n}\right)^{1/2} \geq (1/3)^{1/2}.$$

Then

$$\|g\|^2 = 2A^2 + \frac{1}{27n} = 1.$$

Take

$$f := An^{-1/2}\sum_{k=1}^{3n-1}\varphi_k + \frac{2}{3}\sum_{k=3n}^{6n-1}(k(k+1))^{-1/2}\varphi_k.$$

**First step** We prove that for the dictionary $\mathcal{D} = g \cup \{\varphi_k\}_{k=1}^{\infty}$ we have

$$G^n(f, \mathcal{D}) = u := g + An^{-1/2}\sum_{k=2n+1}^{3n-1}\varphi_k.$$

First of all, it is easy to check that $f - u$ is orthogonal to $g$ and $\varphi_k$, $k = 1, \ldots, 3n - 1$, and

$$\|f - u\|^2 = \frac{1}{9}\sum_{k=3n}^{\infty}\frac{1}{k(k+1)} = \frac{1}{27n}.$$

We shall prove that

$$\sigma_n(f, \mathcal{D})^2 \geq \frac{1}{27n}$$

and that the only approximant which provides equality in this estimate is $u$. We consider two cases.

  **(1)** Assume that $g$ is not among the approximating elements. Then for $\Phi = \{\varphi_k\}_{k=1}^{\infty}$ we have

$$\sigma_n(f, \Phi)^2 = \frac{A^2(2n-1)}{n} + \frac{4}{54n} > \frac{1}{27n}.$$

  **(2)** Assume that $g$ is among the approximating elements; then we should estimate

$$\delta := \inf_a \sigma_{n-1}(f - ag, \Phi)^2.$$

Denote

$$g_s := \sum_{k=s}^{\infty} (k(k+1))^{-1/2} \varphi_k.$$

We have

$$f - ag = (1-a)An^{-1/2} \sum_{k=1}^{2n} \varphi_k + An^{-1/2}$$

$$\times \sum_{k=2n+1}^{3n-1} \varphi_k + (2-a)\frac{(g_{3n} - g_{6n})}{3} - \frac{ag_{6n}}{3}.$$

If $|1-a| \geq 1$ then

$$\sigma_{n-1}(f - ag, \Phi)^2 \geq (1-a)^2 A^2 > \frac{1}{27n}.$$

It remains to consider $0 < a < 2$. In this case the $n-1$ largest in absolute value coefficients of $f - ag$ are those of $\varphi_k$, $k = 2n+1, \ldots, 3n-1$. We have

$$\sigma_{n-1}(f - ag, \Phi)^2 = 2(1-a)^2 A^2 + ((2-a)^2 + a^2)/(54n). \qquad (2.99)$$

It is clear that the right-hand side of (2.99) is greater than or equal to $1/(27n)$ for all $a$, and equals $1/(27n)$ only for $a = 1$. This implies that the best $n$-term approximant to $f$ with regards to $\mathcal{D}$ is unique and coincides with $u$. This concludes the first step.

After the first step we get

$$f_1 := R^n(f) = (g_{3n} - 2g_{6n})/3.$$

**General step** We now prove the following lemma.

**Lemma 2.62** *Consider*

$$h_s := \frac{1}{3} \sum_{k=s}^{\infty} e_k (k(k+1))^{-1/2} \varphi_k, \quad e_k = \pm 1, \quad s \geq 3n.$$

*We have*

$$\sigma_n(h_s, \mathcal{D})^2 = 1/(9(s+n)),$$

*and the best n-term approximant with regard to $\mathcal{D}$ is unique and equals*

$$v_n := \frac{1}{3} \sum_{k=s}^{s+n-1} e_k (k(k+1))^{-1/2} \varphi_k.$$

*Proof* It is easy to verify that

$$\|h_s - v_n\|^2 = 1/(9(s+n)),$$

and that $v_n$ is the unique best $n$-term approximant with regard to $\Phi$. We now prove that for each $a$ we have

$$\sigma_{n-1}(h_s - ag, \Phi)^2 > 1/(9(s+n)).$$

We use the representation

$$h_s - ag = -aAn^{-1/2} \sum_{k=1}^{2n} \varphi_k - a/3 \sum_{k=3n}^{s-1} (k(k+1))^{-1/2} \varphi_k$$

$$+ \frac{1}{3} \sum_{k=s}^{\infty} (e_k - a)(k(k+1))^{-1/2} \varphi_k.$$

Let us assume that an $(n-1)$-term approximant to $h_s - ag$ with regard to $\Phi$ consists of $\mu$, $0 \leq \mu \leq n-1$, elements with indices $k \geq s$ and $n-1-\mu$, with indices $k < s$. Then for the error $e(a, \mu)$ of this approximation we get

$$e(a, \mu)^2 \geq \frac{a^2 A^2(n + \mu + 1)}{n} + \frac{a^2(1/(3n) - 1/s)}{9} + \frac{(1 - |a|)^2}{(9(s+\mu))}. \quad (2.100)$$

Taking into account that

$$\inf_a \sigma_{n-1}(h_s - ag, \Phi)^2 = \inf_{0 \leq \mu \leq n-1} \inf_a e(a, \mu)^2,$$

we conclude that we need to prove the corresponding lower estimate for the right-hand side of (2.100) for all $\mu$ and $a$. We have

$$e(a, \mu)^2 \geq \frac{a^2}{3} + \frac{(1 - |a|)^2}{(9(s+\mu))} \geq \frac{a^2}{3} + \frac{(1 - |a|)^2}{(9(s+n-1))}. \quad (2.101)$$

We now use the following simple relation, for $b, c > 0$ we have

$$\inf_a(a^2 b + (1-a)^2 c) = \frac{bc}{b+c} = c(1 + c/b)^{-1}. \quad (2.102)$$

Specifying $b = 1/3$ and $c = 1/(9(s+n-1))$, we get for all $a$ and $\mu$

$$e(a, \mu)^2 \geq (9(s+n) - 6)^{-1} > (9(s+n))^{-1}.$$

Lemma 2.62 is proved. $\qquad\qquad\qquad \square$

Applying Lemma 2.62 to the second step and to the following steps we obtain that

$$R_m^n(f) = \frac{1}{3} \sum_{k=3n+n(m-1)}^{\infty} e_k (k(k+1))^{-1/2} \varphi_k$$

and

$$\|R_m^n(f)\|^2 = (9n(m+2))^{-1}.$$

This relation and the estimate $\|f\| \leq C$ imply (2.97).

In order to complete the proof of Theorem 2.61 it remains to note that

$$|f|_{A_\beta(\mathcal{D})} \leq C(\beta) n^{1/\beta - 1/2}.$$

$\square$

### 2.7.3 Performance of the $n$-Greedy Algorithm with regard to an incoherent dictionary

In this subsection we demonstrate an advantage of the $n$-Greedy Algorithm over the Pure Greedy Algorithm when they both are run with respect to an incoherent dictionary. For a dictionary $\mathcal{D}$ denote by $\mathcal{D}^n$ a new dictionary that consists of elements

$$\left( \sum_{i \in A} c_i g_i \right) \Big/ \Big\| \sum_{i \in A} c_i g_i \Big\|, \quad |A| = n, \quad g_i \in \mathcal{D}, \quad c_i \in \mathbb{R}.$$

It follows from the definition that the $n$-Greedy Algorithm with regard to $\mathcal{D}$ coincides with the Pure Greedy Algorithm with regard to $\mathcal{D}^n$. Therefore, we can apply the theory developed for the PGA to the $n$-Greedy Algorithm. A typical assumption in that theory is $f \in A_1(\mathcal{D})$. We show how this assumption can be used to prove that $f - p_n(f) \in A_1(\mathcal{D}^n, Bn^{-1/2})$ with an absolute constant $B$, provided $\mathcal{D}$ is $M$-coherent and $Mn \leq 1/4$.

**Lemma 2.63** *Let $\mathcal{D}$ be an $M$-coherent dictionary and let $n$ be such that $Mn \leq 1/2$. Assume that $f$ has a representation $f = \sum_{j=1}^{\infty} b_j \varphi_j$, $\varphi_j \in \mathcal{D}$, with coefficients satisfying the inequalities*

$$|b_j| \leq B/n; \qquad \sum_{j=1}^{\infty} |b_j| \leq B,$$

*with some constant $B$. Then $f \in A_1(\mathcal{D}^n, 3Bn^{-1/2})$.*

*Proof* Without loss of generality, we assume that $|b_1| \geq |b_2| \geq \cdots$. Using the notation $J_l := [ln + 1, (l + 1)n]$ we rewrite the representation of $f$ in the following form:

$$f = \sum_{l=0}^{\infty} \left( \left( \sum_{j \in J_l} b_j \varphi_j \right) \Big/ \Big\| \sum_{j \in J_l} b_j \varphi_j \Big\| \right) \Big\| \sum_{j \in J_l} b_j \varphi_j \Big\|.$$

By Lemma 2.57 we obtain

$$\Big\| \sum_{j \in J_l} b_j \varphi_j \Big\| \leq (1 + Mn)^{1/2} \left( \sum_{j \in J_l} b_j^2 \right)^{1/2}.$$

For $l = 0$ we get

$$\sum_{j \in J_0} b_j^2 \leq n(B/n)^2 = B^2 n^{-1}.$$

For $l \geq 1$,

$$\left( \sum_{j \in J_l} b_j^2 \right)^{1/2} \leq n^{1/2} |b_{ln+1}| \leq n^{1/2} \left( n^{-1} \sum_{j \in J_{l-1}} |b_j| \right)$$

and

$$\sum_{l=1}^{\infty} \left( \sum_{j \in J_l} b_j^2 \right)^{1/2} \leq n^{-1/2} \sum_{l=1}^{\infty} \sum_{j \in J_{l-1}} |b_j| \leq n^{-1/2} \sum_{j=1}^{\infty} |b_j| \leq Bn^{-1/2}.$$

This implies that $f \in \mathcal{A}_1(\mathcal{D}^n, cBn^{-1/2})$ with $c = 2(3/2)^{1/2} < 3$. $\qquad \square$

**Lemma 2.64** *Let $\mathcal{D}$ be an $M$-coherent dictionary and let $n$ be such that $Mn \leq 1/4$. Assume that $f$ has a representation*

$$f = \sum_{i=1}^{\infty} c_i g_i, \quad g_i \in \mathcal{D}, \quad |c_1| \geq |c_2| \geq \cdots, \quad \sum_{i=1}^{\infty} |c_i| \leq 1.$$

*Let elements $\psi_1, \ldots, \psi_n \subset \mathcal{D}$ and coefficients $\beta_1, \ldots, \beta_n$ be such that*

$$\sum_{i=1}^{n} \beta_i \psi_i = P_{\mathrm{span}(\psi_1, \ldots, \psi_n)}(f)$$

*and*

$$\Big\| f - \sum_{i=1}^{n} \beta_i \psi_i \Big\|^2 \leq \sigma_n(f)^2 + n^{-2}.$$

*Then there exists a representation*

$$f_1 := f - \sum_{i=1}^{n} \beta_i \psi_i = \sum_{j=1}^{\infty} b_j \varphi_j, \quad \varphi_j \in \mathcal{D},$$

*such that*

$$|b_j| \le B/n; \qquad \sum_{j=1}^{\infty} |b_j| \le B, \tag{2.103}$$

*with $B = 20$.*

*Proof* Consider

$$h := \sum_{i=1}^{n} c_i g_i - \sum_{j=1}^{n} \beta_j \psi_j = \sum_{k=1}^{m} \alpha_k \eta_k, \quad m \le 2n, \quad \eta_k \in \mathcal{D}.$$

Then

$$f_1 = h + \sum_{i=n+1}^{\infty} c_i g_i, \quad h = f_1 - \sum_{i=n+1}^{\infty} c_i g_i.$$

We begin with a proof of the following bound:

$$\| \sum_{i=n+1}^{\infty} c_i g_i \| \le (5/4)^{1/2} n^{-1/2}. \tag{2.104}$$

Using the above notation $J_l := [ln + 1, (l + 1)n]$ we write

$$\| \sum_{i=n+1}^{\infty} c_i g_i \| \le \sum_{l=1}^{\infty} \| \sum_{i \in J_l} c_i g_i \|.$$

By Lemma 2.57 we obtain

$$\| \sum_{i \in J_l} c_i g_i \| \le \left( \sum_{i \in J_l} c_i^2 \right)^{1/2} (1 + Mn)^{1/2} \le (5/4)^{1/2} \left( \sum_{i \in J_l} c_i^2 \right)^{1/2}. \tag{2.105}$$

The monotonicity assumption on the coefficients $\{c_i\}$ from Lemma 2.64 implies

$$\sum_{l=1}^{\infty} \left( \sum_{i \in J_l} c_i^2 \right)^{1/2} \le \sum_{l=1}^{\infty} n^{1/2} |c_{ln+1}|$$

$$\le \sum_{l=1}^{\infty} n^{1/2} \left( n^{-1} \sum_{i \in J_{l-1}} |c_i| \right) \le n^{-1/2} \sum_{i=1}^{\infty} |c_i| \le n^{-1/2}.$$

$$(2.106)$$

The bound (2.104) follows from (2.105) and (2.106).

Using the inequality $\sigma_n(f) \le \|\sum_{i=n+1}^{\infty} c_i g_i\|$, we obtain from (2.104) for $n \ge 2$

$$\|h\| \le \|f_1\| + \|\sum_{i=n+1}^{\infty} c_i g_i\| \le \left( \frac{5}{4n} + \frac{1}{n^2} \right)^{1/2} + \left( \frac{5}{4n} \right)^{1/2} \le n^{-1/2}(7^{1/2}+5^{1/2})/2.$$

By Lemma 2.57,

$$\sum_{k=1}^{m} \alpha_k^2 \le (1 - 2Mn)^{-1} \|h\|^2$$

and

$$\sum_{k=1}^{m} |\alpha_k| \le m^{1/2} \left( \sum_{k=1}^{m} \alpha_k^2 \right)^{1/2} \le 5. \qquad (2.107)$$

We prove that the representation

$$f_1 = \sum_{k=1}^{m} \alpha_k \eta_k + \sum_{i=n+1}^{\infty} c_i g_i =: \sum_{j=1}^{\infty} b_j \varphi_j$$

satisfies the inequality (2.103) with $B = 20$. We proceed by contradiction. First, we note that

$$\sum_{j=1}^{\infty} |b_j| \le \sum_{k=1}^{m} |\alpha_k| + \sum_{i=n+1}^{\infty} |c_i| \le 6. \qquad (2.108)$$

Second, we assume that (2.103) is not satisfied. Then, taking into account (2.108) and the inequality $|c_{n+1}| < 1/n$, we conclude that there is $v \in [1, m]$ such that $|\alpha_v| > 19/n$. The inequality (2.107) implies

$$\sum_{j=1}^{n} |\beta_j| \le \sum_{i=1}^{n} |c_i| + \sum_{k=1}^{m} |\alpha_k| \le 6.$$

Assuming, without loss of generality, that $|\beta_1| \geq |\beta_2| \geq \cdots \geq |\beta_n|$ we get that $|\beta_n| \leq 6/n$. We now prove that replacing $\beta_n \psi_n$ by $\alpha_\nu \eta_\nu$ in the approximant provides an error better than the best-possible one $\sigma_n(f)$. This contradiction proves the lemma. Using the fact that $\psi_n$ is orthogonal to $f_1$ we obtain

$$\|f_1 + \beta_n \psi_n\|^2 = \|f_1\|^2 + \beta_n^2 \leq \|f_1\|^2 + 36n^{-2}. \tag{2.109}$$

Further,

$$\|f_1 + \beta_n \psi_n - \alpha_\nu \eta_\nu\|^2 = \|f_1 + \beta_n \psi_n\|^2 - 2\alpha_\nu \langle f_1 + \beta_n \psi_n, \eta_\nu \rangle + \alpha_\nu^2. \tag{2.110}$$

We have

$$\langle f_1 + \beta_n \psi_n, \eta_\nu \rangle = \alpha_\nu + \sum_{k \neq \nu} \alpha_k \langle \eta_k, \eta_\nu \rangle + \sum_{i=n+1}^{\infty} c_i \langle g_i, \eta_\nu \rangle + \beta_n \langle \psi_n, \eta_\nu \rangle.$$

The dictionary element $\eta_\nu$ is distinct from all $\eta_k$, $k \neq \nu$, and may coincide with at most one $g_i$. This implies

$$\left| \sum_{k \neq \nu} \alpha_k \langle \eta_k, \eta_\nu \rangle + \sum_{i=n+1}^{\infty} c_i \langle g_i, \eta_\nu \rangle + \beta_n \langle \psi_n, \eta_\nu \rangle \right|$$

$$\leq M \sum_{k \neq \nu} |\alpha_k| + n^{-1} + M \sum_{i=1}^{\infty} |c_i| + |\beta_n| \leq 8.5/n. \tag{2.111}$$

Using (2.109) and (2.111) we get from (2.110) that

$$\|f_1 + \beta_n \psi_n - \alpha_\nu \eta_\nu\|^2 \leq \|f_1\|^2 - 2n^{-2} < \sigma_n(f)^2.$$

$\square$

**Lemma 2.65** *Let $\mathcal{D}$ be an $M$-coherent dictionary and let $n$ be such that $Mn \leq 1/4$. Assume that $f \in A_1(\mathcal{D})$. Let elements $\psi_1, \ldots, \psi_n \subset \mathcal{D}$ and coefficients $\beta_1, \ldots, \beta_n$ be such that*

$$\sum_{i=1}^{n} \beta_i \psi_i = P_{\mathrm{span}(\psi_1,\ldots,\psi_n)}(f)$$

*and*

$$\left\| f - \sum_{i=1}^{n} \beta_i \psi_i \right\|^2 \leq \sigma_n(f)^2 + n^{-2}/2.$$

*Then*

$$f_1 := f - \sum_{i=1}^{n} \beta_i \psi_i \in \mathcal{A}_1(\mathcal{D}^n, 60n^{-1/2}).$$

*Proof* Take arbitrary $\epsilon > 0$. Let $f^\epsilon$ be such that $\| f - f^\epsilon \| \le \epsilon$ and

$$f^\epsilon = \sum_{i=1}^{\infty} c_i g_i, \quad g_i \in \mathcal{D}, \quad |c_1| \ge |c_2| \ge \cdots, \quad \sum_{i=1}^{\infty} |c_i| \le 1.$$

Denote

$$H_n(f) := \text{span}(\psi_1, \dots, \psi_n) \quad \text{and} \quad f_1^\epsilon := f^\epsilon - P_{H_n(f)}(f^\epsilon).$$

Then

$$\| f_1 - f_1^\epsilon \| \le \| f - f^\epsilon \| + \| P_{H_n(f)}(f - f^\epsilon) \| \le 2\epsilon. \tag{2.112}$$

From this and the simple inequality $\sigma_n(f) \le \sigma_n(f^\epsilon) + \epsilon$ we obtain

$$\| f_1^\epsilon \|^2 \le \| f_1 \|^2 + 4\epsilon \| f_1 \| + 4\epsilon^2 \le \sigma_n(f)^2 + n^{-2}/2 + 4\epsilon \| f_1 \| + 4\epsilon^2 \le \sigma_n(f^\epsilon)^2 + n^{-2}$$

for sufficiently small $\epsilon$. By Lemma 2.64 applied to $f^\epsilon$ we see that there exists a representation

$$f_1^\epsilon = \sum_{j=1}^{\infty} b_j \varphi_j, \quad \varphi_j \in \mathcal{D},$$

satisfying (2.103) with $B = 20$. By Lemma 2.63 $f_1^\epsilon \in \mathcal{A}_1(\mathcal{D}^n, 60n^{-1/2})$. This inclusion and (2.112) imply that $f_1 \in \mathcal{A}_1(\mathcal{D}^n, 60n^{-1/2})$. $\quad\square$

**Theorem 2.66** *Let $\mathcal{D}$ be an $M$-coherent dictionary and let $n$ be such that $Mn \le 1/4$. Assume that $f \in A_1(\mathcal{D})$. Then*

$$\| f - G_m^n(f, \mathcal{D}) \| \le 60n^{-1/2} \gamma(m - 1).$$

*Proof* Lemma 2.65 implies that $f_1 := f - p_n(f) \in \mathcal{A}_1(\mathcal{D}^n, 60n^{-1/2})$. This inclusion and the following simple chain of equalities give the proof:

$$f - G_m^n(f, \mathcal{D}) = f_1 - G_{m-1}^n(f_1, \mathcal{D}) = f_1 - G_{m-1}(f_1, \mathcal{D}^n).$$

$\quad\square$

In particular, Theorem 2.66 yields the bound

$$\| f - G_m^n(f, \mathcal{D}) \| \le 60n^{-1/2}(m - 1)^{-1/6}, \quad f \in A_1(\mathcal{D}).$$

We note that $G_m^n(f, \mathcal{D})$ is an $nm$-term approximant. For comparison, the $nm$-term approximant $G_{nm}(f, \mathcal{D})$ provides the error bound

$$\| f - G_{nm}(f, \mathcal{D}) \| \le (nm)^{-1/6}.$$

It is known that the exponent $1/6$ can be improved in the rate of convergence of the PGA. However, it cannot be made close to $1/2$. Recent results show that it is smaller than $0.189$.

## 2.8 Some further remarks

We demonstrated in this chapter that the general theory of greedy approximation with regard to an arbitrary dictionary is well developed. Much less is known about how specific features of a dictionary can be used to our advantage – either to improve rate of convergence results for known algorithms, or to build more efficient algorithms with the same rate of convergence as known general algorithms. A specific feature of a dictionary, $M$-coherence in our case, allows us to build a more efficient greedy algorithm (the Orthogonal Super Greedy Algorithm) than a known algorithm (the Orthogonal Greedy Algorithm) with the same rate of convergence. We study the rate of convergence of greedy algorithms for elements of the closure of the convex hull of the symmetrized dictionary, which is standard in the theory of greedy approximation setting. The presentation in this section is based on Liu and Temlyakov (2010).

Denote as above by $A_1(\mathcal{D})$ the closure of the convex hull of the symmetrized dictionary $\mathcal{D}^{\pm}$. The following theorem is from Temlyakov (2000b) (see Theorem 2.20 in Section 2.3).

**Theorem 2.67** *Let $\mathcal{D}$ be an arbitrary dictionary in $H$. Then for each $f \in A_1(\mathcal{D})$ we have*

$$\|f - G_m^{o,\tau}(f, \mathcal{D})\| \le \left(1 + \sum_{k=1}^{m} t_k^2\right)^{-1/2}. \tag{2.113}$$

We note that in the particular case $t_k = t$, $k = 1, 2, \ldots$, the right-hand side takes the form $(1 + mt^2)^{-1/2}$, which is equal to $(1 + m)^{-1/2}$ for $t = 1$.

We now introduce a new algorithm. Let a natural number $s$ and a sequence $\tau := \{t_k\}_{k=1}^{\infty}$, $t_k \in [0, 1]$, be given. Consider the following Weak Orthogonal Super Greedy Algorithm with parameter $s$.

**WOSGA($s, \tau$)** Initially, $f_0 := f$. Then, for each $m \ge 1$ we inductively define the following.

(1) $\varphi_{(m-1)s+1}, \ldots, \varphi_{ms} \in \mathcal{D}$ are elements of the dictionary $\mathcal{D}$ satisfying the following inequality. Denote $I_m := [(m-1)s+1, ms]$ and assume that

$$\min_{i \in I_m} |\langle f_{m-1}, \varphi_i \rangle| \ge t_m \sup_{g \in \mathcal{D}, g \neq \varphi_i, i \in I_m} |\langle f_{m-1}, g \rangle|.$$

(2) Let $H_m := H_m(f) := \text{span}(\varphi_1, \ldots, \varphi_{ms})$ and let $P_{H_m}$ denote an operator of orthogonal projection onto $H_m$. Define

$$G_m(f) := G_m(f, \mathcal{D}) := G_m^s(f, \mathcal{D}) := P_{H_m}(f).$$

(3) Define the residual after the $m$th iteration of the algorithm as

$$f_m := f_m^s := f - G_m(f, \mathcal{D}).$$

In this section we study the rate of convergence of the WOSGA($s, \tau$) in the case $t_k = t$, $k = 1, 2, \ldots$; in this case we write $t$ instead of $\tau$ in the notations. We assume that the dictionary $\mathcal{D}$ is $M$-coherent and that $f \in A_1(\mathcal{D})$. We begin with the case $t = 1$. In this case we impose an additional assumption, that the $\varphi_i$ from the first step exist. Clearly, it is the case if $\mathcal{D}$ is finite. We call the algorithm WOSGA($s, 1$) the Orthogonal Super Greedy Algorithm with parameter $s$ (OSGA($s$)). We prove the following error bound for the OSGA($s$).

**Theorem 2.68** *Let $\mathcal{D}$ be a dictionary with coherence parameter $M :=$ $M(\mathcal{D})$. Then, for $s \leq (2M)^{-1}$ the OSGA($s$) provides, after $m$ iterations, an approximation of $f \in A_1(\mathcal{D})$ with the following upper bound on the error:*

$$\|f_m\|^2 \leq 40.5(sm)^{-1}, \quad m = 1, 2, \ldots$$

We note that the OSGA($s$) adds $s$ new elements of the dictionary at each iteration and makes one orthogonal projection at each iteration. For comparison, the OGA adds one new element of the dictionary at each iteration and makes one orthogonal projection at each iteration. After $m$ iterations of the OSGA($s$) and after $ms$ iterations of the OGA, both algorithms provide $ms$-term approximants with a guaranteed error bound for $f \in A_1(\mathcal{D})$ of the same order: $O((ms)^{-1/2})$. Both algorithms use the same number, $ms$, of elements of the dictionary. However, the OSGA($s$) makes $m$ iterations and the OGA makes $ms$ ($s$ times more) iterations. Thus, in the sense of number of iterations the OSGA($s$) is $s$ times simpler (more efficient) than the OGA. We gain this simplicity of the OSGA($s$) under an extra assumption of $\mathcal{D}$ being $M$-coherent, and $s \leq (2M)^{-1}$. Therefore, if our dictionary $\mathcal{D}$ is $M$-coherent, then the OSGA($s$) with small enough $s$ approximates with an error whose guaranteed upper bound for $f \in A_1(\mathcal{D})$ is of the same order as that for the OGA.

*Proof of Theorem 2.68.* Denote

$$F_m := \text{span}(\varphi_i, i \in I_m).$$

Then $H_m$ is a direct sum of $H_{m-1}$ and $F_m$. Therefore,

$$f_m = f - P_{H_m}(f) = f_{m-1} + G_{m-1}(f) - P_{H_m}(f_{m-1} + G_{m-1}(f))$$
$$= f_{m-1} - P_{H_m}(f_{m-1}).$$

It is clear that the inclusion $F_m \subset H_m$ implies

$$\|f_m\| \leq \|f_{m-1} - P_{F_m}(f_{m-1})\|. \tag{2.114}$$

Using the notation $p_m := P_{F_m}(f_{m-1})$, we continue

$$\|f_{m-1}\|^2 = \|f_{m-1} - p_m\|^2 + \|p_m\|^2,$$

and, by (2.114),

$$\|f_m\|^2 \le \|f_{m-1}\|^2 - \|p_m\|^2. \tag{2.115}$$

To estimate $\|p_m\|^2$ from below for $f \in A_1(\mathcal{D})$, we first make some auxiliary observations. Let

$$f = \sum_{j=1}^{\infty} c_j g_j, \quad g_j \in \mathcal{D}, \quad \sum_{j=1}^{\infty} |c_j| \le 1, \quad |c_1| \ge |c_2| \ge \cdots \tag{2.116}$$

Every element of $A_1(\mathcal{D})$ can be approximated arbitrarily well by elements of the form (2.116). It will be clear from the following argument that it is sufficient to consider elements $f$ of the form (2.116). Suppose $\nu$ is such that $|c_\nu| \ge 2/s \ge |c_{\nu+1}|$. Then the above assumption on the sequence $\{c_j\}$ implies that $\nu \le s/2$ and $|c_{s+1}| < 1/s$. We claim that elements $g_1, \ldots, g_\nu$ will be chosen among $\varphi_1, \ldots, \varphi_s$ at the first iteration. Indeed, for $j \in [1, \nu]$ we have

$$|\langle f, g_j \rangle| \ge |c_j| - M \sum_{k \ne j}^{\infty} |c_k| \ge 2/s - M(1 - 2/s) > 2/s - M.$$

For all $g$ distinct from $g_1, \ldots, g_s$ we have

$$|\langle f, g \rangle| \le M + 1/s.$$

Our assumption $s \le 1/(2M)$ implies that $M + 1/s \le 2/s - M$. Thus, we do not pick any of $g \in \mathcal{D}$ distinct from $g_1, \ldots, g_s$ until we have chosen all $g_1, \ldots, g_\nu$.

Denote

$$f' := f - \sum_{j=1}^{\nu} c_j g_j = \sum_{j=\nu+1}^{\infty} c_j g_j.$$

It is clear from the above argument that

$$f_1 = f - P_{H_1(f)}(f) = f' - P_{H_1(f)}(f');$$

$$f_m = f - P_{H_m(f)}(f) = f' - P_{H_m(f)}(f').$$

We now estimate $\|p_m\|^2$. For $f_{m-1}$ consider the following quantity

$$q_s := q_s(f_{m-1}) := \sup_{\substack{h_i \in \mathcal{D} \\ i \in [1,s]}} \|P_{H(s)}(f_{m-1})\|,$$

where $H(s) := \text{span}(h_1, \ldots, h_s)$. Then

$$\|P_{H(s)}(f_{m-1})\| = \max_{\psi \in H(s), \|\psi\| \leq 1} |\langle f_{m-1}, \psi \rangle|.$$

Let $\psi = \sum_{i=1}^{s} a_i h_i$. Then by Lemma 2.57 we bound

$$(1 - Ms) \sum_{i=1}^{s} a_i^2 \leq \|\psi\|^2 \leq (1 + Ms) \sum_{i=1}^{s} a_i^2. \tag{2.117}$$

Therefore,

$$(1+Ms)^{-1} \sum_{i=1}^{s} \langle f_{m-1}, h_i \rangle^2 \leq \|P_{H(s)}(f_{m-1})\|^2 \leq (1-Ms)^{-1} \sum_{i=1}^{s} \langle f_{m-1}, h_i \rangle^2,$$

and thus

$$\|p_m\|^2 \geq \frac{1 - Ms}{1 + Ms} q_s^2. \tag{2.118}$$

Using the notation $J_l := [(l-1)s + v + 1, ls + v]$ we write for $m \geq 2$

$$\|f_{m-1}\|^2 = \langle f_{m-1}, f' \rangle = \sum_{l=1}^{\infty} \langle f_{m-1}, \sum_{j \in J_l} c_j g_j \rangle$$

$$\leq q_s (1 + Ms)^{1/2} \sum_{l=1}^{\infty} \left( \sum_{j \in J_l} c_j^2 \right)^{1/2}. \tag{2.119}$$

Since the sequence $\{c_j\}$ has the property

$$|c_{v+1}| \geq |c_{v+2}| \geq \cdots, \quad \sum_{j=v+1}^{\infty} |c_j| \leq 1, \quad |c_{v+1}| \leq 2/s, \tag{2.120}$$

we may apply the simple inequality

$$\left( \sum_{j \in J_l} c_j^2 \right)^{1/2} \leq s^{1/2} |c_{(l-1)s+v+1}|,$$

so that we bound the sum on the right-hand side of (2.119):

$$\sum_{l=1}^{\infty} \left( \sum_{j \in J_l} c_j^2 \right)^{1/2} \leq s^{1/2} \sum_{l=1}^{\infty} |c_{(l-1)s+v+1}|$$

$$\leq s^{1/2} \left( 2/s + \sum_{l=2}^{\infty} s^{-1} \sum_{j \in J_{l-1}} |c_j| \right) \leq 3s^{-1/2}. \tag{2.121}$$

Inequalities (2.119) and (2.121) imply

$$q_s \geq (s^{1/2}/3)(1 + Ms)^{-1/2} \| f_{m-1} \|^2.$$

By (2.118) we have

$$\| p_m \|^2 \geq \frac{s(1 - Ms)}{9(1 + Ms)^2} \| f_{m-1} \|^4. \tag{2.122}$$

Our assumption $Ms \leq 1/2$ implies

$$\frac{1 - Ms}{(1 + Ms)^2} \geq 2/9,$$

and therefore (2.122) gives

$$\| p_m \|^2 \geq (s/A) \| f_{m-1} \|^4, \quad A := 40.5.$$

Thus, by (2.115) we get

$$\| f_m \|^2 \leq \| f_{m-1} \|^2 (1 - (s/A) \| f_{m-1} \|^2). \tag{2.123}$$

Using (2.120) we get for $\| f' \|$

$$\| f' \| \leq \sum_{l=1}^{\infty} \| \sum_{j \in J_l} c_j g_j \| \leq \sum_{l=1}^{\infty} (1 + Ms)^{1/2} \left( \sum_{j \in J_l} c_j^2 \right)^{1/2} \leq (1 + Ms)^{1/2} 3 s^{-1/2}$$

and

$$\| f_1 \|^2 \leq \| f' \|^2 \leq 27/(2s) \leq A/s.$$

By Lemma 2.16 with $a_m := \| f_m \|^2$ we obtain

$$\| f_m \|^2 \leq A(sm)^{-1}, \quad m = 1, 2, \ldots$$

This completes the proof of Theorem 2.68. $\qquad\square$

We now proceed to the case of the WOSGA$(s, t)$ with $t \in (0, 1)$.

**Theorem 2.69** *Let $\mathcal{D}$ be a dictionary with coherence parameter $M := M(\mathcal{D})$. Then, for $s \leq (2M)^{-1}$, the WOSGA$(s, t)$ provides, after $m$ iterations, an approximation of $f \in A_1(\mathcal{D})$ with the following upper bound on the error:*

$$\| f_m \|^2 \leq A(t)(sm)^{-1}, \quad m = 1, 2, \ldots, \quad A(t) := (81/8)(1 + t)^2 t^{-4}.$$

*Proof* Proof of this theorem mimics the proof of Theorem 2.68, except that in the auxiliary observations we choose a threshold $B/s$ with $B := (3 + t)/(2t)$ instead of $2/s$: $|c_\nu| \geq B/s \geq |c_{\nu+1}|$, so that our assumption $Ms \leq 1/2$ implies that $M + 1/s \leq t(B/s - M)$. This, in turn, implies that all $g_1, \ldots, g_\nu$ will be chosen at the first iteration. As a result, the sequence $\{c_j\}$ satisfies the following conditions:

$$|c_{v+1}| \geq |c_{v+2}| \geq \cdots, \qquad \sum_{j=v+1}^{\infty} |c_j| \leq 1, \qquad |c_{v+1}| \leq B/s. \qquad (2.124)$$

To find an analog of inequality (2.118), we begin with the fact that

$$q_s^2 \leq \sup_{\substack{h_i \in \mathcal{D} \\ i \in [1,s]}} (1 - Ms)^{-1} \sum_{i=1}^{s} \langle f_{m-1}, h_i \rangle^2.$$

Now, in order to relate $q_s^2$ to $\|p_m\|^2$, consider an arbitrary set $\{h_i\}_{i=1}^{s}$ of distinct elements of the dictionary $\mathcal{D}$. Let $V$ be a set of all indices $i \in [1, s]$ such that $h_i = \varphi_{k(i)}, k(i) \in I_m$. Denote $V' := \{k(i), i \in V\}$. Then

$$\sum_{i=1}^{s} \langle f_{m-1}, h_i \rangle^2 = \sum_{i \in V} \langle f_{m-1}, h_i \rangle^2 + \sum_{i \in [1,s]\backslash V} \langle f_{m-1}, h_i \rangle^2. \qquad (2.125)$$

From the definition of $\{\varphi_k\}_{k \in I_m}$ we get

$$\max_{i \in [1,s]\backslash V} |\langle f_{m-1}, h_i \rangle| \leq t^{-1} \min_{k \in I_m \backslash V'} |\langle f_{m-1}, \varphi_k \rangle|. \qquad (2.126)$$

Using (2.126) we continue (2.125) as follows:

$$\sum_{i=1}^{s} \langle f_{m-1}, h_i \rangle^2 \leq \sum_{k \in V'} \langle f_{m-1}, \varphi_k \rangle^2 + t^{-2} \sum_{k \in I_m \backslash V'} \langle f_{m-1}, \varphi_k \rangle^2 \leq t^{-2} \sum_{k \in I_m} \langle f_{m-1}, \varphi_k \rangle^2.$$

Therefore,

$$q_s^2 \leq (1 - Ms)^{-1} t^{-2} \sum_{k \in I_m} \langle f_{m-1}, \varphi_k \rangle^2 \leq \frac{1 + Ms}{t^2(1 - Ms)} \|p_m\|^2.$$

This results in the following analog of (2.118):

$$\|p_m\|^2 \geq \frac{t^2(1 - Ms)}{1 + Ms} q_s^2. \qquad (2.127)$$

The use of (2.124) instead of (2.120) gives us the following version of (2.121):

$$\sum_{l=1}^{\infty} \left( \sum_{j \in J_l} c_j^2 \right)^{1/2} \leq (B + 1)s^{-1/2}.$$

The rest of the proof repeats the corresponding part of the proof of Theorem 2.68 with $A := (9(B + 1)^2)/2t^2 = (81/8)(1 + t)^2 t^{-4}$. $\qquad \square$

## 2.9 Open problems

We have already formulated some open problems on greedy approximation in Hilbert spaces. Here we add some more and repeat the problems mentioned above.

**2.1.** Find the order of decay of the sequence

$$\gamma(m) := \sup_{f,\mathcal{D},\{G_m\}} (\|f - G_m(f,\mathcal{D})\|),$$

where the supremum is taken over all dictionaries $\mathcal{D}$, all elements $f \in A_1(\mathcal{D})$ and all possible realizations of $\{G_m\}$.

**2.2.** Let $RB(C_1, C_2)$ denote the collection of all Riesz bases of $H$ with positive constants $C_1, C_2$. A Riesz basis $\mathcal{B}$ with constants $C_1, C_2$ is defined as follows. For any finite number of distinct elements $g_j \in \mathcal{B}$ and any coefficients $c_j$ we have

$$C_1 \left( \sum_j |c_j|^2 \right)^{1/2} \le \|\sum_j c_j g_j\| \le C_2 \left( \sum_j |c_j|^2 \right)^{1/2}.$$

Find the order of decay of the sequence

$$\gamma(m, RB(C_1, C_2)) := \sup_{f,\mathcal{B},\{G_m\}} (\|f - G_m(f,\mathcal{B})\|),$$

where the supremum is taken over all Riesz bases $\mathcal{B} \in RB(C_1, C_2)$, all elements $f \in A_1(\mathcal{B})$ and all possible realizations of $\{G_m\}$.

*Comment* For a normalized dictionary that we consider here we have $C_1 \le 1 \le C_2$. Using the lower bound from the definition of the Riesz basis,

$$C_1 \left( \sum_j |c_j|^2 \right)^{1/2} \le \|\sum_j c_j g_j\|,$$

we deduce that by Definition 2.39 the Riesz basis is $(1, \gamma)$-semistable with $\gamma^{-1/2} = C_1$. Thus by Proposition 2.40 it is $\gamma$-quasi-orthogonal. Therefore, by Theorem 2.33 for $r < (2\gamma)^{-1}$ we have, for $f \in A_1(\mathcal{D})$,

$$\|f - G_m(f,\mathcal{D})\| \le C(r, \gamma) m^{-r}.$$

For example, for $r = 0.2$ it is better than the rate of convergence of the PGA with respect to a general dictionary. We note that we can specify $r = 0.2$ if $C_1 > (0.4)^{1/2}$.

**2.3.** Find the order of decay of the sequence

$$\gamma(m, \lambda) := \sup_{f, \mathcal{D}, \{G_m\}} (\|f - G_m(f, \mathcal{D})\|),$$

where the supremum is taken over all $\lambda$-quasi-orthogonal dictionaries, all elements $f \in A_1(\mathcal{D})$ and all possible realizations of $\{G_m\}$.

**2.4.** Let $\mathcal{R}$ denote the system of ridge functions, i.e. functions $G(x), x \in \mathbb{R}^2$, which can be represented in the form

$$G(x) = g((x, e)),$$

where $g$ is a univariate function and its argument $(x, e)$ is the scalar product of $x$ and a unit vector $e \in \mathbb{R}^2$. Let $D := \{(x_1, x_2) : x_1^2 + x_2^2 \le 1\}$ be the unit disk and let $L_2(D)$ denote the Hilbert space with the norm

$$\|f\|_2 := \|f\|_{L_2(D)} := \left(\frac{1}{\pi} \int_D |f(x)|^2 \, dx\right)^{1/2}.$$

We denote by $\mathcal{R}_2$ the dictionary for $L_2(D)$ which consists of elements of the system $\mathcal{R}$ normalized in $L_2(D)$.

Find the order of decay of the sequence

$$\gamma(m, \mathcal{R}_2) := \sup_{f, \{G_m\}} (\|f - G_m(f, \mathcal{R}_2)\|),$$

where the supremum is taken over all elements $f \in A_1(\mathcal{R}_2)$ and all possible realizations of $\{G_m\}$.

**2.5.** Find the necessary and sufficient conditions on a weakness sequence $\tau$ to guarantee convergence of the Weak Greedy Algorithm with regard to $\mathcal{R}_2$ for each $f \in L_2(D)$.

**2.6.** Let $\Pi_2$ denote the system of functions (bilinear system) of the form $u(x_1)v(x_2) \in L_2([0, 1]^2)$ normalized in $L_2$. Find the necessary and sufficient conditions on a weakness sequence $\tau$ to guarantee convergence of the Weak Greedy Algorithm with regard to $\Pi_2$ for each $f \in L_2([0, 1]^2)$.

**2.7.** Let $0 < r \le 1/2$ be given. Characterize dictionaries $\mathcal{D}$ which possess the property that for any $f \in H$ such that

$$\sigma_m(f, \mathcal{D}) \le m^{-r}, \quad m = 1, 2, \ldots,$$

we have

$$\|f - G_m(f, \mathcal{D})\| \le C(r, \mathcal{D})m^{-r}, \quad m = 1, 2, \ldots$$

# 3

# Entropy

## 3.1 Introduction: definitions and some simple properties

The concept of entropy is also known as *Kolmogorov entropy* and *metric entropy*. This concept allows us to measure the size of a compact set. In the case of finite dimensional compacts it is convenient to compare compact sets by their volumes. In the case of infinite dimensional Banach spaces this method does not work. The concept of entropy is a good replacement for the concept of volume in infinite dimensional Banach spaces. We present some classical basic results and in Section 3.6 give a discussion of a difficult problem that is still unresolved in many important cases.

Let $X$ be a Banach space and let $B_X$ denote the unit ball of $X$ with the center at 0. Denote by $B_X(y, r)$ a ball with center $y$ and radius $r$: $\{x \in X : \|x - y\| \le r\}$. For a compact set $A$ and a positive number $\epsilon$ we define the covering number $N_\epsilon(A)$ as follows:

$$N_\epsilon(A) := N_\epsilon(A, X) := N_\epsilon^1(A, X)$$
$$:= \min\{n : \exists y^1, \ldots, y^n : A \subseteq \cup_{j=1}^n B_X(y^j, \epsilon)\}.$$

Let us list three obvious properties of covering numbers:

$$N_\epsilon(A) \le N_\epsilon(B), \quad \text{provided} \quad A \subseteq B; \tag{3.1}$$
$$N_{\epsilon_1 \epsilon_2}(A, X) \le N_{\epsilon_1}(A, X) N_{\epsilon_2}(B_X, X); \tag{3.2}$$

then let $C = A \oplus B := \{c : c = a + b, a \in A, b \in B\}$, so

$$N_{\epsilon_1 + \epsilon_2}(C) \le N_{\epsilon_1}(A) N_{\epsilon_2}(B). \tag{3.3}$$

For a compact $A$ we define an $\epsilon$-distinguishable set $\{x^1, \ldots, x^m\} \subseteq A$ as a set with the property

$$\|x^i - x^j\| > \epsilon, \quad \text{for all} \quad i, j : i \ne j. \tag{3.4}$$

143

Denote by $M_\epsilon(A) := M_\epsilon(A, X)$ the maximal cardinality of $\epsilon$-distinguishable sets of a compact $A$.

**Theorem 3.1** *For any compact set A we have*

$$M_{2\epsilon}(A) \leq N_\epsilon(A) \leq M_\epsilon(A). \tag{3.5}$$

*Proof* We first prove the second inequality. Let an $\epsilon$-distinguishable set $F$ realise $M_\epsilon(A)$, i.e.

$$F = \{x^1, \ldots, x^{M_\epsilon(A)}\}.$$

By the definition of $M_\epsilon(A)$ as the maximal cardinality of $\epsilon$-distinguishable sets of a compact $A$, we get for any $x \in A$ an index $j := j(x) \in [1, M_\epsilon(A)]$ such that $\|x - x^j\| \leq \epsilon$. Thus we have

$$A \subseteq \cup_{j=1}^{M_\epsilon(A)} B_X(x^j, \epsilon)$$

and the inequality $N_\epsilon(A) \leq M_\epsilon(A)$ follows.

A proof by contradiction gives the first inequality in (3.5). Let $\{y^1, \ldots, y^{N_\epsilon(A)}\}$ be a set such that

$$A \subseteq \cup_{j=1}^{N_\epsilon(A)} B_X(y^j, \epsilon).$$

Assume $M_{2\epsilon}(A) > N_\epsilon(A)$. Then the corresponding $2\epsilon$-distinguishable set $F$ contains two points that are in the same ball $B_X(y^j, \epsilon)$, for some $j \in [1, N_\epsilon(A)]$. This clearly causes a contradiction. $\square$

**Corollary 3.2** *Let $A \subset Y$, and let $Y$ be a subspace of $X$. Then*

$$N_\epsilon(A, X) \geq N_{2\epsilon}(A, Y).$$

Indeed, by Theorem 3.1 we have

$$N_{2\epsilon}(A, Y) \leq M_{2\epsilon}(A, Y) = M_{2\epsilon}(A, X) \leq N_\epsilon(A, X).$$

It is convenient to consider along with the entropy $H_\epsilon(A, X) := \log_2 N_\epsilon(A, X)$ the entropy numbers $\epsilon_k(A, X)$:

$$\epsilon_k(A, X) := \epsilon_k^1(A, X) := \inf\{\epsilon : \exists y^1, \ldots, y^{2^k} \in X : A \subseteq \cup_{j=1}^{2^k} B_X(y^j, \epsilon)\}.$$

## 3.2 Finite dimensional spaces

Let us consider the space $\mathbb{R}^n$ equipped with different norms, say norms $\| \cdot \|_X$ and $\| \cdot \|_Y$. For a Lebesgue measurable set $E \in \mathbb{R}^n$ we denote its Lebesgue measure by $vol(E) := vol_n(E)$.

**Theorem 3.3** *For any two norms X and Y and any $\epsilon > 0$ we have*

$$\frac{1}{\epsilon^n} \frac{vol(B_Y)}{vol(B_X)} \le N_\epsilon(B_Y, X) \le \frac{vol(B_Y(0, 2/\epsilon) \oplus B_X)}{vol(B_X)}. \tag{3.6}$$

*Proof* We begin with the first inequality in (3.6). We have

$$B_Y \subseteq \cup_{j=1}^{N_\epsilon(B_Y,X)} B_X(y^j, \epsilon)$$

and, therefore,

$$vol(B_Y) \le \sum_{j=1}^{N_\epsilon(B_Y,X)} vol(B_X(y^j, \epsilon)) \le N_\epsilon(B_Y, X)\epsilon^n vol(B_X).$$

This gives the required inequality.

We proceed to the second inequality in (3.6). Let $\{x^1, \ldots, x^{M_\epsilon}\}$, $M_\epsilon := M_\epsilon(A)$, be an $\epsilon$-distinguishable set of $B_Y$. Consider the set

$$C := \cup_{j=1}^{M_\epsilon} B_X(x^j, \epsilon/2).$$

Note that the balls $B_X(x^j, \epsilon/2)$ are disjoint. Then

$$C \subseteq B_Y \oplus B_X(0, \epsilon/2) \quad \text{and} \quad M_\epsilon(\epsilon/2)^n vol(B_X) \le vol(B_Y \oplus B_X(0, \epsilon/2)),$$

and the second inequality in (3.6) with $N_\epsilon$ replaced by $M_\epsilon$ follows. It remains to use Theorem 3.1. $\qquad \square$

Let us formulate one immidiate corollary of Theorem 3.3.

**Corollary 3.4** *For any $n$-dimensional Banach space X we have*

$$\epsilon^{-n} \le N_\epsilon(B_X, X) \le (1 + 2/\epsilon)^n,$$

*and, therefore,*

$$\epsilon_k(B_X, X) \le 3(2^{-k/n}).$$

Let us consider some typical $n$-dimensional Banach spaces. These are the spaces $\ell_p^n$: the linear space $\mathbb{R}^n$ equipped with the norms

$$\|x\|_p := \|x\|_{\ell_p^n} := \left( \sum_{j=1}^n |x_j|^p \right)^{1/p}, \quad 1 \le p < \infty,$$

$$\|x\|_\infty := \|x\|_{\ell_\infty^n} := \max_j |x_j|.$$

Denote $B_p^n := B_{\ell_p^n}$. It is obvious that

$$vol(B_\infty^n) = 2^n. \tag{3.7}$$

It is also not difficult to see that

$$vol(B_1^n) = 2^n/n! \tag{3.8}$$

Indeed, consider for $r > 0$

$$O_n(r) := \{x \in \mathbb{R}^n : x_j \geq 0, j = 1, \ldots, n, \sum_{j=1}^{n} x_j \leq r\}, \quad O_n := O_n(1).$$

Then $vol(B_1^n) = 2^n vol(O_n)$ and

$$vol(O_n) = \int_0^1 vol_{n-1}(O_{n-1}(1-t))dt$$

$$= vol_{n-1}(O_{n-1}) \int_0^1 (1-t)^{n-1} dt = vol_{n-1}(O_{n-1})/n.$$

Taking into account that $vol_1(O_1) = 1$, we obtain $vol_n(O_n) = 1/n!$

Let us proceed to the Euclidean case $p = 2$. We will prove the following estimates: there exist two positive absolute constants $C_1$ and $C_2$ such that

$$C_1^n n^{-n/2} \leq vol(B_2^n) \leq C_2^n n^{-n/2}. \tag{3.9}$$

We have

$$vol_n(B_2^n) = \int_{-1}^1 vol_{n-1}(B_2^{n-1}(0, (1-t^2)^{1/2}))dt$$

$$= 2 \int_0^1 (1-t^2)^{(n-1)/2} vol_{n-1}(B_2^{n-1})dt. \tag{3.10}$$

We will estimate the integrals $\int_0^1 (1-t^2)^{(n-1)/2} dt$ and will prove that

$$C_3 n^{-1/2} \leq \int_0^1 (1-t^2)^{(n-1)/2} dt \leq C_4 n^{-1/2}. \tag{3.11}$$

It is clear that identity (3.10) and inequalities (3.11), and $(n/e)^n \leq n! \leq n^n$, imply (3.9). We begin with proving the first inequality in (3.11). We have

$$\int_0^1 (1-t^2)^{(n-1)/2} dt \geq \int_0^{n^{-1/2}} (1-t^2)^{(n-1)/2} dt \geq C_3 n^{-1/2}.$$

We proceed to the second inequality in (3.11). Using the inequality $1 - x \leq e^{-x}$, we get

$$\int_0^1 (1-t^2)^{(n-1)/2} dt \leq \int_0^1 e^{-t^2(n-1)/2} dt \leq e^{1/2} \int_0^1 e^{-nt^2/2} dt$$

$$\leq e^{1/2} \int_0^\infty e^{-nt^2/2} dt \leq n^{-1/2} e^{1/2} \int_0^\infty e^{-y^2/2} dy,$$

where we have made the substitution $t = n^{-1/2}y$. This completes the proof of
(3.11) and (3.9).

**Lemma 3.5** *For $0 < q < \infty$ and $k \le n$ we have*

$$\epsilon_k(B_q^n, \ell_\infty^n) \le C(q)(\ln(en/k)/k)^{1/q}.$$

*Proof* For $x \in B_q^n$ denote

$$I(x, s) := \{i : 2^{-s} < |x_i| \le 2^{-s+1}\} \quad \text{and} \quad n_s(x) := \#I(x, s), \quad s \in \mathbb{N}.$$

It is clear that for $x \in B_q^n$ we have

$$n_s(x) \le [2^{sq}] =: n_s. \tag{3.12}$$

Thus any $x \in B_q^n$ has at most $n_s$ coordinates in the range $(2^{-s}, 2^{-s+1}] \cup [-2^{-s+1}, -2^{-s})$. The number of different coordinate subspaces of $\mathbb{R}^n$ of dimension $\le n_s$ does not exceed

$$K_s := \sum_{v=1}^{n_s} \binom{n}{v} \le C(en/n_s)^{n_s}.$$

We use the inequalities

$$(1 + \frac{1}{v})^v < e; \qquad e^v = \sum_{l=0}^\infty \frac{v^l}{l!} > \frac{v^v}{v!};$$

$$\binom{n}{v} = \frac{n!}{v!(n-v)!} \le \frac{n^v}{v!} < (\frac{en}{v})^v.$$

Take $\epsilon = 2^{-l}$ with $l$ such that $2^{lq} \le n$. In each coordinate subspace $\mathbb{R}^\Lambda := \text{span}(e_i, i \in \Lambda)$ of dimension $v_s := \#\Lambda \le n_s$ consider a cube

$$Cu(\Lambda, s) := \{y : |y_j| \le 2^{-s+1}, j \in \Lambda, y_j = 0, j \notin \Lambda\}.$$

For $s \le l$ we form $\epsilon$-nets for these cubes in $\ell_\infty^{v_s}$ with the number of points $N_\epsilon^\Lambda$ satisfying the inequality

$$N_\epsilon^\Lambda \le 2^{(l-s+1)v_s}.$$

For all $i$ such that $|x_i| \le 2^{-l}$ we replace $x_i$ by 0. Thus we obtain an $\epsilon$-net (in the sense of $\ell_\infty^n$) of $B_q^n$ with the number of points

$$N \le \prod_{s \le l} K_s 2^{(l-s+1)n_s} \le C^l \prod_{s \le l} (en/n_s)^{n_s} 2^{(l-s+1)n_s}$$

and

$$\ln N \le l \ln C + \sum_{s \le l} n_s \ln(en/n_s) + \sum_{s \le l}(l - s + 1)2^{sq}.$$

Taking into account that $n_s \leq n$, we get

$$n_s \ln(en/n_s) \leq 2^{sq} \ln(en2^{-sq}).$$

Thus

$$\ln N \leq C(q) 2^{lq} \ln(en2^{-lq}).$$

Denoting $k := [\log N] + 1$ we get

$$\epsilon_k \leq \epsilon = 2^{-l} \leq C_1(q) \left( \frac{\ln(en/k)}{k} \right)^{1/q}. \tag{3.13}$$

This proves the lemma. □

The following theorem is from Schütt (1984) (see also Höllig (1980) and Maiorov (1978)).

**Theorem 3.6** *For any* $0 < q \leq \infty$ *and* $\max(1, q) \leq p \leq \infty$ *we have*

$$\epsilon_k(B_q^n, \ell_p^n) \leq C(q) \begin{cases} \left( \frac{\ln(en/k)}{k} \right)^{1/q - 1/p}, & k \leq n \\ 2^{-k/n} n^{1/p - 1/q}, & k \geq n. \end{cases}$$

*Proof* Let us first consider the case $k \leq n$. In the case $q < \infty$ and $p = \infty$ the estimate follows from Lemma 3.5. We will deduce the case $q \leq p < \infty$ from the construction in the proof of Lemma 3.5, and keep the notations from that proof. We have constructed an $2^{-l}$-net $\{y^1, \ldots, y^N\}$ such that for each $x \in B_q^n$ there is a $y^j$, $j \in [1, N]$, with the properties $y_i^j = 0$ if $|x_i| \leq 2^{-l}$ and $|x_i - y_i^j| \leq 2^{-l}$ if $|x_i| > 2^{-l}$. Thus we have

$$\|x - y^j\|_p^p \leq 2^{-lp} 2^{lq} + \sum_{i:|x_i| \leq 2^{-l}} |x_i|^p$$

$$\leq 2^{l(q-p)} + 2^{l(q-p)} \sum_i |x_i|^q \leq 2(2^{l(q-p)}).$$

Using the estimate (3.13) for $2^{-l}$ we obtain

$$\epsilon_k(B_q^n, \ell_p^n) \leq C(q) \left( \frac{\ln(en/k)}{k} \right)^{1/q - 1/p}.$$

We will derive the second estimate (for $k \geq n$) from the first one (for $k \leq n$). We use a simple inequality (see (3.2))

$$\epsilon_{k_1 + k_2}(B_Y, X) \leq \epsilon_{k_1}(B_Y, X) \epsilon_{k_2}(B_X, X). \tag{3.14}$$

Therefore

$$\epsilon_k(B_q^n, \ell_p^n) \leq \epsilon_n(B_q^n, \ell_p^n) \epsilon_{k-n}(B_p^n, \ell_p^n). \tag{3.15}$$

We have already proved that

$$\epsilon_n(B_q^n, \ell_p^n) \le C(q)n^{1/p-1/q}. \tag{3.16}$$

Corollary 3.4 implies

$$\epsilon_{k-n}(B_p^n, \ell_p^n) \le 3(2^{-(k-n)/n}) = 6(2^{-k/n}). \tag{3.17}$$

Combining (3.16) with (3.17), we get from (3.15) the second inequality in Theorem 3.6. □

**Remark 3.7** We note that for $q \le p$ the following trivial inequality always holds:

$$\epsilon_k(B_q^n, \ell_p^n) \le 1.$$

Let $|\cdot| := \|\cdot\|_2$ denote the $\ell_2^n$ norm and let $B_2^n$ be a unit ball in $\ell_2^n$. Denote by $S^{n-1}$ the boundary of $B_2^n$. Consider another norm $\|\cdot\|$ on $\mathbb{R}^n$ and denote by $X$ the $\mathbb{R}^n$ equipped with $\|\cdot\|$. We need a formula for the $(n-1)$-dimensional volume of $S^{n-1}$, which we denote by $|S^{n-1}|$. It is convenient for us to write this formula using the Gaussian probability measure on $\mathbb{R}^n$ defined by

$$d\mu := (2\pi)^{-n/2} e^{-|x|^2/2} \, dx.$$

First of all we recall a well known equality that implies $\int_{\mathbb{R}^n} d\mu = 1$:

$$\int_{-\infty}^{\infty} e^{-x^2/2} \, dx = (2\pi)^{1/2}. \tag{3.18}$$

Thus we have

$$\int_{\mathbb{R}^n} e^{-|x|^2/2} \, dx = (2\pi)^{n/2}.$$

Rewriting this integral in polar coordinates yields

$$(2\pi)^{n/2} = \int_0^{\infty} e^{-r^2/2} r^{n-1} \, dr \int_0^{\pi} (\sin \phi_1)^{n-2} \, d\phi_1 \cdots$$

$$\int_0^{\pi} \sin \phi_{n-2} \, d\phi_{n-2} \int_0^{2\pi} d\phi_{n-1}$$

$$= |S^{n-1}| \int_0^{\infty} e^{-r^2/2} r^{n-1} \, dr = \Gamma(n/2) 2^{n/2-1} |S^{n-1}|.$$

Therefore

$$|S^{n-1}| = 2\pi^{n/2} / \Gamma(n/2). \tag{3.19}$$

We define by $d\sigma(x)$ the normalized $(n-1)$-dimensional measure on $S^{n-1}$. Denote

$$M_X := \int_{S^{n-1}} \|x\| d\sigma(x).$$

Let us present a more convenient representation of $M_X$. We have

$$\int_{\mathbb{R}^n} \|x\| e^{-|x|^2/2} \, dx = \int_0^\infty r^n e^{-r^2/2} M_X |S^{n-1}| \, dr$$
$$= M_X (2\pi)^{n/2} 2^{1/2} \Gamma(n/2 + 1/2) \Gamma(n/2)^{-1}.$$

Denote

$$\alpha_n := 2^{-1/2} \Gamma(n/2) \Gamma(n/2 + 1/2)^{-1}.$$

Then

$$M_X = \alpha_n (2\pi)^{-n/2} \int_{\mathbb{R}^n} \|x\| e^{-|x|^2/2} \, dx. \tag{3.20}$$

It is easy to get from the Stirling formula that $\alpha_n \asymp n^{-1/2}$. By (3.20),

$$M_X = \alpha_n \int_{\mathbb{R}^n} \|x\| d\mu.$$

This implies that

$$\mu\{x : \|x\| \le 2M_X/\alpha_n\} \ge 1/2. \tag{3.21}$$

Take $\epsilon > 0$ and consider an $\epsilon$-distinguishable (in the norm $\| \cdot \|$ of $X$) set $\{x_i\}_{i=1}^{M_\epsilon}$ of $B_2^n$. Then the sets $\{x_i + \frac{\epsilon}{2} B_X\}_{i=1}^{M_\epsilon}$ (and their dilations) are disjoint. Therefore,

$$\sum_{i=1}^{M_\epsilon} \mu\{y_i + 2M_X \alpha_n^{-1} B_X\} \le 1, \quad y_i = \frac{4M_X}{\alpha_n \epsilon} x_i. \tag{3.22}$$

Let us estimate from below $\mu\{y_i + 2M_X \alpha_n^{-1} B_X\}$. We have

$$\mu\{y_i + 2M_X \alpha_n^{-1} B_X\} = (2\pi)^{-n/2} \int_{2M_X \alpha_n^{-1} B_X} e^{-|x-y_i|^2/2} \, dx$$
$$= (2\pi)^{-n/2} \int_{2M_X \alpha_n^{-1} B_X} e^{-|x+y_i|^2/2} \, dx$$

(by convexity of $e^{-u}$ we continue)

$$\ge (2\pi)^{-n/2} \int_{2M_X \alpha_n^{-1} B_X} e^{-(|x-y_i|^2 + |x+y_i|^2)/4} \, dx$$
$$= (2\pi)^{-n/2} \int_{2M_X \alpha_n^{-1} B_X} e^{-(|x|^2 + |y_i|^2)/2} \, dx$$

(using (3.21) we arrive at)

$$\ge 0.5 e^{-|y_i|^2/2} \ge 0.5 e^{-(4M_X/\epsilon \alpha_n)^2}.$$

From here and (3.22) we get

$$M_\epsilon \leq 2e^{(4M_X/\epsilon\alpha_n)^2},$$

which implies

$$\epsilon_k \ll (n/k)^{1/2}M_X. \tag{3.23}$$

It makes sense to use this inequality for $k \leq n$. For $k > n$ we get from (3.23), (3.14), and Corollary 3.4 that

$$\epsilon_k \ll M_X 2^{-k/n}, \quad k > n. \tag{3.24}$$

Thus we have proven the following theorem.

**Theorem 3.8** *Let $X$ be $\mathbb{R}^n$ equipped with $\|\cdot\|$ and*

$$M_X = \int_{S^{n-1}} \|x\| d\sigma(x).$$

*Then we have*

$$\epsilon_k(B_2^n, X) \ll M_X \begin{cases} (n/k)^{1/2}, & k \leq n \\ 2^{-k/n}, & k \geq n. \end{cases}$$

Theorem 3.8 is a dual version of the corresponding result from Sudakov (1971); it was proved in Pajor and Tomczak-Yaegermann (1986).

### 3.3 Trigonometric polynomials and volume estimates

#### 3.3.1 Univariate trigonometric polynomials

Functions of the form

$$t(x) = \sum_{|k| \leq n} c_k e^{ikx} = \frac{a_0}{2} + \sum_{k=1}^{n} (a_k \cos kx + b_k \sin kx)$$

(where $c_k, a_k, b_k$ are complex numbers) will be called trigonometric polynomials of order $n$. We shall denote the set of such polynomials by $\mathcal{T}(n)$, and the subset of $\mathcal{T}(n)$ of real polynomials by $\mathcal{RT}(n)$.

We first consider a number of concrete polynomials which play an important role in approximation theory.

**The Dirichlet kernel.** The Dirichlet kernel of order $n$ is as follows:

$$\begin{aligned} \mathcal{D}_n(x) &:= \sum_{|k| \leq n} e^{ikx} = e^{-inx}(e^{i(2n+1)x} - 1)(e^{ix} - 1)^{-1} \\ &= \frac{\sin(n+1/2)x}{\sin(x/2)}. \end{aligned} \tag{3.25}$$

The Dirichlet kernel is an even trigonometric polynomial with the majorant

$$|\mathcal{D}_n(x)| \leq \min(2n + 1, \pi/|x|), \qquad |x| \leq \pi. \tag{3.26}$$

The estimate

$$\|\mathcal{D}_n\|_1 \leq C \ln n, \qquad n = 2, 3, \ldots, \tag{3.27}$$

follows from (3.26).

We mention the well known relation

$$\|\mathcal{D}_n\|_1 = \frac{4}{\pi^2} \ln n + R_n, \qquad |R_n| \leq 3, \qquad n = 1, 2, 3, \ldots.$$

For any trigonometric polynomial $t \in \mathcal{T}(n)$,

$$t * \mathcal{D}_n := (2\pi)^{-1} \int_0^{2\pi} t(x - y)\mathcal{D}_n(y)dy = t.$$

Denote

$$x^l = 2\pi l/(2n + 1), \qquad l = 0, 1, \ldots, 2n.$$

Clearly, the points $x^l$, $l = 1, \ldots, 2n$, are zeros of the Dirichlet kernel $\mathcal{D}_n$ on $[0, 2\pi]$.

For any $|k| \leq n$ we have

$$\sum_{l=1}^{2n} e^{ikx^l} \mathcal{D}_n(x - x^l) = \sum_{|m| \leq n} e^{imx} \sum_{l=0}^{2n} e^{i(k-m)x^l} = e^{ikx}(2n + 1).$$

Consequently, for any $t \in \mathcal{T}(n)$

$$t(x) = (2n + 1)^{-1} \sum_{l=0}^{2n} t(x^l)\mathcal{D}_n(x - x^l). \tag{3.28}$$

Further, it is easy to see that, for any $u, v \in \mathcal{T}(n)$, we have

$$\langle u, v \rangle = (2\pi)^{-1} \int_{-\pi}^{\pi} u(x)\overline{v(x)}dx = (2n + 1)^{-1} \sum_{l=0}^{2n} u(x^l)\overline{v(x^l)} \tag{3.29}$$

and, for any $t \in \mathcal{T}(n)$,

$$\|t\|_2^2 = (2n + 1)^{-1} \sum_{l=0}^{2n} |t(x^l)|^2. \tag{3.30}$$

For $1 < q \leq \infty$ the estimate

$$\|\mathcal{D}_n\|_q \leq C(q)n^{1-1/q} \tag{3.31}$$

follows from (3.26). Applying the Hölder inequality for estimating $\|\mathcal{D}_n\|_2^2$ we get

$$2n + 1 = \|\mathcal{D}_n\|_2^2 \le \|\mathcal{D}_n\|_q \|\mathcal{D}_n\|_{q'}. \tag{3.32}$$

Relations (3.31) and (3.32) imply for $1 < q < \infty$ the relation

$$\|\mathcal{D}_n\|_q \asymp n^{1-1/q}, \tag{3.33}$$

which for $q = \infty$ is obvious.

We denote by $S_n$ the operator of taking the partial sum of order $n$. Then for $f \in L_1$ we have

$$S_n(f) = f * \mathcal{D}_n.$$

We formulate one classical result about operators $S_n$.

**Theorem 3.9** *The operator $S_n$ does not change polynomials from $\mathcal{T}(n)$ and for $p = 1$ or $\infty$ we have*

$$\|S_n\|_{p \to p} \le C \ln n, \qquad n = 2, 3, \ldots,$$

*and for $1 < p < \infty$ for all $n$ we have*

$$\|S_n\|_{p \to p} \le C(p).$$

**The Fejér kernel.** The Fejér kernel of order $n - 1$ is given by

$$\mathcal{K}_{n-1}(x) := n^{-1} \sum_{m=0}^{n-1} \mathcal{D}_m(x) = \sum_{|m| \le n} (1 - |m|/n) e^{imx}$$
$$= \frac{\left(\sin(nx/2)\right)^2}{n\left(\sin(x/2)\right)^2}.$$

The Fejér kernel is an even non-negative trigonometric polynomial in $\mathcal{T}(n - 1)$ with the majorant

$$\left|\mathcal{K}_{n-1}(x)\right| \le \min\left(n, \pi^2/(nx^2)\right), \qquad |x| \le \pi. \tag{3.34}$$

From the obvious relations

$$\|\mathcal{K}_{n-1}\|_1 = 1, \qquad \|\mathcal{K}_{n-1}\|_\infty = n$$

and the inequality

$$\|f\|_q \le \|f\|_1^{1/q} \|f\|_\infty^{1-1/q},$$

we get, in the same way as above,

$$Cn^{1-1/q} \le \|\mathcal{K}_{n-1}\|_q \le n^{1-1/q}, \qquad 1 \le q \le \infty. \tag{3.35}$$

**The de la Vallée Poussin kernels.** These are given by

$$\mathcal{V}_{m,n}(x) = (n - m)^{-1} \sum_{l=m}^{n-1} \mathcal{D}_l(x), \qquad n > m.$$

It is convenient to represent these kernels in terms of the Fejér kernels as follows:

$$\mathcal{V}_{m,n}(x) = (n - m)^{-1}\big(n\mathcal{K}_{n-1}(x) - m\mathcal{K}_{m-1}(x)\big)$$
$$= (\cos mx - \cos nx)\big(2(n - m)(\sin(x/2))^2\big)^{-1}.$$

The de la Vallée Poussin kernels $\mathcal{V}_{m,n}$ are even trigonometric polynomials of order $n - 1$ with the majorant

$$\big|\mathcal{V}_{m,n}(x)\big| \le C \min\big(n, \ 1/|x|, \ 1/((n - m)x^2)\big), \ |x| \le \pi, \qquad (3.36)$$

which implies the estimate

$$\|\mathcal{V}_{m,n}\|_1 \le C \ln\big(1 + n/(n - m)\big).$$

The de la Vallée Poussin kernels with $n = 2m$ are used often; let us denote them by

$$\mathcal{V}_m(x) = \mathcal{V}_{m,2m}(x), \qquad m \ge 1, \qquad \mathcal{V}_0(x) = 1.$$

Then for $m \ge 1$ we have

$$\mathcal{V}_m = 2\mathcal{K}_{2m-1} - \mathcal{K}_{m-1},$$

which with the properties of $\mathcal{K}_n$ implies

$$\|\mathcal{V}_m\|_1 \le 3. \qquad (3.37)$$

In addition,

$$\|\mathcal{V}_m\|_\infty \le 3m.$$

Consequently, in the same way as for the Dirichlet kernels we get

$$\|\mathcal{V}_m\|_q \asymp m^{1-1/q}, \qquad 1 \le q \le \infty. \qquad (3.38)$$

We denote

$$x(l) = \pi l/2m, \qquad l = 1, \ldots, 4m.$$

Then, as in (3.28) for each $t \in T(m)$ we have

$$t(x) = (4m)^{-1} \sum_{l=1}^{4m} t\big(x(l)\big)\mathcal{V}_m\big(x - x(l)\big). \qquad (3.39)$$

The operator $V_m$ defined on $L_1$ by the formula

$$V_m(f) = f * \mathcal{V}_m$$

is called the de la Vallée Poussin operator.

The following theorem is a corollary of the definition of kernels $\mathcal{V}_m$ and (3.37).

**Theorem 3.10** *The operator $V_m$ does not change polynomials from $\mathcal{T}(m)$, and for all $1 \le p \le \infty$ we have*

$$\|V_m\|_{p \to p} \le 3, \qquad m = 1, 2, \dots .$$

In addition, we note two properties of the de la Vallée Poussin kernels.

($1^0$) Relation (3.36) with $n = 2m$ implies the inequality

$$|\mathcal{V}_m(x)| \le C \min(m, 1/(mx^2)), \qquad |x| \le \pi.$$

It is easy to derive from this the following property.

($2^0$) For $h$ satisfying $C_1 \le mh \le C_2$ we have

$$\sum_{0 \le l \le 2\pi/h} |\mathcal{V}(x - lh)| \le Cm.$$

We remark that property ($2^0$) is valid for the Fejér kernel $\mathcal{K}_m$.

**The Rudin–Shapiro polynomials.** We define recursively pairs of trigonometric polynomials $P_j(x)$ and $Q_j(x)$ of order $2^j - 1$:

$$P_0 = Q_0 = 1;$$

$$P_{j+1}(x) = P_j(x) + e^{i2^j x} Q_j(x); \qquad Q_{j+1}(x) = P_j(x) - e^{i2^j x} Q_j(x).$$

Then at each point $x$ we have

$$
\begin{aligned}
|P_{j+1}|^2 + |Q_{j+1}|^2 &= (P_j + e^{i2^j x} Q_j)(\overline{P}_j + e^{-i2^j x}\overline{Q}_j) \\
&\quad + (P_j - e^{i2^j x} Q_j)(\overline{P}_j - e^{-i2^j x}\overline{Q}_j) \\
&= 2(|P_j|^2 + |Q_j|^2).
\end{aligned}
$$

Consequently, for all $x$,

$$|P_j(x)|^2 + |Q_j(x)|^2 = 2^{j+1}.$$

Thus, for example,

$$\|P_n\|_\infty \le 2^{(n+1)/2}. \tag{3.40}$$

It is clear from the definition of the polynomials $P_n$ that

$$P_n(x) = \sum_{k=0}^{2^n-1} \varepsilon_k e^{ikx}, \qquad \varepsilon_k = \pm 1.$$

Let $N$ be a natural number and

$$N = \sum_{j=1}^{m} 2^{n_j}, \qquad n_1 > n_2 > \cdots > n_m \geq 0,$$

its binary representation. We set

$$R'_N(x) = P_{n_1}(x) + \sum_{j=2}^{m} P_{n_j}(x) e^{i(2^{n_1}+\cdots+2^{n_{j-1}})x},$$

$$R_N(x) = R'_N(x) + R'_N(-x) - 1.$$

Then $R_N(x)$ will have the form

$$R_N(x) = \sum_{|k|<N} \varepsilon_k e^{ikx}, \qquad \varepsilon_k = \pm 1,$$

and for this polynomial the estimate

$$\|R_N\|_\infty \leq C N^{1/2}. \tag{3.41}$$

holds.

### 3.3.2 Multivariate trigonometric polynomials

In this subsection we define the analogs of the Dirichlet, Fejér, de la Vallée Poussin and Rudin–Shapiro polynomials (kernels) for $d$-dimensional parallelepipeds of harmonics (frequencies)

$$\Pi(\mathbf{N}, d) := \{\mathbf{a} = (a_1, \ldots, a_d) \in \mathbb{R}^d : |a_j| \leq N_j, \ j = 1, \ldots, d\},$$

where $\mathbf{N} = (N_1, \ldots, N_d)$, $N_j$ are non-negative integers.

Properties of these multidimensional kernels follow easily from the corresponding properties of one-dimensional kernels. In Section 3.3.3 we will prove some estimates of the volumes of sets of Fourier coefficients of bounded polynomials with harmonics from $\Pi(\mathbf{N}, d)$. We will use the volume estimates for studying the entropy numbers of unit balls of subspaces of trigonometric polynomials in the $L_p$ norm. The volume estimates also allow us to prove the existence of polynomials with properties similar to some properties of Rudin–Shapiro polynomials, in each subspace $\Psi \subset \mathcal{T}(\mathbf{N}, d)$ with sufficiently large dimension. Here $\mathcal{T}(\mathbf{N}, d)$ is the set of complex trigonometric polynomials with harmonics from $\Pi(\mathbf{N}, d)$. The set of real trigonometric polynomials with harmonics from $\Pi(\mathbf{N}, d)$ will be denoted by $\mathcal{RT}(\mathbf{N}, d)$.

**The Dirichlet kernels.**

$$\mathcal{D}_{\mathbf{N}}(\mathbf{x}) := \prod_{j=1}^{d} \mathcal{D}_{N_j}(x_j), \quad \mathbf{x} = (x_1, \ldots, x_d), \quad \mathbf{N} = (N_1, \ldots, N_d),$$

have the following properties.

For any trigonometric polynomial $t \in \mathcal{T}(\mathbf{N}, d)$,

$$t * \mathcal{D}_{\mathbf{N}} = t.$$

For $1 < q \leq \infty$,

$$\|\mathcal{D}_{\mathbf{N}}\|_q \asymp v(\mathbf{N})^{1-1/q}, \tag{3.42}$$

where $\overline{N}_j = \max(N_j, 1)$, $v(\mathbf{N}) = \prod_{j=1}^{d} \overline{N}_j$ and

$$\|\mathcal{D}_{\mathbf{N}}\|_1 \asymp \prod_{j=1}^{d} \ln(N_j + 2). \tag{3.43}$$

We denote

$$P(\mathbf{N}) := \Big\{ \mathbf{n} = (n_1, \ldots, n_d), \qquad n_j \text{ are non-negative integers,}$$

$$0 \leq n_j \leq 2N_j, \qquad j = 1, \ldots, d \Big\},$$

and set

$$\mathbf{x}^{\mathbf{n}} = \left( \frac{2\pi n_1}{2N_1 + 1}, \ldots, \frac{2\pi n_d}{2N_d + 1} \right), \qquad \mathbf{n} \in P(\mathbf{N}).$$

Then, for any $t \in \mathcal{T}(\mathbf{N}, d)$,

$$t(\mathbf{x}) = \vartheta(\mathbf{N})^{-1} \sum_{\mathbf{n} \in P(\mathbf{N})} t(\mathbf{x}^{\mathbf{n}}) \mathcal{D}_{\mathbf{N}}(\mathbf{x} - \mathbf{x}^{\mathbf{n}}), \tag{3.44}$$

where $\vartheta(\mathbf{N}) = \prod_{j=1}^{d}(2N_j + 1)$ and, for any $t, u \in \mathcal{T}(\mathbf{N}, d)$,

$$\langle t, u \rangle = \vartheta(\mathbf{N})^{-1} \sum_{\mathbf{n} \in P(\mathbf{N})} t(\mathbf{x}^{\mathbf{n}}) \overline{u}(\mathbf{x}^{\mathbf{n}}), \tag{3.45}$$

$$\|t\|_2^2 = \vartheta(\mathbf{N})^{-1} \sum_{\mathbf{n} \in P(\mathbf{N})} \left| t(\mathbf{x}^{\mathbf{n}}) \right|^2. \tag{3.46}$$

**The Fejér kernels.**

$$\mathcal{K}_{\mathbf{N}}(\mathbf{x}) := \prod_{j=1}^{d} \mathcal{K}_{N_j}(x_j), \quad \mathbf{x} = (x_1, \ldots, x_d), \quad \mathbf{N} = (N_1, \ldots, N_d),$$

are non-negative trigonometric polynomials from $\mathcal{T}(\mathbf{N}, d)$ which have the following properties:

$$\|\mathcal{K}_{\mathbf{N}}\|_1 = 1, \tag{3.47}$$

$$\|\mathcal{K}_{\mathbf{N}}\|_q \asymp \vartheta(\mathbf{N})^{1-1/q}, \quad 1 \le q \le \infty. \tag{3.48}$$

**The de la Vallée Poussin kernels.**

$$\mathcal{V}_{\mathbf{N}}(\mathbf{x}) := \prod_{j=1}^{d} \mathcal{V}_{N_j}(x_j), \quad \mathbf{x} = (x_1, \ldots, x_d), \quad \mathbf{N} = (N_1, \ldots, N_d),$$

have the properties

$$\|\mathcal{V}_{\mathbf{N}}\|_1 \le 3^d, \tag{3.49}$$

$$\|\mathcal{V}_{\mathbf{N}}\|_q \asymp \vartheta(\mathbf{N})^{1-1/q}, \quad 1 \le q \le \infty. \tag{3.50}$$

For any $t \in \mathcal{T}(\mathbf{N}, d)$,

$$\mathcal{V}_{\mathbf{N}}(t) := t * \mathcal{V}_{\mathbf{N}} = t.$$

We denote

$$P'(\mathbf{N}) = \big\{ \mathbf{n} = (n_1, \ldots, n_d), \qquad n_j \text{ are natural numbers,}$$

$$1 \le n_j \le 4N_j, \qquad j = 1, \ldots, d \big\}$$

and set

$$\mathbf{x}(\mathbf{n}) = \left( \frac{\pi n_1}{2N_1}, \ldots, \frac{\pi n_d}{2N_d} \right), \qquad \mathbf{n} \in P'(\mathbf{N}).$$

In the case $N_j = 0$ we assume $x_j(\mathbf{n}) = 0$. Then for any $t \in \mathcal{T}(\mathbf{N}, d)$ we have the representation

$$t(\mathbf{x}) = v(4\mathbf{N})^{-1} \sum_{\mathbf{n} \in P'(\mathbf{N})} t\big(\mathbf{x}(\mathbf{n})\big) \mathcal{V}_{\mathbf{N}}\big(\mathbf{x} - \mathbf{x}(\mathbf{n})\big). \tag{3.51}$$

Relation (3.49) implies that

$$\|V_{\mathbf{N}}\|_{p \to p} \le 3^d, \qquad 1 \le p \le \infty. \tag{3.52}$$

**The Rudin–Shapiro polynomials.**

$$R_{\mathbf{N}}(\mathbf{x}) = \prod_{j=1}^{d} R_{N_j}(x_j), \quad \mathbf{x} = (x_1, \ldots, x_d), \quad \mathbf{N} = (N_1, \ldots, N_d),$$

have the properties

$$R_{\mathbf{N}} \in T(\mathbf{N}, d),$$

$$\hat{R}_{\mathbf{N}}(\mathbf{k}) = \pm 1, \quad |\mathbf{k}| < \mathbf{N}, \quad \|R_{\mathbf{N}}\|_\infty \le C(d)\vartheta(\mathbf{N})^{1/2}, \tag{3.53}$$

where $|\mathbf{k}| = (|k_1|, \ldots, |k_d|)$ and vector inequalities are understood coordinate-wise.

### 3.3.3 Volume estimates; generalized Rudin–Shapiro polynomials

The presentation in this subsection follows Temlyakov (1989b, 1993b). We shall prove the following assertion.

**Theorem 3.11** *Let $\varepsilon > 0$ and let a subspace $\Psi \subset T(\mathbf{m}, d)$ be such that* $\dim \Psi \ge \varepsilon \vartheta(\mathbf{m})$. *Then there is a $t \in \Psi$ such that*

$$\|t\|_\infty = 1, \qquad \|t\|_2 \ge C(\varepsilon, d) > 0.$$

The proof of this theorem is based on the lower estimates of the volumes of the sets of Fourier coefficients of bounded trigonometric polynomials from $T(\mathbf{m}, d)$. We consider arrays of complex numbers $\overline{y} = \{y^{\mathbf{l}}, \mathbf{l} \in P(\mathbf{m})\}$. Then, since $y^{\mathbf{l}} = \operatorname{Re} y^{\mathbf{l}} + i \operatorname{Im} y^{\mathbf{l}}$, we can consider $\overline{y}$ as an element of the space $\mathbb{R}^{2\vartheta(\mathbf{m})}$. We define in the space $\mathbb{R}^{2\vartheta(\mathbf{m})}$ the following linear transformation: we map each $\overline{y}$ to the trigonometric polynomial

$$t(\mathbf{x}, \overline{y}) = B\overline{y} = \vartheta(\mathbf{m})^{-1} \sum_{\mathbf{l} \in P(\mathbf{m})} y^{\mathbf{l}} \mathcal{D}_{\mathbf{m}}(\mathbf{x} - \mathbf{x}^{\mathbf{l}}),$$

and we map each trigonometric polynomial $t \in T(\mathbf{m}, d)$ to the element

$$At = \left\{ (\operatorname{Re} \hat{t}(\mathbf{k}), \operatorname{Im} \hat{t}(\mathbf{k})) \right\}_{|\mathbf{k}| \le \mathbf{m}}$$

of the space $\mathbb{R}^{2\vartheta(\mathbf{m})}$. Then

$$\|t(\cdot, \overline{y})\|_2^2 = \vartheta(\mathbf{m})^{-1} \sum_{\mathbf{l} \in P(\mathbf{m})} |y^{\mathbf{l}}|^2 = \vartheta(\mathbf{m})^{-1} \|\overline{y}\|_2^2$$

and

$$\|t\|_2^2 = \sum_{|\mathbf{k}| \le \mathbf{m}} |\hat{t}(\mathbf{k})|^2 = \|At\|_2^2. \tag{3.54}$$

Thus the operator $AB\vartheta(\mathbf{m})^{1/2}$ is an orthogonal transformation of $\mathbb{R}^{2\vartheta(\mathbf{m})}$. We shall prove the following assertion.

**Lemma 3.12** *Let*

$$S_\infty(\mathbf{m}) := \left\{\overline{y} \in \mathbb{R}^{2\vartheta(\mathbf{m})} : \|t(\cdot, \overline{y})\|_\infty \le 1\right\}.$$

*Then for the volume of this set in the space $\mathbb{R}^{2\vartheta(\mathbf{m})}$ we have the estimate that there is a $C(d) > 0$ such that, for all $\mathbf{m}$,*

$$vol\left(S_\infty(\mathbf{m})\right) \ge C(d)^{-\vartheta(\mathbf{m})}.$$

*Proof* Let $\mathcal{K}_n(t)$ be the Fejér kernel of order $n$, that is

$$\mathcal{K}_n(t) = 1 + 2\sum_{k=1}^{n}\left(1 - \frac{k}{n+1}\right)\cos kt$$

and

$$\mathcal{K}_{\mathbf{m}}(\mathbf{x}) = \prod_{j=1}^{d} \mathcal{K}_{m_j}(x_j).$$

We consider the set $G := \{\overline{y} = K\overline{b}\}$, where $K$ is the linear operator defined as follows:

$$y^{\mathbf{l}} = \vartheta(\mathbf{m})^{-1}\sum_{\mathbf{k}\in P(\mathbf{m})} b^{\mathbf{k}}\mathcal{K}_{\mathbf{m}}(\mathbf{x}^{\mathbf{l}} - \mathbf{x}^{\mathbf{k}}), \qquad \mathbf{l} \in P(\mathbf{m}),$$

and $\overline{b} \in \mathbb{R}^{2\vartheta(\mathbf{m})}$ is such that $|\operatorname{Re} b^{\mathbf{k}}| \le 1$, $|\operatorname{Im} b^{\mathbf{k}}| \le 1$, $\mathbf{k} \in P(\mathbf{m})$. We use the equality

$$(2\pi)^{-d}\int_{[0,2\pi]^d} u(\mathbf{x})w(\mathbf{x})dx = \vartheta(\mathbf{m})^{-1}\sum_{\mathbf{k}\in P(\mathbf{m})} u(\mathbf{x}^{\mathbf{k}})w(\mathbf{x}^{\mathbf{k}}), \tag{3.55}$$

which is valid for any pair of trigonometric polynomials $u, w \in \mathcal{T}(\mathbf{m}, d)$. Then it is not difficult to see that, for all $\mathbf{n}$ such that $|n_j| \le m_j$, $j = 1, \ldots, d$, we have

$$\vartheta(\mathbf{m})^{-1}\sum_{\mathbf{k}\in P(\mathbf{m})} e^{i(\mathbf{n},\mathbf{x}^{\mathbf{k}})}\mathcal{K}_{\mathbf{m}}(\mathbf{x}^{\mathbf{l}} - \mathbf{x}^{\mathbf{k}}) = \hat{\mathcal{K}}_{\mathbf{m}}(\mathbf{n})e^{i(\mathbf{n},\mathbf{x}^{\mathbf{l}})}. \tag{3.56}$$

From (3.56) it follows that the vectors

$$\varepsilon_{\mathbf{n}} = \left\{(\operatorname{Re} e^{i(\mathbf{n},\mathbf{x}^{\mathbf{k}})}, \operatorname{Im} e^{i(\mathbf{n},\mathbf{x}^{\mathbf{k}})})\right\}_{\mathbf{k}\in P(\mathbf{m})} \in \mathbb{R}^{2\vartheta(\mathbf{m})},$$

$$\eta_{\mathbf{n}} = \left\{(\operatorname{Re} ie^{i(\mathbf{n},\mathbf{x}^{\mathbf{k}})}, \operatorname{Im} ie^{i(\mathbf{n},\mathbf{x}^{\mathbf{k}})})\right\}_{\mathbf{k}\in P(\mathbf{m})} \in \mathbb{R}^{2\vartheta(\mathbf{m})}$$

are the eigenvectors of the operator $K$ corresponding to eigenvalues $\hat{\mathcal{K}}_{\mathbf{m}}(\mathbf{n})$, $|\mathbf{n}| \leq \mathbf{m}$. It is not difficult to verify that the vectors $\varepsilon_{\mathbf{n}}$, $\eta_{\mathbf{n}}$, $|\mathbf{n}| \leq \mathbf{m}$, create a set of $2\vartheta(\mathbf{m})$ orthogonal vectors from $\mathbb{R}^{2\vartheta(\mathbf{m})}$. Consequently, the operator $K$ maps the unit cube of the space $\mathbb{R}^{2\vartheta(\mathbf{m})}$ to the set $G$ with volume

$$vol(G) = \prod_{|\mathbf{n}| \leq \mathbf{m}} \hat{\mathcal{K}}_{\mathbf{m}}(\mathbf{n})^2 \geq C_1(d)^{-\vartheta(\mathbf{m})}, \qquad C_1(d) > 0. \tag{3.57}$$

Further, let $\overline{y} \in G$; then

$$\begin{aligned} t(\mathbf{x}, \overline{y}) &= \vartheta(\mathbf{m})^{-1} \sum_{\mathbf{l} \in P(\mathbf{m})} y^{\mathbf{l}} \mathcal{D}_{\mathbf{m}}(\mathbf{x} - \mathbf{x}^{\mathbf{l}}) \\ &= \vartheta(\mathbf{m})^{-1} \sum_{\mathbf{k} \in P(\mathbf{m})} b^{\mathbf{k}} \vartheta(\mathbf{m})^{-1} \sum_{\mathbf{l} \in P(\mathbf{m})} \mathcal{D}_{\mathbf{m}}(\mathbf{x} - \mathbf{x}^{\mathbf{l}}) \mathcal{K}_{\mathbf{m}}(\mathbf{x}^{\mathbf{l}} - \mathbf{x}^{\mathbf{k}}) \\ &= \vartheta(\mathbf{m})^{-1} \sum_{\mathbf{k} \in P(\mathbf{m})} b^{\mathbf{k}} \mathcal{K}_{\mathbf{m}}(\mathbf{x} - \mathbf{x}^{\mathbf{k}}). \end{aligned} \tag{3.58}$$

From the condition $|b^{\mathbf{k}}| \leq 2^{1/2}$, the estimate

$$\mathcal{K}_n(t) \leq C \min\left(n, (nt^2)^{-1}\right)$$

and the representation (3.58) it follows that for some $C_2(d)$ we have

$$\left\| t(\cdot, \overline{y}) \right\|_{\infty} \leq C_2(d). \tag{3.59}$$

The conclusion of Lemma 3.12 follows from (3.57) and (3.59). $\qquad\square$

This lemma and the property of orthogonality of the operator $AB\vartheta(\mathbf{m})^{1/2}$ imply the following statement.

**Lemma 3.13** *Let*

$$A_{\infty}(\mathbf{m}) := \left\{ At \in \mathbb{R}^{2\vartheta(\mathbf{m})}, t \in \mathcal{T}(\mathbf{m}, d), \|t\|_{\infty} \leq 1 \right\}.$$

*Then for the volume of this set in the space $\mathbb{R}^{2\vartheta(\mathbf{m})}$ the following estimate holds: there is a $C(d) > 0$ such that, for all $\mathbf{m}$,*

$$vol\left(A_{\infty}(\mathbf{m})\right) \geq \vartheta(\mathbf{m})^{-\vartheta(\mathbf{m})} C(d)^{-\vartheta(\mathbf{m})}.$$

Lemma 3.13 in the case $d = 1$ was proved by Kashin (1980).

*Proof of Theorem 3.11.* Let $\varepsilon > 0$ and let a subspace $\Psi \in \mathcal{T}(\mathbf{m}, d)$ with $\dim \Psi \geq \varepsilon\vartheta(\mathbf{m})$ be given. Let $U \subset \mathbb{R}^{2\vartheta(\mathbf{m})}$ be the image of $\Psi$ under the transformation $A$ defined above. Then $\dim U = 2 \dim \Psi \geq 2\varepsilon\vartheta(\mathbf{m})$.

The conclusion of the theorem will follow from Lemma 3.13 and the following lemma.

**Lemma 3.14** *Let a set $B$ be contained in the unit ball $B_2^N$ of the Euclidean space $\mathbb{R}^N$ and be a convex and centrally symmetric set with*

$$vol(B) \geq vol(B_2^N)C^{-N},$$

*where $C > 0$ is a constant independent of $N$. Then for any hyperplane $U(0 \in U)$ of dimension $n \geq \varepsilon N, \varepsilon > 0$, there is an element $b \in B \cap U$ for which*

$$\|b\|_2 \geq C(\varepsilon) > 0.$$

Indeed, to prove this it suffices to use the relation (see (3.9))

$$vol(B_2^N) = \pi^{N/2}\Gamma(1 + N/2)^{-1} \leq N^{-N/2}C^{-N}, \qquad C > 0. \qquad (3.60)$$

$\square$

It remains to prove Lemma 3.14.

*Proof of Lemma 3.14.* The proof is based on the following corollary of the Brun theorem.

**Theorem 3.15** *Let $B \subset \mathbb{R}^n$ be a convex, centrally symmetric set, let $\mathbf{u}$ be some unit vector from $\mathbb{R}^n$ and let $B_\alpha = \{\mathbf{x} \in B : (\mathbf{x}, \mathbf{u}) = \alpha\}$ be sections of $B$ by hyperplanes of the dimension $n - 1$ orthogonal to $\mathbf{u}$. Then the $(n - 1)$-dimensional volume of $B_\alpha$ is non-increasing in $[0, +\infty)$ as a function of $\alpha$.*

Let $U$ be from Lemma 3.14, let $U^\perp$ denote the orthogonal complement of $U$ with respect to $\mathbb{R}^N$ and let $B_2(U^\perp)$ be the Euclidean unit ball in $U^\perp$. For any $\mathbf{h} \in U^\perp$ we denote $S_\mathbf{h} = B \cap L_\mathbf{h}$, where $L_\mathbf{h}$ is the linear manifold of the space $\mathbb{R}^N$ such that

$$L_\mathbf{h} = \{\mathbf{l} \in \mathbb{R}^N : \mathbf{l} = \mathbf{h} + \mathbf{u}, \mathbf{u} \in U\},$$

that is, a hyperplane which is parallel to $U$ and passes through $\mathbf{h}$. Since $B \subset B_2^N$ we have that for $\mathbf{h} \notin B_2(U^\perp)$, $S_\mathbf{h} = \varnothing$. Along with the set $B$ we consider the set

$$B' = \{\mathbf{y} \in \mathbb{R}^N : \mathbf{y} = \mathbf{u} + \mathbf{h}, \mathbf{u} \in S_0, \mathbf{h} \in B_2(U^\perp)\}.$$

Then by Theorem 3.15 for any $\mathbf{h} \in B_2(U^\perp)$,

$$|S_\mathbf{h}'| = |S_0| \geq |S_\mathbf{h}|,$$

where $S_\mathbf{h}' = B' \cap L_\mathbf{h}$. This implies the estimate

$$vol(B) \leq vol(B'). \qquad (3.61)$$

Further,

$$vol(B') = |S_0| |B_2(U^\perp)|. \tag{3.62}$$

Let

$$a = \max_{\mathbf{u} \in S_0} \|\mathbf{u}\|_2.$$

Then from (3.60)–(3.62) we get

$$vol(B) \le a^n n^{-n/2} (N - n)^{-1/2(N-n)} C^{-N}. \tag{3.63}$$

Due to the boundedness on $(0, 1)$ of the function $x^{-x}(1 - x)^{x-1}$ we obtain, from (3.63),

$$vol(B) \le a^n N^{-N/2} C^{-N}. \tag{3.64}$$

From the hypotheses of Lemma 3.14 and the relations (3.60) and (3.64) we get

$$a \ge C(\varepsilon) > 0,$$

which implies the conclusion of Lemma 3.14. $\qquad\qquad \square$

Let us use now the volume estimates from Lemma 3.13 for studying the entropy numbers. We will prove the following theorem.

**Theorem 3.16** *There exists a positive constant $C(d)$ such that for the unit $L_\infty$-ball $\mathcal{T}(\mathbf{m}, d)_\infty$ of $\mathcal{T}(\mathbf{m}, d)$ we have the following lower bound:*

$$\epsilon_k(\mathcal{T}(\mathbf{m}, d)_\infty, L_1) \ge C(d), \quad k \le \vartheta(\mathbf{m}).$$

*Proof* We begin this proof with a lemma on upper estimates of the volume of the $L_1$-analog of $A_\infty(\mathbf{m})$.

**Lemma 3.17** *Let*

$$A_1(\mathbf{m}) := \{ At \in \mathbb{R}^{2\vartheta(\mathbf{m})}, t \in \mathcal{T}(\mathbf{m}, d), \|t\|_1 \le 1 \}.$$

*Then for the volume of this set in the space $\mathbb{R}^{2\vartheta(\mathbf{m})}$ the following estimate holds: there is a $C(d) > 0$ such that, for all $\mathbf{m}$,*

$$vol(A_1(\mathbf{m})) \le \vartheta(\mathbf{m})^{-\vartheta(\mathbf{m})} C(d)^{\vartheta(\mathbf{m})}.$$

*Proof* Using property $(2^0)$ of the de la Vallée Poussin kernels we can prove the inequality: for $t \in \mathcal{T}(m, 1)$

$$(2m + 1)^{-1} \sum_{l=0}^{2m} |t(x^l)| \le C \|t\|_1.$$

This inequality implies its $d$-dimensional analog.

**Lemma 3.18** *For any* $t \in \mathcal{T}(\mathbf{m}, d)$ *we have*

$$\vartheta(\mathbf{m})^{-1} \sum_{\mathbf{l} \in P(\mathbf{m})} |t(\mathbf{x}^{\mathbf{l}})| \leq C(d) \|t\|_1.$$

Therefore,

$$A_1(\mathbf{m}) \subseteq \{ AB\overline{y}, \|\overline{y}\|_{\ell_1^{2\vartheta(\mathbf{m})}} \leq C(d)\vartheta(\mathbf{m}) \}$$

with $C(d)$ the constant from Lemma 3.18. Using the fact that the operator $AB\vartheta(\mathbf{m})^{1/2}$ is an orthogonal transformation of $\mathbb{R}^{2\vartheta(\mathbf{m})}$, we get

$$vol(A_1(\mathbf{m})) \leq \vartheta(\mathbf{m})^{-\vartheta(\mathbf{m})} vol(O_{2\vartheta(\mathbf{m})}(C(d)\vartheta(\mathbf{m}))).$$

By (3.8) we obtain from here that

$$vol(A_1(\mathbf{m})) \leq \vartheta(\mathbf{m})^{-\vartheta(\mathbf{m})} C(d)^{\vartheta(\mathbf{m})}.$$

Lemma 3.17 is now proved. □

Let us apply Theorem 3.3 with $n = 2\vartheta(\mathbf{m})$:

$$\|z\|_X := \|A^{-1}z\|_{L_1}; \quad \|z\|_Y := \|A^{-1}z\|_{L_\infty}, \quad z \in \mathbb{R}^n.$$

Then $B_X = A_1(\mathbf{m})$ and $B_Y = A_\infty(\mathbf{m})$. Combining Lemma 3.13 with Lemma 3.17 we get from (3.6) that

$$N_\epsilon(B_Y, B_X) \geq (C(d)/\epsilon)^n. \tag{3.65}$$

In particular, (3.65) implies

$$\epsilon_n(B_Y, B_X) \geq C(d) > 0.$$

This inequality can be rewritten in the form

$$\epsilon_n(\mathcal{T}(\mathbf{m}, d)_\infty, L_1 \cap \mathcal{T}(\mathbf{m}, d)) \geq C(d). \tag{3.66}$$

We want to replace $L_1 \cap \mathcal{T}(\mathbf{m}, d)$ by $L_1$ in (3.66). There are different ways of doing this. The simplest way is to use Corollary 3.2.

The proof of Theorem 3.16 is now complete. □

**Theorem 3.19** *For any* $1 \leq q \leq p \leq \infty$ *we have*

$$\epsilon_k(\mathcal{T}(\mathbf{m}, d)_q, L_p)$$

$$\leq C(q, d)v(4\mathbf{m})^{1/q-1/p} \begin{cases} \left( \frac{\ln(e2v(4\mathbf{m})/k)}{k} \right)^{1/q-1/p}, & k \leq 2v(4\mathbf{m}) \\ 2^{-k/2v(4\mathbf{m})} v(4\mathbf{m})^{1/p-1/q}, & k \geq 2v(4\mathbf{m}). \end{cases}$$

*Proof* We consider arrays of complex numbers $z = \{z^{\mathbf{n}}, \mathbf{n} \in P'(\mathbf{m})\}$. Then, since $z^{\mathbf{n}} = \operatorname{Re} z^{\mathbf{n}} + i \operatorname{Im} z^{\mathbf{n}}$, we interpret $z$ as an element of the space $\mathbb{R}^{2v(4\mathbf{m})}$.

With each polynomial $t \in T(\mathbf{m}, d)$ we associate $z(t) := \{t(\mathbf{x}(\mathbf{n})), \mathbf{n} \in P'(\mathbf{m})\}$ and with any $z = \{z^{\mathbf{n}}, \mathbf{n} \in P'(\mathbf{m})\}$ we associate

$$t(\mathbf{x}, z) := \nu(4\mathbf{m})^{-1} \sum_{\mathbf{n} \in P'(\mathbf{m})} z^{\mathbf{n}} \mathcal{V}_{\mathbf{m}}(\mathbf{x} - \mathbf{x}(\mathbf{n})).$$

We need a multivariate version of the Marcinkiewicz theorem (see Temlyakov (1993a)).

**Lemma 3.20** *There exist two positive constants $C_1(d)$ and $C_2(d)$ such that for any $t \in T(\mathbf{m}, d)$ one has, for $1 \le p \le \infty$,*

$$C_1(d)\|t\|_p \le \nu(4\mathbf{m})^{-1/p} \left( \sum_{\mathbf{n} \in P'(\mathbf{m})} |t(\mathbf{x}(\mathbf{n}))|^p \right)^{1/p} \le C_2(d)\|t\|_p.$$

Thus, if

$$t \in T(\mathbf{m}, d)_q, \quad \text{then} \quad \|z(t)\|_{\ell_q^{2\nu(4\mathbf{m})}} \le C_2(d)\nu(4\mathbf{m})^{1/q}.$$

Next, by Temlyakov (1993a), chap. 2, lemma 2.6, we get

$$\|t(\cdot, z)\|_p \le C(d)\nu(4\mathbf{m})^{-1/p}\|z\|_{\ell_p^{2\nu(4\mathbf{m})}}.$$

Therefore

$$\epsilon_k(T(\mathbf{m}, d)_q, L_p) \le C(d)\nu(4\mathbf{m})^{1/q-1/p}\epsilon_k(B_q^{2\nu(4\mathbf{m})}, \ell_p^{2\nu(4\mathbf{m})}).$$

Applying Theorem 3.6 we complete the proof of Theorem 3.19. $\qquad\square$

## 3.4 The function classes

We illustrate in this section how the finite dimensional results from Sections 3.2 and 3.3 can be applied to the smoothness classes. We begin with a definition of two standard univariate smoothness classes $W_{q,\alpha}^r$ and $H_q^r$.

For $r > 0$ and $\alpha \in \mathbb{R}$ the functions

$$F_r(x, \alpha) = 1 + 2\sum_{k=1}^{\infty} k^{-r} \cos(kx - \alpha\pi/2)$$

will be called Bernoulli kernels.

We define the following operator in the space $L_1$:

$$(I_\alpha^r \phi)(x) := (2\pi)^{-1} \int_0^{2\pi} F_r(x - y, \alpha)\phi(y)dy. \tag{3.67}$$

Denote by $W_{q,\alpha}^r B$, $r > 0$, $\alpha \in \mathbb{R}$, $1 \le q \le \infty$, the class of functions $f(x)$ representable in the form

$$f = I_\alpha^r \phi, \qquad \|\phi\|_q \le B. \tag{3.68}$$

Let us define the classes $H_q^r B, r > 0, 1 \le q \le \infty$, as follows:

$$H_q^r B = \left\{ f \in L_q : \|f\|_q \le B, \ \left\|\Delta_t^a f(x)\right\|_q \le B|t|^r, \ a = [r] + 1 \right\},$$

$$\Delta_t f(x) = f(x) - f(x+t), \qquad \Delta_t^a = (\Delta_t)^a.$$

When $B = 1$, we shall not include it in our notation.

It is well known that the classes $W_{q,\alpha}^r B$ are embedded into the classes $H_q^r B$:

$$W_{q,\alpha}^r B \subset H_q^r B'.$$

The classes $H_q^r B$ have a convenient representation that we now describe. Let us consider special trigonometric polynomials. Let $s$ be a non-negative integer. We define

$$\mathcal{A}_0(x) := 1, \quad \mathcal{A}_1(x) := \mathcal{V}_1(x) - 1, \quad \mathcal{A}_s(x) = \mathcal{V}_{2^{s-1}}(x) - \mathcal{V}_{2^{s-2}}(x), \quad s \ge 2,$$

where $\mathcal{V}_m$ are the de la Vallée Poussin kernels. Then $\mathcal{A}_s \in \mathcal{T}(2^s)$ and, by (3.37),

$$\|\mathcal{A}_s\|_1 \le 6. \tag{3.69}$$

Let

$$A_s(f) := \mathcal{A}_s * f.$$

We formulate one classical representation theorem for classes $H_q^r$.

**Theorem 3.21** *Let $f \in L_q$, $1 \le q \le \infty$, $\|f\|_q \le 1$. For $\left\|\Delta_t^a f\right\|_q \le |t|^r$, $a = [r] + 1$, it is necessary and sufficient that the following conditions be satisfied:*

$$\left\|A_s(f)\right\|_q \le C(r,q)2^{-rs}, \qquad s = 0, 1, \ldots .$$

*(The constants $C(r,q)$ in the cases of necessity and sufficiency may be different.)*

The following theorem is a classical result in approximation theory.

**Theorem 3.22** *For any $1 \le q, p \le \infty$ and $r > (1/q - 1/p)_+$ we have*

$$\epsilon_k(W_{q,\alpha}^r, L_p) \asymp \epsilon_k(H_q^r, L_p) \asymp k^{-r}.$$

*Proof* The embedding $W_{q,\alpha}^r \subset H_q^r B$ (see above) and monotonicity of the $L_p$ norms imply that it is sufficient to prove the lower estimate for the class $W_{\infty,\alpha}^r$ in the $L_1$ norm. By Bernstein's inequality for trigonometric polynomials we get

$$\mathcal{T}(m)_\infty \subset W_{\infty,\alpha}^r C(r) m^r.$$

This embedding and Theorem 3.16 imply

$$\epsilon_m(W^r_{\infty,\alpha}, L_1) \geq C'(r)m^{-r}.$$

The lower estimate in Theorem 3.22 is proved.

We proceed to the upper estimate. It is clear that it suffices to consider the $H^r_q$ classes (see the embedding mentioned above). By Theorem 3.21 we get

$$H^r_q \subseteq \bigoplus_{s=1}^{\infty} T(r, s, q),$$

where $T(r, s, q) \subset T(2^s)$ and

$$\sup_{t \in T(r,s,q)} \|t\|_q \ll 2^{-rs}.$$

Thus by (3.3) we write for any $n$, $\{n_s\}$ satisfying $\sum_s n_s \leq n$

$$\epsilon_n(H^r_q, L_p) \leq \sum_{s=1}^{\infty} \epsilon_{n_s}(T(r, s, q), L_p) \ll \sum_{s=1}^{\infty} 2^{-rs} \epsilon_{n_s}(T(2^s)_q, L_p). \quad (3.70)$$

It remains to estimate $\epsilon_k(T(m)_q, L_p)$ and specify a sequence $\{n_s\}$. To estimate $\epsilon_k(T(m)_q, L_p)$ we use Theorem 3.19. Let $a > 0$ be such that $r > (1/q - 1/p)(1 + a)$. We take any $l \in \mathbb{N}$ and set $n_s := [(r + a)(l - s)2^{s+3}]$ for $s < l$ and $n_s := [2^{l-a(s-l)}]$ for $s \geq l$. Then

$$n := \sum_s n_s \ll \sum_{s=1}^{l}(l - s)2^s + \sum_{s=l}^{\infty} 2^{l-a(s-l)} \ll 2^l. \quad (3.71)$$

By Theorem 3.19 we obtain from (3.70)

$$\epsilon_n(H^r_q, L_p) \ll \sum_{s=1}^{l-1} 2^{-rs} 2^{-n_s/2^{s+3}}$$

$$+ \sum_{s=l}^{\infty} 2^{-(r-1/q+1/p)s} \left( \frac{\ln(e 2^{s+3}/n_s)}{n_s} \right)^{1/q-1/p}$$

$$\ll \sum_{s=1}^{l-1} 2^{-rs-(r+a)(l-s)}$$

$$+ \sum_{s=l}^{\infty} 2^{-(r-1/q+1/p)s-(1/q-1/p)(l-a(s-l))}(s - l + 1)^{1/q-1/p}$$

$$\ll 2^{-rl}. \quad (3.72)$$

Combining (3.71) and (3.72) we get the required upper estimates.

Theorem 3.22 is now proved. $\qquad \square$

### 3.5 General inequalities

A number of different widths are being studied in approximation theory: Kolmogorov widths, linear widths, Fourier widths, Gel'fand widths, Alexandrov widths and others. All these widths were introduced in approximation theory as characteristics of function classes (more generally compact sets) which give the best possible accuracy of algorithms with certain restrictions. For instance, Kolmogorov's $n$-width for a centrally symmetric compact set $F$ in a Banach space $X$ is defined as follows:

$$d_n(F, X) := \inf_L \sup_{f \in F} \inf_{g \in L} \| f - g \|_X,$$

where $\inf_L$ is taken over all $n$-dimensional subspaces of $X$. In other words, the Kolmogorov $n$-width gives the best possible error in approximating a compact set $F$ by $n$-dimensional linear subspaces.

There has been an increasing interest since 2000 in nonlinear $m$-term approximation with regard to different systems. In this section we generalize the concept of the classical Kolmogorov width in order to use it in estimating the best $m$-term approximation (see Temlyakov (1998d)). For this purpose we introduce a nonlinear Kolmogorov $(N, m)$-width as follows:

$$d_m(F, X, N) := \inf_{\Lambda_N, \#\Lambda_N \le N} \sup_{f \in F} \inf_{L \in \Lambda_N} \inf_{g \in L} \| f - g \|_X,$$

where $\Lambda_N$ is a set of at most $N$ $m$-dimensional subspaces $L$. It is clear that

$$d_m(F, X, 1) = d_m(F, X).$$

The new feature of $d_m(F, X, N)$ is that we allow one to choose a subspace $L \in \Lambda_N$ depending on $f \in F$. It is clear that the bigger the $N$ the more flexibility we have to approximate $f$. It turns out that from the point of view of our applications the two cases

$$N \asymp K^m, \tag{3.73}$$

where $K > 1$ is a constant, and

$$N \asymp m^{am}, \tag{3.74}$$

where $a > 0$ is a fixed number, play an important role.

It is known (see Temlyakov (2003a)) that the $(N, m)$-widths can be used for estimating from below the best $m$-term approximations. There are several general results (see Carl (1981)) which give lower estimates of the Kolmogorov widths $d_n(F, X)$ in terms of the entropy numbers $\epsilon_k(F, X)$. We will generalize the Carl (see Carl (1981)) inequality: for any $r > 0$ we have

$$\max_{1\leq k\leq n} k^r \epsilon_k(F, X) \leq C(r) \max_{1\leq m\leq n} m^r d_{m-1}(F, X). \tag{3.75}$$

It is clear that

$$d_1(F, X, 2^n) \leq \epsilon_n(F, X).$$

We prove here the inequality

$$\max_{1\leq k\leq n} k^r \epsilon_k(F, X) \leq C(r, K) \max_{1\leq m\leq n} m^r d_{m-1}(F, X, K^m), \tag{3.76}$$

where we denote

$$d_0(F, X, N) := \sup_{f\in F} \|f\|_X.$$

This inequality is a generalization of inequality (3.75). We also prove the inequality

$$\max_{1\leq k\leq n} k^r \epsilon_{(a+r)k\log k}(F, X) \leq C(r, a) \max_{1\leq m\leq n} m^r d_{m-1}(F, X, m^{am}) \tag{3.77}$$

and give an example showing that $k \log k$ in this inequality cannot be replaced by a slower growing function on $k$. For non-integer $k$ we set $\epsilon_k(F, X) := \epsilon_{[k]}(F, X)$, where $[k]$ is the integral part of the number $k$.

**Theorem 3.23** *For any compact $F \subset X$ and any $r > 0$ we have, for all $n \in \mathbb{N}$,*

$$\max_{1\leq k\leq n} k^r \epsilon_k(F, X) \leq C(r, K) \max_{1\leq m\leq n} m^r d_{m-1}(F, X, K^m).$$

*Proof* Let $X(N, m)$ denote the union of some $N$ subspaces $L$ with dim $L = m$. Consider a collection $\mathcal{K}(K, l) := \{X(K^{2^{s+1}}, 2^{s+1})\}_{s=1}^l$ and denote

$$H^r(\mathcal{K}(K, l)) := \{f \in X : \exists L_1(f), \dots, L_l(f) : L_s(f) \in X(K^{2^{s+1}}, 2^{s+1})$$

and $\exists t_s(f) \in L_s(f)$ such that

$$\|t_s(f)\|_X \leq 2^{-r(s-1)}, \quad s = 1, \dots, l; \quad \|f - \sum_{s=1}^l t_s(f)\|_X \leq 2^{-rl}\}.$$

**Lemma 3.24** *We have for $r > 0$*

$$\epsilon_{2^l}(H^r(\mathcal{K}(K, l)), X) \leq C(r, K)2^{-rl}.$$

*Proof* We use Corollary 3.4 to estimate $\epsilon_n(B_X, X)$ of the unit ball $B_X$ in the $d$-dimensional space $X$:

$$\epsilon_n(B_X, X) \leq 3(2^{-n/d}). \tag{3.78}$$

Take any sequence $\{n_s\}_{s=1}^{l(r)}$ of $l(r) \leq l$ non-negative integers. Construct $\epsilon_{n_s}$-nets consisting of $2^{n_s}$ points each for all unit balls of the spaces in

$X(K^{2^{s+1}}, 2^{s+1})$. Then the total number of the elements $y_j^s$ in these $\epsilon_{n_s}$-nets does not exceed

$$M_s := K^{2^{s+1}} 2^{n_s}.$$

We now consider the set $A$ of elements of the form

$$y_{j_1}^1 + 2^{-r} y_{j_2}^2 + \cdots + 2^{-r(l(r)-1)} y_{j_{l(r)}}^{l(r)}, \quad j_s = 1, \ldots, M_s, \quad s = 1, \ldots, l(r).$$

The total number of these elements does not exceed

$$M = \prod_{s=1}^{l(r)} M_s \leq K^{2^{l(r)+2}} 2^{\sum_{s=1}^{l(r)} n_s}.$$

It is clear that it suffices to consider the case of large $l \geq l(r, K)$. We take now

$$n_s := [(r+1)(l-s)2^{s+1}], \quad s = 1, \ldots, l(r),$$

where $[x]$ denotes the integer part of a number $x$. We choose $l(r) \leq l$ as a maximal natural number satisfying

$$\sum_{s=1}^{l(r)} n_s \leq 2^{l-1}$$

and

$$2^{l(r)+2} \log K \leq 2^{l-1}.$$

It is clear that

$$l(r) \geq l - C(r, K).$$

Then we have

$$M \leq 2^{2^l}.$$

For the error $\epsilon(f)$ of approximation of $f \in H^r(\mathcal{K}(K, l))$ by elements of $A$ we have

$$\epsilon(f) \leq 2^{-rl} + \sum_{s=1}^{l(r)} \|t_s(f) - 2^{-r(s-1)} y_{j_s}^s\|_X + \sum_{s=l(r)+1}^{l} \|t_s(f)\|_X$$

$$\leq C(r, K) 2^{-rl} + \sum_{s=1}^{l(r)} 2^{-r(s-1)} \epsilon_{n_s}(B_{L_s(f)}, X)$$

$$\leq C(r, K) 2^{-rl} + 3 \sum_{s=1}^{l(r)} 2^{-r(s-1)} 2^{-n_s/2^{s+1}} \leq C(r, K) 2^{-rl}.$$

Lemma 3.24 is now proved. $\qquad\qquad\qquad\qquad\qquad\qquad\qquad\qquad\square$

We continue the proof of Theorem 3.23. Without loss of generality, assume

$$\max_{1 \leq m \leq n} m^r d_{m-1}(F, X, K^m) < 1/2.$$

Then for $s = 1, 2, \ldots, l; l \leq [\log(n-1)]$ we have

$$d_{2^s}(F, X, K^{2^s}) < 2^{-rs-1}.$$

This means that, for each $s = 1, 2, \ldots, l$, there is a collection $\Lambda_{K^{2^s}}$ of $K^{2^s}$ $2^s$-dimensional spaces $L^s_j$, $j = 1, \ldots, K^{2^s}$, such that, for each $f \in F$, there exists a subspace $L^s_{j_s}(f)$ and an approximant $a_s(f) \in L^s_{j_s}(f)$ such that

$$\|f - a_s(f)\| \leq 2^{-rs-1}.$$

Consider

$$t_s(f) := a_s(f) - a_{s-1}(f), \quad s = 2, \ldots, l. \tag{3.79}$$

Then we have

$$t_s(f) \in L^s_{j_s}(f) \oplus L^{s-1}_{j_{s-1}}(f), \quad \dim(L^s_{j_s}(f) \oplus L^{s-1}_{j_{s-1}}(f)) \leq 2^s + 2^{s-1} < 2^{s+1}.$$

Let $X(K^{2^{s+1}}, 2^{s+1})$ denote the collection of all $L^s_{j_s} \oplus L^{s-1}_{j_{s-1}}$ over various $1 \leq j_s \leq K^{2^s}; 1 \leq j_{s-1} \leq K^{2^{s-1}}$. For $t_s(f)$ defined by (3.79) we have

$$\|t_s(f)\| \leq 2^{-rs-1} + 2^{-r(s-1)-1} \leq 2^{-r(s-1)}.$$

Next, for $a_1(f) \in L^1(f)$ we have

$$\|f - a_1(f)\| \leq 1/2$$

and from $d_0(F, X) \leq 1/2$ we get

$$\|a_1(f)\| \leq 1.$$

Take $t_1(f) = a_1(f)$. Then we have $F \subset H^r(\mathcal{K}(K, l))$ and Lemma 3.24 gives the required bound

$$\epsilon_{2^l}(F) \leq C(r, K)2^{-rl}, \quad 1 \leq l \leq [\log(n-1)].$$

It is clear that these inequalities imply the conclusion of Theorem 3.23. $\square$

**Remark 3.25** On examining the proof of Theorem 3.23, one can check that the inequality holds for $K^m$ replaced by a larger function. For example, we have

$$\max_{1 \leq k \leq n} k^r \epsilon_k(F, X) \leq C(r, K) \max_{1 \leq m \leq n} m^r d_{m-1}(F, X, (Kn/m)^m).$$

We proceed now to the case (3.74) when $N \asymp m^{am}$. We prove a lemma which will imply the inequality (3.77).

**Lemma 3.26** *For any compact set $F \subset X$ and any real numbers $0 < a < b$ we have*

$$\epsilon_{bm \log m}(F, X) \leq C(d_0(F, X)m^{a-b} + d_m(F, X, m^{am})).$$

*Proof* Let $N := [m^{am}]$. For a given $\delta > 0$, denote by $\Lambda_N$ a collection of $m$-dimensional subspaces $L_j$, $j = 1, \ldots, N$, such that for each $f \in F$ there exists $j(f) \in [1, N]$ and an element $g(f) \in L_{j(f)}$ with the approximating property

$$\|f - g(f)\| \leq d_m(F, X, m^{am}) + \delta. \tag{3.80}$$

Then

$$\|g(f)\|_X \leq \|f\|_X + \|f - g(f)\|_X \leq d_0(F, X) + d_m(F, X, m^{am}) + \delta =: \alpha.$$

Thus we must estimate the $\epsilon$-entropy of the union $U$ of $m$-dimensional balls of radius $\alpha$ in $L_j$ over $j \in [1, N]$. By (3.78) we have

$$\epsilon_{n+[\log N]+1}(U, X) \leq \alpha 3(2^{-n/m}).$$

If $(b - a)m \log m < 1$ the statement of Lemma 3.26 is trivial. Assume $(b - a)m \log m \geq 1$ and choose $n$ such that $n = [(b-a)m \log m] - 1$; then we have

$$n + [\log N] + 1 \leq bm \log m$$

and

$$\alpha 3(2^{-n/m}) \leq C\alpha m^{a-b}. \tag{3.81}$$

Combining (3.80) and (3.81) we get the statement of Lemma 3.26. $\qquad \square$

It is easy to see that this lemma implies the inequality (3.77). We give now an example showing that we cannot get rid of $\log k$ in (3.77).

**Example** Let $r > 0$ and $a > 0$ be given. We specify $X := L_\infty := L_\infty([0, 1])$. Consider a partition of $[0, 1]$ into $N = [n^{a+1}/3]$ segments $I_j := [\frac{j-1}{N}, \frac{j}{N})$, $j = 1, \ldots, N$, and form the set of all $n$-dimensional subspaces of the form

$$X_Q := \text{span}\{\chi_{I_j}\}_{j \in Q}, \quad Q \subset \{1, 2, \ldots, N\}, \quad \#Q = n,$$

where $\chi_I$ denotes the characteristic function of a segment $I$. The number of these subspaces is given by

$$N(n) = \binom{N}{n}; \quad (n^a/3 - 1)^n \leq \binom{N}{n} \leq n^{an}.$$

Consider

$$F_n := \cup_Q (B_{L_\infty} \cap X_Q) n^{-r}.$$

Then we have $F_n \subset n^{-r} B_{L_\infty}$, which implies for all $m$

$$d_m(F_n, L_\infty) \leq n^{-r}.$$

We also have

$$d_n(F_n, L_\infty, n^{an}) \leq d_n(F_n, L_\infty, N(n)) = 0.$$

This implies that for any $s \in \mathbb{N}$ we have, for the right-hand side of (3.75),

$$\max_{1 \leq m \leq s} m^r d_m(F_n, L_\infty, m^{am}) \leq 1.$$

Next, consider the set of functions $\chi_{G_Q}$, $G_Q = \cup_{j \in Q} I_j$, where $\chi_G$ is the characteristic function of $G$. Then for any $Q \neq Q'$ we have

$$\|\chi_{G_Q} - \chi_{G_{Q'}}\|_\infty \geq 1.$$

The number of functions $\{\chi_{G_Q}\}$ is equal to $N(n)$. This implies that

$$\epsilon_{[\log N(n)]}(F_n, L_\infty) \geq n^{-r}/2.$$

Assume we can replace $\log k$ in (3.75) by a slower growing function $\phi(k)$. Take any $n \in \mathbb{N}$ and let $m_n \in \mathbb{N}$ be the largest number satisfying the inequality

$$(a + r)m_n\phi(m_n) \leq [\log N(n)].$$

Then our assumption implies that

$$\lim_{n \to \infty} m_n/n = \infty.$$

Thus, for the left-hand side of (3.77) we get

$$\max_{1 \leq k \leq m_n} k^r \epsilon_{(a+r)k\phi(k)}(F_n, X) \geq m_n^r \epsilon_{(a+r)m_n\phi(m_n)}(F_n, X) \geq (m_n/n)^r/2 \to \infty$$

as $n \to \infty$, which is a contradiction to (3.77).

We now proceed to two multiplicative inequalities for the $L_p$ spaces. Let $D$ be a domain in $\mathbb{R}^d$ and let $L_p := L_p(D)$ denote the corresponding $L_p$ space, $1 \leq p \leq \infty$, with respect to the Lebesgue measure. We note that the inequalities below hold for any measure $\mu$ on $D$.

**Theorem 3.27** *Let $A \subset L_1 \cap L_\infty$. Then for any $1 \leq p \leq \infty$ we have $A \subset L_p$ and*

$$\epsilon_{n+m}(A, L_p) \leq 2\epsilon_n(A, L_1)^{1/p}\epsilon_m(A, L_\infty)^{1-1/p}.$$

*Proof* The simple inequality $\|f\|_p \leq \|f\|_1^{1/p}\|f\|_\infty^{1-1/p}$ implies $A \subset L_p$. Let $a$ and $b$ be any positive numbers satisfying $a > \epsilon_n(A, L_1)$ and $b > \epsilon_m(A, L_\infty)$. By the definition of entropy numbers (see Section 3.1) there exist $g_1, \ldots, g_{2^n}$ in $L_1$ and $h_1, \ldots, h_{2^m}$ in $L_\infty$ such that

$$A \subset \cup_{k=1}^{2^n} B_{L_1}(g_k, a), \qquad A \subset \cup_{l=1}^{2^m} B_{L_\infty}(h_l, b).$$

We now set $\epsilon := 2a^{1/p}b^{1-1/p}$ and bound from above the $M_\epsilon(A)$. We want to prove that $M_\epsilon(A) \leq 2^{n+m}$. We take any set $f_1, \ldots, f_N$ of elements of $A$ with $N > 2^{n+m}$ and prove that for some $i$ and $j$ we have $\|f_i - f_j\|_p \leq \epsilon$. Indeed, the total number of sets $G_{k,l} := B_{L_1}(g_k, a) \cap B_{L_\infty}(h_l, b)$ is less than or equal to $2^{n+m} < N$. Therefore, there exist two indices $i$ and $j$ such that for some $k$ and $l$ we have $f_i \in G_{k,l}$ and $f_j \in G_{k,l}$. This means that

$$\|f_i - f_j\|_1 \leq 2a, \qquad \|f_i - f_j\|_\infty \leq 2b$$

and

$$\|f_i - f_j\|_p \leq \|f_i - f_j\|_1^{1/p}\|f_i - f_j\|_\infty^{1-1/p} \leq \epsilon.$$

This implies $M_\epsilon(A) \leq 2^{n+m}$ and, by Theorem 3.1, $N_\epsilon(A) \leq 2^{n+m}$. This completes the proof. $\square$

It will be convenient for us to formulate one more inequality in terms of entropy numbers of operators. Let $S$ be a linear operator from $X$ to $Y$. We define the $n$th entropy number of $S$ as

$$\epsilon_n(S : X \to Y) := \epsilon_n(S(B_X), Y),$$

where $S(B_X)$ is the image of $B_X$ under the mapping $S$.

**Theorem 3.28** *For any $1 \leq p \leq \infty$ and any Banach space $Y$ we have*

$$\epsilon_{n+m}(S : L_p \to Y) \leq 2\epsilon_n(S : L_1 \to Y)^{1/p}\epsilon_m(S : L_\infty \to Y)^{1-1/p}.$$

*Proof* We begin with a simple well known lemma.

**Lemma 3.29** *For any $f \in L_p$, $\|f\|_p \leq 1$, and any positive numbers $a$, $b$ there exists a representation $f = f_1 + f_\infty$ such that*

$$a\|f_1\|_1 \leq a^{1/p}b^{1-1/p}, \qquad b\|f_\infty\|_\infty \leq a^{1/p}b^{1-1/p}.$$

*Proof* Let $f_T$ denote the $T$ cut off of $f$: $f_T(x) = f(x)$ if $|f(x)| \leq T$ and $f_T(x) = 0$ otherwise. Clearly, $\|f_T\|_\infty \leq T$. Set $f^T := f - f_T$. We now estimate the $L_1$ norm of the $f^T$. Let $E := \{x : f^T(x) \neq 0\}$. First, we bound from above the measure of $E$. We have

$$1 \geq \int_D |f(x)|^p \, dx \geq \int_E T^p \, dx = T^p|E|.$$

Second, we bound the $\|f^T\|_1$:

$$\|f^T\|_1 = \int_E |f^T(x)|dx \le \left(\int_E |f^T(x)|^p \, dx\right)^{1/p} |E|^{1-1/p} \le T^{1-p}.$$

Specifying $T = (a/b)^{1/p}$ we get

$$a\|f^T\|_1 \le a^{1/p}b^{1-1/p}, \qquad b\|f_T\|_\infty \le a^{1/p}b^{1-1/p}.$$

This proves the lemma. $\qquad\qquad\qquad\qquad\qquad\qquad\qquad\qquad\qquad\qquad\square$

We continue the proof of Theorem 3.28. Let $a$ and $b$ be such that

$$a > \epsilon_n(S : L_1 \to Y), \qquad b > \epsilon_m(S : L_\infty \to Y).$$

Find $y_1, \ldots, y_{2^n}$ and $z_1, \ldots, z_{2^m}$ such that

$$S(B_{L_1}) \subset \cup_{k=1}^{2^n} B_Y(y_k, a), \qquad S(B_{L_\infty}) \subset \cup_{l=1}^{2^m} B_Y(z_l, b).$$

Take any $f \in L_p$, $\|f\|_p \le 1$. Set $\epsilon := a^{1/p}b^{1-1/p}$ and by Lemma 3.29 find $f_1$ and $f_\infty$ such that $f = f_1 + f_\infty$ and

$$a\|f_1\|_1 \le \epsilon, \qquad b\|f_\infty\|_\infty \le \epsilon.$$

Clearly, for some $k$

$$S(af_1/\epsilon) \in B_Y(y_k, a) \Rightarrow S(f_1) \in B_Y(\epsilon y_k/a, \epsilon) \qquad (3.82)$$

and for some $l$

$$S(bf_\infty/\epsilon) \in B_Y(z_l, b) \Rightarrow S(f_\infty) \in B_Y(\epsilon z_l/b, \epsilon). \qquad (3.83)$$

Consider the sets $G_{i,j} := B_Y(\epsilon y_i/a + \epsilon z_j/b, 2\epsilon)$, $i = 1, \ldots, 2^n$, $j = 1, \ldots, 2^m$. Relations (3.82) and (3.83) imply $S(f) \in G_{k,l}$. Thus

$$\epsilon_{n+m}(S : L_p \to Y) \le 2\epsilon.$$

$\qquad\qquad\qquad\qquad\qquad\qquad\qquad\qquad\qquad\qquad\qquad\qquad\square$

## 3.6 Some further remarks

In Section 3.4 we discussed the rate of decay of $\epsilon_k(W_{q,\alpha}^r, L_p)$ as a function on $k$. Theorem 3.22 gives the rate of decay

$$\epsilon_k(W_{q,\alpha}^r, L_p) \asymp k^{-r} \qquad (3.84)$$

that holds for all $1 \le q, p \le \infty$ and $r > (1/q - 1/p)_+$. We note that the condition $r > (1/q - 1/p)_+$ is a necessary and sufficient condition for the compact embedding of $W_{q,\alpha}^r$ into $L_p$. Thus (3.84) provides a complete description of the rate of $\epsilon_k(W^r_{q,\alpha}, L_p)$ in the univariate case. We point out that (3.84) shows that the rate of decay of $\epsilon_k(W^r_{q,\alpha}, L_p)$ depends only on $r$ and does not depend on $q$ and $p$. In this sense the strongest upper bound (for $r > 1$) is $\epsilon_k(W^r_{1,\alpha}, L_\infty) \ll k^{-r}$ and the strongest lower bound is $\epsilon_k(W^r_{\infty,\alpha}, L_1) \gg k^{-r}$.

There are different generalizations of classes $W^r_{q,\alpha}$ to the case of multivariate functions. In this section we only discuss classes $MW^r_{q,\alpha}$ of functions with bounded mixed derivative ($M$ in the notation refers to *mixed*). We decided to discuss these classes for several reasons. (1) The problem of the rate of decay of $\epsilon_k(MW^r_{q,\alpha}, L_p)$ in the particular case $r = 1$, $q = 2$, $p = \infty$ is equivalent (see Kuelbs and Li (1993)) to a fundamental problem of probability theory (the small ball problem). Both of these problems are still open for $d > 2$. (2) The problem of the rate of decay of $\epsilon_k(MW^r_{q,\alpha}, L_p)$ turns out to be a very rich and difficult problem. There are still many open problems. Those problems that have been resolved required different non-trivial methods for different pairs $(q, p)$.

Let $F_r(x, \alpha)$ be the Bernoulli kernels defined in Section 3.4. For $\mathbf{x} = (x_1, \ldots, x_d)$ and $\alpha = (\alpha_1, \ldots, \alpha_d)$ we define

$$F_r(\mathbf{x}, \alpha) := \prod_{i=1}^{d} F_r(x_i, \alpha_i)$$

and

$$MW^r_{q,\alpha} := \{f : f = F_r(\cdot, \alpha) * \varphi, \quad \|\varphi\|_q \le 1\},$$

where $*$ means convolution.

The problem of estimating $\epsilon_k(MW^r_{q,\alpha}, L_p)$ has a long history. The first result on the right order of $\epsilon_k(MW^r_{2,\alpha}, L_2)$ was obtained by Smolyak (1960). Later (see Temlyakov (1988b, 1989b)) it was established that

$$\epsilon_k(MW^r_{q,\alpha}, L_p) \asymp k^{-r}(\log k)^{r(d-1)} \tag{3.85}$$

holds for all $1 < q, p < \infty, r > 1$. The case $1 < q = p < \infty, r > 0$ was established by Ding Dung (1985). Belinsky (1998) extended (3.85) to the case $r > (1/q - 1/p)_+$ when $1 < q, p < \infty$.

It is known in approximation theory (see Temlyakov (1993a)) that investigation of asymptotic characteristics of classes $MW^r_{q,\alpha}$ in $L_p$ becomes more difficult when $q$ or $p$ takes the value either 1 or $\infty$ than when $1 < q, p < \infty$. It turns out to be the case for $\epsilon_k(MW^r_{q,\alpha}, L_p)$ too. It was discovered that in some of these extreme cases ($q$ or $p$ equals 1 or $\infty$) relation (3.85) holds and in other cases it does not hold. We describe the picture in detail. It was proved by Temlyakov (1989b) that (3.85) holds for $p = 1, 1 < q < \infty, r > 0$. It was also proved that (3.85) holds for $p = 1, q = \infty$ (see Belinsky (1998) for $r > 1/2$ and Kashin and Temlyakov (1995) for $r > 0$). Summarizing, we state that (3.85) holds for $1 < q$, $p < \infty$ and $p = 1$, $1 < q \leq \infty$ for all $d$ (with appropriate restrictions on $r$). This easily implies that (3.85) also holds for $q = \infty$, $1 \leq p < \infty$. For all other pairs $(q, p)$, namely for $p = \infty$, $1 \leq q \leq \infty$ and $q = 1$, $1 \leq p \leq \infty$, the rate of $\epsilon_k(MW_{q,\alpha}^r, L_p)$ is not known in the case $d > 2$. It is an outstanding open problem.

In the case $d = 2$ this problem is essentially solved. We now cite the corresponding results. The first result on the right order of $\epsilon_k(MW_{q,\alpha}^r, L_p)$ in the case $p = \infty$ was obtained by Kuelbs and Li (1993) for $q = 2$, $r = 1$. It was proved by Temlyakov (1995a) that

$$\epsilon_k(MW_{q,\alpha}^r, L_\infty) \asymp k^{-r}(\log k)^{r+1/2} \tag{3.86}$$

holds for $1 < q < \infty$, $r > 1$. We note that the upper bound in (3.86) was proved under the condition $r > 1$ and the lower bound in (3.86) was proved under the condition $r > 1/q$. Belinsky (1998) proved the upper bound in (3.86) for $1 < q < \infty$ under the condition $r > \max(1/q, 1/2)$. Temlyakov (1998e) proved (3.86) for $q = \infty$ under the assumption $r > 1/2$.

The case $q = 1$, $1 \leq p \leq \infty$ was settled by Kashin and Temlyakov (2003), who proved that

$$\epsilon_k(MW_{1,\alpha}^r, L_p) \asymp k^{-r}(\log k)^{r+1/2} \tag{3.87}$$

holds for $1 \leq p < \infty$, $r > \max(1/2, 1 - 1/p)$ and

$$\epsilon_k(MW_{1,0}^r, L_\infty) \asymp k^{-r}(\log k)^{r+1}, \quad r > 1. \tag{3.88}$$

Let us make an observation on the basis of the above discussion. In the univariate case the entropy numbers $\epsilon_k(W_{q,\alpha}^r, L_p)$ have the same order of decay with respect to $k$ for all pairs $(q, p)$, $1 \leq q$, $p \leq \infty$. In the case $d = 2$ we have three different orders of decay of $\epsilon_k(MW_{q,\alpha}^r, L_p)$ which depend on the pair $(q, p)$. For instance, in the case $1 < q$, $p < \infty$ it is $k^{-r}(\log k)^r$; in the case $q = 1$, $1 < p < \infty$ it is $k^{-r}(\log k)^{r+1/2}$ and in the case $q = 1$, $p = \infty$ it is $k^{-r}(\log k)^{r+1}$.

We have discussed the results on the right order of decay of the entropy numbers. Clearly, each order relation $\asymp$ is a combination of the upper bound $\ll$ and the matching lower bound $\gg$. We now briefly discuss methods that were used for proving upper and lower bounds. The upper bounds in (3.85) were proved by the standard method of reduction by discretization to estimates of the entropy numbers of finite dimensional sets. We used this method in Section 3.3 when we deduced Theorem 3.19 from Theorem 3.6. Theorem 3.6 plays a key role in this method. It is clear from the above discussion that it was sufficient to prove the lower bound in (3.85) in the case $p = 1$. The proof of this lower bound is more difficult and is based on non-trivial estimates of the volumes of the sets of Fourier coefficients of bounded trigonometric polynomials. Lemma 3.13 plays a key role in this method.

An analog of the upper bound in (3.86) for any $d$ was obtained by Belinsky (1998):

$$\epsilon_k(MW^r_{q,\alpha}, L_\infty) \ll k^{-r}(\log k)^{(d-1)r+1/2}, \quad r > \max(1/q, 1/2). \quad (3.89)$$

That proof is based on Theorem 3.8.

We now proceed to the case $p = \infty$. We already pointed out that the case $r = 1, q = 2, p = \infty$ is equivalent to the small ball problem from probability theory. We discuss related results in detail. Consider the centered Gaussian process $\mathbf{B}_d := (B_x)_{x \in [0,1]^d}$ with covariance

$$E(B_x B_y) = \prod_{i=1}^{d} \min(x_i, y_i), \quad x = (x_1, \ldots, x_d), \quad y = (y_1, \ldots, y_d).$$

This process is called the Brownian sheet. It is known that the sample paths of $\mathbf{B}_d$ are almost surely continuous. We consider them as random elements of the space $C([0, 1]^d)$. The small ball problem is the problem of the asymptotic behavior of the small ball probabilities

$$P(\sup_{x \in [0,1]^d} |B_x| \le \epsilon)$$

as $\epsilon$ tends to zero. We introduce the notation

$$\phi(\epsilon) := -\ln P(\sup_{x \in [0,1]^d} |B_x| \le \epsilon).$$

The following relation is a fundamental result of probability theory: for $d = 2$ and $\epsilon < 1/2$

$$\phi(\epsilon) \asymp \epsilon^{-2}(\ln(1/\epsilon))^3. \quad (3.90)$$

The upper bound in (3.90) was obtained by Lifshits and Tsirelson (1986) and by Bass (1988). The lower bound in (3.90) was obtained by Talagrand (1994).

Kuelbs and Li (1993) discovered the fact that there is a tight relationship between $\phi(\epsilon)$ and the entropy $H_\epsilon(MW^1_{2,\alpha}, L_\infty)$. We note that they considered the general setting of a Gaussian measure on a Banach space. We formulate a special case of their result here, in convenient terms. They proved the equivalence relations: for any $d$

$$\phi(\epsilon) \ll \epsilon^{-2}(\ln(1/\epsilon))^\beta \iff \epsilon_k(MW^1_{2,\alpha}, L_\infty) \ll k^{-1}(\ln k)^{\beta/2};$$

$$\phi(\epsilon) \gg \epsilon^{-2}(\ln(1/\epsilon))^\beta \iff \epsilon_k(MW^1_{2,\alpha}, L_\infty) \gg k^{-1}(\ln k)^{\beta/2}.$$

These relations and (3.90) imply, for $d = 2$,

$$\epsilon_k(MW_{2,\alpha}^1, L_\infty) \asymp k^{-1}(\ln k)^{3/2}. \tag{3.91}$$

Proof of the most difficult part of (3.90) – the lower bound – is based on a special inequality for the Haar polynomials proved by Talagrand (1994) (see Temlyakov (1995b) for a simple proof). We formulate this inequality using dyadic enumeration of the Haar system (see Section 1.3)

$$H_I(x) = H_{I_1}(x_1)H_{I_2}(x_2), \quad x = (x_1, x_2), \quad I = I_1 \times I_2.$$

Talagrand's inequality claims that for any coefficients $\{c_I\}$

$$\Big\| \sum_{I:|I|=2^{-n}} c_I H_I(x) \Big\|_\infty \geq \frac{1}{2} \sum_{m=0}^n \Big\| \sum_{I:|I_1|=2^{-m},|I_2|=2^{m-n}} c_I H_I(x) \Big\|_1, \tag{3.92}$$

where $|I|$ means the measure of $I$.

We note that the lower bound in (3.91) can be deduced directly from (3.92). However, this does not work for deducing the lower bound in (3.86) for general $r$. This difficulty was overcome in Temlyakov (1995a) by proving an analog of (3.92) for the trigonometric system. Let $s = (s_1, s_2)$ be a vector whose coordinates are non-negative integers and

$$\rho(s) := \{k = (k_1, k_2) \in \mathbb{Z}^2 : [2^{s_j-1}] \leq |k_j| < 2^{s_j}, j = 1, 2\}.$$

For an even number $n$ define

$$Y_n := \{s = (2n_1, 2n_2), \quad n_1 + n_2 = n/2\}.$$

Then for any coefficients $\{c_k\}$

$$\Big\| \sum_{s \in Y_n} \sum_{k \in \rho(s)} c_k e^{i(k,x)} \Big\|_\infty \geq C \sum_{s \in Y_n} \Big\| \sum_{k \in \rho(s)} c_k e^{i(k,x)} \Big\|_1, \tag{3.93}$$

where $C$ is a positive number. Inequality (3.93) plays a key role in the proof of lower bounds in (3.86).

We now proceed to the case $q = 1$, $d = 2$. For a natural number $n$ let us denote

$$Q_n := \cup_{s:s_1+s_2 \leq n} \rho(s); \qquad \Delta Q_n := Q_n \setminus Q_{n-1} = \cup_{s:s_1+s_2=n} \rho(s).$$

We call a set $\Delta Q_n$ a *hyperbolic layer*. For a finite set $\Lambda \subset \mathbb{Z}^2$ define

$$\mathcal{T}(\Lambda) := \{f \in L_1 : \hat{f}(k) = 0, k \in \mathbb{Z}^2 \setminus \Lambda\}.$$

For a finite set $\Lambda$ as in Section 3.3.3 we assign to each $f = \sum_{k \in \Lambda} \hat{f}(k) e^{i(k,x)} \in \mathcal{T}(\Lambda)$ a vector

$$A(f) := \{(\mathrm{Re}\,\hat{f}(k), \mathrm{Im}\,\hat{f}(k)), \quad k \in \Lambda\} \in \mathbb{R}^{2|\Lambda|}$$

where $|\Lambda|$ denotes the cardinality of $\Lambda$, and define

$$B_\Lambda(L_p) := \{A(f) : f \in T(\Lambda), \quad \|f\|_p \le 1\}.$$

The volume estimates of the sets $B_\Lambda(L_p)$ and related questions have been studied in a number of papers: $\Lambda = [-n, n]$, $p = \infty$ in Kashin (1980); $\Lambda = [-N_1, N_1] \times [-N_2, N_2]$, $p = \infty$ in Temlyakov (1989b, 1993b) (see also Lemma 3.13 above); arbitrary $\Lambda$ and $p = 1$ in Kashin and Temlyakov (1994). In particular, the results of Kashin and Temlyakov (1994) imply for $d = 2$ and $1 \le p < \infty$ that

$$(vol(B_{\Delta Q_n}(L_p)))^{(2|\Delta Q_n|)^{-1}} \asymp |\Delta Q_n|^{-1/2} \asymp (2^n n)^{-1/2}.$$

It was proved in Kashin and Temlyakov (2003) that in the case $p = \infty$ the volume estimate is different:

$$(vol(B_{\Delta Q_n}(L_\infty)))^{(2|\Delta Q_n|)^{-1}} \asymp (2^n n^2)^{-1/2}. \tag{3.94}$$

We note that in the case $\Lambda = [-N_1, N_1] \times [-N_2, N_2]$ the volume estimate is the same for all $1 \le p \le \infty$. The volume estimate (3.94) plays the key role in the proof of (3.87) and (3.88).

We have previously discussed known results on the rate of decay of $\epsilon_k(MW^r_{q,\alpha}, L_p)$. In the case $d = 2$ the picture is almost complete, whereas for $d > 2$ the situation is fundamentally different. The problem of the right order of decay of $\epsilon_k(MW^r_{q,\alpha}, L_p)$ is still open for $q = 1, 1 \le p \le \infty$ and $p = \infty, 1 \le q \le \infty$. In particular, it is open in the case $q = 2, p = \infty, r = 1$, which is related to the small ball problem. We discuss in more detail the case $p = \infty, 1 \le q \le \infty$. We pointed out above that in the case $d = 2$ the proof of lower bounds (the most difficult part) was based on inequalities (3.92) $(r = 1)$ and (3.93) (all $r$). The existing conjecture is that

$$\epsilon_k(MW^r_{q,\alpha}, L_\infty) \asymp k^{-r}(\ln k)^{(d-1)r+1/2}, \quad 1 < q < \infty, \tag{3.95}$$

for large enough $r$. The upper bound in (3.95) follows from (3.89). It is known that the corresponding lower bound in (3.95) would follow from the $d$-dimensional version of (3.93), which we formulate below. For $s \in \mathbb{Z}^d_+$ define

$$\rho(s) := \{k = (k_1, \ldots, k_d) \in \mathbb{Z}^d : [2^{s_j-1}] \le |k_j| < 2^{s_j}, j = 1, \ldots, d\}.$$

For even $n$, put

$$Y^d_n := \{s = (2l_1, \ldots, 2l_d), l_1 + \cdots + l_d = n/2, l_j \in \mathbb{Z}_+, j = 1, \ldots, d\}.$$

It is conjectured (see, for instance, Kashin and Temlyakov (2008)) that the following inequality holds for any coefficients $\{c_k\}$:

$$\|\sum_{s \in Y_n^d} \sum_{k \in \rho(s)} c_k e^{i(k,x)}\|_\infty \geq C(d) n^{-(d-2)/2} \sum_{s \in Y_n^d} \|\sum_{k \in \rho(s)} c_k e^{i(k,x)}\|_1. \quad (3.96)$$

We note that a weaker version of (3.96) with exponent $(d-2)/2$ replaced by $(d-1)/2$ is a direct corollary of the Parseval identity, the Cauchy inequality and the monotonicity of the $L_p$ norms.

The $d$-dimensional version of (3.92) is similar to (3.96):

$$\|\sum_{I:|I|=2^{-n}} c_I H_I(x)\|_\infty$$

$$\geq C(d) n^{-(d-2)/2} \sum_{s:s_1+\cdots+s_d=n} \|\sum_{I:|I_j|=2^{-s_j},j=1,\ldots,d} c_I H_I(x)\|_1. \quad (3.97)$$

Recently, Bilyk and Lacey (2008) and Bilyk, Lacey and Vagharshakyan (2008) proved (3.97) with the exponent $(d-1)/2 - \delta(d)$ with some positive $\delta(d)$ instead of $(d-2)/2$. There is no progress in proving (3.96).

Kashin and Temlyakov (2008) considered a new norm – the $QC$ norm, which we now briefly discuss. For a periodic univariate function $f \in L_1$ with the Fourier series

$$f \sim \sum_k \hat{f}(k) e^{ikx},$$

we define

$$\delta_0(f,x) := \hat{f}(0), \quad \delta_s(f,x) := \sum_{k:2^{s-1} \leq |k| < 2^s} \hat{f}(k) e^{ikx}, \quad s = 1, 2, \ldots.$$

Let $\{r_s(w)\}_{s=0}^\infty$ be the Rademaher system. Set

$$\|f\|_{QC} := \int_0^1 \|\sum_{s=0}^\infty r_s(w) \delta_s(f,\cdot)\|_\infty \, dw. \quad (3.98)$$

The space of quasi-continuous functions is defined as the closure of the set of trigonometric polynomials in the $QC$ norm (3.98). There are different ways to define the $QC$ norm in the multivariate case (see Kashin and Temlyakov (2008)). We discuss one of them here. For $x = (x_1, \ldots, x_d) \in \mathbb{T}^d$, we set $x^1 = (x_2, \ldots, x_d) \in \mathbb{T}^{d-1}$. Then for $d \geq 2$ we define

$$\|f\|_{QC} := \|\|f(\cdot, x^1)\|_{QC}\|_\infty,$$

where the $QC$ norm is with respect to $x_1$ and the $L_\infty$ norm is with respect to the remaining variables. Kashin and Temlyakov (2008) proved an analog of

(3.96) for the $QC$ norm instead of the $L_\infty$ norm. This allowed the authors to find the right order of decay of the corresponding entropy numbers:

$$\epsilon_k(MW^r_{q,\alpha}, QC) \asymp k^{-r}(\ln k)^{r(d-1)+1/2}, \quad d \geq 2,$$

for $1 < q \leq \infty$ and $r > \max(1/q, 1/2)$.

## 3.7 Open problems

We have already mentioned some open problems in Section 3.6. Problems 3.1–3.4 concern the order in parameter $k$.

**3.1.** Find the order of $\epsilon_k(MW^r_{\infty,\alpha}, L_\infty)$ in the case $d = 2$ and $0 < r \leq 1/2$.
**3.2.** Find the order of $\epsilon_k(MW^r_{1,\alpha}, L_\infty)$ in the case $d = 2$ and $\alpha \neq 0$.
**3.3.** For $d > 2$ find the order of $\epsilon_k(MW^r_{1,\alpha}, L_p)$, $1 \leq p \leq \infty$.
**3.4.** For $d > 2$ find the order of $\epsilon_k(MW^r_{q,\alpha}, L_\infty)$, $1 \leq q \leq \infty$.
**3.5.** Prove (3.97) for $d > 2$.
**3.6.** Prove (3.96) for $d > 2$.

# 4

# Approximation in learning theory

## 4.1 Introduction

This chapter is devoted to some mathematical aspects of recent results on supervised learning. Supervised learning, or learning from examples, refers to a process that builds on the base of available data of inputs $x_i$ and outputs $y_i, i = 1, \ldots, m$, a function that best represents the relation between the inputs $x \in X$ and the corresponding outputs $y \in Y$. This is a big area of research, both in non-parametric statistics and in learning theory. In this chapter we confine ourselves to recent further developments in settings and results obtained from Cucker and Smale (2001). In this chapter we illustrate how methods of approximation theory, in particular greedy algorithms, can be used in learning theory. We begin our discussion with a very brief survey of different settings that are close to the setting of our main interest.

### 4.1.1 Approximation theory; recovery of functions

We discuss the following deterministic model: given

$$\mathbf{z} := ((x_1, y_1), \ldots, (x_m, y_m)) \quad : \quad y_i = f(x_i), \quad i = 1, \ldots, m, \quad f \in \Theta,$$

we recover $f \in \Theta$ (find an approximant of $f$). The error of approximation is measured in some norm $\| \cdot \|$. Usually it is the $L_p$ norm, $1 \leq p \leq \infty$, with respect to the Lebesgue measure on a given domain $X$.

### 4.1.2 Statistics; regression theory

(a) Fixed design model. Given

$$\mathbf{z} := ((x_1, y_1), \ldots, (x_m, y_m)) \quad : \quad y_i = f(x_i) + \epsilon_i,$$

183

$x_1, \ldots, x_m$ fixed; $\epsilon_i$ independent identically distributed (i.i.d.), $E\epsilon_i = 0$, $f \in \Theta$, we find an approximant for $f$ (estimator $\hat{f}$). The unknown function $f$ is called the regression function. Error is measured by expectation $E(\|f - \hat{f}\|^2)$ of some of the standard norms.

(b)  Random design model. Given

$$\mathbf{z} := ((x_1, y_1), \ldots, (x_m, y_m)) \quad : \quad y_i = f(x_i) + \epsilon_i,$$

$x_1, \ldots, x_m$ random, i.i.d.; $\epsilon_i$ i.i.d. (independent of $x_i$), $E\epsilon_i = 0$, $f \in \Theta$, we find an estimator $\hat{f}$ for $f$. The error is measured by the expectation $E(\|f - \hat{f}\|^2)$.

(c)  Distribution-free theory of regression. Let $X \subset \mathbb{R}^d$, $Y \subset \mathbb{R}$ be Borel sets, and let $\rho$ be a Borel probability measure on $Z = X \times Y$. For $f : X \to Y$ define *the error*

$$\mathcal{E}(f) := \int_Z (f(x) - y)^2 \, d\rho.$$

Consider $\rho_X$, the marginal probability measure on $X$ (for $S \subset X$, $\rho_X(S) = \rho(S \times Y)$). Define

$$f_\rho(x) := E(y|x)$$

to be a conditional expectation of $y$. The function $f_\rho$ is known in statistics as the *regression function* of $\rho$. It is clear that if $f_\rho \in L_2(\rho_X)$ then it minimizes the error $\mathcal{E}(f)$ over all $f \in L_2(\rho_X)$: $\mathcal{E}(f_\rho) \leq \mathcal{E}(f)$, $f \in L_2(\rho_X)$. Thus, in the sense of error $\mathcal{E}(\cdot)$ the regression function $f_\rho$ is the best one to describe the relation between inputs $x \in X$ and outputs $y \in Y$. Given: $(x_i, y_i)$, $i = 1, \ldots, m$, independent identically distributed according to $\rho$, $|y| \leq M$ a.e. Find an estimator $\hat{f}$ for $f_\rho$. Error: $E(\|f_\rho - \hat{f}\|^2_{L_2(\rho_X)})$.

### 4.1.3  Learning theory

This is a vast area of research with a wide range of different settings. In this chapter we only discuss a development of a setting from Cucker and Smale (2001). For results in other settings we recommend the fundamental book by Vapnik (1998) and a nice survey on the classification problem by Lugosi (2002). Our setting is similar to the setting of the distribution-free regression problem. The goal is to find an estimator $f_\mathbf{z}$, on the basis of given data $\mathbf{z} = ((x_1, y_1), \ldots, (x_m, y_m))$ that approximates $f_\rho$ (or its projection) well with high probability. We assume that $(x_i, y_i)$, $i = 1, \ldots, m$, are independent and distributed according to $\rho$. As in the distribution-free theory of regression we measure the error in the $L_2(\rho_X)$ norm. This differentiates between distribution-free theory of regression and our setting of learning theory from classical non-parametric statistics. One can find a discussion of the relationship between the fixed design model, the random design model, and the distribution-free theory of regression in Györfy *et al.* (2002) (see also Barron, Birgé and Massart (1999) and Van de Geer (2000)). Here we only mention that the problem of learning theory that we discuss in this chapter can be rewritten in the form

$$y_i = f_\rho(x_i) + \epsilon_i, \quad \epsilon := y - f_\rho(x),$$

close to the form of the random design model. However, in our setting we are not assuming that $\epsilon$ and $x$ are independent. While the theories of fixed and random design models do not directly apply to our setting, they utilize several of the same techniques we shall encounter, such as the use of entropy and the construction of estimators through minimal risk.

We note that a standard setting in the distribution-free theory of regression (see Györfy *et al.* (2002)) involves the expectation as a measure of quality of an estimator. An important new feature of the setting in learning theory formulated in Cucker and Smale (2001) comprises the following. They propose to study systematically the probability distribution function

$$\rho^m \{\mathbf{z} : \| f_\rho - f_\mathbf{z} \|_{L_2(\rho_X)} \geq \eta \}$$

instead of the expectation.

There are several important ingredients in the mathematical formulation of the learning problem. In our formulation we follow the way that has become standard in approximation theory and is based on the concept of *optimal method*.

We begin with a class $\mathcal{M}$ of admissible measures $\rho$. Usually, we impose restrictions on $\rho$ in the form of restrictions on the regression function $f_\rho$: $f_\rho \in \Theta$. Then the first step is to find an optimal estimator for a given class $\Theta$ of priors (we assume $f_\rho \in \Theta$). In regression theory a usual way to evaluate performance of an estimator $f_\mathbf{z}$ is by studying its convergence in expectation, i.e. the rate of decay of the quantity $E(\| f_\rho - f_\mathbf{z} \|^2_{L_2(\rho_X)})$ as the sample size $m$ increases. Here the expectation is taken with respect to the product measure $\rho^m$ defined on $Z^m$. We note that $\mathcal{E}(f_\mathbf{z}) - \mathcal{E}(f_\rho) = \| f_\mathbf{z} - f_\rho \|^2_{L_2(\rho_X)}$ (see Proposition 4.31 below). As we have already mentioned above a more accurate and more delicate way of evaluating the performance of $f_\mathbf{z}$ has been described in Cucker and Smale (2001). In this chapter we concentrate on a discussion of the results on the probability distribution function.

An important question in finding an optimal $f_z$ is: How should we describe the class $\Theta$ of priors? In other words, what characteristics of $\Theta$ govern, say, the optimal rate of decay of $E(\| f_\rho - f_z \|_{L_2(\rho_X)}^2)$ for $f_\rho \in \Theta$? Previous and recent work in statistics and learning theory (see Barron (1991), Barron, Birgé and Massart (1999), Barron *et al.* (2008), Binev *et al.* (2005), Cucker and Smale (2001), DeVore *et al.* (2004, 2006), Györfy *et al.* (2002), Konyagin and Temlyakov (2004, 2007), Lugosi (2002), Temlyakov (2008a), Van de Geer (2000), Vapnik (1998)) indicate that the compactness characteristics of $\Theta$ play a fundamental role in the above problem. It is convenient for us to express the compactness of $\Theta$ in terms of the entropy numbers. In this chapter we discuss the classical concept of entropy and the concept of tight entropy. We note that some other concepts of entropy, for example entropy with bracketing, prove to be useful in the theory of empirical processes and non-parametric statistics (see Van de Geer (2000) and Vapnik (1998)). There is a concept of $VC$ dimension that plays a fundamental role in the problem of pattern recognition and classification (Vapnik, 1998). This concept is also useful in describing compactness characteristics of sets.

For a compact subset $\Theta$ of a Banach space $B$ we define the entropy numbers as follows (see Chapter 3):

$$\epsilon_n(\Theta, B) := \epsilon_n^2(\Theta, B) := \inf\left\{\epsilon : \exists f_1, \ldots, f_{2^n} \in \Theta : \Theta \subset \cup_{j=1}^{2^n}(f_j + \epsilon U(B))\right\},$$

where $U(B)$ is the unit ball of a Banach space $B$. We denote by $N(\Theta, \epsilon, B)$ the covering number that is the minimal number of balls of radius $\epsilon$ with centers in $\Theta$ needed for covering $\Theta$. The corresponding $\epsilon$-net is denoted by $\mathcal{N}_\epsilon(\Theta, B)$. In Cucker and Smale (2001), DeVore *et al.* (2006) and Konyagin and Temlyakov (2004), in the most cases the space $\mathcal{C} := \mathcal{C}(X)$ of continuous functions on a compact $X \subset \mathbb{R}^d$ has been taken as a Banach space $B$. This allows us to formulate all results with assumptions on $\Theta$ independent of $\rho$. In Konyagin and Temlyakov (2007) and Binev *et al.* (2005) some results are obtained for $B = L_2(\rho_X)$. On the one hand, we weaken assumptions on the class $\Theta$, and on the other hand this results in the use of $\rho_X$ in the construction of an estimator. Thus, we have a tradeoff between treating wider classes and building estimators that are independent of $\rho_X$. We note that in practice we often do not know the $\rho_X$. Thus, it is very desirable to build estimators independent of $\rho_X$. In statistics this type of regression problem is referred to as *distribution-free*. A recent survey on distribution-free regression theory is provided in Györfy *et al.* (2002).

In this chapter we always assume that the unknown measure $\rho$ satisfies the condition $|y| \leq M$ (or, a little weaker, $|y| \leq M$ a.e. with respect to $\rho_X$) with some fixed $M$. Then it is clear that for $f_\rho$ we have $|f_\rho(x)| \leq M$ for all $x$ (for almost all $x$). Therefore, it is natural to assume that a class $\Theta$ of priors where $f_\rho$ belongs is embedded into the $C(X)$-ball ($L_\infty$-ball) of radius $M$.

In DeVore *et al.* (2006) and Konyagin and Temlyakov (2004) the restrictions on a class $\Theta$ were imposed in the following forms:

$$\epsilon_n(\Theta, \mathcal{C}) \leq Dn^{-r}, \quad n = 1, 2, \ldots, \quad \Theta \subset DU(\mathcal{C}), \quad (4.1)$$

or

$$d_n(\Theta, \mathcal{C}) \leq Kn^{-r}, \quad n = 1, 2, \ldots, \quad \Theta \subset KU(\mathcal{C}). \quad (4.2)$$

Here, $d_n(\Theta, B)$ is the Kolmogorov width. Kolmogorov's $n$-width for the centrally symmetric compact set $\Theta$ in the Banach space $B$ is defined as follows:

$$d_n(\Theta, B) := \inf_L \sup_{f \in \Theta} \inf_{g \in L} \|f - g\|_B,$$

where $\inf_L$ is taken over all $n$-dimensional linear subspaces of $B$. In Konyagin and Temlyakov (2007) a weaker restriction,

$$\epsilon_n(\Theta, L_2(\rho_X)) \leq Dn^{-r}, \quad n = 1, 2, \ldots, \quad \Theta \subset DU(L_2(\rho_X)), \quad (4.3)$$

was imposed.

We have already mentioned that the study of the probability distribution function $\rho^m\{\mathbf{z} : \|f_\rho - f_\mathbf{z}\|_{L_2(\rho_X)} \geq \eta\}$ is a more difficult and delicate problem than the study of the expectation $E(\|f_\rho - f_\mathbf{z}\|^2_{L_2(\rho_X)})$. We encounter this difficulty even at the level of the formulation of a problem. The reason for this is that the probability distribution function controls two characteristics: $\eta$, the error of estimation, and $1 - \rho^m\{\mathbf{z} : \|f_\rho - f_\mathbf{z}\|_{L_2(\rho_X)} \geq \eta\}$, the confidence of the error $\eta$. Therefore, we need a mathematical formulation of the above discussed problems of optimal estimators.

We propose (see DeVore *et al.* (2006)) to study the following function that we call the *accuracy confidence function*. Let a set $\mathcal{M}$ of admissible measures $\rho$ and a sequence $\mathbb{E} := \{\mathbb{E}(m)\}_{m=1}^\infty$ of allowed classes $\mathbb{E}(m)$ of estimators be given. For $m \in \mathbb{N}$, $\eta > 0$, we define

$$\mathbf{AC}_m(\mathcal{M}, \mathbb{E}, \eta) := \inf_{E_m \in \mathbb{E}(m)} \sup_{\rho \in \mathcal{M}} \rho^m\{\mathbf{z} : \|f_\rho - f_\mathbf{z}\|_{L_2(\rho_X)} \geq \eta\},$$

where $E_m$ is an estimator that maps $\mathbf{z} \to f_\mathbf{z}$. For example, $\mathbb{E}(m)$ could be a class of all estimators, a class of linear estimators of the form

$$f_\mathbf{z} = \sum_{i=1}^m w_i(x_1, \ldots, x_m, x)y_i,$$

or a specific estimator. In the case where $\mathbb{E}(m)$ is the set of all estimators, $m = 1, 2, \ldots,$ we write $\mathbf{AC}_m(\mathcal{M}, \eta)$.

We discuss results on $\mathbf{AC}_m(\mathcal{M}, \mathbb{E}, \eta)$ with $\mathcal{M} = \mathcal{M}(\Theta) := \{\rho : f_\rho \in \Theta\}$. In this case we write $\mathbf{AC}_m(\mathcal{M}(\Theta), \mathbb{E}, \eta) =: \mathbf{AC}_m(\Theta, \mathbb{E}, \eta)$. Section 4.4 is devoted to the study of priors on $f_\rho$ in the form $f_\rho \in \Theta$. This setting is referred to as *the proper function learning problem*. In Section 4.4 (see Theorem 4.74) we obtain a right behavior of the **AC** function for classes satisfying the entropy condition

$$\epsilon_n(\Theta, L_2(\mu)) \asymp n^{-r}.$$

This is one of the main results of this chapter. We give detailed proof of this theorem here. The upper bounds are proved in Section 4.4 and the corresponding lower bounds are proved in Section 4.5. It is interesting to note that the proof uses only well known classical results from statistics: the Bernstein inequality in the proof of the upper bounds and the Fano lemma in the proof of the lower bounds.

We point out that results on the expectation can be obtained as a corollary of the results on the **AC** function. For a class $\Theta$ consider

$$E(\Theta, m, \hat{f}) := \sup_{f_\rho \in \Theta} E\left(\|f_\rho - \hat{f}\|_{L_2(\rho_X)}^2\right),$$

$$E(\Theta, m) := \inf_{\hat{f}} E(\Theta, m, \hat{f}).$$

It is clear from the definition of $E(\Theta, m)$ and $\mathbf{AC}_m(\Theta, \eta)$ that

$$\int_0^\infty \mathbf{AC}_m(\Theta, \eta^{1/2})d\eta \leq E(\Theta, m), \qquad (4.4)$$

and, for $\rho, \Theta$ satisfying $|y| \leq M$, $\Theta \subset MU(\mathcal{C}(X))$,

$$E(\Theta, m) \leq \min_\eta(\eta^2 + 4M^2 \mathbf{AC}_m(\Theta, \eta)). \qquad (4.5)$$

One of the important variants of the learning problem formulated in Cucker and Smale (2001) is the following. We now do not impose any restrictions on $\rho$, except $|y| \leq M$ a.e., and instead of estimating the regression function $f_\rho$ we estimate a projection $(f_\rho)_W$ of $f_\rho$ onto a compact set $W$ of our choice. This setting is referred to as *the improper function learning problem* or *the projection learning problem*. As in the above case ($f_\rho \in \Theta$), we introduce the corresponding accuracy confidence function

$$\mathbf{AC}_m^p(W, \mathbb{E}, \eta) := \inf_{E_m \in \mathbb{E}(m)} \sup_\rho \rho^m \{\mathbf{z} : \mathcal{E}(f_{\mathbf{z}}) - \mathcal{E}((f_\rho)_W) \geq \eta^2\}.$$

In the case where $\mathbb{E}(m)$, $m = 1, 2, \ldots$, is a collection of all estimators $E_m$ : $\mathbf{z} \to f_{\mathbf{z}} \in W$ we drop $\mathbb{E}$ from the notation. We note (see Lemma 4.47) that, in the case of convex $W$, we have for any $f \in W$

$$\|f - (f_\rho)w\|^2_{L_2(\rho_X)} \le \mathcal{E}(f) - \mathcal{E}((f_\rho)w).$$

We discuss related results in Section 4.3. In Sections 4.3 and 4.4 we provide a more detailed discussion of the upper bounds for the corresponding accuracy confidence functions. The upper bounds from Sections 4.3 and 4.4 are complemented by the lower bounds from Section 4.5. We formulate one important conclusion of results from Sections 4.4 and 4.5 here: the entropy numbers $\epsilon_n(\Theta, L_2(\rho_X))$ are the right characteristic of the class of priors $\Theta$ in the estimation problem. Further discussion can be found in Section 4.4.

## 4.2 Some basic concepts of probability theory

### 4.2.1 The measure theory and integration

We begin by recalling that a $\sigma$-algebra $\Sigma$ of subsets of a given set $X$ contains $\emptyset$ and is closed with respect to complements and with respect to countable unions. By a set function $\mu$ we mean a function which assigns an extended real number to certain sets.

**Definition 4.1** By a measurable space we mean a couple $(X, \Sigma)$ consisting of a set $X$ and $\sigma$-algebra $\Sigma$ of subsets of $X$. A subset $A$ of $X$ is called measurable (or measurable with respect to $\Sigma$) if $A \in \Sigma$.

**Definition 4.2** By a measure $\mu$ on a measurable space $(X, \Sigma)$ we mean a non-negative set function defined for all sets of $\Sigma$ and satisfying $\mu(\emptyset) = 0$ and

$$\mu\left(\bigcup_{i=1}^{\infty} A_i\right) = \sum_{i=1}^{\infty} \mu(A_i) \tag{4.6}$$

for any sequence $\{A_i\}$ of disjoint measurable sets. By a measure space $(X, \Sigma, \mu)$ we mean a measurable space $(X, \Sigma)$ together with a measure $\mu$ defined on $\Sigma$.

The property (4.6) of $\mu$ is referred to as *countable additivity*. We also have that $\mu$ is *finitely additive*:

$$\mu\left(\bigcup_{i=1}^{N} A_i\right) = \sum_{i=1}^{N} \mu(A_i)$$

for disjoint sets $A_i$ belonging to $\Sigma$.

One example of a measure space is $(\mathbb{R}, \mathcal{L}, \mu)$, where $\mathbb{R}$ is the set of real numbers, $\mathcal{L}$ is the Lebesgue measurable sets of real numbers and $\mu$ is the Lebesgue measure.

We formulate three basic properties of the measures.

**Proposition 4.3** *If $A \in \Sigma$, $B \in \Sigma$ and $A \subseteq B$, then*

$$\mu(A) \leq \mu(B).$$

**Proposition 4.4** *If $A_i \in \Sigma$, $\mu(A_1) < \infty$ and $A_{i+1} \subseteq A_i$, $i = 1, 2, \ldots$, then*

$$\mu\left(\bigcap_{i=1}^{\infty} A_i\right) = \lim_{i \to \infty} \mu(A_i).$$

**Proposition 4.5** *If $A_i \in \Sigma$, $i = 1, 2, \ldots$, then*

$$\mu\left(\bigcup_{i=1}^{\infty} A_i\right) \leq \sum_{i=1}^{\infty} \mu(A_i).$$

A measure space $(X, \Sigma, \mu)$ is said to be *complete* if $\Sigma$ contains all subsets of sets of measure zero; that is, if $B \in \Sigma$, $\mu(B) = 0$ and $A \subseteq B$, then $A \in \Sigma$. Thus the Lebesgue measure is complete, while the Lebesgue measure restricted to the $\sigma$-algebra of Borel sets is not complete.

**Proposition 4.6** *If $(X, \Sigma, \mu)$ is a measure space, then we can find a complete measure space $(X, \Sigma_0, \mu_0)$ such that*

$$\Sigma \subseteq \Sigma_0; \quad A \in \Sigma \Rightarrow \mu(A) = \mu_0(A);$$

$$D \in \Sigma_0 \iff D = A \cup B, \quad B \in \Sigma, \quad A \subseteq C, \quad C \in \Sigma, \quad \mu(C) = 0.$$

**Definition 4.7** A real-valued function $f$ defined on $X$ is called measurable (or measurable with respect to $\Sigma$) if for any $a$

$$\{x \in X : f(x) \geq a\} \in \Sigma.$$

The following theorem shows that the property of being measurable is stable under many operations.

**Theorem 4.8** *If $c$ is a constant and the functions $f$ and $g$ are measurable, then so are the functions $cf$, $f + g$, $fg$ and $\max(f, g)$. Moreover, if $\{f_n\}_{n=1}^{\infty}$ is a sequence of measurable functions, then the functions $\sup_n\{f_n\}$, $\inf_n\{f_n\}$, $\limsup_{n \to \infty} f_n$ and $\liminf_{n \to \infty} f_n$ are all measurable.*

By a simple function we mean a finite linear combination

$$g(x) = \sum_{i=1}^{n} c_i \chi_{A_i}(x) \tag{4.7}$$

of characteristic functions $\chi_{A_i}(x)$ of measurable sets $A_i$.

**Proposition 4.9** *Let $f$ be a non-negative measurable function. Then there is a sequence $\{g_n\}$ of simple functions with $g_{n+1} \geq g_n$ such that $f(x) = \lim_{n\to\infty} g_n(x)$ at each point of $X$.*

If $A$ is a mesurable set and $g$ is a non-negative simple function, we define

$$\int_A g \, d\mu := \sum_{i=1}^{n} c_i \mu(A_i \cap A),$$

where $g$ is given by (4.7).

It is easily seen that the value of this integral is independent of the representation of $g$ which we use.

**Definition 4.10** Let $f$ be a non-negative real-valued measurable function on the measure space $(X, \Sigma, \mu)$. Then $\int_X f \, d\mu := \int f \, d\mu$ is the supremum of the integrals $\int g \, d\mu$ as $g$ ranges over all simple functions with $0 \leq g \leq f$.

We formulate some theorems on the properties of the integral.

**Theorem 4.11** (**Fatou's lemma**) *Let $\{f_n\}$ be a sequence of non-negative measurable functions that converge almost everywhere on a measurable set $A$ to a function $f$. Then*

$$\int_A f \, d\mu \leq \liminf_{n\to\infty} \int_A f_n \, d\mu.$$

**Theorem 4.12** (**Monotone convergence theorem**) *Let $\{f_n\}$ be a sequence of non-negative measurable functions that converge almost everywhere to a function $f$ and suppose that $f_n \leq f_{n+1}$ for all $n$. Then*

$$\int f \, d\mu = \lim_{n\to\infty} \int f_n \, d\mu.$$

**Corollary 4.13** *Let $\{f_n\}$ be a sequence of non-negative measurable functions. Then*

$$\int \sum_{n=1}^{\infty} f_n \, d\mu = \sum_{n=1}^{\infty} \int f_n \, d\mu.$$

A non-negative function $f$ is called *integrable* (over a mesurable set $A$ with respect to $\mu$) if it is measurable and

$$\int_A f \, d\mu < \infty.$$

An arbitrary function $f$ is said to be integrable if both $f^+(x) := \max(f(x), 0)$ and $f^- := f^+ - f$ are integrable. In this case we define

$$\int_A f \, d\mu := \int_A f^+ \, d\mu - \int_A f^- \, d\mu.$$

**Theorem 4.14** (**Lebesgue convergence theorem**) *Let g be integrable over A, and suppose that* $\{f_n\}$ *is a sequence of measurable functions such that on A*

$$|f_n(x)| \leq g(x)$$

*and such that almost everywhere on A*

$$f_n(x) \to f(x).$$

*Then*

$$\int_A f \, d\mu = \lim_{n \to \infty} \int_A f_n \, d\mu.$$

A measure $\nu$ is said to be *absolutely continuous* with respect to the measure $\mu$ if $\nu(A) = 0$ for each set $A$ for which $\mu(A) = 0$. Whenever we are dealing with more than one measure on a measurable space $(X, \Sigma)$, the term "almost everywhere" becomes ambiguous, and we must specify almost everywhere with respect to $\mu$ or almost everywhere with respect to $\nu$, etc. These are usually abbreviated as a.e. $[\mu]$ and a.e. $[\nu]$.

Let $\mu$ be a measure and let $f$ be a non-negative measurable function on $X$. For $A \in \Sigma$, set

$$\nu(A) := \int_A f \, d\mu.$$

Then $\nu$ is a set function defined on $\Sigma$, and it follows from Corollary 4.13 that $\nu$ is countably additive and hence a measure. The measure $\nu$ will be finite if and only if $f$ is integrable. Since the integral over a set of $\mu$-measure zero is zero, we have $\nu$ absolutely continuous with respect to $\mu$. The next theorem shows that every absolutely continuous measure $\nu$ is obtained in this fashion.

**Theorem 4.15** (**Radon–Nikodim theorem**) *Let* $(X, \Sigma, \mu)$ *be a finite measure space, and let* $\nu$ *be a measure defined on* $\Sigma$ *which is absolutely continuous with respect to* $\mu$. *Then there is a non-negative measurable function* $f$ *such that for each set A in* $\Sigma$ *we have*

$$\nu(A) = \int_A f \, d\mu.$$

*The function f is unique in the sense that if g is any measurable function with this property then* $g = f$ *a.e.* $[\mu]$.

If $(X, \Sigma, \mu)$ is a complete measure space, for $1 \leq p < \infty$ we denote by $L_p(\mu) := L_p(X, \mu)$ the space of all measurable functions on $X$ for which

$$\|f\|_p := \left( \int |f|^p \, d\mu \right)^{1/p} < \infty,$$

and for $p = \infty$ we set

$$\|f\|_\infty := \text{esssup}|f|.$$

For $1 \le p \le \infty$ the spaces $L_p(\mu)$ are Banach spaces.

**Theorem 4.16 (Hölder's inequality)** *If $f \in L_p(\mu)$ and $g \in L_q(\mu)$, with $1/p + 1/q = 1$, then $fg \in L_1(\mu)$ and*

$$\int |fg|d\mu \le \|f\|_p \|g\|_q.$$

**Theorem 4.17 (Riesz representation theorem)** *Let $F$ be a bounded linear functional on $L_p(\mu)$ with $1 \le p < \infty$, where $\mu$ is a finite measure. Then there is a unique element $g$ in $L_q(\mu)$, where $1/q + 1/p = 1$, such that*

$$F(f) = \int fg\,d\mu.$$

*We have also $\|F\| = \|g\|_q$.*

We now describe a way to obtain the product measure. Let $(X_1, \Sigma_1, \mu_1)$ and $(X_2, \Sigma_2, \mu_2)$ be two complete measure spaces, and consider the direct product $X := X_1 \times X_2$. If $A \subseteq X_1$ and $B \subseteq X_2$, we call $A \times B$ a rectangle. If $A \in \Sigma_1$ and $B \in \Sigma_2$, we call $A \times B$ a measurable rectangle. If $A \times B$ is a measurable rectangle, we set

$$\lambda(A \times B) := \mu_1(A)\mu_2(B).$$

The measure $\lambda$ can be extended to be a complete measure on a $\sigma$-algebra $\Sigma$ containing all measurable rectangles. This extended measure $\mu$ is called the product measure of $\mu_1$ and $\mu_2$ and is denoted by $\mu := \mu_1 \times \mu_2$. In a particular case when $(X_1, \Sigma_1, \mu_1) = (X_2, \Sigma_2, \mu_2)$ we write $\mu = \mu_1^2$. If $\mu_i$, $i = 1, 2$, are finite, so is $\mu$. If $X_1$ and $X_2$ are the real line and $\mu_1$ and $\mu_2$ are both Lebesgue measures, then $\mu$ is called the two-dimensional Lebesgue measure for the plane.

**Theorem 4.18 (Fubini theorem)** *Using the above notations let $f$ be an integrable function on $X$ with respect to $\mu$. Then we have the following.*

(i) *For almost all $x_1 \in X_1$ the function $f(x_1, \cdot)$ is an integrable function on $X_2$ with respect to $\mu_2$. For almost all $x_2 \in X_2$ the function $f(\cdot, x_2)$ is an integrable function on $X_1$ with respect to $\mu_1$.*

(ii) *The functions*

$$\int_{X_2} f(x_1, x_2)d\mu_2, \quad \int_{X_1} f(x_1, x_2)d\mu_1$$

*are integrable functions on $X_1$ and $X_2$, respectively.*

*(iii) The following equalities hold:*

$$\int_{X_1} \left( \int_{X_2} f \, d\mu_2 \right) d\mu_1 = \int_X f \, d\mu = \int_{X_2} \left( \int_{X_1} f \, d\mu_1 \right) d\mu_2.$$

Let us introduce some concepts of probability theory. The measure space $(X, \Sigma, \mu)$ is called the *probability space* if $\mu(X) = 1$ and the measure $\mu$ on a probability space is called a *probability measure*. Another way of saying that $\mu$ is a probability measure on $X$ is to say that a sample $x$ is distributed on $X$ according to $\mu$. The elements of $\Sigma$ are called the *events*. For an event $A \in \Sigma$ the number $\mu(A)$ is called the *probability of the event A*. It is clear that for all events $A$ one has $0 \leq \mu(A) \leq 1$. The term *almost everywhere* is replaced by *almost surely*. A real-valued measurable function $f$ defined on $X$ is called a *random variable*. For a random variable $f$ that is integrable we define the *expectation* by the formula

$$Ef := E(f) := \int_X f \, d\mu.$$

We define the conditional probability of an event $A$, given that $B$ has already occurred, in symbols $\mu(A|B)$, as the ratio

$$\mu(A|B) := \mu(A \cap B)/\mu(B) \tag{4.8}$$

provided $\mu(B) \neq 0$. We say that two events $A$ and $B$ are *independent* (*stochastically independent*) if

$$\mu(A|B) = \mu(A) \tag{4.9}$$

provided $\mu(B)$ is non-zero. It is clear from the definition (4.8) that (4.9) is equivalent to

$$\mu(A \cap B) = \mu(A)\mu(B). \tag{4.10}$$

We use (4.10) as a definition of the stochastic independence for any two events $A$ and $B$. It now follows from this definition that an event $A$ such that $\mu(A) = 0$ is independent with any event $B$.

The events of the class **A** are *independent* if

$$\mu \left( \bigcap_{i=1}^n A_i \right) = \prod_{i=1}^n \mu(A_i)$$

for every finite class $\{A_i\}_{i=1}^n$ of distinct events in **A**. The random variables of the class **F** are *independent* if

$$\mu \left( \bigcap_{i=1}^n \{x : f_i(x) \in M_i\} \right) = \prod_{i=1}^n \mu(\{x : f_i(x) \in M_i\})$$

for every finite subset $\{f_i\}_{i=1}^n$ of distinct random variables of $\mathbf{F}$ and for every class $\{M_i\}_{i=1}^n$ of Borel sets on the real line. An equivalent way of expressing this condition is to say that if, for each $f \in \mathbf{F}$, $M_f$ is a Borel set on the real line, then, for every possible choice of the Borel sets $M_f$, the events (sets) of the class $\mathbf{A} := \{f^{-1}(M_f) : f \in \mathbf{F}\}$ are independent.

For integrable independent random variables $f_1$ and $f_2$ their product $f_1 f_2$ is also integrable and

$$E(f_1 f_2) = E(f_1)E(f_2). \tag{4.11}$$

The variance of an integrable random variable $f$, denoted $\sigma^2(f)$, is defined by

$$\sigma^2(f) := \int (f - Ef)^2 \, d\mu. \tag{4.12}$$

**Theorem 4.19** *If $f$ and $g$ are integrable independent random variables with a finite variance, then*

$$\sigma^2(f + g) = \sigma^2(f) + \sigma^2(g).$$

*Proof* First of all, we note that our assumption implies that $f, g \in L_2(\mu)$. The definition (4.12) implies

$$\sigma^2(f) = E(f^2) - (Ef)^2. \tag{4.13}$$

Therefore,

$$\sigma^2(f + g) = E((f + g)^2) - (Ef + Eg)^2. \tag{4.14}$$

Using (4.11) we obtain

$$E((f + g)^2) = E(f^2) + 2E(f)E(g) + E(g^2). \tag{4.15}$$

Combining (4.14) and (4.15) we get

$$\sigma^2(f + g) = E(f^2) - (Ef)^2 + E(g^2) - (Eg)^2 = \sigma^2(f) + \sigma^2(g).$$

$\square$

We now proceed to a more delicate concept of the *conditional expectation*. There are different equivalent ways to define this concept; we have chosen the way that is most convenient for our applications, albeit not the most general one. Let $Z = X \times Y$ and $(Z, \Sigma, \rho)$ be a probability space. First, we define a "projection" of this probability space onto $X$. We let

$$\Sigma_X := \{A \subseteq X : A \times Y \in \Sigma\},$$

and for $A \in \Sigma_X$ we define

$$\rho_X(A) := \rho(A \times Y).$$

It is clear that the $(X, \Sigma_X, \rho_X)$ forms a probability space. The measure $\rho_X$ is called the *marginal probability*.

Let us consider a random variable $f \in L_2(\rho) := L_2(Z, \rho)$. We will define the conditional expectation $E(f(x, y)|x)$. We begin with the bounded linear functional $\varphi_f$ defined on the $L_2(\rho_X) := L_2(X, \rho_X)$ by the formula

$$\varphi_f(g) := \int_Z f(x, y)g(x)d\rho, \quad g \in L_2(\rho_X).$$

The boundedness of this functional follows from the following observation. The condition $g \in L_2(\rho_X)$ implies that $g \in L_2(Z, \rho)$. The boundedness of $\varphi_f$ on the $L_2(\rho)$ follows from Theorem 4.16. Therefore, by Theorem 4.17 with $p = 2$ there is a unique element $E(f|x)$ in $L_2(\rho_X)$ such that

$$\varphi_f(g) = \int_X E(f|x)g \, d\rho_X.$$

We call the $E(f|x)$ the *conditional expectation* of $f$ (the conditional expectation of $f$ with respect to $x$).

Let us list some properties of the conditional expectation. First of all, it is clear that

$$E(f_1 + f_2|x) = E(f_1|x) + E(f_2|x).$$

Next, for any bounded measurable with respect to $\rho_X$ function $u(x)$ we have

$$E(f(x, y)u(x)|x) = u(x)E(f|x). \tag{4.16}$$

Indeed,

$$\varphi_{fu}(g) = \varphi_f(ug) = \int_X E(f|x)u(x)g(x)d\rho_X.$$

By the uniqueness in the Riesz representation theorem we get (4.16).

### 4.2.2 The concentration of measure inequalities

We begin this section with some classical inequalities that follow directly from the definition of the expectation. Let $(X, \Sigma, \mu)$ be a probability space.

If $\xi$ is a non-negative random variable, then the Markov inequality says

$$\mu\{\omega : \xi(\omega) \geq t\} \leq E\xi/t, \quad t > 0. \tag{4.17}$$

If $\xi$ is a random variable, then the Chebyshev inequality says

$$\mu\{\omega : |\xi(\omega)| \geq t\} \leq E((\xi)^2)/t^2, \quad t > 0. \tag{4.18}$$

The Chernoff inequality gives, for $s > 0$,

$$\mu\{\omega : \xi(\omega) \geq t\} = \mu\{\omega : e^{s\xi(\omega)} \geq e^{st}\} \leq e^{-st}E(e^{s\xi}). \tag{4.19}$$

**Lemma 4.20** *Let $\xi$ be a random variable with $E\xi = 0$ and $|\xi| \leq b$ almost surely. Then for $s \geq 0$*

$$E(e^{s\xi}) \leq \cosh(sb) \leq \exp((sb)^2/2).$$

*Proof* By convexity of $e^x$ we have for $|\xi| \leq b$

$$e^{s\xi} \leq \frac{1}{2}\left(1 + \frac{\xi}{b}\right)e^{sb} + \frac{1}{2}\left(1 - \frac{\xi}{b}\right)e^{-sb}$$

and

$$E(e^{s\xi}) \leq \cosh(sb).$$

Next,

$$\cosh x = \sum_{k=0}^{\infty} \frac{x^{2k}}{(2k)!}$$

and

$$e^{x^2/2} = \sum_{k=0}^{\infty} \frac{x^{2k}}{2^k k!}.$$

Then the inequality $(2k)! \geq 2^k k!$ implies $\cosh x \leq e^{x^2/2}$. $\qquad\square$

**Theorem 4.21 (Hoeffding's inequality)** *Let $\xi_i$ be random variables on $(X, \Sigma, \mu)$ such that $|\xi_i - E\xi_i| \leq b_i$, $i = 1, \ldots, m$, almost surely. Consider a new random variable $\zeta$ on $(X^m, \Sigma^m, \mu^m)$ defined as*

$$\zeta(\omega) := \sum_{i=1}^{m} \xi_i(\omega_i), \quad \omega = (\omega_1, \ldots, \omega_m).$$

*Then for $t > 0$*

$$\mu^m\{\omega : |\zeta(\omega) - E\zeta| \geq mt\} \leq 2\exp\left(-\frac{m^2 t^2}{2\|b\|_2^2}\right).$$

*Proof* By Chernoff's inequality

$$\mu^m\{\omega : \zeta(\omega) - E\zeta \geq mt\} \leq \inf_{s>0} e^{-mst} E(e^{s(\zeta - E\zeta)}). \qquad (4.20)$$

Next, we have

$$E(e^{s(\zeta - E\zeta)}) = \prod_{i=1}^{m} E(e^{s(\xi_i(\omega_i) - E\xi_i)}). \qquad (4.21)$$

By Lemma 4.20 we continue

$$E(e^{s(\zeta - E\zeta)}) \le \prod_{i=1}^{m} \exp((sb_i)^2/2).$$

Thus we need to minimize $(s\|b\|_2)^2/2 - mst$ over $s > 0$. The minimum is attained at $s = mt/\|b\|_2^2$ and is equal to $-m^2t^2/2\|b\|_2^2$. Therefore,

$$\mu^m\{\omega : \zeta(\omega) - E\zeta \ge mt\} \le \exp\left(-\frac{m^2t^2}{2\|b\|_2^2}\right). \tag{4.22}$$

In the same way we obtain

$$\mu^m\{\omega : E\zeta - \zeta(\omega) \ge mt\} \le \exp\left(-\frac{m^2t^2}{2\|b\|_2^2}\right). \tag{4.23}$$

Combining (4.22) and (4.23) we complete the proof of Theorem 4.21.    $\square$

**Theorem 4.22 (Bernstein's inequality)** *Let $\xi$ be a random variable on $(X, \Sigma, \mu)$ such that $|\xi - E\xi| \le b$ almost surely. Denote $\sigma^2 := E(\xi - E\xi)^2$. For $m \in \mathbb{N}$ consider a new random variable $\zeta$ on $(X^m, \Sigma^m, \mu^m)$ defined as*

$$\zeta(\omega) := \sum_{i=1}^{m} \xi(\omega_i), \quad \omega = (\omega_1, \ldots, \omega_m).$$

*Then for $t > 0$*

$$\mu^m\{\omega : |\zeta(\omega) - E\zeta| \ge mt\} \le 2\exp\left(-\frac{mt^2}{2(\sigma^2 + bt/3)}\right).$$

*Proof* Denote $\beta := \xi - E\xi$. As in the proof of Theorem 4.21 we use the Chernoff inequality (4.19). Thus, by (4.21) we need to estimate $E(e^{s\beta})$, $s > 0$. Writing

$$e^{s\beta} = 1 + s\beta + \frac{s^2\beta^2}{2!} + \cdots + \frac{s^k\beta^k}{k!} + \cdots$$

we get

$$E(e^{s\beta}) = 1 + \frac{s^2\sigma^2}{2} + \sum_{k=3}^{\infty} \frac{s^k E(\beta^k)}{k!}.$$

Using the inequality $E(\beta^k) \le b^{k-2}\sigma^2$ we continue

$$E(e^{s\beta}) \le 1 + s^2\sigma^2 \sum_{k=2}^{\infty} \frac{(sb)^{k-2}}{k!} \le 1 + \frac{s^2\sigma^2}{2} \sum_{k=2}^{\infty} \left(\frac{sb}{3}\right)^{k-2}. \tag{4.24}$$

Later we will specify $s$ in such a way that $sb < 3$. Therefore, we get from (4.24)

$$E(e^{s\beta}) \leq 1 + \frac{s^2\sigma^2}{2} \frac{1}{1 - sb/3}.$$

Next, we use the inequality $1 + x \leq e^x$ and minimize

$$\frac{s^2\sigma^2}{2} \frac{1}{1 - sb/3} - st$$

over $s \geq 0$. We choose $s$ from the equation

$$\frac{s^2\sigma^2}{1 - sb/3} = st.$$

With this $s = t/(\sigma^2 + tb/3)$ we obtain

$$\min_{s \geq 0} \left( \frac{s^2\sigma^2}{2} \frac{1}{1 - sb/3} - st \right) \leq \frac{1}{2} \frac{t^2}{(\sigma^2 + tb/3)^2} \frac{\sigma^2}{\sigma^2} (\sigma^2 + tb/3) - \frac{t^2}{\sigma^2 + tb/3}$$

$$= -\frac{1}{2} \frac{t^2}{\sigma^2 + tb/3}.$$

Therefore, in the same way as in (4.22) and (4.23) we get

$$\mu^m \{ \omega : \zeta(\omega) - E\zeta \geq mt \} \leq \exp \left( -\frac{mt^2}{2(\sigma^2 + bt/3)} \right) \qquad (4.25)$$

and

$$\mu^m \{ \omega : E\zeta - \zeta(\omega) \geq mt \} \leq \exp \left( -\frac{mt^2}{2(\sigma^2 + bt/3)} \right). \qquad (4.26)$$

Combining (4.25) and (4.26) we complete the proof of Theorem 4.22. $\qquad \square$

**Remark 4.23** The condition $0 \leq \xi \leq b$ implies $|\xi - E\xi| \leq b$.

We note that in the proof of Theorem 4.22 we used the assumption $|\beta| \leq b$, $\beta := \xi - E\xi$, to estimate $E\beta^k$. The above proof gives the following analog of Theorem 4.22.

**Theorem 4.24** *Let $\xi$ be a random variable on $(X, \Sigma, \mu)$ such that $E\beta^k \leq C_0(bk)^k$, $\beta := \xi - E\xi$, $k = 2, \ldots$ . For $m \in \mathbb{N}$ consider a new random variable $\zeta$ on $(X^m, \Sigma^m, \mu^m)$ defined as*

$$\zeta(\omega) := \sum_{i=1}^m \xi(\omega_i), \quad \omega = (\omega_1, \ldots, \omega_m).$$

*Then for $t \in (0, 1)$*

$$\mu^m\{\omega : |\zeta(\omega) - E\zeta| \geq mt\} \leq 2\exp(-c_0 mt^2)$$

*with* $c_0 := \min((8C_0)^{-1}(be)^{-2}, (4be)^{-1})$.

**Theorem 4.25 (Bennett's inequality)** *Let* $\xi$ *be a random variable on* $(X, \Sigma, \mu)$ *such that* $|\xi - E\xi| \leq b$ *almost surely. Denote* $\sigma^2 := E(\xi - E\xi)^2$. *For* $m \in \mathbb{N}$ *consider a new random variable* $\zeta$ *on* $(X^m, \Sigma^m, \mu^m)$ *defined as*

$$\zeta(\omega) := \sum_{i=1}^{m} \xi(\omega_i), \quad \omega = (\omega_1, \ldots, \omega_m).$$

*Then for* $t > 0$

$$\mu^m\{\omega : |\zeta(\omega) - E\zeta| \geq mt\} \leq 2\exp\left(-\frac{m\sigma^2}{b^2}h\left(\frac{bt}{\sigma^2}\right)\right),$$

*where the function* $h$ *is defined by* $h(u) := (1+u)\ln(1+u) - u$, $u \geq 0$.

*Proof* The beginning of the proof is the same as in the proof of Theorem 4.22. It deviates at the estimate (4.24) for $E(e^{s\beta})$. This time we write

$$E(e^{s\beta}) \leq 1 + s^2\sigma^2 \sum_{k=2}^{\infty} \frac{(sb)^{k-2}}{k!} = 1 + \frac{\sigma^2}{b^2}\sum_{k=2}^{\infty} \frac{(sb)^k}{k!}$$

$$= 1 + \frac{\sigma^2}{b^2}(e^{sb} - 1 - sb) \leq \exp\left(\frac{\sigma^2}{b^2}(e^{sb} - 1 - sb)\right).$$

Thus we need to minimize over $s \geq 0$ the expression

$$F(s) := \frac{\sigma^2}{b^2}(e^{sb} - 1 - sb) - st.$$

It is clear that $F(s)$ takes a min value at

$$s_0 = \frac{1}{b}\ln\left(1 + \frac{tb}{\sigma^2}\right).$$

We have

$$F(s_0) = \frac{\sigma^2}{b^2}\left(1 + \frac{tb}{\sigma^2} - 1 - \ln\left(1 + \frac{tb}{\sigma^2}\right) - \frac{tb}{\sigma^2}\ln\left(1 + \frac{tb}{\sigma^2}\right)\right)$$

$$= -\frac{\sigma^2}{b^2}h\left(\frac{tb}{\sigma^2}\right).$$

Using the Chernoff inequality (4.19) we complete the proof in the same way as in Theorems 4.21 and 4.22. $\square$

**Corollary 4.26** *Theorem 4.25 implies Theorem 4.22.*

*Proof* This follows from the inequality

$$h(u) \geq \frac{u^2}{2(1 + u/3)} =: g(u), \quad u \geq 0. \tag{4.27}$$

The following relations imply (4.27):

$$h(0) = g(0) = 0, \quad h'(0) = g'(0) = 0$$

and

$$h''(u) = \frac{1}{1 + u} \geq g''(u) = \frac{1}{(1 + u/3)^3}, \quad u \geq 0.$$

$\square$

### 4.2.3 The Kullback–Leibler information and the Hellinger distance

Let two probability measures $\mu$ and $\nu$ be absolutely continuous with respect to the third probability measure $w$ defined on a measurable space $(X, \Sigma)$. Then, by the Radon–Nikodim theorem,

$$d\mu = u \, dw, \quad d\nu = v \, dw. \tag{4.28}$$

Denote

$$S(\mu) := \{x \in X : u(x) > 0\}.$$

We define the Kullback–Leibler information as follows:

$$\mathcal{K}(\mu, \nu) := \int_{S(\mu)} \ln(u/v) d\mu = \int_{S(\mu)} u \ln(u/v) dw = \int_X u \ln(u/v) dw. \tag{4.29}$$

One can prove that $\mathcal{K}(\mu, \nu)$ does not depend on the measure $w$. In particular, if $\mu$ is absolutely continuous with respect to $\nu$, $d\mu = g \, d\nu$, then we have

$$\mathcal{K}(\mu, \nu) = \int_{S(\mu)} \ln g \, d\mu = \int_{S(\mu)} g \ln g \, d\nu. \tag{4.30}$$

We write

$$\mathcal{K}(\mu, \nu) = - \int_{S(\mu)} u \ln(v/u) dw$$

$$= - \int_{S(\mu)} (\ln(v/u) - (v/u - 1)) u \, dw + \int_{S(\mu)} (1 - v/u) u \, dw.$$

By the inequality $\ln(1 + y) \leq y$ we obtain

$$\ln(v/u) - (v/u - 1) \leq 0.$$

Therefore,

$$-\int_{S(\mu)} (\ln(v/u) - (v/u - 1))u\,dw \geq 0.$$

Using the fact that $\mu$ and $v$ are the probability measures

$$\mu(X) = \int_X u\,dw = 1, \quad v(X) = \int_X v\,dw = 1,$$

we obtain

$$\int_{S(\mu)} (1 - v/u)u\,dw = 1 - \int_{S(\mu)} v\,dw \geq 0.$$

Thus,

$$\mathcal{K}(\mu, v) \geq 0.$$

It is clear that $\mathcal{K}(\mu, \mu) = 0$. The quantity $\mathcal{K}(\mu, v)$ indicates how close the measures $\mu$ and $v$ are to each other. However, in general $\mathcal{K}(\mu, v) \neq \mathcal{K}(v, \mu)$; therefore, it is not a metric.

We proceed to the Hellinger distance that is a metric. For $\mu$, $v$ satisfying (4.28) we define

$$h(\mu, v) := \int_X (u^{1/2} - v^{1/2})^2\,dw = \|u^{1/2} - v^{1/2}\|_{L_2(w)}^2.$$

The Kulback–Leibler information and the Hellinger distance satisfy certain relations. Again, using the inequality $\ln(1 + y) \leq y$ we get

$$\ln(v/u) = 2\ln(1 + ((v/u)^{1/2} - 1)) \leq 2((v/u)^{1/2} - 1)$$

and

$$\mathcal{K}(\mu, v) = -\int_X u\ln(v/u)dw \geq -2\left(\int_X (uv)^{1/2}\,dw - 1\right) = h(\mu, v).$$

Thus,

$$\mathcal{K}(\mu, v) \geq h(\mu, v).$$

The Kullback–Leibler information is convenient for working with measures $\mu^m$ and $v^m$:

$$\mathcal{K}(\mu^m, v^m) = m\mathcal{K}(\mu, v). \tag{4.31}$$

We note that if $\mu$ and $v$ are of the form (4.28) and $\mu$ is not absolutely continuous with respect to $v$ (there exists an $A \in \Sigma$ such that $\mu(A) > 0$, $v(A) = 0$) then $\mathcal{K}(\mu, v) = \infty$. So, we complement the relation (4.30) that gives the $\mathcal{K}(\mu, v)$ in the case where $\mu$ is absolutely continuous with respect to $v$ by the relation $\mathcal{K}(\mu, \nu) = \infty$ if $\mu$ is not absolutely continuous with respect to $\nu$. This defines $\mathcal{K}(\mu, \nu)$ for any pair of probability measures $\mu$, $\nu$.

We will prove a duality property of the $\mathcal{K}(\mu, \nu)$ that will be used later on.

**Lemma 4.27** *Let $\mu$ and $\nu$ be two probability measures defined on $(X, \Sigma)$. Denote*

$$V := \left\{ f : \int e^f \, d\nu = 1 \right\}.$$

*Then*

$$\mathcal{K}(\mu, \nu) = \sup_{f \in V} \int f \, d\mu.$$

*Proof* In the case where $\mu$ is not absolutely continuous with respect to $\nu$ we take an $A$ such that $\mu(A) > 0$ and $\nu(A) = 0$. Then for $f = \lambda\chi_A$, $\lambda > 0$, we have

$$\int e^f \, d\nu = \int_{X \setminus A} d\nu = 1.$$

Also,

$$\int f \, d\mu = \lambda\mu(A) \to \infty \quad \text{as} \quad \lambda \to \infty.$$

Thus, in this case,

$$\sup_{f \in V} \int f \, d\mu = \infty.$$

Now let $d\mu = g \, d\nu$. First, we note that the function $f = \ln g$ belongs to $V$ and

$$\mathcal{K}(\mu, \nu) = \int \ln g \, d\mu.$$

Therefore,

$$\mathcal{K}(\mu, \nu) \leq \sup_{f \in V} \int f \, d\mu.$$

We will prove that, for any $f \in V$,

$$\int f \, d\mu \leq \mathcal{K}(\mu, \nu).$$

Denote $\Phi(x) := x \ln x$, $x > 0$. Then

$$\mathcal{K}(\mu, \nu) - \int f \, d\mu = \int g \ln g \, d\nu - \int f \, d\mu = \int g(\ln g - f)d\nu$$

$$= \int (ge^{-f} \ln(ge^{-f}))e^f \, d\nu = \int \Phi(ge^{-f})e^f \, d\nu.$$

The function $\Phi$ is a convex function and $\int e^f \, dv = 1$. Therefore, using the Jensen inequality, we continue

$$\int \Phi(ge^{-f})e^f \, dv \geq \Phi\left(\int g \, dv\right) = 0.$$

This completes the proof of Lemma 4.27. $\qquad\square$

**Corollary 4.28** *Let $\mu$ and $v$ be two probability measures defined on $(X, \Sigma)$. Then for any $A \in \Sigma$*

$$\mathcal{K}(\mu, v) \geq \lambda \mu(A) - \ln[(e^\lambda - 1)v(A) + 1].$$

*Proof* Indeed, it is easy to verify that

$$f = \lambda \chi_A - \ln[(e^\lambda - 1)v(A) + 1] \in V.$$

It remains to apply Lemma 4.27. $\qquad\square$

**Lemma 4.29** (**Fano's inequality**) *Let $(X, \Sigma)$ be a measurable space and let $A_i \in \Sigma$, $i \in \{0, 1, \ldots, M\}$, be such that $\forall i \neq j$, $A_i \cap A_j = \emptyset$. Assume that $\mu_i$, $i \in \{0, 1 \ldots, M\}$, are probability measures on $(X, \Sigma)$. Denote*

$$p := \sup_{0 \leq i \leq M} \mu_i(X \setminus A_i).$$

*Then either $p > M/(M + 1)$ or*

$$\inf_{j \in \{0, 1, \ldots, M\}} \frac{1}{M} \sum_{i: i \neq j} \mathcal{K}(\mu_i, \mu_j) \geq \Psi_M(p),$$

*where*

$$\Psi_M(p) := (1 - p) \ln\left[\left(\frac{1-p}{p}\right)\left(\frac{M-p}{p}\right)\right] - \ln\frac{M-p}{Mp}.$$

*Proof* Let us define

$$a := \inf_{0 \leq i \leq M} \mu_i(A_i).$$

Then $p = 1 - a$. Using Corollary 4.28 we write

$$\mathcal{K}(\mu_i, \mu_0) \geq \lambda \mu_i(A_i) - \ln[(e^\lambda - 1)\mu_0(A_i) + 1].$$

Next,

$$\frac{1}{M} \sum_{i=1}^{M} \mathcal{K}(\mu_i, \mu_0) \geq \lambda \frac{1}{M} \sum_{i=1}^{M} \mu_i(A_i) - \frac{1}{M} \sum_{i=1}^{M} \ln[(e^\lambda - 1)\mu_0(A_i) + 1].$$

Using the Jensen inequality for the concave function $\ln x$, we continue

$$\frac{1}{M} \sum_{i=1}^{M} \mathcal{K}(\mu_i, \mu_0) \geq \lambda a - \ln \left[ (e^\lambda - 1) \frac{1}{M} \sum_{i=1}^{M} \mu_0(A_i) + 1 \right]$$

$$= \lambda a - \ln \left[ (e^\lambda - 1) \frac{1}{M} \mu_0(\cup_{i=1}^{M} A_i) + 1 \right].$$

By the inequality

$$\mu_0(\cup_{i=1}^{M} A_i) \leq (1 - \mu_0(A_0)) \leq 1 - a,$$

we get

$$\frac{1}{M} \sum_{i=1}^{M} \mathcal{K}(\mu_i, \mu_0) \geq \lambda a - \ln \left[ (e^\lambda - 1) \frac{1}{M} \mu_0(\cup_{i=1}^{M} A_i) + 1 \right]$$

$$\geq \lambda a - \ln \left[ (e^\lambda - 1) \frac{1 - a}{M} + 1 \right].$$

It is clear that we obtain the same inequality if we replace $\mu_0$ by $\mu_j$ with some $j \in [1, M]$ and replace the summation over $[1, M]$ by summation over $[0, M] \setminus \{j\}$. Therefore, we get

$$\inf_{j \in \{0,1,\ldots,M\}} \frac{1}{M} \sum_{i:i \neq j} \mathcal{K}(\mu_i, \mu_j) \geq \lambda a - \ln \left[ (e^\lambda - 1) \frac{1 - a}{M} + 1 \right].$$

We now want to maximize the right-hand side over $\lambda \geq 0$. If $0 \leq x \leq a \leq 1$, then the maximum value of the function $\lambda a - \ln[(e^\lambda - 1)x + 1]$ is attained at

$$\lambda = \ln \left[ \left( \frac{a}{1-a} \right) \left( \frac{1-x}{x} \right) \right]$$

and we have

$$\sup_{\lambda \geq 0}(\lambda a - \ln[(e^\lambda - 1)x + 1]) = a \ln \left[ \left( \frac{a}{1-a} \right) \left( \frac{1-x}{x} \right) \right] - \ln \frac{1-x}{1-a}.$$

So, if $1/(M + 1) \leq a \leq 1$ then we have

$$\sup_{\lambda \geq 0} \left( \lambda a - \ln \left[ (e^\lambda - 1) \left( \frac{1-a}{M} \right) + 1 \right] \right)$$

$$= a \ln \left[ \left( \frac{a}{1-a} \right) \left( \frac{M - 1 + a}{1 - a} \right) \right] - \ln \frac{M - 1 + a}{M(1 - a)}.$$

Therefore, we have either $0 \leq a < 1/(M+1)$, (equivalently $M/(M+1) < p$), or $1/(M + 1) \leq a \leq 1$ and then

$$\inf_{j \in \{0,\ldots,M\}} \frac{1}{M} \sum_{i:i \neq j} \mathcal{K}(\mu_i, \mu_j) \geq a \ln \left[ \left( \frac{a}{1-a} \right) \left( \frac{M - 1 + a}{1 - a} \right) \right] - \ln \frac{M - 1 + a}{M(1 - a)}$$

or

$$\inf_{j \in \{0,1,\dots,M\}} \frac{1}{M} \sum_{i: i \neq j} \mathcal{K}(\mu_i, \mu_j) \geq (1-p) \ln \left[ \left( \frac{1-p}{p} \right) \left( \frac{M-p}{p} \right) \right] - \ln \frac{M-p}{Mp}.$$

This completes the proof of Lemma 4.29. □

**Corollary 4.30** *Under the assumptions of Lemma 4.29 we have*

$$p \geq \min \left( \frac{1}{2}, e^{-3/e} \exp \left( - \left( \inf_{j \in \{0,1,\dots,M\}} \frac{1}{M} \sum_{i: i \neq j} \mathcal{K}(\mu_i, \mu_j) - \frac{1}{2} \ln M \right) \right) \right).$$

*Proof* Let $p \in [0, M/(M+1)]$. We consider

$$\Psi_M(p) = \ln M + (1-p) \ln \left( \frac{1-p}{p} \right) + p \ln \left( \frac{p}{M-p} \right)$$

$$= \ln M + (1-p) \ln(1-p) - (1-p) \ln p + p \ln p - p \ln(M-p).$$

Using the inequality $x \ln x \geq -1/e$ for $x \in [0, 1]$ we continue

$$\geq -\ln p - \frac{3}{e} + (1-p) \ln M.$$

Thus, by Lemma 4.29,

$$\inf_{j \in \{0,1,\dots,M\}} \frac{1}{M} \sum_{i: i \neq j} \mathcal{K}(\mu_i, \mu_j) \geq -\ln p - \frac{3}{e} + (1-p) \ln M.$$

In the case $p \leq 1/2$ this gives

$$\inf_{j \in \{0,1,\dots,M\}} \frac{1}{M} \sum_{i: i \neq j} \mathcal{K}(\mu_i, \mu_j) \geq -\ln p - \frac{3}{e} + \frac{1}{2} \ln M.$$

Therefore, either $p \geq 1/2$ or

$$p \geq e^{-3/e} \exp \left( - \left( \inf_{j \in \{0,1,\dots,M\}} \frac{1}{M} \sum_{i: i \neq j} \mathcal{K}(\mu_i, \mu_j) - \frac{1}{2} \ln M \right) \right).$$

□

## 4.3 Improper function learning; upper estimates

### 4.3.1 Introduction

As above, let $X \subset \mathbb{R}^d$, $Y \subset \mathbb{R}$ be Borel sets and let $\rho$ be a Borel probability measure on $Z = X \times Y$. As we pointed out in Section 4.1 the regression function $f_\rho \in L_2(\rho_X)$ minimizes the error $\mathcal{E}(f)$ over all $f \in L_2(\rho_X)$: $\mathcal{E}(f_\rho) \leq$

$\mathcal{E}(f)$, $f \in L_2(\rho_X)$. In the sense of an error $\mathcal{E}(\cdot)$ the regression function $f_\rho$ is the best to describe the relationship between inputs $x \in X$ and outputs $y \in Y$. Now our goal is to find an estimator $f_\mathbf{z}$, on the basis of the given data $\mathbf{z} = ((x_1, y_1), \ldots, (x_m, y_m))$, that approximates $f_\rho$ well with high probability. We assume that $(x_i, y_i), i = 1, \ldots, m$ are independent and distributed according to $\rho$. There are several important ingredients in the mathematical formulation of this problem. We follow the method that has become standard in approximation theory and has been used, for example, in DeVore *et al.* (2006) and Konyagin and Temlyakov (2007). In this approach we first choose a function class $W$ (a hypothesis space $\mathcal{H}$ in Cucker and Smale (2001)) to work with. After selecting a class $W$, there are two routes we can take. The first (see Cucker and Smale (2001), Konyagin and Temlyakov (2007) and Poggio and Smale (2003)) is based on the idea of studying the approximation of a projection $f_W$ of $f_\rho$ onto $W$. This setting is known as *the improper function learning problem (the projection learning problem)*. In this case we do not assume that the regression function $f_\rho$ comes from a specific (say, smoothness) class of functions. We study this problem in this section. The second way (see Cucker and Smale (2001), DeVore *et al.* (2006), Konyagin and Temlyakov (2007) and Poggio and Smale (2003)) is based on the assumption $f_\rho \in W$. This setting is known as *the proper function learning problem*. For instance, we may assume that $f_\rho$ has some smoothness. We study this problem in Section 4.4, where, following a tradition from non-parametric statistics, we denote a class of priors by $\Theta$ (instead of $W$ as in this section).

The next step in both settings is to find a method for constructing an estimator $f_\mathbf{z}$ that provides a good (optimal, near optimal in a certain sense) approximation with high probability with respect to $\rho$. A problem of optimization is naturally broken into two parts: upper estimates and lower estimates. In order to prove the upper estimates we need to decide upon the form of an estimator $f_\mathbf{z}$. In other words, we need to specify the *hypothesis space* $\mathcal{H}$ (see Cucker and Smale (2001), Konyagin and Temlyakov (2007) and Poggio and Smale (2003)) (*approximation space* (DeVore *et al.*, 2006; Konyagin and Temlyakov, 2007)) where an estimator $f_\mathbf{z}$ comes from.

The next question is how to build $f_\mathbf{z} \in \mathcal{H}$. In this section we discuss a method, which is standard in statistics, of *empirical risk minimization* that takes

$$f_{\mathbf{z},\mathcal{H}} = \arg \min_{f \in \mathcal{H}} \mathcal{E}_\mathbf{z}(f),$$

where

$$\mathcal{E}_\mathbf{z}(f) := \frac{1}{m} \sum_{i=1}^{m} (f(x_i) - y_i)^2$$

is the *empirical error (risk)* of $f$. This $f_{\mathbf{z},\mathcal{H}}$ is called the *empirical optimum* or the *least squares estimator (LSE)*.

In Section 4.1 we discussed the importance of the characteristics of a class $W$, closely related to the concept of entropy numbers. In this section it will be convenient for us to define the entropy numbers in the same way as in Section 4.1. This definition differs slightly from the definition used in Chapter 3. For a compact subset $W$ of a Banach space $B$ we define the entropy numbers as follows:

$$\epsilon_n(W, B) := \epsilon_n^2(W, B) := \inf\{\epsilon : \exists f_1, \ldots, f_{2^n} \in W : W \subset \cup_{j=1}^{2^n}(f_j + \epsilon U(B))\},$$

where $U(B)$ is the unit ball of Banach space $B$. A set $\{f_1, \ldots, f_N\} \subset W$ satisfying the condition $W \subset \cup_{j=1}^{N}(f_j + \epsilon U(B))$ is called an $\epsilon$-net of $W$. We denote by $N(W, \epsilon, B)$ the covering number, that is the minimal number of points in $\epsilon$-nets of $W$. We note that $N(W, \epsilon_n(W, B), B) \leq 2^n$ and

$$N_\epsilon(W, B) \leq N(W, \epsilon, B) \leq N_{\epsilon/2}(W, B).$$

This implies that

$$\epsilon_n^1(W, B) \leq \epsilon_n^2(W, B) \leq 2\epsilon_n^1(W, B),$$

where $N_\epsilon(W, B)$ and $\epsilon_n^1(W, B)$ are defined in Chapter 3. Therefore, all the conditions that we will impose on $\{\epsilon_n(W, B)\}$ can be expressed in an equivalent way in both $\{\epsilon_n^1(W, B)\}$ and $\{\epsilon_n^2(W, B)\}$.

We mentioned in Section 4.1 that in DeVore *et al.* (2004) and Konyagin and Temlyakov (2004) the restrictions on a class $W$ were imposed in the following form:

$$\epsilon_n(W, C) \leq Dn^{-r}, \quad n = 1, 2, \ldots, \quad W \subset DU(C). \tag{4.32}$$

We denote by $S^r$ the collection of classes satisfying (4.32). In Konyagin and Temlyakov (2007) weaker restrictions were imposed, i.e.

$$\epsilon_n(W, L_2(\rho_X)) \leq Dn^{-r}, \quad n = 1, 2, \ldots, \quad W \subset DU(L_2(\rho_X)). \tag{4.33}$$

We denote by $S_2^r$ the collection of classes satisfying (4.33).

After building $f_{\mathbf{z}}$ we need to choose an appropriate way to measure the descrepancy between $f_{\mathbf{z}}$ and the target function $f_W$. In Cucker and Smale (2001) the quality of approximation is measured by $\mathcal{E}(f_{\mathbf{z}}) - \mathcal{E}(f_W)$. It is easy to see (Proposition 4.31 below) that for any $f \in L_2(\rho_X)$

$$\mathcal{E}(f) - \mathcal{E}(f_\rho) = \|f - f_\rho\|_{L_2(\rho_X)}^2. \tag{4.34}$$

Thus the choice $\|\cdot\| = \|\cdot\|_{L_2(\rho_X)}$ seems natural. This norm was also used in DeVore *et al.* (2006) and Konyagin and Temlyakov (2007) for measuring the error. The use of the $L_2(\rho_X)$ norm in measuring the error is one of the reasons for us to consider restictions (4.33) instead of (4.32).

One of important questions discussed in Cucker and Smale (2001), DeVore *et al.* (2006) and Konyagin and Temlyakov (2007) is to estimate the *defect function* $L_{\mathbf{z}}(f) := \mathcal{E}(f) - \mathcal{E}_{\mathbf{z}}(f)$ of $f \in W$. We discuss this question in detail in this section.

If $\xi$ is a random variable (a real-valued function on a probability space $Z$) then denote

$$E(\xi) := \int_Z \xi \, d\rho; \quad \sigma^2(\xi) := \int_Z (\xi - E(\xi))^2 \, d\rho. \quad (4.35)$$

We will assume that $\rho$ and $W$ satisfy the following condition:

for all $f \in W$, $f : X \to Y$ is such that $|f(x) - y| \le M$ a.e. (4.36)

We complete the introduction by including the formulation of the classical Bernstein's inequalities (see Theorem 4.22) that we use systematically in this section. For a single function $f$, we have the following: if $|\xi(z) - E(\xi)| \le M$ a.e. then, for any $\epsilon > 0$,

$$\rho^m \left\{ \mathbf{z} : |\frac{1}{m} \sum_{i=1}^m \xi(z_i) - E(\xi)| \ge \epsilon \right\} \le 2\exp\left(-\frac{m\epsilon^2}{2(\sigma^2(\xi) + M\epsilon/3)}\right); \quad (4.37)$$

$$\rho^m \left\{ \mathbf{z} : \frac{1}{m} \sum_{i=1}^m \xi(z_i) - E(\xi) \ge \epsilon \right\} \le \exp\left(-\frac{m\epsilon^2}{2(\sigma^2(\xi) + M\epsilon/3)}\right); \quad (4.38)$$

$$\rho^m \left\{ \mathbf{z} : \frac{1}{m} \sum_{i=1}^m \xi(z_i) - E(\xi) \le -\epsilon \right\} \le \exp\left(-\frac{m\epsilon^2}{2(\sigma^2(\xi) + M\epsilon/3)}\right). \quad (4.39)$$

### 4.3.2 First estimates for classes from $\mathcal{S}^r$

Let $\rho$ be a Borel probability measure on $Z = X \times Y$. The following proposition gives a relation between $\mathcal{E}(f) - \mathcal{E}(f_\rho)$ and $\|f - f_\rho\|_{L_2(\rho_X)}$.

**Proposition 4.31** *For every* $f : X \to Y$, $f \in L_2(\rho_X)$,

$$\mathcal{E}(f) - \mathcal{E}(f_\rho) = \int_X (f(x) - f_\rho(x))^2 \, d\rho_X.$$

*Proof* We have for $f \in L_2(\rho_X)$

$$\mathcal{E}(f) = \int_Z (f(x) - y)^2 \, d\rho = \int_Z (f(x) - f_\rho(x) + f_\rho(x) - y)^2 \, d\rho$$

$$= \int_Z (f(x) - f_\rho(x))^2 \, d\rho + 2 \int_Z (f(x) - f_\rho(x))(f_\rho(x) - y) \, d\rho$$

$$+ \int_Z (f_\rho(x) - y)^2 \, d\rho.$$

Next, by the definition of $f_\rho(x) := E(y|x)$, we obtain

$$\int_Z (f(x) - f_\rho(x)) y \, d\rho = \int_X E(y|x)(f(x) - f_\rho(x)) d\rho_X$$

$$= \int_X f_\rho(x)(f(x) - f_\rho(x)) d\rho_X.$$

Combining the above two relations, we complete the proof.                    $\square$

We define as above the *empirical error* of $f$ as

$$\mathcal{E}_{\mathbf{z}}(f) := \frac{1}{m} \sum_{i=1}^m (f(x_i) - y_i)^2.$$

Let $f \in L_2(\rho_X)$. The *defect function* of $f$ is given by

$$L_{\mathbf{z}}(f) := L_{\mathbf{z},\rho}(f) := \mathcal{E}(f) - \mathcal{E}_{\mathbf{z}}(f); \quad \mathbf{z} = (z_1, \ldots, z_m), \quad z_i = (x_i, y_i).$$

We are interested in estimating $L_{\mathbf{z}}(f)$ for functions $f$ coming from a given class $W$. We begin with a formulation of a corresponding result in the case when $W$ consists of only one element.

**Theorem 4.32** *Let $M > 0$ and $f : X \to Y$ be such that $|f(x) - y| \leq M$ a.e. Then, for all $\epsilon > 0$,*

$$\rho^m \{\mathbf{z} : |L_{\mathbf{z}}(f)| \leq \epsilon\} \geq 1 - 2 \exp\left(-\frac{m\epsilon^2}{2(\sigma^2 + M^2\epsilon/3)}\right), \tag{4.40}$$

*where $\sigma^2 := \sigma^2((f(x) - y)^2)$.*

This theorem is a direct corollary of the Bernstein inequality (Theorem 4.22) and Remark 4.23. Indeed, taking $\xi(z) := (f(x) - y)^2$ and noting that $E(\xi) = \mathcal{E}(f)$, we get (4.40) from Theorem 4.22.

We proceed to a function class $W$. It is clear how Theorem 4.32 can be directly applied in the case of $W$ consisting of a finite number of elements. The main idea of studying infinite classes $W$ is in the approximate representation of $W$ by a finite set. This leads to the idea of using $\epsilon$-nets of $W$ for such an approximate representation, and we follow this idea in our study. An important question in this regard is the following: What is an appropriate Banach space $B$ for building $\epsilon$-nets? We already mentioned in Section 4.1 that there are different natural options for the corresponding Banach spaces. We begin with the case for which $B$ is $C(X)$, the space of functions continuous on a compact subset $X$ of $\mathbb{R}^d$ with the norm

$$\|f\|_\infty := \sup_{x \in X} |f(x)|.$$

We use the abbreviated notations

$$N(W, \epsilon) := N(W, \epsilon, C); \quad \epsilon_n(W) := \epsilon_n(W, C).$$

**Theorem 4.33** *Let $W$ be a compact subset of $C(X)$. Assume that $\rho$ and $W$ satisfy (4.36). Then, for all $\epsilon > 0$,*

$$\rho^m\{\mathbf{z} : \sup_{f \in W} |L_\mathbf{z}(f)| \leq \epsilon\} \geq 1 - N(W, \epsilon/(8M))2 \exp\left(-\frac{m\epsilon^2}{8(\sigma^2 + M^2\epsilon/6)}\right).$$

(4.41)

*Here $\sigma^2 := \sigma^2(W) := \sup_{f \in W} \sigma^2((f(x) - y)^2)$.*

**Remark 4.34** In general we cannot guarantee that the set $\{\mathbf{z} : \sup_{f \in W} | L_\mathbf{z}(f)| \geq \eta\}$ is $\rho^m$-measurable. In such a case the relation (4.41) and further relations of this type are understood in the sense of outer measure associated with the $\rho^m$. For instance, for (4.41) this means that there exists a $\rho^m$-measurable set $G$ such that $\{\mathbf{z} : \sup_{f \in W} |L_\mathbf{z}(f)| \geq \eta\} \subset G$ and (4.41) holds for $G$.

We note that the above theorem is related to the concept of the Glivenko–Cantelli sample complexity of a class $\Phi$ with accuracy $\epsilon$ and confidence $\delta$:

$$S_\Phi(\epsilon, \delta) := \min\{n : \quad \text{for all} \quad m \geq n, \quad \text{for all} \quad \rho,$$

$$\rho^m\{\mathbf{z} = (z_1, \ldots, z_m) : \sup_{\phi \in \Phi} |\int_Z \phi \, d\rho - \frac{1}{m}\sum_{i=1}^m \phi(z_i)| \geq \epsilon\} \leq \delta\}.$$

In order to see that, we define $z_i := (x_i, y_i), i = 1, \ldots, m; \phi(x, y) := (f(x) - y)^2; \Phi := \{(f(x) - y)^2, f \in W\}$. A survey of recent results on the Glivenko–Cantelli sample complexity may be found in Mendelson (2003) and results and the corresponding historical remarks related to Theorem 4.33 may be found in Györfy *et al.* (2002).

In the proof of Theorem 4.33 we will use the following simple relation.

**Proposition 4.35** *If $|f_j(x) - y| \leq M$ a.e. for $j = 1, 2$, then*

$$|L_\mathbf{z}(f_1) - L_\mathbf{z}(f_2)| \leq 4M\|f_1 - f_2\|_\infty.$$

We leave a detailed proof of this proposition to the reader. We only point out that in the proof of this proposition we use

$$|(f_1(x) - y)^2 - (f_2(x) - y)^2| \leq 2M\|f_1 - f_2\|_\infty.$$

*Proof of Theorem 4.33.* Let $f_1, \ldots, f_N$ be the $\epsilon/(8M)$-net of $W$, $N :=$ $N(W, \epsilon/(8M))$. Then for any $f \in W$ there is an $f_j$ such that $\|f - f_j\|_\infty \leq$ $\epsilon/(8M)$ and, by Proposition 4.35,

$$|L_{\mathbf{z}}(f) - L_{\mathbf{z}}(f_j)| \leq \epsilon/2.$$

Therefore, $|L_{\mathbf{z}}(f)| \geq \epsilon$ implies that there is a $j \in [1, N]$ such that $|L_{\mathbf{z}}(f_j)| \geq$ $\epsilon/2$. Using Theorem 4.32 we obtain from here that

$$\rho^m\{\mathbf{z} : \sup_{f \in W} |L_{\mathbf{z}}(f)| \geq \epsilon\} \leq \sum_{j=1}^N \rho^m\{\mathbf{z} : |L_{\mathbf{z}}(f_j)| \geq \epsilon/2\}$$

$$\leq 2N \exp\left(-\frac{m\epsilon^2}{8(\sigma^2 + M^2\epsilon/6)}\right). \qquad \square$$

For a compact $W \subset L_2(\rho_X)$ we denote by $f_W$ a projection of $f_\rho$ onto $W$, that is a function from $W$ that minimizes the error $\mathcal{E}(f)$:

$$f_W = \arg\min_{f \in W} \mathcal{E}(f).$$

We now illustrate how a result on a defect function (Theorem 4.33) can be used in estimating the error between the projection $f_W$ and the least squares estimator $f_{\mathbf{z}, W}$.

**Theorem 4.36** *Let $W$ be a compact subset of $C(X)$. Assume that $\rho$ and $W$ satisfy (4.36). Then, for all $\epsilon > 0$,*

$$\rho^m\{\mathbf{z} : \mathcal{E}(f_{\mathbf{z}, W}) - \mathcal{E}(f_W) \leq \epsilon\} \geq 1 - N(W, \epsilon/(16M))2\exp\left(-\frac{m\epsilon^2}{8(4\sigma^2 + M^2\epsilon/3)}\right).$$

*Here $\sigma^2 := \sigma^2(W) := \sup_{f \in W} \sigma^2((f(x) - y)^2).$*

This theorem follows from Theorem 4.33 and the chain of inequalities

$$0 \leq \mathcal{E}(f_{\mathbf{z}, W}) - \mathcal{E}(f_W) = \mathcal{E}(f_{\mathbf{z}, W}) - \mathcal{E}_{\mathbf{z}}(f_{\mathbf{z}, W}) + \mathcal{E}_{\mathbf{z}}(f_{\mathbf{z}, W}) - \mathcal{E}_{\mathbf{z}}(f_W)$$

$$+ \mathcal{E}_{\mathbf{z}}(f_W) - \mathcal{E}(f_W) \leq \mathcal{E}(f_{\mathbf{z}, W}) - \mathcal{E}_{\mathbf{z}}(f_{\mathbf{z}, W}) + \mathcal{E}_{\mathbf{z}}(f_W) - \mathcal{E}(f_W).$$

Assume $W \in \mathcal{S}^r$ such that

$$\epsilon_n(W) \leq Dn^{-r}, \quad n = 1, 2, \ldots, \quad W \subset DU(\mathcal{C}). \tag{4.42}$$

Then

$$N(W, \epsilon) \leq 2^{(C_1/\epsilon)^{1/r}+1} \leq 2^{(C_2/\epsilon)^{1/r}}. \tag{4.43}$$

Substituting this into Theorem 4.36 and optimizing over $\epsilon$ we get for $\epsilon = Am^{-r/(1+2r)}$, $A \geq A_0(M, D, r)$,

$$\rho^m \left\{ \mathbf{z} : \mathcal{E}(f_{\mathbf{z},W}) - \mathcal{E}(f_W) \leq Am^{-r/(1+2r)} \right\} \geq 1 - \exp\left( -c(M)A^2 m^{1/(1+2r)} \right).$$

We have proved the following theorem.

**Theorem 4.37** *Assume that $W \in \mathcal{S}^r$ and that $\rho$, $W$ satisfy (4.36). Then for $\eta \geq \eta_m := A_0(M, D, r)m^{-r/(1+2r)}$*

$$\rho^m \{ \mathbf{z} : \mathcal{E}(f_{\mathbf{z},W}) - \mathcal{E}(f_W) \leq \eta \} \geq 1 - \exp(-c(M)m\eta^2). \tag{4.44}$$

### 4.3.3 Further estimates for classes from $\mathcal{S}^r$; chaining technique

In Section 4.3.2 we demonstrated what can be obtained for $W \in \mathcal{S}^r$ by using the simplest technique – a direct application of the Bernstein concentration of measure theorem. In this subsection we develop a technique that uses the ideas from Section 4.3.2 and adds a new ingredient. We work here with a chain of $\epsilon$-nets of $W$.

**Lemma 4.38** *If $\delta > \eta/(8M)$, $|f_j(x) - y| \leq M$ a.e. for $j = 1, 2$ and $\| f_1 - f_2 \|_\infty \leq \delta$, then*

$$\rho^m \{ \mathbf{z} : |L_{\mathbf{z}}(f_1) - L_{\mathbf{z}}(f_2)| \leq \eta \} \geq 1 - 2 \exp\left( -\frac{m\eta^2}{30M^2\delta^2} \right).$$

*Proof* Consider the random variable $\xi = (f_1(x) - y)^2 - (f_2(x) - y)^2$. We use

$$|\xi| \leq 2M\| f_1 - f_2 \|_\infty \quad \text{a.e.}$$

Therefore, $|\xi - E\xi| \leq 4M\delta$ a.e. and the variance $V$ of $\xi$ is at most $4M^2\delta^2$. Applying the Bernstein inequality (4.37) to $\xi$ we get

$$\rho^m \{ \mathbf{z} : |L_{\mathbf{z}}(f_1) - L_{\mathbf{z}}(f_2)| \geq \eta \} = \rho^m \left\{ \mathbf{z} : \left| \frac{1}{m} \sum_{i=1}^{m} \xi(z_i) - E(\xi) \right| \geq \eta \right\}$$

$$\leq 2 \exp\left( -\frac{m\eta^2}{2(4M^2\delta^2 + 4M\delta\eta/3)} \right)$$

$$\leq 2 \exp\left( -\frac{m\eta^2}{2(44M^2\delta^2/3)} \right), \tag{4.45}$$

and Lemma 4.38 follows. $\qquad\square$

**Theorem 4.39** *Assume that $\rho$, $W$ satisfy (4.36) and $W$ is such that*

$$\sum_{n=1}^{\infty} n^{-1/2} \epsilon_n(W) < \infty. \tag{4.46}$$

*Then for $m\eta^2 \geq 1$ we have*

$$\rho^m\{\mathbf{z} : \sup_{f \in W} |L_{\mathbf{z}}(f)| \geq \eta\} \leq C(M, \epsilon(W)) \exp(-c(M)m\eta^2);$$

*$C(M, \epsilon(W))$ may depend on $M$ and $\epsilon(W) := \{\epsilon_n(W, C)\}$, and $c(M)$ may depend only on $M$.*

We note that the condition on the entropy numbers is similar to the corresponding Dudley entropy condition (see Dudley (1967)) expressed in terms of the covering numbers. It is more convenient for us to formulate the results in terms of the entropy numbers. Talagrand (2005) also prefers to work with the entropy numbers rather than with the covering numbers. Theorem 4.39 was proved in Konyagin and Temlyakov (2007). These types of results have been developed in the study of the central limit theorem in probability theory (see, for instance, Gine and Zinn (1984)). Condition (4.46) is equivalent to the Pollard entropy condition from Pollard (1984). The reader can find further results on the chaining technique in Talagrand (2005).

*Proof* We use a convenient variant of the chaining technique that is a standard technique in stochastic processes (see, for instance, Pollard (1984)). It is clear that (4.46) implies that

$$\sum_{j=0}^{\infty} 2^{j/2} \epsilon_{2^j}(W) < \infty. \tag{4.47}$$

Denote $\delta_j := \epsilon_{2^j}$, $j = 0, 1, \ldots$, and consider minimal $\delta_j$-nets $\mathcal{N}_j \subset W$ of $W$. We will use the notation $N_j := |\mathcal{N}_j|$. Let $J$ be the minimal $j$ satisfying $\delta_j \leq \eta/(8M)$. For $j = 1, \ldots, J$ we define a mapping $A_j$ that associates with a function $f \in W$ a function $A_j(f) \in \mathcal{N}_j$ closest to $f$ in the $C$ norm. Then, clearly,

$$\|f - A_j(f)\|_C \leq \delta_j.$$

We use the mappings $A_j$, $j = 1, \ldots, J$, to associate with a function $f \in W$ a sequence (a chain) of functions $f_J, f_{J-1}, \ldots, f_1$ in the following way:

$$f_J := A_J(f), \quad f_j := A_j(f_{j+1}), \quad j = 1, \ldots, J-1.$$

We introduce an auxiliary sequence

$$\eta_j := (30)^{1/2} M\eta 2^{(j+1)/2} \epsilon_{2^{j-1}}, \quad j = 1, 2, \ldots, \tag{4.48}$$

and define $I := I(M, \epsilon(W))$ to be the minimal satisfying

$$\sum_{j \geq I} \eta_j \leq \eta/4. \tag{4.49}$$

We now proceed to the estimate of $\rho^m\{\mathbf{z} : \sup_{f \in W} |L_{\mathbf{z}}(f)| \geq \eta\}$ with $m, \eta$ satisfying $m\eta^2 \geq 1$. First of all by Proposition 4.35 the assumption $\delta_J \leq \eta/(8M)$ implies that if $|L_{\mathbf{z}}(f)| \geq \eta$ then $|L_{\mathbf{z}}(f_J)| \geq \eta/2$. Using this, (4.49) and rewriting

$$L_{\mathbf{z}}(f_J) = L_{\mathbf{z}}(f_J) - L_{\mathbf{z}}(f_{J-1}) + \cdots + L_{\mathbf{z}}(f_{I+1}) - L_{\mathbf{z}}(f_I) + L_{\mathbf{z}}(f_I),$$

we conclude that if $|L_{\mathbf{z}}(f)| \geq \eta$ then at least one of the following events occurs:

$$|L_{\mathbf{z}}(f_j) - L_{\mathbf{z}}(f_{j-1})| \geq \eta_j \quad \text{for some} \quad j \in (I, J] \quad \text{or} \quad |L_{\mathbf{z}}(f_I)| \geq \eta/4.$$

Therefore

$$\rho^m\{\mathbf{z} : \sup_{f \in W} |L_{\mathbf{z}}(f)| \geq \eta\} \leq \rho^m\{\mathbf{z} : \sup_{f \in \mathcal{N}_I} |L_{\mathbf{z}}(f)| \geq \eta/4\}$$

$$+ \sum_{j \in (I,J]} \sum_{f \in \mathcal{N}_j} \rho^m\{\mathbf{z} : |L_{\mathbf{z}}(f) - L_{\mathbf{z}}(A_{j-1}(f))| \geq \eta_j\}$$

$$\leq \rho^m\{\mathbf{z} : \sup_{f \in \mathcal{N}_I} |L_{\mathbf{z}}(f)| \geq \eta/4\}$$

$$+ \sum_{j \in (I,J]} N_j \sup_{f \in W} \rho^m\{\mathbf{z} : |L_{\mathbf{z}}(f) - L_{\mathbf{z}}(A_{j-1}(f))| \geq \eta_j\}. \tag{4.50}$$

By our choice of $\delta_j = \epsilon_{2^j}$, we get $N_j \leq 2^{2^j} < e^{2^j}$. Applying Lemma 4.38, we obtain

$$\sup_{f \in W} \rho^m\{\mathbf{z} : |L_{\mathbf{z}}(f) - L_{\mathbf{z}}(A_{j-1}(f))| \geq \eta_j\} \leq 2 \exp\left(-\frac{m\eta_j^2}{30M^2\delta_{j-1}^2}\right).$$

From the definition (4.48) of $\eta_j$ we get

$$\frac{m\eta_j^2}{30M^2\delta_{j-1}^2} = m\eta^2 2^{j+1}$$

and

$$N_j \exp\left(-\frac{m\eta_j^2}{30M^2\delta_{j-1}^2}\right) \leq \exp(-m\eta^2 2^j).$$

Therefore

$$\sum_{j \in (I,J]} N_j \exp\left(-\frac{m\eta_j^2}{30M^2\delta_{j-1}^2}\right) \leq 2\exp(-m\eta^2 2^I). \tag{4.51}$$

By Theorem 4.21,

$$\rho^m\{\mathbf{z}: \sup_{f\in\mathcal{N}_I} |L_{\mathbf{z}}(f)| \geq \eta/4\} \leq 2N_I \exp\left(-\frac{m\eta^2}{C(M)}\right). \tag{4.52}$$

Combining (4.51) and (4.52) we obtain

$$\rho^m\{\mathbf{z}: \sup_{f\in W} |L_{\mathbf{z}}(f)| \geq \eta\} \leq C(M, \epsilon(W)) \exp(-c(M)m\eta^2).$$

This completes the proof of Theorem 4.39.                                          □

Theorem 4.39 shows that if the sequence $\{\epsilon_n(W)\}$ decays fast enough to satisfy (4.46) then we guarantee that the defect function is small with high probability for all $f \in W$. It is interesting to note that Theorem 4.39 does not give a better result for sequences $\{\epsilon_n(W)\}$ that decay very fast. We will give a result for sequences $\{\epsilon_n(W)\}$ that do not satisfy (4.46). In this case the result depends on how fast the sequence $\{\epsilon_n(W)\}$ decays.

**Theorem 4.40** *Assume that $\rho$, $W$ satisfy (4.36) and that $W$ is such that*

$$\sum_{n=1}^{\infty} n^{-1/2}\epsilon_n(W) = \infty.$$

*For $\eta > 0$ define $J := J(\eta/M)$ as the minimal $j$ satisfying $\epsilon_{2^j} \leq \eta/(8M)$ and*

$$S_J := \sum_{j=1}^{J} 2^{(j+1)/2}\epsilon_{2^{j-1}}.$$

*Then for $m$, $\eta$ satisfying $m(\eta/S_J)^2 \geq 480M^2$ we have*

$$\rho^m\{\mathbf{z}: \sup_{f\in W} |L_{\mathbf{z}}(f)| \geq \eta\} \leq C(M, \epsilon(W)) \exp(-c(M)m(\eta/S_J)^2).$$

*Proof* This proof differs from the above proof of Theorem 4.39 only in the choice of an auxiliary sequence $\{\eta_j\}$. Thus we keep notations from the proof of Theorem 4.39. Now, instead of (4.48) we define $\{\eta_j\}$ as follows:

$$\eta_j := \frac{\eta}{4}\frac{2^{(j+1)/2}\epsilon_{2^{j-1}}}{S_J}.$$

Proceeding as in the proof of Theorem 4.39 with $I = 1$, we need to check that

$$2^j - \frac{m\eta_j^2}{30M^2\delta_{j-1}^2} \leq -2^j\frac{m(\eta/S_J)^2}{480M^2}.$$

Indeed, using the assumption $m(\eta/S_J)^2 \geq 480M^2$ we obtain

$$\frac{m\eta_j^2}{30M^2\delta_{j-1}^2} - 2^j = \frac{m(\eta/S_J)^2}{480M^2}2^{j+1} - 2^j \geq \frac{m(\eta/S_J)^2}{480M^2}2^j.$$

We complete the proof in the same way as in Theorem 4.39. $\qquad\square$

**Remark 4.41** Let $a = \{a_n\}$ be a majorant sequence for $\{\epsilon_n(W)\}$: $\epsilon_n(W) \leq a_n$, $n = 1, 2, \ldots$. It is clear that Theorem 4.40 holds with $J$ replaced by $J(a)$ – the minimal $j$ satisfying $a_{2^j} \leq \eta/(8M)$ and with $S_J$ replaced by

$$S_{J(a)} := \sum_{j=1}^{J(a)} 2^{(j+1)/2}a_{2^{j-1}}.$$

We formulate three corollaries of Theorem 4.40. All the proofs are similar. We only prove Corollary 4.44 here.

**Corollary 4.42** *Assume $\rho$, $W$ satisfy (4.36) and $\epsilon_n(W) \leq Dn^{-1/2}$. Then for $m$, $\eta$ satisfying $m(\eta/(1 + \log(M/\eta)))^2 \geq C_1(M, D)$ we have*

$$\rho^m\{\mathbf{z} : \sup_{f\in W} |L_{\mathbf{z}}(f)| \geq \eta\} \leq C(M, D)\exp(-c(M, D)m(\eta/(1+\log(M/\eta)))^2).$$

**Corollary 4.43** *Assume $\rho$, $W$ satisfy (4.36) and $\epsilon_n(W) \leq Dn^{-r}$, $r \in (0, 1/2)$. Then for $m$, $\eta$ satisfying $m\eta^{1/r} \geq C_1(M, D, r)$ we have*

$$\rho^m\{\mathbf{z} : \sup_{f\in W} |L_{\mathbf{z}}(f)| \geq \eta\} \leq C(M, D, r)\exp(-c(M, D, r)m\eta^{1/r}).$$

Denote by $\mathcal{N}_\delta(W)$ the $\delta$-net of $W$ in the $\mathcal{C}$ norm.

**Corollary 4.44** *Assume $\rho$, $W$ satisfy (4.36) and $\epsilon_n(W) \leq Dn^{-r}$, $r \in (0, 1/2)$. Then for $m$, $\eta$, $\delta \geq \eta/(8M)$ satisfying $m\eta^2\delta^{1/r-2} \geq C_1(M, D, r)$ we have*

$$\rho^m\{\mathbf{z} : \sup_{f\in\mathcal{N}_\delta(W)} |L_{\mathbf{z}}(f)| \geq 2\eta\} \leq C(M, D, r)\exp(-c(M, D, r)m\eta^2\delta^{1/r-2}).$$

*Proof* We apply Theorem 4.40 to $\mathcal{N}_\delta(W)$. First of all we note that for $n$ such that $\epsilon_n(W) \leq \delta$ we have $\epsilon_n(\mathcal{N}_\delta(W)) = 0$. Also, for $n$ such that $\epsilon_n(W) > \delta$ we have

$$\epsilon_n(\mathcal{N}_\delta(W)) \leq \epsilon_n(W) + \delta \leq 2\epsilon_n(W).$$

We now estimate the $S_J$ from Theorem 4.40. Denote by $J_\delta$ the minimal $j$ satisfying $\epsilon_{2^j}(W) \leq \delta$ and keep the notation $J$ for the minimal $j$ satisfying

$\epsilon_{2^j}(W) \leq \eta/(8M)$. Then it is clear from our assumption $\delta \geq \eta/(8M)$ that $J_\delta \leq J$ and $\epsilon_{2^{j-1}}(\mathcal{N}_\delta(W)) = 0$ for $j > J_\delta$. Therefore,

$$S_J \leq 2 \sum_{j=1}^{J_\delta} 2^{(j+1)/2} \epsilon_{2^{j-1}}(W) \leq 2^{3/2+r} D \sum_{j=1}^{J_\delta} 2^{j(1/2-r)} \leq C_1(r) D 2^{J_\delta(1/2-r)}.$$

Next,

$$D2^{-r(J_\delta-1)} \geq \epsilon_{2^{J_\delta-1}} > \delta \quad \text{implies} \quad 2^{J_\delta} \leq 2(D/\delta)^{1/r}.$$

Thus

$$S_J \leq C_1(D, r)(1/\delta)^{(1/2r)-1}.$$

It remains to apply Theorem 4.40. $\qquad\square$

The above results on the defect function imply the corresponding results for $\mathcal{E}(f_{\mathbf{z},W}) - \mathcal{E}(f_W)$. We formulate these results as one theorem.

**Theorem 4.45** *Assume that $\rho$ and $W$ satisfy (4.32) and (4.36). Then we have the following estimates:*

$$\rho^m\{\mathbf{z} : \mathcal{E}(f_{\mathbf{z},W}) - \mathcal{E}(f_W) \leq \eta\} \geq 1 - C(M, D, r)\exp(-c(M)m\eta^2), \quad (4.53)$$

$$\rho^m\{\mathbf{z} : |\mathcal{E}_{\mathbf{z}}(f_{\mathbf{z},W}) - \mathcal{E}(f_W)| \leq 2\eta\} \geq 1 - C(M, D, r)\exp(-c(M)m\eta^2),$$

*provided $r > 1/2$, $m\eta^2 \geq 1$;*

$$\rho^m\{\mathbf{z} : \mathcal{E}(f_{\mathbf{z},W}) - \mathcal{E}(f_W) \leq \eta\}$$
$$\geq 1 - C_1(M, D)\exp(-c(M, D)m(\eta/(1 + \log(M/\eta)))^2),$$

$$\rho^m\{\mathbf{z} : |\mathcal{E}_{\mathbf{z}}(f_{\mathbf{z},W}) - \mathcal{E}(f_W)| \leq 2\eta\}$$
$$\geq 1 - C_1(M, D)\exp(-c(M, D)m(\eta/(1 + \log(M/\eta)))^2),$$

*provided $r = 1/2$, $m(\eta/(1 + \log(M/\eta)))^2 \geq C_2(M, D)$;*

$$\rho^m\{\mathbf{z} : \mathcal{E}(f_{\mathbf{z},W}) - \mathcal{E}(f_W) \leq \eta\} \geq 1 - C_1(M, D, r)\exp(-c(M, D, r)m\eta^{1/r}),$$

$$\rho^m\{\mathbf{z} : |\mathcal{E}_{\mathbf{z}}(f_{\mathbf{z},W}) - \mathcal{E}(f_W)| \leq 2\eta\} \geq 1 - C_1(M, D, r)\exp(-c(M, D, r)m\eta^{1/r}),$$

*provided $r \in (0, 1/2)$, $m\eta^{1/r} \geq C_2(M, D, r)$.*

*Proof* This theorem follows from Theorem 4.39, Corollaries 4.42 and 4.43 and the chain of inequalities:

$$0 \leq \mathcal{E}(f_{\mathbf{z},W}) - \mathcal{E}(f_W) = \mathcal{E}(f_{\mathbf{z},W}) - \mathcal{E}_{\mathbf{z}}(f_{\mathbf{z},W}) + \mathcal{E}_{\mathbf{z}}(f_{\mathbf{z},W}) - \mathcal{E}_{\mathbf{z}}(f_W) + \mathcal{E}_{\mathbf{z}}(f_W)$$
$$- \mathcal{E}(f_W)$$
$$\leq \mathcal{E}(f_{\mathbf{z},W}) - \mathcal{E}_{\mathbf{z}}(f_{\mathbf{z},W}) + \mathcal{E}_{\mathbf{z}}(f_W) - \mathcal{E}(f_W). \qquad\square$$

We note that in (4.53) we can take $\eta$ as small as $\eta = Am^{-1/2}$ and that $m^{-1/2}$ is the best rate we can achieve using (4.53).

### 4.3.4 Least squares estimators for convex hypothesis spaces

In Sections 4.3.4 and 4.3.5 we continue to study the projection learning problem. Results from these subsections are oriented for applications in the proper function learning problem. In this case the class $W$ plays an intermediate role in the estimation process. The standard name for this class in that context is the hypothesis space and the standard notation for it is $\mathcal{H}$. Thus, in Sections 4.3.4 and 4.3.5 we denote the class of our interest by $\mathcal{H}$ instead of $W$ used in the preceding sections. We begin with the following theorem.

**Theorem 4.46** *Suppose that $\mathcal{H}$ is a compact and convex subset of $C(X)$. Assume that $\rho$ and $\mathcal{H}$ satisfy (4.36). Then, for all $\epsilon > 0$,*

$$\rho^m \{\mathbf{z} : \mathcal{E}(f_{\mathbf{z},\mathcal{H}}) - \mathcal{E}(f_{\mathcal{H}}) \le \epsilon\} \ge 1 - N(\mathcal{H}, \epsilon/(16M)) \exp\left(-\frac{m\epsilon}{80M^2}\right).$$

We will prove some lemmas and theorems in preparation for the proof of Theorem 4.46. From Proposition 4.31 we know that $\|f - f_\rho\|^2_{L_2(\rho_X)} = \mathcal{E}(f) - \mathcal{E}(f_\rho)$. The following lemma gives an inequality relating the corresponding quantities with $f_{\mathcal{H}}$ instead of $f_\rho$.

**Lemma 4.47** *Let $\mathcal{H}$ be a convex subset of $C(X)$ such that $f_{\mathcal{H}}$ exists. Then, for all $f \in \mathcal{H}$,*

$$\|f - f_{\mathcal{H}}\|^2_{L_2(\rho_X)} \le \mathcal{E}(f) - \mathcal{E}(f_{\mathcal{H}}).$$

*Proof* By the convexity assumption for any $f \in \mathcal{H}$ and $g := f - f_{\mathcal{H}}$, we have $(1 - \epsilon)f_{\mathcal{H}} + \epsilon f = f_{\mathcal{H}} + \epsilon g$ is in $\mathcal{H}$ and, therefore,

$$0 \le \|f_\rho - f_{\mathcal{H}} - \epsilon g\|^2_{L_2(\rho_X)} - \|f_\rho - f_{\mathcal{H}}\|^2_{L_2(\rho_X)}$$
$$= -2\epsilon \int_X (f_\rho - f_{\mathcal{H}})g \, d\rho_X + \epsilon^2 \int_X g^2 \, d\rho_X.$$

Letting $\epsilon \to 0$, we obtain the following inequality:

$$\int_X (f_\rho - f_{\mathcal{H}})(f - f_{\mathcal{H}}) d\rho_X \le 0, \quad f \in \mathcal{H}. \tag{4.54}$$

Then, letting $\epsilon = 1$, we see that $\|f_\rho - f\|_{L_2(\rho_X)} > \|f_\rho - f_{\mathcal{H}}\|_{L_2(\rho_X)}$ whenever $f \ne f_{\mathcal{H}}$ and so $f_{\mathcal{H}}$ is unique. Also, (4.54) gives

$$\|f - f_{\mathcal{H}}\|^2_{L_2(\rho_X)} \le \|f - f_\rho\|^2_{L_2(\rho_X)} - \|f_\rho - f_{\mathcal{H}}\|^2_{L_2(\rho_X)} = \mathcal{E}(f) - \mathcal{E}(f_{\mathcal{H}}).$$

$\square$

We will use the following notations:

$$\delta(\mathcal{H}) := \|f_{\mathcal{H}} - f_\rho\|^2_{L_2(\rho_X)};$$

$$\mathcal{E}_{\mathcal{H}}(f) := \mathcal{E}(f) - \mathcal{E}(f_{\mathcal{H}}); \qquad \mathcal{E}_{\mathcal{H},\mathbf{z}}(f) := \mathcal{E}_{\mathbf{z}}(f) - \mathcal{E}_{\mathbf{z}}(f_{\mathcal{H}});$$

$$\ell(f) := \ell(f, z) := (f(x) - y)^2 - (f_{\mathcal{H}}(x) - y)^2, \quad z = (x, y).$$

We note that

$$\mathcal{E}_{\mathcal{H}}(f) = E_\rho(\ell(f, z)); \qquad \mathcal{E}_{\mathcal{H},\mathbf{z}}(f) = \frac{1}{m} \sum_{i=1}^{m} \ell(f, z_i).$$

**Lemma 4.48** *Assume that $\mathcal{H}$ is convex and that $\rho$ and $\mathcal{H}$ satisfy (4.36). Then we have, for $f \in \mathcal{H}$,*

$$\sigma^2 := \sigma^2(\ell(f)) \le 4M^2 \mathcal{E}_{\mathcal{H}}(f).$$

*Proof* We have

$$\sigma^2(\ell(f)) \le E(\ell(f)^2) = E((f(x) - f_{\mathcal{H}}(x))^2 (f(x) + f_{\mathcal{H}}(x) - 2y)^2)$$
$$\le 4M^2 E((f(x) - f_{\mathcal{H}}(x))^2) = 4M^2 \|f - f_{\mathcal{H}}\|^2_{L_2(\rho_X)} \le 4M^2 \mathcal{E}_{\mathcal{H}}(f).$$

At the last step we have used Lemma 4.47. $\qquad\square$

**Lemma 4.49** *Assume that $\mathcal{H}$ is convex and that $\rho$ and $\mathcal{H}$ satisfy (4.36). Let $f \in \mathcal{H}$. For all $\epsilon > 0$, $\alpha \in (0, 1]$, one has*

$$\rho^m\{\mathbf{z} : \mathcal{E}_{\mathcal{H}}(f) - \mathcal{E}_{\mathcal{H},\mathbf{z}}(f) \ge \alpha(\mathcal{E}_{\mathcal{H}}(f) + \epsilon)\} \le \exp\left(-\frac{\alpha^2 m\epsilon}{5M^2}\right);$$

$$\rho^m\{\mathbf{z} : \mathcal{E}_{\mathcal{H}}(f) - \mathcal{E}_{\mathcal{H},\mathbf{z}}(f) \le -\alpha(\mathcal{E}_{\mathcal{H}}(f) + \epsilon)\} \le \exp\left(-\frac{\alpha^2 m\epsilon}{5M^2}\right).$$

*Proof* Denote $a := \mathcal{E}_{\mathcal{H}}(f)$. The proofs of both inequalities are the same, so we will only carry out the proof of the first one. Using the one-sided Bernstein's inequality (4.39) for $\ell(f)$, we obtain

$$\rho^m\{\mathbf{z} : \mathcal{E}_{\mathcal{H}}(f) - \mathcal{E}_{\mathcal{H},\mathbf{z}}(f) \ge \alpha(a + \epsilon)\} \le \exp\left(-\frac{m\alpha^2(a + \epsilon)^2}{2(\sigma^2 + M^2\alpha(a + \epsilon)/3)}\right).$$

It remains to check that

$$\frac{(a + \epsilon)^2}{2(\sigma^2 + M^2\alpha(a + \epsilon)/3)} \ge \frac{\epsilon}{5M^2}. \tag{4.55}$$

Using Lemma 4.48 we get, on the one hand,

$$2\epsilon(\sigma^2 + M^2\alpha(a + \epsilon)/3) \le M^2\epsilon(9a + 2\epsilon/3); \tag{4.56}$$

on the other hand,

$$5M^2(a + \epsilon)^2 \geq M^2\epsilon(10a + 5\epsilon). \tag{4.57}$$

Comparing (4.56) and (4.57), we obtain (4.55). □

**Lemma 4.50** *Assume that $\rho$ and $\mathcal{H}$ satisfy (4.36). Let $\alpha \in (0, 1)$, $\epsilon > 0$, and let $f \in \mathcal{H}$ be such that*

$$\frac{\mathcal{E}_{\mathcal{H}}(f) - \mathcal{E}_{\mathcal{H},\mathbf{z}}(f)}{\mathcal{E}_{\mathcal{H}}(f) + \epsilon} < \alpha. \tag{4.58}$$

*Then for all $g \in \mathcal{H}$ such that $\|f - g\|_{\mathcal{C}(X)} \leq \alpha\epsilon/4M$ we have*

$$\frac{\mathcal{E}_{\mathcal{H}}(g) - \mathcal{E}_{\mathcal{H},\mathbf{z}}(g)}{\mathcal{E}_{\mathcal{H}}(g) + \epsilon} < 2\alpha. \tag{4.59}$$

*Proof* Denote

$$a := \mathcal{E}_{\mathcal{H}}(f), \quad a' := \mathcal{E}_{\mathcal{H}}(g), \quad b := \mathcal{E}_{\mathcal{H},\mathbf{z}}(f), \quad b' := \mathcal{E}_{\mathcal{H},\mathbf{z}}(g).$$

Then our assumption $\|f - g\|_{\mathcal{C}(X)} \leq \alpha\epsilon/4M$ implies

$$|a - a'| \leq \alpha\epsilon/2, \qquad |b - b'| \leq \alpha\epsilon/2. \tag{4.60}$$

By (4.58) and (4.60) we get ($a \geq 0$)

$$a(1 - \alpha) < b + \alpha\epsilon \leq b' + 3\alpha\epsilon/2. \tag{4.61}$$

Also, by (4.60),

$$a(1 - \alpha) \geq (a' - \alpha\epsilon/2)(1 - \alpha) \geq a' - \alpha a' - \alpha\epsilon/2. \tag{4.62}$$

Combining (4.61) and (4.62) we obtain

$$a' - b' < \alpha a' + 2\alpha\epsilon \leq 2\alpha(a' + \epsilon),$$

which implies (4.59). □

A combination of Lemmas 4.49 and 4.50 yields the following theorem.

**Theorem 4.51** *Assume that $\mathcal{H}$ is convex and that $\rho$, $\mathcal{H}$ satisfy (4.36). Then for all $\epsilon > 0$ and $\alpha \in (0, 1)$*

$$\rho^m \left\{ \mathbf{z} : \sup_{f \in \mathcal{H}} \frac{\mathcal{E}_{\mathcal{H}}(f) - \mathcal{E}_{\mathcal{H},\mathbf{z}}(f)}{\mathcal{E}_{\mathcal{H}}(f) + \epsilon} \geq 2\alpha \right\} \leq N \left( \mathcal{H}, \frac{\alpha\epsilon}{4M}, \mathcal{C}(X) \right) \exp \left( -\frac{\alpha^2 m\epsilon}{5M^2} \right).$$

*Proof* Let $f_1, \ldots, f_N$ be the $(\alpha\epsilon/4M)$-net of $\mathcal{H}$ in $\mathcal{C}(X)$, $N := N(\mathcal{H}, \alpha\epsilon/4M, \mathcal{C}(X))$. Let $\Lambda$ be the set of $\mathbf{z}$ such that for all $j = 1, \ldots, N$ we have

$$\frac{\mathcal{E}_{\mathcal{H}}(f_j) - \mathcal{E}_{\mathcal{H},\mathbf{z}}(f_j)}{\mathcal{E}_{\mathcal{H}}(f_j) + \epsilon} < \alpha.$$

Then, by Lemma 4.49,

$$\rho^m(\Lambda) \geq 1 - N \exp\left(-\frac{\alpha^2 m\epsilon}{5M^2}\right). \tag{4.63}$$

We take any $\mathbf{z} \in \Lambda$ and any $g \in \mathcal{H}$. Let $f_j$ be such that $\|g - f_j\|_{C(X)} \leq \alpha\epsilon/4M$. By Lemma 4.50 we obtain that

$$\frac{\mathcal{E}_{\mathcal{H}}(g) - \mathcal{E}_{\mathcal{H},\mathbf{z}}(g)}{\mathcal{E}_{\mathcal{H}}(g) + \epsilon} < 2\alpha.$$

It remains to use (4.63).                                                    □

**Theorem 4.52** *Let $\mathcal{H}$ be a compact and convex subset of $C(X)$ and let $\rho$, $\mathcal{H}$ satisfy (4.36). Then for all $\epsilon > 0$ with probability at least*

$$p(\mathcal{H}, \epsilon) := 1 - N\left(\mathcal{H}, \frac{\epsilon}{16M}, C(X)\right) \exp\left(-\frac{m\epsilon}{80M^2}\right),$$

*one has for all $f \in \mathcal{H}$*

$$\mathcal{E}(f) \leq 2\mathcal{E}_{\mathbf{z}}(f) + \epsilon - \mathcal{E}(f_{\mathcal{H}}) + 2(\mathcal{E}(f_{\mathcal{H}}) - \mathcal{E}_{\mathbf{z}}(f_{\mathcal{H}})). \tag{4.64}$$

*Proof* Using Theorem 4.51 with $\alpha = 1/4$ we get, with probability at least $p(\mathcal{H}, \epsilon)$,

$$\mathcal{E}_{\mathcal{H}}(f) \leq 2\mathcal{E}_{\mathcal{H},\mathbf{z}}(f) + \epsilon. \tag{4.65}$$

Substituting

$$\mathcal{E}_{\mathcal{H}}(f) := \mathcal{E}(f) - \mathcal{E}(f_{\mathcal{H}}); \qquad \mathcal{E}_{\mathcal{H},\mathbf{z}}(f) := \mathcal{E}_{\mathbf{z}}(f) - \mathcal{E}_{\mathbf{z}}(f_{\mathcal{H}}),$$

we obtain (4.64).                                                            □

The definition of $f_{\mathcal{H}}$ and the inequality (4.65) imply that for any $f \in \mathcal{H}$ we have $2\mathcal{E}_{\mathcal{H},\mathbf{z}}(f) + \epsilon \geq 0$. Applying this inequality to $f = f_{\mathbf{z},\mathcal{H}}$ we obtain the following corollary.

**Corollary 4.53** *Under the assumptions of Theorem 4.52 we have*

$$\mathcal{E}_{\mathbf{z}}(f_{\mathcal{H}}) - \mathcal{E}_{\mathbf{z}}(f_{\mathbf{z},\mathcal{H}}) \leq \epsilon/2$$

*with probability at least $p(\mathcal{H}, \epsilon)$.*

*Proof of Theorem 4.46* We use (4.65) with $f = f_{\mathbf{z},\mathcal{H}}$. From the definition of $f_{\mathbf{z},\mathcal{H}}$ we obtain that $\mathcal{E}_{\mathbf{z},\mathcal{H}}(f_{\mathbf{z},\mathcal{H}}) \leq 0$. This completes the proof of Theorem 4.46.                                                            □

The following theorem is a direct corollary of Theorem 4.46.

**Theorem 4.54** *Assume that $\mathcal{H} \in \mathcal{S}^r$ is convex and that $\rho$, $\mathcal{H}$ satisfy (4.36). Then for $\eta \geq \eta_m := A_0(M, D, r)m^{-r/(1+r)}$ one has*

$$\rho^m\{\mathbf{z} : \mathcal{E}(f_{\mathbf{z},\mathcal{H}}) - \mathcal{E}(f_{\mathcal{H}}) \geq \eta\} \leq \exp(-c(M)m\eta).$$

*Proof* As in (4.43) we get from the assumption $\mathcal{H} \in \mathcal{S}^r$ that

$$N(\mathcal{H}, \eta/(16M)) \leq 2^{(CDM/\eta)^{1/r}}.$$

Expressing $\eta$ in the form $\eta = Am^{-r/(1+r)}$ we obtain that for $A \geq A_0(M, D, r)$

$$2^{(CDM/\eta)^{1/r}} \exp\left(-\frac{m\eta}{80M}\right) \leq \exp(-c(M)m\eta). \qquad \square$$

### 4.3.5 Least squares estimators for non-convex hypothesis spaces

In Section 4.3.4 the assumption that $\mathcal{H}$ is convex was important. Clearly, this assumption holds when $\mathcal{H}$ is a ball in the $L_2(\rho_X)$. In the case of linear methods of approximation used for building an estimator, the convexity assumption for the hypothesis space $\mathcal{H}$ is usually satisfied. However, this assumption is not satisfied when we opt for a nonlinear method of approximation. In this subsection we develop a technique parallel to the one from Section 4.3.4 with the convexity assumption replaced by an assumption on the deviation of $\mathcal{E}(f_{\mathcal{H}})$ from $\mathcal{E}(f_\rho)$. The following result has been proved in DeVore *et al.* (2004).

**Theorem 4.55** *Let $\mathcal{H}$ be a compact subset of $C(X)$. Assume that $\rho$ and $\mathcal{H}$ satisfy (4.36). Then, for all $\epsilon > 0$,*

$$\rho^m\{\mathbf{z} : \mathcal{E}(f_{\mathbf{z},\mathcal{H}}) - \mathcal{E}(f_{\mathcal{H}}) \leq \epsilon\} \geq 1 - N(\mathcal{H}, \epsilon/(24M))2\exp\left(-\frac{m\epsilon}{C(M, K)}\right)$$

*under the assumption $\mathcal{E}(f_{\mathcal{H}}) - \mathcal{E}(f_\rho) \leq K\epsilon$.*

Theorem 4.55 shows that we obtain an analog of Theorem 4.46 with the convexity assumption replaced by the assumption $\delta(\mathcal{H}) := \mathcal{E}(f_{\mathcal{H}}) - \mathcal{E}(f_\rho) \leq K\epsilon$. In this subsection we will develop further the idea of replacing the convexity assumption by an estimate for $\delta(\mathcal{H})$. The motivation for this is that applications of results of the type of Theorem 4.55 in the construction of universal estimators require bounds in a more general situation than $\delta(\mathcal{H}) \leq K\epsilon$. The following theorem, from Temlyakov (2008a), provides bounds for $\rho^m\{\mathbf{z} : \mathcal{E}(f_{\mathbf{z},\mathcal{H}}) - \mathcal{E}(f_{\mathcal{H}}) \geq \epsilon\}$ in the case of arbitrary $\epsilon$ and $\delta(\mathcal{H})$.

**Theorem 4.56** *Suppose $\mathcal{H}$ is a compact subset of $C(X)$ and $\mathcal{E}(f_{\mathcal{H}}) - \mathcal{E}(f_\rho) \leq \delta$. Assume that $\rho$, $\mathcal{H}$ satisfy (4.36). Then, for all $\epsilon > 0$,*

$$\rho^m\{\mathbf{z} : \mathcal{E}(f_{\mathbf{z},\mathcal{H}}) - \mathcal{E}(f_{\mathcal{H}}) \geq \epsilon\} \leq N\left(\mathcal{H}, \frac{\epsilon}{16M}, C(X)\right) \exp\left(-\frac{m\epsilon^2}{2^9 M^2(\epsilon + \delta)}\right).$$

The proof of this theorem is similar to the proof of Theorem 4.46. As above, we begin with some lemmas.

**Lemma 4.57** *For any f we have*

$$\|f - f_{\mathcal{H}}\|^2_{L_2(\rho_X)} \le 2\left(\mathcal{E}(f) - \mathcal{E}(f_{\mathcal{H}}) + 2\|f_{\mathcal{H}} - f_{\rho}\|^2_{L_2(\rho_X)}\right).$$

*Proof* We have

$$\|f - f_{\mathcal{H}}\|_{L_2(\rho_X)} \le \|f - f_{\rho}\|_{L_2(\rho_X)} + \|f_{\rho} - f_{\mathcal{H}}\|_{L_2(\rho_X)}.$$

Next,

$$\|f - f_{\rho}\|^2_{L_2(\rho_X)} = \mathcal{E}(f) - \mathcal{E}(f_{\rho}) = \mathcal{E}(f) - \mathcal{E}(f_{\mathcal{H}}) + \mathcal{E}(f_{\mathcal{H}}) - \mathcal{E}(f_{\rho}).$$

Combining the above two relations, we get

$$\begin{aligned}
\|f - f_{\mathcal{H}}\|^2_{L_2(\rho_X)} &\le 2(\|f - f_{\rho}\|^2_{L_2(\rho_X)} + \|f_{\mathcal{H}} - f_{\rho}\|^2_{L_2(\rho_X)}) \\
&\le 2(\mathcal{E}(f) - \mathcal{E}(f_{\mathcal{H}}) + 2\|f_{\mathcal{H}} - f_{\rho}\|^2_{L_2(\rho_X)}). \qquad \square
\end{aligned}$$

**Lemma 4.58** *Assume that $\rho$ and $\mathcal{H}$ satisfy (4.36). Then we have, for $f \in \mathcal{H}$,*

$$\sigma^2 := \sigma^2(\ell(f)) \le 8M^2(\mathcal{E}_{\mathcal{H}}(f) + 2\delta(\mathcal{H})).$$

*Proof* It is clear that $|\ell(f)| \le M^2$. Therefore,

$$\begin{aligned}
\sigma^2(\ell(f)) &\le E(\ell(f)^2) = E\left((f(x) - f_{\mathcal{H}}(x))^2(f(x) + f_{\mathcal{H}}(x) - 2y)^2\right) \\
&\le 4M^2 E\left((f(x) - f_{\mathcal{H}}(x))^2\right) \\
&= 4M^2 \|f - f_{\mathcal{H}}\|^2_{L_2(\rho_X)} \le 8M^2(\mathcal{E}_{\mathcal{H}}(f) + 2\delta(\mathcal{H})).
\end{aligned}$$

At the last step we have used Lemma 4.57. $\qquad \square$

**Lemma 4.59** *Assume that $\rho$ and $\mathcal{H}$ satisfy (4.36). Let $f \in \mathcal{H}$. For all $\epsilon > 0$, $\alpha \in (0, 1]$, one has*

$$\rho^m\{\mathbf{z} : \mathcal{E}_{\mathcal{H}}(f) - \mathcal{E}_{\mathcal{H},\mathbf{z}}(f) \ge \alpha(\mathcal{E}_{\mathcal{H}}(f) + \epsilon)\} \le \exp\left(-\frac{\alpha^2 m\epsilon^2}{32M^2(\epsilon + \delta(\mathcal{H}))}\right);$$

$$\rho^m\{\mathbf{z} : \mathcal{E}_{\mathcal{H}}(f) - \mathcal{E}_{\mathcal{H},\mathbf{z}}(f) \le -\alpha(\mathcal{E}_{\mathcal{H}}(f) + \epsilon)\} \le \exp\left(-\frac{\alpha^2 m\epsilon^2}{32M^2(\epsilon + \delta(\mathcal{H}))}\right).$$

*Proof* Denote $a := \mathcal{E}_{\mathcal{H}}(f)$. The proofs of both inequalities are the same. We will carry out only the proof of the first one. Using the one-sided Bernstein's inequality (4.39) for $\ell(f)$ we obtain

$$\rho^m\{\mathbf{z} : \mathcal{E}_{\mathcal{H}}(f) - \mathcal{E}_{\mathcal{H},\mathbf{z}}(f) \ge \alpha(a + \epsilon)\} \le \exp\left(-\frac{m\alpha^2(a + \epsilon)^2}{2(\sigma^2 + M^2\alpha(a + \epsilon)/3)}\right).$$

It remains to check that

$$\frac{(a+\epsilon)^2}{2(\sigma^2 + M^2\alpha(a+\epsilon)/3)} \geq \frac{\epsilon^2}{32M^2(\epsilon + \delta(\mathcal{H}))}. \tag{4.66}$$

Using Lemma 4.58 we get, on the one hand,

$$\epsilon^2(\sigma^2 + M^2\alpha(a+\epsilon)/3) \leq M^2\epsilon^2(9a + 16\delta(\mathcal{H}) + \epsilon/3). \tag{4.67}$$

On the other hand,

$$16M^2(\epsilon + \delta(\mathcal{H}))(a+\epsilon)^2 \geq M^2\epsilon^2(32a + 16\delta(\mathcal{H}) + 16\epsilon). \tag{4.68}$$

Comparing (4.67) and (4.68) we obtain (4.66). $\qquad\square$

A combination of Lemmas 4.59 and 4.50 yields the following theorem.

**Theorem 4.60** *Assume that $\rho$ and $\mathcal{H}$ satisfy (4.36) and are such that $\mathcal{E}(f_\mathcal{H}) - \mathcal{E}(f_\rho) \leq \delta$. Then, for all $\epsilon > 0$ and $\alpha \in (0, 1)$,*

$$\rho^m \left\{ \mathbf{z} : \sup_{f\in\mathcal{H}} \frac{\mathcal{E}_\mathcal{H}(f) - \mathcal{E}_{\mathcal{H},\mathbf{z}}(f)}{\mathcal{E}_\mathcal{H}(f) + \epsilon} \geq 2\alpha \right\}$$
$$\leq N\left(\mathcal{H}, \frac{\alpha\epsilon}{4M}, C(X)\right) \exp\left(-\frac{\alpha^2 m\epsilon^2}{32M^2(\epsilon + \delta)}\right).$$

*Proof* Let $f_1, \ldots, f_N$ be the $(\alpha\epsilon/4M)$-net of $\mathcal{H}$ in $C(X)$, $N := N(\mathcal{H}, \alpha\epsilon/4M, C(X))$. Let $\Lambda$ be the set of $\mathbf{z}$ such that for all $j = 1, \ldots, N$ we have

$$\frac{\mathcal{E}_\mathcal{H}(f_j) - \mathcal{E}_{\mathcal{H},\mathbf{z}}(f_j)}{\mathcal{E}_\mathcal{H}(f_j) + \epsilon} < \alpha.$$

Then, by Lemma 4.59,

$$\rho^m(\Lambda) \geq 1 - N\exp\left(-\frac{\alpha^2 m\epsilon^2}{32M^2(\epsilon + \delta)}\right). \tag{4.69}$$

We take any $\mathbf{z} \in \Lambda$ and any $g \in \mathcal{H}$. Let $f_j$ be such that $\|g - f_j\|_{C(X)} \leq \alpha\epsilon/4M$. By Lemma 4.50 we obtain

$$\frac{\mathcal{E}_\mathcal{H}(g) - \mathcal{E}_{\mathcal{H},\mathbf{z}}(g)}{\mathcal{E}_\mathcal{H}(g) + \epsilon} < 2\alpha.$$

It remains to use (4.69). $\qquad\square$

**Theorem 4.61** *Let $\mathcal{H}$ be a compact subset of $C(X)$ such that $\mathcal{E}(f_\mathcal{H}) - \mathcal{E}(f_\rho) \leq \delta$. Then, for all $\epsilon > 0$ with probability at least*

$$p(\mathcal{H}, \epsilon, \delta) := 1 - N\left(\mathcal{H}, \frac{\epsilon}{16M}, C(X)\right) \exp\left(-\frac{m\epsilon^2}{2^9 M^2(\epsilon + \delta)}\right),$$

*one has for all $f \in \mathcal{H}$*

$$\mathcal{E}(f) \le 2\mathcal{E}_{\mathbf{z}}(f) + \epsilon - \mathcal{E}(f_{\mathcal{H}}) + 2(\mathcal{E}(f_{\mathcal{H}}) - \mathcal{E}_{\mathbf{z}}(f_{\mathcal{H}})). \tag{4.70}$$

*Proof* Using Theorem 4.60 with $\alpha = 1/4$ we get, with probability at least $p(\mathcal{H}, \epsilon, \delta)$,

$$\mathcal{E}_{\mathcal{H}}(f) \le 2\mathcal{E}_{\mathcal{H},\mathbf{z}}(f) + \epsilon. \tag{4.71}$$

Substituting

$$\mathcal{E}_{\mathcal{H}}(f) := \mathcal{E}(f) - \mathcal{E}(f_{\mathcal{H}}); \qquad \mathcal{E}_{\mathcal{H},\mathbf{z}}(f) := \mathcal{E}_{\mathbf{z}}(f) - \mathcal{E}_{\mathbf{z}}(f_{\mathcal{H}}),$$

we obtain (4.70). $\qquad\qquad\qquad\qquad\qquad\qquad\qquad\qquad\qquad\qquad\qquad\square$

**Corollary 4.62** *Under the assumptions of Theorem 4.61 we have*

$$\mathcal{E}_{\mathbf{z}}(f_{\mathcal{H}}) - \mathcal{E}_{\mathbf{z}}(f_{\mathbf{z},\mathcal{H}}) \le \epsilon/2$$

*with probability at least $p(\mathcal{H}, \epsilon, \delta)$.*

*Proof of Theorem 4.56.* The statement of the theorem follows immediately from (4.71) with $f = f_{\mathbf{z},\mathcal{H}}$ because $\mathcal{E}_{\mathcal{H},\mathbf{z}}(f_{\mathbf{z},\mathcal{H}}) \le 0$ from the definition of $f_{\mathbf{z},\mathcal{H}}$. $\qquad\qquad\qquad\qquad\qquad\qquad\qquad\qquad\qquad\qquad\qquad\square$

### 4.3.6 Estimates for classes from $S_2^r$

We presented the chaining technique in Section 4.3.3, where it was used for the study of the defect function of sets from $S^r$. In this subsection we demonstrate how the chaining technique works for sets from $S_2^r$. Sets from $S_2^r$ are larger than sets from $S^r$. This results in the following modification of the corresponding results from Section 4.3.3. For instance, in Theorem 4.64 – an analog of Theorem 4.39 – we prove the probability bound for the supremum over an appropriate $\delta$-net of $W$, instead of $W$ itself as in Theorem 4.39. We present results from Konyagin and Temlyakov (2007).

**Lemma 4.63** *If $|f_j(x) - y| \le M$ a.e. for $j = 1, 2$ and $\|f_1 - f_2\|_{L_2(\rho_X)} \le \delta$, then for $\delta^2 \ge \eta$*

$$\rho^m\{\mathbf{z} : |L_{\mathbf{z}}(f_1) - L_{\mathbf{z}}(f_2)| \le \eta\} \ge 1 - 2\exp\left(-\frac{m\eta^2}{9M^2\delta^2}\right),$$

*and for $\delta^2 < \eta$*

$$\rho^m\{\mathbf{z} : |L_{\mathbf{z}}(f_1) - L_{\mathbf{z}}(f_2)| \le \eta\} \ge 1 - 2\exp\left(-\frac{m\eta}{9M^2}\right).$$

*Proof* Consider the random variable $\xi = (f_1(x) - y)^2 - (f_2(x) - y)^2$. We use

$$|\xi| \leq M^2, \quad \sigma(\xi) \leq 2M\delta.$$

Applying the Bernstein inequality (4.37) to $\xi$ we get

$$\rho^m\{\mathbf{z} : |L_{\mathbf{z}}(f_1) - L_{\mathbf{z}}(f_2)| \geq \eta\} = \rho^m \left\{ \mathbf{z} : \left| \frac{1}{m} \sum_{i=1}^{m} \xi(z_i) - E(\xi) \right| \geq \eta \right\}$$

$$\leq 2 \exp \left( -\frac{m\eta^2}{2(4M^2\delta^2 + M^2\eta/3)} \right), \quad (4.72)$$

and Lemma 4.63 follows. $\square$

**Theorem 4.64** *Assume that $\rho$, $W$ satisfy (4.36) and that $W$ is such that*

$$\sum_{n=1}^{\infty} n^{-1/2} \epsilon_n(W, L_2(\rho_X)) < \infty. \quad (4.73)$$

*Let $m\eta^2 \geq 1$. Then for any $\delta$ satisfying $\delta^2 \geq \eta$ we have for a minimal $\delta$-net $\mathcal{N}_\delta(W)$ of $W$ in the $L_2(\rho_X)$ norm*

$$\rho^m\{\mathbf{z} : \sup_{f \in \mathcal{N}_\delta(W)} |L_{\mathbf{z}}(f)| \geq \eta\} \leq C(M, \epsilon(W)) \exp(-c(M)m\eta^2).$$

*Proof* It is clear that (4.73) implies that

$$\sum_{j=0}^{\infty} 2^{j/2} \epsilon_{2^j}(W, L_2(\rho_X)) < \infty. \quad (4.74)$$

Denote $\delta_j := \epsilon_{2^j}(W, L_2(\rho_X))$, $j = 0, 1, \ldots$, and consider minimal $\delta_j$-nets $\mathcal{N}_j := \mathcal{N}_{\delta_j}(W) \subset W$ of $W$. We will use the notation $N_j := |\mathcal{N}_j|$. Let $J$ be the minimal $j$ satisfying $\delta_j \leq \delta$. We modify $\delta_J$ by setting $\delta_J = \delta$. Then $\mathcal{N}_J = \mathcal{N}_\delta(W)$. For $j = 1, \ldots, J$ we define a mapping $A_j$ that associates with a function $f \in W$ a function $A_j(f) \in \mathcal{N}_j$ closest to $f$ in the $L_2(\rho_X)$ norm. Then, clearly,

$$\|f - A_j(f)\|_{L_2(\rho_X)} \leq \delta_j.$$

We use the mappings $A_j$, $j = 1, \ldots, J$, to associate with a function $f \in W$ a sequence of functions $f_J, f_{J-1}, \ldots, f_1$ in the following way:

$$f_J := A_J(f), \quad f_j := A_j(f_{j+1}), \quad j = 1, \ldots, J - 1.$$

We introduce an auxiliary sequence

$$\eta_j := 3M\eta 2^{(j+1)/2} \epsilon_{2^{j-1}}, \quad j = 1, 2, \ldots, \quad (4.75)$$

and define $I := I(M, \epsilon(W))$ to be the minimal number satisfying

$$\sum_{j \geq I} M2^{(j+1)/2} \epsilon_{2j-1} \leq 1/6 \quad \text{or} \quad \sum_{j \geq I} \eta_j \leq \eta/2. \tag{4.76}$$

We now proceed to the estimate of $\rho^m\{\mathbf{z} : \sup_{f \in \mathcal{N}_\delta(W)} |L_\mathbf{z}(f)| \geq \eta\}$ with $m, \eta$ satisfying $m\eta^2 \geq 1$. If $J \leq I$ then the statement of Theorem 4.64 follows from Theorem 4.32. We consider the case $J > I$. Assume $|L_\mathbf{z}(f_J)| \geq \eta$. Then rewriting

$$L_\mathbf{z}(f_J) = L_\mathbf{z}(f_J) - L_\mathbf{z}(f_{J-1}) + \cdots + L_\mathbf{z}(f_{I+1}) - L_\mathbf{z}(f_I) + L_\mathbf{z}(f_I)$$

we conclude that at least one of the following events occurs:

$$|L_\mathbf{z}(f_j) - L_\mathbf{z}(f_{j-1})| \geq \eta_j \quad \text{for some} \quad j \in (I, J] \quad \text{or} \quad |L_\mathbf{z}(f_I)| \geq \eta/2.$$

Therefore

$$\rho^m\{\mathbf{z} : \sup_{f \in \mathcal{N}_\delta(W)} |L_\mathbf{z}(f)| \geq \eta\} \leq \rho^m\{\mathbf{z} : \sup_{f \in \mathcal{N}_I} |L_\mathbf{z}(f)| \geq \eta/2\}$$

$$+ \sum_{j \in (I,J]} \sum_{f \in \mathcal{N}_j} \rho^m\{\mathbf{z} : |L_\mathbf{z}(f) - L_\mathbf{z}(A_{j-1}(f))| \geq \eta_j\}$$

$$\leq \rho^m\{\mathbf{z} : \sup_{f \in \mathcal{N}_I} |L_\mathbf{z}(f_I)| \geq \eta/2\}$$

$$+ \sum_{j \in (I,J]} N_j \sup_{f \in W} \rho^m\{\mathbf{z} : |L_\mathbf{z}(f) - L_\mathbf{z}(A_{j-1}(f))| \geq \eta_j\}. \tag{4.77}$$

By our choice of $\delta_j = \epsilon_{2j}(W, L_2(\rho_X))$ we get $N_j \leq 2^{2^j} < e^{2^j}$. Let $\eta, \delta$ be such that $m\eta^2 \geq 1$ and $\eta \leq \delta^2$. It is clear that $\delta_j^2 \geq \eta_j$, $j = I, \ldots, J$. Applying Lemma 4.63 we obtain for $j \in [I, J]$

$$\sup_{f \in W} \rho^m\{\mathbf{z} : |L_\mathbf{z}(f) - L_\mathbf{z}(A_{j-1}(f))| \geq \eta_j\} \leq 2\exp\left(-\frac{m\eta_j^2}{9M^2\delta_{j-1}^2}\right).$$

From the definition (4.75) of $\eta_j$ we get

$$\frac{m\eta_j^2}{9M^2\delta_{j-1}^2} = m\eta^2 2^{j+1}$$

and

$$N_j \exp\left(-\frac{m\eta_j^2}{9M^2\delta_{j-1}^2}\right) \leq \exp(-m\eta^2 2^j).$$

Therefore

$$\sum_{j \in (I,J]} N_j \exp\left(-\frac{m\eta_j^2}{9M^2\delta_{j-1}^2}\right) \leq 2\exp(-m\eta^2 2^I). \tag{4.78}$$

By Theorem 4.32,

$$\rho^m\{\mathbf{z} : \sup_{f \in \mathcal{N}_I} |L_{\mathbf{z}}(f)| \geq \eta/2\} \leq 2N_I \exp\left(-\frac{m\eta^2}{C(M)}\right). \tag{4.79}$$

Combining (4.78) and (4.79) we obtain

$$\rho^m\{\mathbf{z} : \sup_{f \in \mathcal{N}_\delta(W)} |L_{\mathbf{z}}(f)| \geq \eta\} \leq C(M, \epsilon(W)) \exp(-c(M)m\eta^2).$$

This completes the proof of Theorem 4.64. $\qquad\square$

We get the following error estimates for $\mathcal{E}(f_{\mathbf{z}}) - \mathcal{E}(f_W)$ from Theorem 4.64.

**Theorem 4.65** *Assume that $\rho$, $W$ satisfy (4.36), and $W \in \mathcal{S}_2^r$ with $r > 1/2$. Let $m\eta^{1+\max(1/r,1)} \geq A_0(M, D, r) \geq 1$. Then there exists an estimator $f_{\mathbf{z}} \in W$ such that*

$$\rho^m\{\mathbf{z} : \mathcal{E}(f_{\mathbf{z}}) - \mathcal{E}(f_W) \leq 5\eta\} \geq 1 - C_1(M, D, r) \exp(-c_1(M)m\eta^2).$$

*Proof* It suffices to prove the theorem for $r \in (1/2, 1]$. Let us take $\delta_0 := \eta^{1/2}$ and $\mathcal{H}_0 := \mathcal{N}_{\delta_0}(W)$ to be a minimal $\delta_0$-net for $W$. Let $\delta := \eta/(2M)$ and $\mathcal{H} := \mathcal{N}_\delta(W)$ to be a minimal $\delta$-net for $W$. Denote $f_{\mathbf{z}} := f_{\mathbf{z},\mathcal{H}}$. For any $f \in \mathcal{H}$ there is $A(f) \in \mathcal{H}_0$ such that $\|f - A(f)\|_{L_2(\rho_X)} \leq \delta_0$. By Lemma 4.63,

$$\rho^m\{\mathbf{z} : |L_{\mathbf{z}}(f) - L_{\mathbf{z}}(A(f))| \leq \eta\} \geq 1 - 2\exp\left(-\frac{m\eta}{9M^2}\right).$$

Using the above inequality and Theorem 4.64 ($m\eta^2 \geq 1$) we get

$$\rho^m\{\mathbf{z} : \sup_{f \in \mathcal{H}} |L_{\mathbf{z}}(f)| \geq 2\eta\} \leq \rho^m\{\mathbf{z} : \sup_{f \in \mathcal{H}} |L_{\mathbf{z}}(f) - L_{\mathbf{z}}(A(f))| \geq \eta\}$$

$$+ \rho^m\{\mathbf{z} : \sup_{f \in \mathcal{H}_0} |L_{\mathbf{z}}(f)| \geq \eta\} \leq 2\#\mathcal{H} \exp\left(-\frac{m\eta}{9M^2}\right)$$

$$+ C(M, D, r) \exp(-c(M)m\eta^2)$$

$$\leq 4\exp\left((\eta^{-1/r})(2MD)^{1/r}\right) \exp\left(-\frac{m\eta}{9M^2}\right)$$

$$+ C(M, D, r) \exp(-c(M)m\eta^2). \tag{4.80}$$

Let us specify $A_0(M, D, r) := \max(18M^2(2MD)^{1/r}, 1)$, $r \in (1/2, 1]$. Then

$$m\eta^{1+1/r} \geq 18M^2(2MD)^{1/r} \tag{4.81}$$

and (4.80) imply

$$\rho^m\{\mathbf{z} : \sup_{f \in \mathcal{H}} |L_{\mathbf{z}}(f)| \geq 2\eta\} \leq 4\exp\left(-\frac{m\eta}{18M^2}\right) + C(M, D, r)\exp(-c(M)m\eta^2).$$

Further, we can assume that $\eta < M^2$ (otherwise, the statement of Theorem 4.65 is trivial). Therefore, we deduce from the last estimate that

$$\rho^m\{\mathbf{z} : \sup_{f \in \mathcal{H}} |L_{\mathbf{z}}(f)| \geq 2\eta\} \leq C_1(M, D, r)\exp(-c_1(M)m\eta^2).$$

Let $g \in \mathcal{H}$ be such that $\|f_W - g\|_{L_2(\rho_X)} \leq \delta$. We now observe that, by the choice of $\delta$,

$$\mathcal{E}(f_{\mathcal{H}}) - \mathcal{E}(f_W) \leq \mathcal{E}(g) - \mathcal{E}(f_W)$$

$$= \int_Z ((g(x) - y)^2 - (f_W(x) - y)^2)d\rho$$

$$\leq \|f_W - g\|_{L_1(\rho_X)} 2M \leq \eta. \tag{4.82}$$

Using

$$\begin{aligned}
\mathcal{E}(f_{\mathbf{z},\mathcal{H}}) - \mathcal{E}(f_W) &= \mathcal{E}(f_{\mathcal{H}}) - \mathcal{E}(f_W) + \mathcal{E}(f_{\mathbf{z},\mathcal{H}}) - \mathcal{E}_{\mathbf{z}}(f_{\mathbf{z},\mathcal{H}}) \\
&\quad + \mathcal{E}_{\mathbf{z}}(f_{\mathbf{z},\mathcal{H}}) - \mathcal{E}_{\mathbf{z}}(f_{\mathcal{H}}) + \mathcal{E}_{\mathbf{z}}(f_{\mathcal{H}}) - \mathcal{E}(f_{\mathcal{H}}) \\
&\leq \mathcal{E}(f_{\mathcal{H}}) - \mathcal{E}(f_W) + \mathcal{E}(f_{\mathbf{z},\mathcal{H}}) - \mathcal{E}_{\mathbf{z}}(f_{\mathbf{z},\mathcal{H}}) + \mathcal{E}_{\mathbf{z}}(f_{\mathcal{H}}) \\
&\quad - \mathcal{E}(f_{\mathcal{H}}), \tag{4.83}
\end{aligned}$$

we see that the following inequality holds:

$$\mathcal{E}(f_{\mathbf{z},\mathcal{H}}) - \mathcal{E}(f_W) \leq \eta + \mathcal{E}(f_{\mathbf{z},\mathcal{H}}) - \mathcal{E}_{\mathbf{z}}(f_{\mathbf{z},\mathcal{H}}) + \mathcal{E}_{\mathbf{z}}(f_{\mathcal{H}}) - \mathcal{E}(f_{\mathcal{H}}). \tag{4.84}$$

Hence, if $\sup_{f \in \mathcal{H}} |L_{\mathbf{z}}(f)| \leq 2\eta$, then $\mathcal{E}(f_{\mathbf{z},\mathcal{H}}) - \mathcal{E}(f_W) \leq 5\eta$. This completes the proof of Theorem 4.65. $\qquad\square$

We now proceed to the case $r \in (0, 1/2]$. We prove the following analogs of the results from Section 4.3.3 with restrictions imposed in the $L_2(\rho_X)$ norm.

**Theorem 4.66** *Assume that $\rho$, $W$ satisfy (4.36) and*

$$\sum_{n=1}^{\infty} n^{-1/2}\epsilon_n = \infty, \quad \epsilon_n := \epsilon_n(W, L_2(\rho_X)).$$

*Let $\eta, \delta$ be such that $\delta^2 \geq \eta$. Define $J := J(\delta)$ as the minimal $j$ satisfying $\epsilon_{2^j} \leq \delta$ and*

$$S_J := \sum_{j=1}^{J} 2^{(j+1)/2}\epsilon_{2^{j-1}}, \quad J \geq 1; \quad S_0 := 1.$$

*Then for m, η satisfying* $m(\eta/S_J)^2 \geq 36M^2$ *we have*

$$\rho^m\{\mathbf{z}: \sup_{f\in\mathcal{N}_\delta(W)} |L_{\mathbf{z}}(f)| \geq \eta\} \leq C(M, \epsilon(W))\exp(-c(M)m(\eta/S_J)^2),$$

*where* $\mathcal{N}_\delta(W)$ *is a minimal* $\delta$-*net of* $W$ *in the* $L_2(\rho_X)$.

*Proof* In the case $J = 0$ the statement of Theorem 4.66 follows from Theorem 4.32. In the case $J \geq 1$ the proof differs from the proof of Theorem 4.64 only in the choice of an auxiliary sequence $\{\eta_j\}$. Thus we retain the notations from the proof of Theorem 4.64. Now, instead of (4.75) we define $\{\eta_j\}$ as follows:

$$\eta_j := \frac{\eta}{2}\frac{2^{(j+1)/2}\epsilon_{2^{j-1}}}{S_J}.$$

Proceeding as in the proof of Theorem 4.64 with $I = 1$ we need to check that

$$2^j - \frac{m\eta_j^2}{9M^2\delta_{j-1}^2} \leq -2^j\frac{m(\eta/S_J)^2}{36M^2}.$$

Indeed, using the assumption $m(\eta/S_J)^2 \geq 36M^2$ we obtain

$$\frac{m\eta_j^2}{9M^2\delta_{j-1}^2} - 2^j = \frac{m(\eta/S_J)^2}{36M^2}2^{j+1} - 2^j \geq \frac{m(\eta/S_J)^2}{36M^2}2^j.$$

We complete the proof in the same way as in Theorem 4.64.  □

**Corollary 4.67** *Assume* $\rho$, $W$ *satisfy* (4.36) *and* $\epsilon_n(W, L_2(\rho_X)) \leq Dn^{-1/2}$. *Then for m, η satisfying* $m\eta^2/(1 + (\log(M/\eta))^2) \geq C_1(M, D)$ *we have for* $\delta^2 \geq \eta$

$$\rho^m\{\mathbf{z}: \sup_{f\in\mathcal{N}_\delta(W)} |L_{\mathbf{z}}(f)| \geq \eta\}$$

$$\leq C(M, D)\exp(-c(M, D)m\eta^2/(1 + (\log(M/\eta))^2)).$$

**Corollary 4.68** *Assume* $\rho$, $W$ *satisfy* (4.36) *and* $\epsilon_n(W, L_2(\rho_X)) \leq Dn^{-r}$, $r \in (0, 1/2)$. *Then for m, η,* $\delta^2 \geq \eta$ *satisfying* $m\eta^2\delta^{1/r-2} \geq C_1(M, D, r)$ *we have*

$$\rho^m\{\mathbf{z}: \sup_{f\in\mathcal{N}_\delta(W)} |L_{\mathbf{z}}(f)| \geq \eta\} \leq C(M, D, r)\exp(-c(M, D, r)m\eta^2\delta^{1/r-2}).$$

The proofs of both corollaries are the same. We present here only the proof of Corollary 4.68.

*Proof of Corollary 4.68* We use Theorem 4.66. Similarly to the proof of Theorem 4.66 it is sufficient to consider the case $J \geq 1$. We estimate the $S_J$ from Theorem 4.66:

$$S_J = \sum_{j=1}^{J} 2^{(j+1)/2} \epsilon_{2^{j-1}} \leq 2^{1/2+r} D \sum_{j=1}^{J} 2^{j(1/2-r)} \leq C_1(r) D 2^{J(1/2-r)}.$$

Next,

$$D 2^{-r(J-1)} \geq \epsilon_{2^{J-1}} > \delta \quad \text{implies} \quad 2^J \leq 2(D/\delta)^{1/r}.$$

Thus

$$S_J \leq C_1(D, r)(1/\delta)^{(1/2r)-1}.$$

It remains to apply Theorem 4.66. □

We now prove an analog of Theorem 4.65.

**Theorem 4.69** *Assume that $\rho$, $W$ satisfy (4.36) and that $W \in S_2^r$ with $r \in (0, 1/2]$. Let $m\eta^{1+1/r} \geq A_0(M, D, r) \geq 1$. Then there exists an estimator $f_{\mathbf{z}} \in W$ such that*

$$\rho^m \{\mathbf{z} : \mathcal{E}(f_{\mathbf{z}}) - \mathcal{E}(f_W) \leq 5\eta\}$$

$$\geq 1 - C(M, D) \exp(-c(M, D) m\eta^2 / (1 + (\log(M/\eta))^2))$$

*provided $r = 1/2$,*

$$\rho^m \{\mathbf{z} : \mathcal{E}(f_{\mathbf{z}}) - \mathcal{E}(f_W) \leq 5\eta\} \geq 1 - C(M, D, r) \exp\left(-c(M, D, r) m\eta^{1+1/(2r)}\right)$$

*provided $r \in (0, 1/2)$.*

*Proof* The proof in both cases $r = 1/2$ and $r \in (0, 1/2)$ is similar to the proof of Theorem 4.65. We will sketch the proof only in the case $r \in (0, 1/2)$, $\eta \leq 1$. We use the notations from the proof of Theorem 4.65. We choose $A_0(M, D, r) \geq C_1(M, D, r)$ – the constant from Corollary 4.68. Then we can use Corollary 4.68 with $\delta = \eta^{1/2}$ because

$$m\eta^2 \delta^{1/r-2} = m\eta^{1+1/(2r)} \geq m\eta^{1+1/r} \geq A_0(M, D, r) \geq C_1(M, D, r).$$

We obtain the following analog of (4.80):

$$\rho^m \{\mathbf{z} : \sup_{f \in \mathcal{H}} |L_{\mathbf{z}}(f)| \geq 2\eta\} \leq 4 \exp\left((\eta^{-1/r})(2MD)^{1/r}\right) \exp\left(-\frac{m\eta}{9M^2}\right)$$

$$+ C(M, D, r) \exp\left(-c(M, D, r) m\eta^{1+1/(2r)}\right).$$

We complete the proof in the same way as in the proof of Theorem 4.65. □

### 4.3.7 Estimates for classes from $\mathcal{S}_1^r$

In this subsection we demonstrate how a combination of a geometric assumption (convexity) and a complexity assumption ($W \in \mathcal{S}_1^r$) results in an improved bound for the accuracy of estimation. We first prove the corresponding theorem and then give a discussion.

**Theorem 4.70** *Let $W$ be convex and compact in the $L_1(\rho_X)$ set and let $\rho$, $W$ satisfy (4.36). Assume $W \in \mathcal{S}_1^r$, that is*

$$\epsilon_n(W, L_1(\rho_X)) \leq Dn^{-r}, \quad n = 1, 2, \ldots, \quad W \subset DU(L_1(\rho_X)). \quad (4.85)$$

*Then there exists an estimator $f_\mathbf{z}$ such that for $\eta \geq \eta_m := (6M + 4)\epsilon_0$, $\epsilon_0 := C(M, D)m^{-r/(1+r)}$, $m \geq 60(M/D)^2$, we have*

$$\rho^m\{\mathbf{z} : \mathcal{E}(f_\mathbf{z}) - \mathcal{E}(f_W) \geq \eta\} \leq \exp(-c(M)m\eta).$$

*Proof* Let $\mathcal{N} := \mathcal{N}_{\epsilon_0}(W, L_1(\rho_X))$ be a minimal $\epsilon_0$-net of $W$ in the $L_1(\rho_X)$ norm. The constant $C(M, D)$ will be chosen later. Then (4.85) implies that

$$|\mathcal{N}| \leq 2^{(D/\epsilon_0)^{1/r}+1}. \quad (4.86)$$

As an estimator $f_\mathbf{z}$ we take

$$f_\mathbf{z} := f_{\mathbf{z},\mathcal{N}} := \arg \min_{f \in \mathcal{N}} \mathcal{E}_\mathbf{z}(f).$$

We take $\epsilon \geq \epsilon_0$ and apply the first inequality of Lemma 4.49 with $\alpha = 1/2$ to each $f \in \mathcal{N}$. In such a way we obtain a set $\Lambda_1$ with

$$\rho^m(\Lambda_1) \geq 1 - |\mathcal{N}| \exp\left(-\frac{m\epsilon}{20M^2}\right)$$

with the following property: For all $f \in \mathcal{N}$ and all $\mathbf{z} \in \Lambda_1$ one has

$$\mathcal{E}_W(f) \leq 2\mathcal{E}_{W,\mathbf{z}}(f) + \epsilon. \quad (4.87)$$

Therefore, for $\mathbf{z} \in \Lambda_1$,

$$\mathcal{E}_W(f_\mathbf{z}) \leq 2\mathcal{E}_{W,\mathbf{z}}(f_\mathbf{z}) + \epsilon \leq 2\mathcal{E}_{W,\mathbf{z}}(f_\mathcal{N}) + \epsilon. \quad (4.88)$$

Let $\Lambda_2$ be the set of all $\mathbf{z}$ such that

$$\mathcal{E}_W(f_\mathcal{N}) - \mathcal{E}_{W,\mathbf{z}}(f_\mathcal{N}) \leq -\frac{1}{2}(\mathcal{E}_W(f_\mathcal{N}) + \epsilon). \quad (4.89)$$

By the second inequality of Lemma 4.49 with $\alpha = 1/2$,

$$\rho^m(\Lambda_2) \leq \exp\left(-\frac{m\epsilon}{20M^2}\right).$$

Consider $\Lambda := \Lambda_1 \setminus \Lambda_2$. Then

$$\rho^m(\Lambda) \geq 1 - (|\mathcal{N}| + 1) \exp\left(-\frac{m\epsilon}{20M^2}\right).$$

Using the inequality opposite to (4.89) we continue (4.88) for $\mathbf{z} \in \Lambda$:

$$\mathcal{E}_W(f_{\mathbf{z}}) \leq 2\mathcal{E}_{W,\mathbf{z}}(f_{\mathcal{N}}) + \epsilon \leq 3\mathcal{E}_W(f_{\mathcal{N}}) + 2\epsilon \leq 6M\epsilon_0 + 2\epsilon.$$

We choose $\epsilon_0 \leq D$ from the equation

$$3(D/\epsilon_0)^{1/r} = \frac{m\epsilon_0}{20M^2}.$$

We get

$$\epsilon_0 = (60M^2)^{r/(1+r)} D^{1/(1+r)} m^{-r/(1+r)}.$$

For $m \geq 60M^2/D$ we have $\epsilon_0 \leq D$. We let $\eta = 6M\epsilon_0 + 2\epsilon$. Then our assumption $\eta \geq (6M + 4)\epsilon_0$ implies $\epsilon \geq 2\epsilon_0$ and

$$\rho^m(Z^m \setminus \Lambda) \leq (|\mathcal{N}| + 1) \exp\left(-\frac{m\epsilon_0}{20M^2}\right) \exp\left(-\frac{m(\epsilon - \epsilon_0)}{20M^2}\right)$$

$$\leq \exp\left(-\frac{m\epsilon}{40M^2}\right) \leq \exp\left(-\frac{m\eta}{40M^2(3M+2)}\right).$$

This completes the proof of Theorem 4.70. □

We note that Theorem 4.70 with the assumption $W \in \mathcal{S}^r$ (instead of $W \in \mathcal{S}_1^r$) has been proved in Cucker and Smale (2001) and DeVore *et al.* (2004) with $f_{\mathbf{z}} = f_{\mathbf{z},W}$ (see Theorem 4.54 above). It is interesting to compare Theorem 4.70 with the corresponding results when we do not assume that $W$ is convex. Let us compare only the accuracy thresholds $\eta_m$. Theorem 4.70 says that for a convex $W$ the assumption $W \in \mathcal{S}_1^r$ implies

$$\eta_m \ll m^{-r/(1+r)}.$$

The results of Section 4.3.6 (see Theorems 4.65 and 4.69) state that $W \in \mathcal{S}_2^r$ (the no convexity assumption) implies

$$\eta_m \ll m^{-r/(1+r)}, \quad r \in (0, 1],$$

$$\eta_m \ll m^{-1/2}, \quad r \geq 1.$$

The results of Section 4.3.3 (see Theorem 4.45) give the following estimates for $W \in \mathcal{S}^r$:

$$\eta_m \ll m^{-r}, \quad r \in (0, 1/2),$$

$$\eta_m \ll m^{-1/2}, \quad r > 1/2.$$

It will be proved in Section 4.5 that the above bounds cannot be improved. Therefore, even under a strong assumption $W \in \mathcal{S}^r$ the best we can get is

$\eta_m \ll m^{-1/2}$. Theorem 4.70 shows that the convexity combined with a weaker assumption $W \in \mathcal{S}_1^r$ provide better estimates for large $r$. Section 4.5 contains further comments on studying the accuracy confidence function for the projection learning problem (improper function learning problem) that was defined in the Introduction to this chapter.

## 4.4 Proper function learning; upper estimates

### 4.4.1 Introduction

In this section we continue a discussion from Section 4.1 and the Introduction from Section 4.3. The main question of non-parametric regression theory and learning theory is how to choose an estimator $f_{\mathbf{z}}$. There are several different approaches to this problem, some of which we now discuss. Recently, driven by ideas from approximation theory, the following general approach to this problem has been developed. The idea of this approach is to choose an estimator $f_{\mathbf{z}}$ as a solution (approximate solution) of an optimization problem (minimax problem). So, in this approach we should begin with a formulation of an optimization problem. A standard formulation of such a problem is the following. We begin with a fixed class $\Theta$ of priors (or a fixed class $W$ where we project $f_{\rho}$). That means we impose a restriction on an unknown measure $\rho$, which we want to study, in the form $f_{\rho} \in \Theta$. Developing this approach we encounter three immediate questions.

(1) What classes $\Theta$ of priors (or classes $W$) should we choose?
(2) What should be the form of $f_{\mathbf{z}}$?
(3) How should we measure the quality of estimation (approximation)?

We will not discuss these questions in detail here. We only note that the following partial answers to the above questions are widely accepted.

(1) A very important characteristic of $\Theta$ that governs the quality of estimation is a sequence of the entropy numbers $\epsilon_n(\Theta, B)$ of $\Theta$ in a suitable Banach space $B$.
(2) The following way of building $f_{\mathbf{z}}$ provides a near optimal estimator in many cases. First, choose a right hypothesis space $\mathcal{H}$ (that may depend on $\Theta$). Second, construct $f_{\mathbf{z},\mathcal{H}} \in \mathcal{H}$ as the empirical optimum (least squares estimator).
(3) It seems natural (see Binev *et al.* (2005), Cucker and Smale (2001), DeVore *et al.* (2006), Györfy *et al.* (2002) and Konyagin and Temlyakov (2004, 2007)) to measure the quality of approximation by

$$\mathcal{E}(f) - \mathcal{E}(f_{\rho}) = \|f - f_{\rho}\|_{L_2(\rho_X)}^2. \tag{4.90}$$

In this section we address the following important issue. In many cases we do not know exactly the form of the class $\Theta$ of priors where an unknown $f_\rho$ comes from. Therefore, we try to construct an estimator that provides a good estimation (near optimal) not for a single class of priors $\Theta$ but for a collection of classes of priors. Clearly, in order to claim that an estimator $f_z$ is near optimal for a class $\Theta$ we need to compare the upper estimates of approximation by $f_z$ with the corresponding lower bounds of optimal estimation for $\Theta$.

The concept of universality is rigorously defined and well understood in approximation theory (see, for example, Temlyakov (1988a, 2003a)). We propose to study the universality of estimation in learning theory in three steps. At the first step we solve an optimization problem for classes of our interest. Then, at the second step, using results on the optimization problem, we study the universality problem from a theoretical point of view based on accuracy and confidence. Finally, the problem of building practical universal algorithms with good theoretical properties should be addressed. In this book we discuss only the first two steps of the above program. The problem of constructing practical universal algorithms has been considered in Binev *et al.* (2005).

We begin with some lower bounds from Section 4.5. We formulate Theorem 4.101 from Section 4.5.4 for convenience. We let $\mu$ be any Borel probability measure defined on $X$ and let $\mathcal{M}(\Theta, \mu)$ denote the set of all $\rho \in \mathcal{M}(\Theta)$ such that $\rho_X = \mu$, $|y| \leq 1$. As above, $\mathcal{M}(\Theta) = \{\rho : f_\rho \in \Theta\}$.

**Theorem 4.71** *Assume $\Theta$ is a compact subset of $L_2(\mu)$ such that $\Theta \subset (1/4)U(\mathcal{C}(X))$ and*

$$\epsilon_n(\Theta, L_2(\mu)) \asymp n^{-r}. \tag{4.91}$$

*Then there exist $\delta_0 > 0$ and $\eta_m := \eta_m(r) \asymp m^{-r/(1+2r)}$ such that*

$$\mathbf{AC}_m(\mathcal{M}(\Theta, \mu), \eta) \geq \delta_0 \quad \text{for} \quad \eta \leq \eta_m \tag{4.92}$$

*and*

$$\mathbf{AC}_m(\mathcal{M}(\Theta, \mu), \eta) \geq C e^{-c(r)m\eta^2} \quad \text{for} \quad \eta \geq \eta_m. \tag{4.93}$$

**Remark 4.72** Theorem 4.71 holds in the case $\Theta \subset (M/4)U(\mathcal{C}(X))$, $|y| \leq M$, with constants allowed to depend on $M$.

The lower estimates from Theorem 4.71 will serve as a benchmark for the performance of particular estimators. Let us formulate a condition on a measure $\rho$ and a class $\Theta$ that we will often use:

for all $f \in \Theta$, we have $|f(x)-y| \leq M$ a.e. with respect to $\rho$. (4.94)

Clearly, (4.94) is satisfied if $|y| \leq M/2$ and $|f(x)| \leq M/2$, $f \in \Theta$.

The results of this section are from Temlyakov (2008a). In this section we prove the following result, complementary to Theorem 4.71.

**Theorem 4.73** *Let $f_\rho \in \Theta$ and let $\rho$, $\Theta$ satisfy (4.94). Assume*

$$\epsilon_n(\Theta, L_2(\rho_X)) \leq Dn^{-r}, \quad n = 1, 2, \ldots, \quad \Theta \subset DU(L_2(\rho_X)).$$

*Then there exists an estimator $f_{\mathbf{z}}$ such that, for $\eta^2 \geq 7\epsilon_0$, $\epsilon_0 := C(M, D, r)m^{-2r/(1+2r)}$, $m \geq 60(M/D)^2$, we have*

$$\rho^m\{\mathbf{z} : \|f_{\mathbf{z}} - f_\rho\|_{L_2(\rho_X)} \geq \eta\} \leq \exp\left(-\frac{m\eta^2}{140M^2}\right).$$

In the case when $\Theta$ satisfies the assumption,

$$\epsilon_n(\Theta, \mathcal{C}(X)) \leq Dn^{-r}, \quad n = 1, 2, \ldots, \quad \Theta \subset DU(\mathcal{C}(X)).$$

Theorem 4.73 was obtained in Konyagin and Temlyakov (2004).

A combination of Theorem 4.71 and Theorem 4.73 completes the study of the behavior (in the sense of order) of the **AC**-function of classes satisfying (4.91). We formulate this as a theorem (see Temlyakov (2008a)).

**Theorem 4.74** *Let $\mu$ be a Borel probability measure on $X$. Assume that $r > 0$ and that $\Theta$ is a compact subset of $L_2(\mu)$ such that $\Theta \subset (1/4)U(\mathcal{C}(X))$ and*

$$\epsilon_n(\Theta, L_2(\mu)) \asymp n^{-r}.$$

*Then there exist $\delta_0 > 0$ and $\eta_m^- \leq \eta_m^+$, $\eta_m^- \asymp \eta_m^+ \asymp m^{-r/(1+2r)}$ such that*

$$\mathbf{AC}_m(\mathcal{M}(\Theta, \mu), \eta) \geq \delta_0 \quad \text{for} \quad \eta \leq \eta_m^-$$

*and*

$$C_1 e^{-c_1(r)m\eta^2} \leq \mathbf{AC}_m(\mathcal{M}(\Theta, \mu), \eta) \leq e^{-c_2 m\eta^2}$$

*for $\eta \geq \eta_m^+$.*

Theorem 4.74 solves the optimization problem. Let us now make some conclusions. First of all, Theorem 4.74 shows that the entropy numbers $\epsilon_n(\Theta, L_2(\mu))$ are the right characteristic of the class $\Theta$ in the estimation problem. The behavior of the sequence $\{\epsilon_n(\Theta, L_2(\mu))\}$ determines the behavior of the sequence $\{\mathbf{AC}_m(\mathcal{M}(\Theta, \mu), \eta)\}$ of the **AC**-functions. Second, the proof of Theorem 4.73 points out that the optimal (in the sense of order) estimator can be always constructed as a least squares estimator. Theorem 4.74 discovers a new phenomenon – *sharp phase transition*. The behavior of the accuracy confidence function changes dramatically within the *critical interval* $[\eta_m^-, \eta_m^+]$. It drops from a constant $\delta_0$ to an exponentially small quantity $\exp(-cm^{1/(1+2r)})$. One may also call the interval $[\eta_m^-, \eta_m^+]$ *the interval of phase transition*.

Let us make a general remark on the technique that we use in this section: it usually consists of a combination of results from non-parametric statistics with results from approximation theory. Both the results from non-parametric statistics and the results from approximation theory that we use are either known or are very close to known results. For example, in the proof of Theorem 4.73 we have used a statistical technique that was used in many papers (for example, Barron, Birgé and Massart (1999), Cucker and Smale (2001) and Lee, Bartlett and Williamson (1998)) and goes back to Barron's seminal paper, Barron (1991). We also used some elementary results on the entropy numbers from approximation theory.

We now proceed to the results from this section on the construction of universal (adaptive) estimators. Let $a$, $b$, be two positive numbers. Consider a collection $\mathcal{K}(a, b)$ of compacts $K_n$ in $C(X)$ satisfying

$$N(K_n, \epsilon, C(X)) \leq (a(1 + 1/\epsilon))^n n^{bn}, \quad n = 1, 2, \ldots. \tag{4.95}$$

The following two theorems that we prove here form a basis for constructing universal estimators. We begin with the definition of our estimator. As above, let $\mathcal{K} := \mathcal{K}(a, b)$ be a collection of compacts $K_n$ in $C(X)$ satisfying (4.95).

We take a parameter $A \geq 1$ and consider the following estimator that we call the penalized least squares estimator (PLSE):

$$f_{\mathbf{z}}^A := f_{\mathbf{z}}^A(\mathcal{K}) := f_{\mathbf{z}, K_{n(\mathbf{z})}}$$

with

$$n(\mathbf{z}) := \arg \min_{1 \leq j \leq m} \left( \mathcal{E}_{\mathbf{z}}(f_{\mathbf{z}, K_j}) + \frac{Aj \ln m}{m} \right).$$

Denote for a set $L$ of a Banach space $B$

$$d(\Theta, L)_B := \sup_{f \in \Theta} \inf_{g \in L} \| f - g \|_B.$$

**Theorem 4.75** *For* $\mathcal{K} := \{K_n\}_{n=1}^{\infty}$ *satisfying (4.95) and* $M > 0$ *there exists* $A_0 := A_0(a, b, M)$ *such that, for any* $A \geq A_0$ *and any* $\rho$ *such that* $\rho$, $K_n$, $n = 1, 2, \ldots$ *satisfy (4.94), we have*

$$\| f_{\mathbf{z}}^A - f_\rho \|_{L_2(\rho_X)}^2 \leq \min_{1 \leq j \leq m} \left( 3d(f_\rho, K_j)_{L_2(\rho_X)}^2 + \frac{4Aj \ln m}{m} \right)$$

*with probability* $\geq 1 - m^{-c(M)A}$.

Theorem 4.75 is from Temlyakov (2008a). The reader can find the results in the style of Theorem 4.75 with bounds of the expectation $E_{\rho^m}(\| f_{\mathbf{z}}^A - f_\rho \|_{L_2(\rho_X)}^2)$ in Györfy *et al.* (2002), chap. 12.

**Theorem 4.76** *Let compacts $\{K_n\}$ satisfy (4.95) and $M > 0$ be given. There exists $A_0 := A_0(a, b, M) \geq 1$ such that for any $A \geq A_0$ and any $\rho$ satisfying*

$$d(f_\rho, K_n)_{L_2(\rho_X)} \leq A^{1/2} n^{-r}, \quad n = 1, 2, \ldots, \tag{4.96}$$

*and such that $\rho$, $K_n$, $n = 1, 2, \ldots$, satisfy (4.94) we have for $\eta \geq A^{1/2} (\ln m/m)^{r/(1+2r)}$*

$$\rho^m \{\mathbf{z} : \|f_{\mathbf{z}}^A - f_\rho\|_{L_2(\rho_X)} \geq 4A^{1/2} \eta\} \leq C e^{-c(M)m\eta^2}.$$

Let us make some comments on the construction of the estimator $f_{\mathbf{z}}^A$. In approximation theory we have the following setting of the universality problem (see Temlyakov (1988a, 2003a)). We begin with a collection $\mathcal{F} := \{F\}$ of classes of interest to us and try to build a universal approximation method of a certain type (for instance, approximation by elements of a sequence of $n$-dimensional subspaces, or $m$-term approximations with regard to a basis). Usually, the classes $F \in \mathcal{F}$ are smoothness classes with smoothness defined in the $L_q$ spaces with respect to the Lebesgue measure (see a discussion in DeVore *et al.* (2006)). It is well known in approximation theory that smoothness classes (Besov classes, Hölder–Nikol'skii classes) can be equivalently described as approximation classes (see, for instance, DeVore (1998) or DeVore *et al.* (2006)). In this case, approximation takes place in the $L_p$ space with respect to the Lebesgue measure. In a learning theory setting we want to approximate in the $L_2(\rho_X)$ norm with a rather general probability measure $\rho_X$. There is no theory of smoothness classes in such a general setting. Therefore, we describe our classes of interest as approximation classes, where approximation takes place in the $L_2(\rho_X)$ norm. As a result we transform the universality setting in learning theory to the following setting. We now begin with a sequence $\{\mathcal{H}_n\}$ that plays a double role: $\{\mathcal{H}_n\}$ is a sequence of hypothesis spaces where our estimators are supposed to come from and, also, $\{\mathcal{H}_n\}$ is a sequence of sets of approximants that is used in the definition of approximation classes. We stress that it is important that our estimators are of the same form as the approximants used in the definition of classes. In Theorem 4.76 the approximation classes are given by (4.96) and the estimator $f_{\mathbf{z}}^A \in K_{n(\mathbf{z})}$ is exactly in the form of an approximant. In other words, the sparseness of the regression function $f_\rho$ described in terms of the approximation classes is reflected by the sparseness of the estimator. We note that this property of our universal estimator $f_{\mathbf{z}}^A$ distinguishes it from the estimator considered in Györfy *et al.* (2002). We will give a brief description of their construction. A standard assumption we make in supervised learning is that $|y| \leq M$. This assumption implies that $|f_\rho(x)| \leq M$. Thus, it is natural to restrict a search of the estimator to functions bounded (in absolute value) by $M$. Now, the question is how to incorporate this into a construction of an estimator. Györfy *et al.* (2002) use the truncation operator $T_M$ for this purpose ($T_M(u) = u$ if $|u| \leq M$ and $T_M(u) = M \text{sign} u$ if $|u| > M$). Their approach, in general terms, is as follows. Let a sequence of hypothesis spaces $\{\mathcal{H}_n\}$ be given. For given $m$ and $\mathbf{z}$ they build an estimator $\hat{f}_{\mathbf{z}} \in \mathcal{H}_{n(m,\mathbf{z})}$ and then use the estimator $f_{\mathbf{z}} := T_M(\hat{f}_{\mathbf{z}})$. Next, they study the expectation of the error $\|f_\rho - f_{\mathbf{z}}\|^2_{L_2(\rho_X)}$ for $f_\rho$ satisfying

$$d(f_\rho, \mathcal{H}_n)_{L_2(\rho_X)} \leq Cn^{-r}.$$

Usually, an $\mathcal{H}_n$ is a linear subspace or a nonlinear manifold that is not a bounded subset of the space $\mathcal{B}(X)$ of bounded functions with $\| \cdot \|_{\mathcal{B}(X)} := \sup_{x \in X} |f(x)|$. This is why the truncation operator $T_M$ is used in the construction of the $f_{\mathbf{z}}$. Results of Györfy *et al.* (2002) show that the estimator $f_{\mathbf{z}}$ is universal in the sense of expectation. The above approach has a drawback in that the use of the truncation operator entails that the estimator $f_{\mathbf{z}}$ is (in general) not an element of $\mathcal{H}_n$. Thus, they describe a class using approximation by elements of one form and build an estimator of another, more complex, form.

Let us make a comment on the condition (4.95). This condition is written in a form that allows us to treat simultaneously the case of linear approximation and the case of nonlinear approximation within the framework of a unified general theory. The term $(a(1 + 1/\epsilon))^n$ corresponds to the covering number for an $a$-ball in an $n$-dimensional Banach space and an extra factor $n^{bn}$ takes care of nonlinear $n$-term approximation (see Section 4.4.3 for details).

In this section we formulate assumptions on a class $W$ in the following form:

$$\epsilon_n(W, B) \leq Dn^{-r}, \quad W \subset DU(B). \tag{4.97}$$

We denote by $\mathcal{S}^r := \mathcal{S}^r(D) := \mathcal{S}^r_\infty(D)$ a collection of classes $W$ that satisfy (4.97) with $B = \mathcal{C}(X)$. The notation $\mathcal{S}^r_p := \mathcal{S}^r_p(D)$ is used for a collection of classes $W$ satisfying (4.97) with $B = L_p(\rho_X)$, $1 \leq p < \infty$.

We often have error estimates of the form $(\ln m/m)^\alpha$ that hold for $m \geq 2$. We could write these estimates in the form, say, $(\ln(m+1)/m)^\alpha$ to make them valid for all $m \in \mathbb{N}$. However, we use the first variant throughout the book for the following two reasons: (i) simpler notations may be used and (ii) we are looking for the asymptotic behavior of the error.

### 4.4.2 The least squares estimators

The technique from Section 4.3 can also be used in the following situation. Define

$$\mathcal{E}_\rho(f) := \mathcal{E}(f) - \mathcal{E}(f_\rho); \qquad \mathcal{E}_{\rho,\mathbf{z}}(f) := \mathcal{E}_{\mathbf{z}}(f) - \mathcal{E}_{\mathbf{z}}(f_\rho);$$
$$\ell_\rho(f) := (f(x) - y)^2 - (f_\rho(x) - y)^2.$$

Then we have the following analogs of the lemmas from Section 4.3 (see Section 4.3.4).

**Lemma 4.77** *Let $f$, $f_\rho$ be such that $|f(x) - y| \le M$, $|f_\rho(x) - y| \le M$ a.s. Then we have*

$$\sigma^2(\ell_\rho(f)) \le 4M^2 \mathcal{E}_\rho(f).$$

**Lemma 4.78** *Let $f$, $f_\rho$ be such that $|f(x) - y| \le M$, $|f_\rho(x) - y| \le M$ a.s. Then for all $\epsilon > 0$, $\alpha \in (0, 1]$ one has*

$$\rho^m\{\mathbf{z} : \mathcal{E}_\rho(f) - \mathcal{E}_{\rho,\mathbf{z}}(f) \ge \alpha(\mathcal{E}_\rho(f) + \epsilon)\} \le \exp\left(-\frac{\alpha^2 m\epsilon}{5M^2}\right),$$

$$\rho^m\{\mathbf{z} : \mathcal{E}_\rho(f) - \mathcal{E}_{\rho,\mathbf{z}}(f) \le -\alpha(\mathcal{E}_\rho(f) + \epsilon)\} \le \exp\left(-\frac{\alpha^2 m\epsilon}{5M^2}\right).$$

**Lemma 4.79** *Assume that $\rho$, $\mathcal{H}$ satisfy (4.94) and that $f, g \in \mathcal{H}$ are such that $\|f - g\|_{C(X)} \le \alpha\epsilon/(4M)$. Let $\alpha \in (0, 1)$, $\epsilon > 0$, and let $f$ be such that*

$$\frac{\mathcal{E}_\rho(f) - \mathcal{E}_{\rho,\mathbf{z}}(f)}{\mathcal{E}_\rho(f) + \epsilon} < \alpha.$$

*Then we have*

$$\frac{\mathcal{E}_\rho(g) - \mathcal{E}_{\rho,\mathbf{z}}(g)}{\mathcal{E}_\rho(g) + \epsilon} < 2\alpha.$$

These lemmas imply the following analog of Theorem 4.52.

**Theorem 4.80** *Let $\mathcal{H}$ be a compact subset of $C(X)$. Assume that $\rho$ and $\mathcal{H}$ satisfy (4.94). Then for all $\epsilon > 0$ with probability at least*

$$p(\mathcal{H}, \rho, \epsilon) := 1 - N\left(\mathcal{H}, \frac{\epsilon}{16M}, C(X)\right) \exp\left(-\frac{m\epsilon}{80M^2}\right)$$

*one has, for all $f \in \mathcal{H}$,*

$$\mathcal{E}_\rho(f) \le 2\mathcal{E}_{\rho,\mathbf{z}}(f) + \epsilon. \tag{4.98}$$

We first demonstrate how Lemma 4.78 can be used in proving optimal upper estimates.

**Theorem 4.81** *Let $f_\rho \in \Theta$ and let $\rho$, $\Theta$ satisfy (4.94). Assume*

$$\epsilon_n(\Theta, L_2(\rho_X)) \le Dn^{-r}, \quad n = 1, 2, \ldots, \quad \Theta \subset DU(L_2(\rho_X)). \tag{4.99}$$

*Then there exists an estimator $f_{\mathbf{z}}$ such that for $\eta \ge \eta_m := 7\epsilon_0$, $\epsilon_0 := C(M, D, r)m^{-2r/(1+2r)}$, $m \ge 60(M/D)^2$, we have*

$$\rho^m\{\mathbf{z} : \|f_{\mathbf{z}} - f_\rho\|^2_{L_2(\rho_X)} \ge \eta\} \le \exp\left(-\frac{m\eta}{140M^2}\right).$$

*Proof* Let $\mathcal{N} := \mathcal{N}_{\epsilon_0^{1/2}}(\Theta, L_2(\rho_X))$ be a minimal $\epsilon_0^{1/2}$-net of $\Theta$ in the $L_2(\rho_X)$ norm. The constant $C(M, D, r)$ will be chosen later. Then (4.99) implies that

$$|\mathcal{N}| \leq 2^{(D^2/\epsilon_0)^{1/(2r)}+1}. \tag{4.100}$$

As an estimator $f_{\mathbf{z}}$ we take

$$f_{\mathbf{z}} := f_{\mathbf{z},\mathcal{N}} := \arg \min_{f \in \mathcal{N}} \mathcal{E}_{\mathbf{z}}(f).$$

We take $\epsilon \geq \epsilon_0$ and apply the first inequality of Lemma 4.78 with $\alpha = 1/2$ to each $f \in \mathcal{N}$. In such a way we obtain a set $\Lambda_1$ with

$$\rho^m(\Lambda_1) \geq 1 - |\mathcal{N}| \exp\left(-\frac{m\epsilon}{20M^2}\right)$$

with the following property: For all $f \in \mathcal{N}$ and all $\mathbf{z} \in \Lambda_1$ one has

$$\mathcal{E}_\rho(f) \leq 2\mathcal{E}_{\rho,\mathbf{z}}(f) + \epsilon. \tag{4.101}$$

Therefore, for $\mathbf{z} \in \Lambda_1$,

$$\mathcal{E}_\rho(f_{\mathbf{z}}) \leq 2\mathcal{E}_{\rho,\mathbf{z}}(f_{\mathbf{z}}) + \epsilon \leq 2\mathcal{E}_{\rho,\mathbf{z}}(f_{\mathcal{N}}) + \epsilon. \tag{4.102}$$

Let $\Lambda_2$ be the set of all $\mathbf{z}$ such that

$$\mathcal{E}_\rho(f_{\mathcal{N}}) - \mathcal{E}_{\rho,\mathbf{z}}(f_{\mathcal{N}}) \leq -\frac{1}{2}(\mathcal{E}_\rho(f_{\mathcal{N}}) + \epsilon). \tag{4.103}$$

By the second inequality of Lemma 4.78 with $\alpha = 1/2$,

$$\rho^m(\Lambda_2) \leq \exp\left(-\frac{m\epsilon}{20M^2}\right).$$

Consider $\Lambda := \Lambda_1 \setminus \Lambda_2$. Then

$$\rho^m(\Lambda) \geq 1 - (|\mathcal{N}| + 1) \exp\left(-\frac{m\epsilon}{20M^2}\right).$$

Using the inequality opposite to (4.103) we continue (4.102) for $\mathbf{z} \in \Lambda$:

$$\mathcal{E}_\rho(f_{\mathbf{z}}) \leq 2\mathcal{E}_{\rho,\mathbf{z}}(f_{\mathcal{N}}) + \epsilon \leq 3\mathcal{E}_\rho(f_{\mathcal{N}}) + 2\epsilon \leq 3\epsilon_0 + 2\epsilon.$$

We choose $\epsilon_0 \leq D^2$ from the equation

$$3(D^2/\epsilon_0)^{1/2r} = \frac{m\epsilon_0}{20M^2}.$$

We get

$$\epsilon_0 = (60M^2)^{2r/(1+2r)} D^{2/(1+2r)} m^{-2r/(1+2r)}.$$

For $m \geq 60(M/D)^2$ we have $\epsilon_0 \leq D^2$. We let $\eta = 3\epsilon_0 + 2\epsilon$. Then our assumption $\eta \geq 7\epsilon_0$ implies $\epsilon \geq 2\epsilon_0$ and

$$\rho^m(Z^m \setminus \Lambda) \leq (|\mathcal{N}| + 1) \exp\left(-\frac{m\epsilon_0}{20M^2}\right) \exp\left(-\frac{m(\epsilon - \epsilon_0)}{20M^2}\right)$$
$$\leq \exp\left(-\frac{m\epsilon}{40M^2}\right) \leq \exp\left(-\frac{m\eta}{140M^2}\right).$$

This completes the proof of Theorem 4.81. □

We proceed to the universal estimators. For convenience we repeat the definition of the estimator $f_{\mathbf{z}}^A$ and the formulations of the corresponding theorems from the Introduction. Let $a, b$ be two positive numbers. Consider a collection $\mathcal{K} := \mathcal{K}(a, b)$ of compacts $K_n$ in $C(X)$ satisfying

$$N(K_n, \epsilon, C(X)) \leq (a(1 + 1/\epsilon))^n n^{bn}, \quad n = 1, 2, \ldots \qquad (4.104)$$

We take a parameter $A \geq 1$ and consider the following estimator

$$f_{\mathbf{z}}^A := f_{\mathbf{z}}^A(\mathcal{K}) := f_{\mathbf{z}, K_{n(\mathbf{z})}}$$

with

$$n(\mathbf{z}) := \arg \min_{1 \leq j \leq m} \left(\mathcal{E}_{\mathbf{z}}(f_{\mathbf{z}, K_j}) + \frac{Aj \ln m}{m}\right).$$

**Theorem 4.82** *For* $\mathcal{K} := \{K_n\}_{n=1}^{\infty}$ *satisfying (4.104) there exists* $A_0 := A_0(a, b, M)$ *such that, for any* $A \geq A_0$ *and any* $\rho$ *such that* $\rho$, $K_n$, $n = 1, 2, \ldots$, *satisfy (4.94), we have*

$$\|f_{\mathbf{z}}^A - f_\rho\|_{L_2(\rho_X)}^2 \leq \min_{1 \leq j \leq m} \left(3d(f_\rho, K_j)_{L_2(\rho_X)}^2 + \frac{4Aj \ln m}{m}\right)$$

*with probability* $\geq 1 - m^{-c(M)A}$.

*Proof* We set $\epsilon_j := (2Aj \ln m)/m$ for all $j \in [1, m]$. Applying Theorem 4.80 and Lemma 4.78 we find a set $\Lambda$ with

$$\rho^m(\Lambda) \geq 1 - m^{-c(M)A}$$

such that for all $\mathbf{z} \in \Lambda$, $j \in [1, m]$ we have

$$\mathcal{E}_\rho(f) \leq 2\mathcal{E}_{\rho, \mathbf{z}}(f) + \epsilon_j, \quad \forall f \in K_j,$$
$$\mathcal{E}_{\rho, \mathbf{z}}(f_{K_j}) \leq \frac{3}{2}\mathcal{E}_\rho(f_{K_j}) + \epsilon_j/2.$$

We get from here, for $\mathbf{z} \in \Lambda$,

$$\mathcal{E}_\rho(f_{\mathbf{z}}^A) \leq 2\mathcal{E}_{\rho,\mathbf{z}}(f_{\mathbf{z},K_{n(\mathbf{z})}}) + \epsilon_{n(\mathbf{z})} = 2\left(\mathcal{E}_{\rho,\mathbf{z}}(f_{\mathbf{z},K_{n(\mathbf{z})}}) + \frac{An(\mathbf{z})\ln m}{m}\right)$$

$$= 2\min_{1\leq j\leq m}\left(\mathcal{E}_{\rho,\mathbf{z}}(f_{\mathbf{z},K_j}) + \frac{Aj\ln m}{m}\right)$$

$$\leq 2\min_{1\leq j\leq m}\left(\mathcal{E}_{\rho,\mathbf{z}}(f_{K_j}) + \frac{Aj\ln m}{m}\right)$$

$$\leq 2\min_{1\leq j\leq m}\left(\frac{3}{2}\mathcal{E}_\rho(f_{K_j}) + \frac{2Aj\ln m}{m}\right). \qquad \square$$

**Theorem 4.83** *Let compacts* $\{K_n\}$ *satisfy (4.104). There exists* $A_0 := A_0(a, b, M) \geq 1$ *such that for any* $A \geq A_0$ *and any* $\rho$ *satisfying*

$$d(f_\rho, K_n)_{L_2(\rho_X)} \leq A^{1/2}n^{-r}, \quad n = 1, 2, \ldots,$$

*and such that* $\rho$, $K_n$, $n = 1, 2, \ldots$, *satisfy (4.94) we have for* $\eta \geq A^{1/2}(\ln m/m)^{r/(1+2r)}$

$$\rho^m\{\mathbf{z} : \|f_{\mathbf{z}}^A - f_\rho\|_{L_2(\rho_X)} \geq 4A^{1/2}\eta\} \leq Ce^{-c(M)m\eta^2}.$$

*Proof* Let $\eta \geq A^{1/2}(\ln m/m)^{r/(1+2r)}$. We define $n$ as the smallest integer such that $2n \geq m\eta^2/\ln m$. Denote $\epsilon_j := (2Aj\ln m)/m$, $j \in (n, m]$; $\epsilon_j := A\eta^2$, $j \in [1, n]$. We apply Theorem 4.80 to $K_j$ with $\epsilon = \epsilon_j$. Denote by $\Lambda_j$ the set of all $\mathbf{z}$ such that, for any $f \in K_j$,

$$\mathcal{E}_\rho(f) \leq 2\mathcal{E}_{\rho,\mathbf{z}}(f) + \epsilon_j. \qquad (4.105)$$

By Theorem 4.80,

$$\rho^m(\Lambda_j) \geq p(K_j, \rho, \epsilon_j).$$

For estimating $p(K_j, \rho, \epsilon_j)$ we write ($j \in [n, m]$):

$$\ln\left(N(K_j, \epsilon_j/(16M), \mathcal{C}(X))\exp\left(-\frac{m\epsilon_j}{80M^2}\right)\right)$$

$$\leq j\ln(a(1 + 16M/\epsilon_j)) + bj\ln j - \frac{m\epsilon_j}{80M^2}$$

$$\leq j\ln(a(1 + 8M)) + j(1 + b)\ln m - Ac_2(M)j\ln m$$

$$\leq -Ac_3(M)j\ln m \leq -Ac_3(M)n\ln m \leq -Ac_4(M)m\eta^2$$

for $A \geq C_1(a, b, M)$. A similar estimate for $j \in [1, n]$ follows from the above estimate with $j = n$.

Thus (4.105) holds for all $1 \leq j \leq m$ on the set $\Lambda' := \cap_{j=1}^m \Lambda_j$ with

$$\rho^m(\Lambda') \geq 1 - e^{-c_5(M)m\eta^2}.$$

For $j \in [1, m]$ we have, by the assumption of Theorem 4.83, that

$$\mathcal{E}_\rho(f_{K_j}) = \|f_{K_j} - f_\rho\|^2_{L_2(\rho_X)} \leq Aj^{-2r}. \tag{4.106}$$

We apply the second inequality of Lemma 4.78 to each $f_{K_j}$ with $\alpha = 1/2$ and $\epsilon_j$ chosen above, $j = 1, \ldots, m$. Then we obtain a set $\Lambda''$ of $\mathbf{z}$ such that

$$\mathcal{E}_{\rho,\mathbf{z}}(f_{K_j}) \leq \frac{3}{2}\mathcal{E}_\rho(f_{K_j}) + \epsilon_j/2, \quad j = 1, \ldots, m, \tag{4.107}$$

and

$$\rho^m(\Lambda'') \geq 1 - \sum_{j=1}^m \exp\left(-\frac{m\epsilon_j}{20M^2}\right) \geq 1 - e^{-c_6(M)m\eta^2}. \tag{4.108}$$

For the set $\Lambda := \Lambda' \cap \Lambda''$ we have the inequalities (4.105) and (4.107) for all $j \in [1, m]$. Let $\mathbf{z} \in \Lambda$. We apply (4.105) to $f_{\mathbf{z}}^A = f_{\mathbf{z}, K_{n(\mathbf{z})}}$. We consider separately two cases: (i) $n(\mathbf{z}) > n$; (ii) $n(\mathbf{z}) \leq n$. In the first case we obtain

$$\mathcal{E}_\rho(f_{\mathbf{z}}^A) \leq 2\left(\mathcal{E}_{\rho,\mathbf{z}}(f_{\mathbf{z}}^A) + \frac{An(\mathbf{z})\ln m}{m}\right). \tag{4.109}$$

Using the definition of $f_{\mathbf{z}}^A$ and the inequality (4.107) we get

$$\begin{aligned}
\mathcal{E}_{\rho,\mathbf{z}}(f_{\mathbf{z}}^A) + \frac{An(\mathbf{z})\ln m}{m} &= \min_{1 \leq j \leq m}\left(\mathcal{E}_{\rho,\mathbf{z}}(f_{\mathbf{z}, K_j}) + \frac{Aj\ln m}{m}\right) \\
&\leq \min_{1 \leq j \leq m}\left(\mathcal{E}_{\rho,\mathbf{z}}(f_{K_j}) + \frac{Aj\ln m}{m}\right) \\
&\leq \min\left(\min_{n < j \leq m}\left(\frac{3}{2}\mathcal{E}(f_{K_j}) + \frac{2Aj\ln m}{m}\right),\right. \\
&\quad \left.\min_{1 \leq j \leq n}\left(\frac{3}{2}\mathcal{E}(f_{K_j}) + \frac{Aj\ln m}{m}\right) + A\eta^2/2\right) \\
&\leq \min_{1 \leq j \leq m}\left(\frac{3}{2}\mathcal{E}(f_{K_j}) + \frac{2Aj\ln m}{m}\right) + A\eta^2/2 \\
&\leq \min_{1 \leq j \leq m}\left(\frac{3}{2}Aj^{-2r} + \frac{2Aj\ln m}{m}\right) + A\eta^2/2.
\end{aligned} \tag{4.110}$$

Substituting $j = [(m/\ln m)^{1/(1+2r)}] + 1$ and using the inequalities

$$(m/\ln m)^{1/(1+2r)} \leq j \leq 2(m/\ln m)^{1/(1+2r)},$$

we obtain from (4.109) and (4.110)

$$\mathcal{E}_\rho(f_{\mathbf{z}}^A) \leq 11A\left(\frac{\ln m}{m}\right)^{2r/(1+2r)} + A\eta^2 \leq 12A\eta^2.$$

This gives the required bound

$$\|f_{\mathbf{z}}^A - f_\rho\|_{L_2(\rho_X)} \leq 4A^{1/2}\eta.$$

In the second case we obtain

$$\mathcal{E}_\rho(f_{\mathbf{z}}^A) \leq 2\mathcal{E}_{\rho,\mathbf{z}}(f_{\mathbf{z}}^A) + A\eta^2.$$

Next, we have

$$\mathcal{E}_{\rho,\mathbf{z}}(f_{\mathbf{z}}^A) \leq \min_{1 \leq j \leq m} \left( \mathcal{E}_{\rho,\mathbf{z}}(f_{K_j}) + \frac{Aj \ln m}{m} \right).$$

Using Lemma 4.78 we continue:

$$\leq \min_{1 \leq j \leq m} \left( \frac{3}{2}\mathcal{E}_\rho(f_{K_j}) + A\eta^2/2 + \frac{2Aj \ln m}{m} \right)$$

$$\leq \min_{1 \leq j \leq m} \left( \frac{3}{2}Aj^{-2r} + \frac{2Aj \ln m}{m} \right) + A\eta^2/2 \leq 6A\eta^2.$$

Therefore,

$$\mathcal{E}_\rho(f_{\mathbf{z}}^A) \leq 13A\eta^2.$$

The proof of Theorem 4.81 is complete.                                      □

As mentioned in the Introduction in DeVore *et al.* (2006) we propose to study the **AC**-function. In the discussion that follows it will be more convenient for us to express the results in terms of the following variant of the accuracy confidence function:

$$\mathbf{ac}_m(\mathcal{M}, \mathbb{E}, \eta) := \mathbf{AC}_m(\mathcal{M}, \mathbb{E}, \eta^{1/2}).$$

We may study the **ac**-function in two steps.

**Step (1)** For given $\mathcal{M}$, $m$, $E(m)$ find for $\delta \in (0, 1)$ the smallest $t_m(\mathcal{M}, \delta) := t_m(\mathcal{M}, E(m), \delta)$ such that

$$\mathbf{ac}_m(\mathcal{M}, \mathbb{E}, t_m(\mathcal{M}, \delta)) \leq \delta.$$

It is clear that for $\eta > t_m(\mathcal{M}, \delta)$ we have $\mathbf{ac}_m(\mathcal{M}, \mathbb{E}, \eta) \leq \delta$ and for $\eta < t_m(\mathcal{M}, \delta)$ we have $\mathbf{ac}_m(\mathcal{M}, \mathbb{E}, \eta) > \delta$.

The following modification of the above $t_m(\mathcal{M}, \delta)$ is also of interest. We now look for the smallest $t_m(\mathcal{M}, \delta, c)$ such that

$$\mathbf{ac}_m(\mathcal{M}, \mathbb{E}, t_m(\mathcal{M}, \delta, c)) \leq \delta m^{-c}, \quad c > 0.$$

It is clear that $t_m(\mathcal{M}, \delta) \leq t_m(\mathcal{M}, \delta, c)$. We call the $t_m(\mathcal{M}, \delta)$ and $t_m(\mathcal{M}, \delta, c)$ the *approximation threshold for the proper function learning*.

**Step (2)** Find the right order of $\mathbf{ac}_m(\mathcal{M}, \mathbb{E}, \eta)$ for $\eta \geq t_m(\mathcal{M}, \delta)$ as a function on $m$ and $\eta$.

It was proved in DeVore *et al.* (2006) and Temlyakov (2006d) (see Theorem 4.71 above) that for a compact $\Theta \subset L_2(\mu)$ such that $\Theta \subset (1/4)U(L_\infty(\mu))$ and

$$\epsilon_n(\Theta, L_2(\mu)) \asymp n^{-r}, \tag{4.111}$$

there exists $\delta_0 > 0$ such that for any $\delta \in (0, \delta_0]$

$$t_m(\mathcal{M}(\Theta, \mu), \delta) \gg m^{-2r/(1+2r)}.$$

Theorem 4.81 implies that, under the above assumptions,

$$t_m(\mathcal{M}(\Theta, \mu), \delta) \ll m^{-2r/(1+2r)}.$$

Therefore, for any $\Theta$ satisfying the above assumptions we have

$$t_m(\mathcal{M}(\Theta, \mu), \delta) \asymp m^{-2r/(1+2r)}, \quad \delta \in (0, \delta_0]. \tag{4.112}$$

We now proceed to the concept of universal (universally optimal) estimators. Let a collection $\mathbb{M} := \{\mathcal{M}\}$ of classes $\mathcal{M}$ of measures and a sequence $\mathbb{E}$ of allowed classes of estimators be given.

**Definition 4.84** An estimator $f_{\mathbf{z}} \in E(m)$ is a *universal (universally optimal) in a weak sense for the pair* $(\mathbb{M}, \mathbb{E})$ if, for any $\rho \in \mathcal{M} \in \mathbb{M}$, we have

$$\rho^m\{\mathbf{z} : \|f_\rho - f_{\mathbf{z}}\|^2_{L_2(\rho_X)} \geq C_1(\mathbb{M}, \mathbb{E})(\ln m)^w t_m(\mathcal{M}, \delta, c)\} \leq C_2 m^{-c_1},$$

where $C_1, c_1$ and $C_2$ do not depend on $\rho$ and $m$.

In the case $w = 0$ in the above definition we replace *in a weak sense* by *in a strong sense*.

We now discuss an application of Theorem 4.83 for the construction of universal estimators. Let $\mathcal{L} := \{L_n\}_{n=1}^\infty$ be a sequence of $n$-dimensional subspaces of $\mathcal{C}(X)$. Consider the $\mathcal{C}(X)$-balls in $L_n$ of radius $D$:

$$K_n := DU(\mathcal{C}(X)) \cap L_n, \quad n = 1, 2, \ldots$$

Then the sequence $\mathcal{K} := \{K_n\}$ satisfy (4.104) with $a = 2D$. Consider the classes

$$\Theta^r(\mathcal{K}, \mu) := \{f : d(f, K_n)_{L_2(\mu)} \leq C_3 n^{-r}, \quad n = 1, 2, \ldots\}, \tag{4.113}$$

with $C_3$ a fixed positive number. We also consider a set $V$ of Borel probability measures $v$ defined on $X$ such that

$$\epsilon_n(\Theta^r(\mathcal{K}, v), L_2(v)) \geq C_4 n^{-r}, \quad v \in V, \quad r \in [\alpha, \beta]. \tag{4.114}$$

We consider a class $\mathcal{M}(r, v)$ of measures $\rho$ such that $|y| \leq M$ a.e. with respect to $\rho$ and $\rho_X = v$, $f_\rho \in \Theta^r(\mathcal{K}, v)$. Finally, we define a collection

$$\mathbb{M} := \{\rho : \rho \in \mathcal{M}(r, v), \quad v \in V, \quad r \in [\alpha, \beta]\}.$$

Then for any $v \in V, r \in [\alpha, \beta]$ our assumptions (4.113) and (4.114) imply (by Carl's inequality, Carl (1981)) that

$$\epsilon_n(\Theta^r(\mathcal{K}, v), L_2(v)) \asymp n^{-r}, \quad v \in V, \quad r \in [\alpha, \beta]. \tag{4.115}$$

Therefore, by Theorem 4.71,

$$t_m(\mathcal{M}(r, v), \delta, c) \gg m^{-2r/(1+2r)}.$$

Choosing $w = 1$, we get from Theorem 4.83 that for any $\rho \in \mathcal{M}(r, v) \in \mathbb{M}$

$$\rho^m\{\mathbf{z} : \|f_{\mathbf{z}}^A - f_\rho\|_{L_2(\rho_X)}^2 \geq C_1(\mathbb{M})(\ln m)t_m(\mathcal{M}(r, v), \delta, c)\} \leq C_2(\mathbb{M})m^{-c_1},$$

provided that $A$ is big enough. This indicates that the estimator $f_{\mathbf{z}}^A$ is a universal estimator in a weak sense for the collection $\mathbb{M}$.

### 4.4.3 Some examples

In Section 4.4.2 we presented a way of constructing universal estimators: the penalized least squares estimators. This method is based on a given sequence of compacts in $\mathcal{C}(X)$. In Section 4.4.2 we considered a collection $\mathcal{K}(a, b)$ of compacts $K_n$ in $\mathcal{C}(X)$ satisfying

$$N(K_n, \epsilon, \mathcal{C}(X)) \leq (an^b(1 + 1/\epsilon))^n, \quad n = 1, 2, \ldots \tag{4.116}$$

We begin with a construction based on the concept of the Kolmogorov width. This construction has been used in DeVore *et al.* (2006).

**Example 1** Let $\mathcal{L} = \{L_n\}_{n=1}^{\infty}$ be a sequence of $n$-dimensional subspaces of $\mathcal{C}(X)$. For $Q > 0$ we define

$$K_n := QU(\mathcal{C}(X)) \cap L_n = \{f \in L_n : \|f\|_{\mathcal{C}(X)} \leq Q\}, \quad n = 1, 2, \ldots \tag{4.117}$$

Then it is well known (Pisier (1989)) that

$$N(K_n, \epsilon, \mathcal{C}(X)) \leq (1 + 2Q/\epsilon)^n.$$

We note that $\{K_n\}_{n=1}^{\infty} = \mathcal{K}(\max(1, 2Q), 0)$. Therefore, Theorem 4.83 applies to this sequence of compacts. Let us discuss the condition

$$d(f_\rho, K_n)_{L_2(\rho_X)} \leq A^{1/2}n^{-r}, \quad n = 1, 2, \ldots, \tag{4.118}$$

from Theorem 4.83. We compare (4.118) with a standard condition in approximation theory,

$$d(f_\rho, L_n)_{\mathcal{C}(X)} \leq Dn^{-r}, \quad n = 1, 2, \ldots, \quad f_\rho \in DU(\mathcal{C}(X)). \quad (4.119)$$

First of all we observe that (4.119) implies that there exists $\varphi_n \in L_n$, $\|\varphi_n\|_{\mathcal{C}(X)} \leq 2D$, such that

$$\|f_\rho - \varphi_n\|_{\mathcal{C}(X)} \leq Dn^{-r}.$$

Thus (4.119) implies

$$d(f_\rho, K_n)_{\mathcal{C}(X)} \leq Dn^{-r}, \quad n = 1, 2, \ldots \quad (4.120)$$

provided $Q \geq 2D$. Also, (4.120) implies (4.118) provided $A^{1/2} \geq D$. Therefore, Theorem 4.83 can be used for $f_\rho$ satisfying (4.119). We formulate this result as a theorem.

**Theorem 4.85** Let $\mathcal{L} = \{L_n\}_{n=1}^\infty$ be a sequence of n-dimensional subspaces of $\mathcal{C}(X)$. For given positive numbers $D$, $M_1$, $M := M_1 + D$ there exists $A_0 := A_0(D, M)$ with the following property: For any $A \geq A_0$ there exists an estimator $f_{\mathbf{z}}^A$ such that, for any $\rho$ with the properties $|y| \leq M_1$ a.e. with respect to $\rho$ and

$$d(f_\rho, L_n)_{\mathcal{C}(X)} \leq Dn^{-r}, \quad n = 1, 2, \ldots, \quad f_\rho \in DU(\mathcal{C}(X)),$$

we have for $\eta \geq A^{1/2} (\ln m/m)^{r/(1+2r)}$

$$\rho^m\{\mathbf{z} : \|f_{\mathbf{z}}^A - f_\rho\|_{L_2(\rho_X)} \geq 4A^{1/2}\eta\} \leq Ce^{-c(M)m\eta^2}. \quad (4.121)$$

Theorem 4.85 is an extension of theorem 4.10 from DeVore *et al.* (2006); which yields (4.121) with $e^{-c(M)m\eta^2}$ replaced by $e^{-c(M)m\eta^4}$ under an extra restriction $r \leq 1/2$.

**Example 2** In the previous example we worked in the $\mathcal{C}(X)$ space. We now want to replace (4.119) by a weaker condition, i.e.

$$d(f_\rho, L_n)_{L_2(\rho_X)} \leq Dn^{-r}, \quad n = 1, 2, \ldots, \quad f_\rho \in DU(L_2(\rho_X)). \quad (4.122)$$

This condition is compatible with condition (4.118) (from Theorem 4.83) in the sense of approximation in the $L_2(\rho_X)$ norm. However, conditions (4.122) and (4.118) differ in the sense of the approximation set: it is a linear subspace $L_n$ in (4.122) and a compact subset of $\mathcal{C}(X)$ in (4.118). In Example 1, the approximation (4.119) by a linear subspace automatically provided the approximation (4.118) by a suitable compact of $\mathcal{C}(X)$. It is clear that, similarly to Example 1, approximation (4.122) by a linear subspace $L_n$ provides approximation (4.118) by a compact $K_n \subset L_n$ of the $L_2(\rho_X)$ instead of the $\mathcal{C}(X)$. We cannot apply Theorem 4.83 in such a situation. In order to overcome this difficulty we impose extra restrictions on the sequence $\mathcal{L}$ and on the measure $\rho$. We discuss the setting from Konyagin and Temlyakov (2007). Let $\mathcal{B}(X)$ be a Banach space with the norm $\|f\|_{\mathcal{B}(X)} := \sup_{x \in X} |f(x)|$. Let $\{L_n\}_{n=1}^{\infty}$ be a given sequence of $n$-dimensional linear subspaces of $\mathcal{B}(X)$ such that $L_n$ is also a subspace of each $L_{\infty}(\mu)$, where $\mu$ is a Borel probability measure on $X$, $n = 1, 2, \ldots$. Assume that $n$-dimensional linear subspaces $L_n$ have the following property: For any Borel probability measure $\mu$ on $X$ one has

$$\|P_{L_n}^{\mu}\|_{\mathcal{B}(X) \to \mathcal{B}(X)} \le K, \quad n = 1, 2, \ldots, \tag{4.123}$$

where $P_L^{\mu}$ is the operator of the $L_2(\mu)$ projection onto $L$. Then our standard assumption $|y| \le M_1$ implies $\|f_{\rho}\|_{L_{\infty}(\rho_X)} \le M_1$, and (4.122) and (4.113) yield

$$d(f_{\rho}, K_n)_{L_2(\rho_X)} \le Dn^{-r}, \quad n = 1, 2, \ldots,$$

where

$$K_n := (K + 1)M_1 U(\mathcal{B}(X)) \cap L_n.$$

We note that Theorem 4.83 holds for compacts satisfying (4.104) in the $\mathcal{B}(X)$ norm instead of the $\mathcal{C}(X)$ norm. Thus, as a corollary of Theorem 4.83 we obtain the following result.

**Theorem 4.86** *Let $\mathcal{L} = \{L_n\}_{n=1}^{\infty}$ be a sequence of $n$-dimensional subspaces of $\mathcal{B}(X)$ satisfying (4.123). For given positive numbers $D, M_1, M := M_1 + D$ there exists $A_0 := A_0(K, D, M)$ with the following property. For any $A \ge A_0$ there exists an estimator $f_{\mathbf{z}}^A$ such that for any $\rho$ with the properties $|y| \le M_1$ a.e. with respect to $\rho$ and*

$$d(f_{\rho}, L_n)_{L_2(\rho_X)} \le Dn^{-r}, \quad n = 1, 2, \ldots,$$

*we have, for $\eta \ge \eta_m := A^{1/2} (\ln m/m)^{r/(1+2r)}$,*

$$\rho^m\{\mathbf{z} : \|f_{\mathbf{z}}^A - f_{\rho}\|_{L_2(\rho_X)} \ge 4A^{1/2}\eta\} \le Ce^{-c(M)m\eta^2}. \tag{4.124}$$

Theorem 4.86 is an extension of theorem 4.3 from Konyagin and Temlyakov (2007); which yields (4.124) with $4A^{1/2}\eta$ replaced by $C(D)\eta_m$ and $e^{-c(M)m\eta^2}$ replaced by $m^{-c(M)A}$ under an extra restriction $r \le 1/2$.

**Remark 4.87** In Theorem 4.86 we can replace the assumption that $\mathcal{L}$ satisfies (4.123) for all Borel probability measures $\mu$ by the assumption that (4.123) is satisfied for $\mu \in \mathcal{M}$ and add the assumption $\rho_X \in \mathcal{M}$.

**Example 3** Our construction here is based on the concept of nonlinear Kolmogorov's $(N, n)$-width (Temlyakov (1998d)):

$$d_n(F, B, N) := \inf_{\mathcal{L}_N, \#\mathcal{L}_N \leq N} \sup_{f \in F} \inf_{L \in \mathcal{L}_N} \inf_{g \in L} \|f - g\|_B,$$

where $\mathcal{L}_N$ is a set of at most $N$ $n$-dimensional subspaces $L$. It is clear that

$$d_n(F, B, 1) = d_n(F, B).$$

The new feature of $d_n(F, B, N)$ is that we may now choose a subspace $L \in \mathcal{L}_N$ depending on $f \in F$. It is clear that the larger the $N$ the more flexibility we have to approximate $f$.

Let $\mathbb{L} := \{\mathcal{L}_n\}_{n=1}^{\infty}$ be a sequence of collections $\mathcal{L}_n := \{L_n^j\}_{j=1}^{N_n}$ of $n$-dimensional subspaces $L_n^j$ of $C(X)$. Assume $N_n \leq n^{bn}$. For $Q > 0$ we now consider

$$K_n := \bigcup_{j=1}^{N_n} (QU(C(X)) \cap L_n^j).$$

Then $\{K_n\}_{n=1}^{\infty} = \mathcal{K}(\max(1, 2Q), b)$. It is also clear that the condition

$$\min_{1 \leq j \leq N_n} d(f_\rho, L_n^j)_{C(X)} \leq Dn^{-r}, \quad n = 1, 2, \ldots, \quad f_\rho \in DU(C(X)),$$

$$(4.125)$$

implies

$$d(f_\rho, K_n)_{C(X)} \leq Dn^{-r}, \quad n = 1, 2, \ldots,$$

provided $Q \geq 2D$.

We have the following analog of Theorem 4.85.

**Theorem 4.88** *Let* $\mathbb{L} := \{\mathcal{L}_n\}_{n=1}^{\infty}$ *be a sequence of collections* $\mathcal{L}_n := \{L_n^j\}_{j=1}^{N_n}$ *of $n$-dimensional subspaces* $L_n^j$ *of $C(X)$. Assume $N_n \leq n^{bn}$. For given positive numbers $D$, $M_1$, $M := M_1 + D$ there exists $A_0 := A_0(b, D, M)$ with the following property: For any $A \geq A_0$ there exists an estimator $f_\mathbf{z}^A$ such that for any $\rho$ with the properties: $|y| \leq M_1$ a.e. with respect to $\rho$ and*

$$\min_{1 \leq j \leq N_n} d(f_\rho, L_n^j)_{C(X)} \leq Dn^{-r}, \quad n = 1, 2, \ldots, \quad f_\rho \in DU(C(X)),$$

*we have, for $\eta \geq A^{1/2} (\ln m/m)^{r/(1+2r)}$,*

$$\rho^m\{\mathbf{z} : \|f_\mathbf{z}^A - f_\rho\|_{L_2(\rho_X)} \geq 4A^{1/2}\eta\} \leq Ce^{-c(M)m\eta^2}.$$

**Example 4** In this example we apply the ideas of Examples 2 and 3 for nonlinear $m$-term approximation with regard to a given countable dictionary. Let $\Psi := \{\psi_n\}_{n=1}^{\infty}$ be a system of functions $\psi_n$ from $\mathcal{B}(X)$. Let $\gamma \geq 0$

and let $\mathcal{M}(\gamma)$ be a set of Borel probability measures $\mu$ such that all $\psi_n$ are $\mu$-measurable and

$$\left\| \sum_{n=1}^{N} a_n \psi_n \right\|_{\mathcal{B}(X)} \leq C_1 N^\gamma \left\| \sum_{n=1}^{N} a_n \psi_n \right\|_{L_2(\mu)}. \tag{4.126}$$

We fix a parameter $q \geq 1$ and define the best $m$-term approximation with depth $m^q$ as follows:

$$\sigma_{m,q}(f, \Psi)_{L_2(\mu)} := \inf_{c_i ; n_i \leq m^q} \left\| f - \sum_{i=1}^{m} c_i \psi_{n_i} \right\|_{L_2(\mu)}.$$

For a fixed $Q > 0$ that will be chosen later we now consider

$$K_n(Q) := \{ f : f = \sum_{i=1}^{n} a_i \psi_{n_i}, \quad n_i \leq n^q, \quad i = 1, \ldots, n, \quad \|f\|_{\mathcal{B}(X)} \leq Q \}.$$

Then

$$N(K_n(Q), \epsilon, \mathcal{B}(X)) \leq (1 + 2Q/\epsilon))^n \binom{n^q}{n} \leq (n^q (1 + 2Q/\epsilon))^n. \tag{4.127}$$

Suppose we have for $\mu \in \mathcal{M}(\gamma)$

$$\sigma_{m,q}(f, \Psi)_{L_2(\mu)} \leq D n^{-r}, \quad f \in DU(L_2(\mu)).$$

Then there exists $\varphi_n$ of the form

$$\varphi_n = \sum_{i=1}^{n} a_i \psi_{n_i}, \quad n_i \leq n^q, \quad i = 1, \ldots, n, \quad \|\varphi_n\|_{L_2(\mu)} \leq 2D$$

such that

$$\|f - \varphi_n\|_{L_2(\mu)} \leq D n^{-r}.$$

Next, by our assumption (4.126) we get

$$\|\varphi_n\|_{\mathcal{B}(X)} \leq 2D C_1 n^{\gamma q}.$$

Therefore $\varphi_n \in K_n(2D C_1 n^{\gamma q})$. The inequality (4.127) implies that

$$\{K_n(2D C_1 n^{\gamma q})\} = \mathcal{K}(\max(1, 4D C_1), (1 + \gamma)q).$$

Consequently, Theorem 4.83 applies in this situation. We formulate the result as a theorem.

**Theorem 4.89** *Let $\Psi$ and $\mathcal{M}(\gamma)$ be as above. For given positive numbers $q$, $D$, $M_1$, $M := M_1 + D$, there exists $A_0 := A_0(\gamma, C_1, q, D, M)$ with the* *following property: For any $A \geq A_0$ there exists an estimator $f_{\mathbf{z}}^A$ such that for any $\rho$ with the properties $|y| \leq M_1$ a.e. with respect to $\rho$, $\rho_X \in \mathcal{M}(\gamma)$, and*

$$\sigma_{n,q}(f_\rho, \Psi)_{L_2(\rho_X)} \leq Dn^{-r}, \quad n = 1, 2, \ldots, \quad f_\rho \in DU(\mathcal{B}(X)),$$

*we have, for $\eta \geq A^{1/2} (\ln m/m)^{r/(1+2r)}$,*

$$\rho^m\{\mathbf{z} : \|f_{\mathbf{z}}^A - f_\rho\|_{L_2(\rho_X)} \geq 4A^{1/2}\eta\} \leq Ce^{-c(M)mn^2}.$$

**Remark 4.90** In Theorem 4.89 the condition (4.126) can be replaced by the following weaker condition. Let $m_i \in [1, n^q]$, $i = 1, \ldots, n$, and

$$L_n := \mathrm{span}\{\psi_{m_i}\}_{i=1}^n.$$

Assume that for $\mu \in \mathcal{M}(\gamma)$

$$\|P_{L_n}^\mu\|_{\mathcal{B}(X) \to \mathcal{B}(X)} \leq C_1 n^{qn}.$$

## 4.5 The lower estimates

### 4.5.1 Introduction

We recall that we study the two variants of an optimization problem: the projection learning problem and the proper function learning problem. We formulated the corresponding optimization problems in Section 4.1. In both cases these problems are formulated in terms of the accuracy confidence functions $\mathbf{AC}_m(\mathcal{M}, \mathbb{E}, \eta)$ and $\mathbf{AC}_m^p(W, \mathbb{E}, \eta)$ (the superscript $p$ stands for "projection"). We note that the right behavior of the accuracy confidence functions is known only in some special cases. As we pointed out above, the results in these special cases show that the behavior of the accuracy confidence function exibits a mixture of two different features of the problem: (1) deterministic, expressed in terms of the size of a class of priors $\Theta$, or a class $W$; (2) stochastic.

We begin with some simple remarks that illustrate the above observation. Let a class of priors $\Theta$ be given and let $\mathcal{M}[\Theta]$ be a set of measures $\rho$ such that $f_\rho \in \Theta$ and $\rho_X$ is the Lebesgue measure on $X$. Assume that each class $\mathbb{E}(m)$ of allowed estimators is a linear subspace $L_{n(m)}$ of the $L_2(X)$ of dimension $n(m)$. Then it follows from the definition of the Kolmogorov widths that for each $m$ there is a $f_{\rho_m} \in \Theta$ such that

$$d(f_{\rho_m}, L_{n(m)})_{L_2(X)} \geq d_{n(m)}(\Theta, L_2(X)).$$

Therefore, for any $E_m \in \mathbb{E}(m)$ such that $E_m : \mathbf{z} \to f_{\mathbf{z}} \in L_{n(m)}$ we have

$$\rho_m^m\{\mathbf{z} : \|f_{\rho_m} - f_{\mathbf{z}}\|_{L_2(X)} \geq d_{n(m)}(\Theta, L_2(X))\} = 1.$$

Thus

$$\mathbf{AC}_m(\mathcal{M}[\Theta], \mathbb{E}, \eta) = 1, \quad \eta \le d_{n(m)}(\Theta, L_2(X)). \tag{4.128}$$

The above argument uses only the deterministic feature of the problem: the size of $\Theta$, expressed in terms of the sequence of the Kolmogorov widths. For instance, if we know that $d_n(\Theta, L_2(X)) \asymp n^{-r}$, then (4.128) says that in choosing estimators from $n(m)$-dimensional subspaces we are limited to the accuracy $n(m)^{-r}$ (at best). All our upper estimates from Sections 4.3 and 4.4 provide accuracy not better than $m^{-1/2}$. The following simple example indicates that such a saturation property is caused by the stochastic feature of the problem.

Let $X = [0, 1]$. Consider $\Theta := \{\pm f_s\}_{s=1}^{\infty}$, $f_s := (\ln(s + 1))^{-r}$, if $x \in [0, 1/s]$ and $f_s := 0$ otherwise. Now let $\mathcal{M}[\Theta]$ be the set of measures $\rho$ such that $\rho_X$ is the Lebesgue measure on $[0, 1]$ and $f_\rho \in \Theta$. We assume that $\rho(y|x)$ is the Dirac measure $\delta(f_\rho(x))$. Then it is easy to check that

$$\epsilon_n(\Theta, \mathcal{B}(X)) \le (n \ln 2)^{-r}, \quad n = 1, 2, \dots$$

For a fixed $m$ the probability of the event $\{x_i \in (1/m, 1], i = 1, \dots, m\}$ is equal to $(1 - 1/m)^m \ge \delta > 0$. In the case $x_i \in (1/m, 1], i = 1, \dots, m$, the functions $f_m$ and $-f_m$ are indistinguishable from the data. Therefore, for any estimator $f_\mathbf{z}$ with probability at least $\delta$ one has

$$\max(\|f_m - f_\mathbf{z}\|_2, \|f_m + f_\mathbf{z}\|_2) \ge \|f_m\|_2 = m^{-1/2}(\ln(m + 1))^{-r}.$$

Thus for $\eta \le m^{-1/2}(\ln(m + 1))^{-r}$

$$\mathbf{AC}_m(\mathcal{M}[\Theta], \eta) \ge \delta.$$

Theorem 4.74 provides the right estimates (in the sense of order) of the **AC**-function of a class $\Theta$ satisfying the condition

$$\epsilon_n(\Theta, L_2(\mu)) \asymp n^{-r}.$$

Let us make a comment on studying the accuracy confidence function for the projection learning problem. We recall that, similarly to the case of the proper function learning problem, we introduced the corresponding accuracy confidence function

$$\mathbf{AC}_m^P(W, \mathbb{E}, \eta) := \inf_{E_m \in \mathbb{E}(m)} \sup_\rho \rho^m \left\{ \mathbf{z} : \mathcal{E}(f_\mathbf{z}) - \mathcal{E}((f_\rho)w) \ge \eta^2 \right\},$$

where $\sup_\rho$ is taken over $\rho$ such that $\rho$, $W$ satisfy (4.94). We note that in the case of convex $W$ we have by Lemma 4.47, for any $f \in W$,

$$\|f - (f_\rho)w\|_{L_2(\rho_X)}^2 \le \mathcal{E}(f) - \mathcal{E}((f_\rho)w).$$

Theorem 4.70 provides an upper estimate for the $\mathbf{AC}^p$-function in the case of convex $W$ from $\mathcal{S}_1^r$:

$$\mathbf{AC}_m^p(W, \eta^{1/2}) \leq \exp(-c(M)m\eta), \quad \eta \geq \eta_m \gg m^{-r/(1+r)}.$$

We note that the behavior of the $\mathbf{AC}^p$-function is well understood only in the following special cases. Let $r > 1/2$ then (see Konyagin and Temlyakov (2004), Temlyakov (2006d))

$$C_1 \exp(-c_1(M)m\eta^4) \leq \sup_{W \in \mathcal{S}^r(D)} \mathbf{AC}_m^p(W, \eta) \leq C(M, D, r) \exp(-c_2(M)m\eta^4)$$

for $\eta \geq m^{-1/4}$. Also for $r \geq 1$ (see Konyagin and Temlyakov (2007))

$$C_1 \exp(-c_1(M)m\eta^4) \leq \sup_{W \in \mathcal{S}_2^r(D)} \mathbf{AC}_m^p(W, \eta) \leq C(M, D, r) \exp(-c_3(M)m\eta^4)$$

provided $\eta \gg m^{-1/4}$.

It would be interesting to find the behavior of

$$\sup_{W \in \mathcal{S}} \mathbf{AC}_m^p(W, \eta)$$

in the following cases: (i) $\mathcal{S} = \mathcal{S}^r(D)$, $r \leq 1/2$; (ii) $\mathcal{S} = \mathcal{S}_2^r(D)$, $r < 1$; (iii) $\mathcal{S} = \{W : W \in \mathcal{S}_q^r(D), W \text{ is convex}\}$, $q = 1, 2, \infty$.

### 4.5.2 The projection learning

We now prove that in general we cannot estimate $\mathcal{E}(f_{\mathbf{z},W}) - \mathcal{E}(f_W)$ with a better rate than $m^{-1/2}$.

**Theorem 4.91** *There exist two positive constants* $c_1$, $c_2$ *and a class* $W$ *consisting of two functions* $1$ *and* $-1$ *such that, for every* $m = 2, 3, \ldots$ *and* $m^{-1/4} \leq \eta \leq 1$, *there are two measures* $\rho_0$ *and* $\rho_1$ *such that, for any estimator* $f_{\mathbf{z}} \in W$, *for one of* $\rho = \rho_0$ *or* $\rho = \rho_1$ *we have*

$$\rho^m\{\mathbf{z} : \mathcal{E}(f_{\mathbf{z}}) - \mathcal{E}(f_W) \geq \eta^2\} \geq c_1 \exp(-c_2 m\eta^4).$$

*Proof* Let $X = [0, 1]$, $Y = [-1, 1]$. For a given $m \in \mathbb{N}$ we define $\rho_0$, $\rho_1$ as follows. For both $\rho_0$, $\rho_1$ the $\rho_X$ is the Lebesgue measure on $[0, 1]$ (the proof below works for any $\rho_X$), and for $x \in [0, 1]$ we define

$$\rho_0(1|x) = \rho_1(-1|x) = p; \quad \rho_0(-1|x) = \rho_1(1|x) = 1 - p$$

with $p = (1 + \eta^2)/2$. Then

$$f_{\rho_0} = \eta^2; \quad f_{\rho_1} = -\eta^2.$$

and

$$(f_{\rho_0})w = 1; \quad (f_{\rho_1})w = -1.$$

Let $\mathbf{z} = ((x_1, y_1), \ldots, (x_m, y_m)) =: (x, y)$, $x = (x_1, \ldots, x_m)$, $y = (y_1, \ldots, y_m)$. For a fixed $x \in X^m$ we will prove the lower estimate for the probability in $Y^m$. For a subset $e \subset \{1, \ldots, m\}$ we denote by $\chi_e$ the vector $y = (y_1, \ldots, y_m)$ such that $y_j = 1$ for $j \in e$ and $y_j = -1$ otherwise. For a given estimator $f_{\mathbf{z}}$ consider the following two sets:

$$E_1 := \{e \subset \{1, \ldots, m\} \quad : \quad f_{\mathbf{z}} = 1 \text{ if } \mathbf{z} = (x, \chi_e)\},$$

$$E_{-1} := \{e \subset \{1, \ldots, m\} \quad : \quad f_{\mathbf{z}} = -1 \text{ if } \mathbf{z} = (x, \chi_e)\}.$$

Then for the measure $\rho_0$ we have

$$\mathcal{E}(f_{\mathbf{z}}) - \mathcal{E}((f_{\rho_0})w) = 0 \quad \text{for} \quad \mathbf{z} = (x, \chi_e), \quad e \in E_1,$$

$$\mathcal{E}(f_{\mathbf{z}}) - \mathcal{E}((f_{\rho_0})w) = 4\eta^2 \quad \text{for} \quad \mathbf{z} = (x, \chi_e), \quad e \in E_{-1}.$$

Similarly for the measure $\rho_1$ we have

$$\mathcal{E}(f_{\mathbf{z}}) - \mathcal{E}((f_{\rho_1})w) = 4\eta^2 \quad \text{for} \quad \mathbf{z} = (x, \chi_e), \quad e \in E_1,$$

$$\mathcal{E}(f_{\mathbf{z}}) - \mathcal{E}((f_{\rho_1})w) = 0 \quad \text{for} \quad \mathbf{z} = (x, \chi_e), \quad e \in E_{-1}.$$

The conditional probability of realization of $y = \chi_e$ for a fixed $x$ in the case of measure $\rho_0$ is equal to $p^{|e|}(1 - p)^{m-|e|}$ and in the case of measure $\rho_1$ is equal to $p^{m-|e|}(1 - p)^{|e|}$. Therefore in the case of $\rho_0$ we have

$$\text{Prob}_{y \in Y^m}\{\mathcal{E}(f_{\mathbf{z}}) - \mathcal{E}((f_{\rho_0})w) = 4\eta^2|x\} = \sum_{e \in E_{-1}} p^{|e|}(1 - p)^{m-|e|}$$

and, in the case of $\rho_1$,

$$\text{Prob}_{y \in Y^m}\{\mathcal{E}(f_{\mathbf{z}}) - \mathcal{E}((f_{\rho_1})w) = 4\eta^2|x\} = \sum_{e \in E_1} p^{m-|e|}(1 - p)^{|e|}.$$

We will prove that for $p = (1 + \eta^2)/2$ we have

$$\Sigma := \sum_{e \in E_{-1}} p^{|e|}(1 - p)^{m-|e|} + \sum_{e \in E_1} p^{m-|e|}(1 - p)^{|e|} \geq 2c_1 \exp\left(-c_2 m \eta^4\right)$$

$$(4.129)$$

with absolute constants $c_1, c_2$. This implies Theorem 4.91. We restrict summation in both sums from (4.129) to those $e$ with $m/2 - m^{1/2} \leq |e| \leq m/2 + m^{1/2}$. For such an $e$ we have

$$p^{|e|}(1-p)^{m-|e|} = 2^{-m}(1+\eta^2)^{|e|}(1-\eta^2)^{m-|e|}$$
$$\geq 2^{-m}(1-\eta^4)^{m/2}(1-\eta^2)^{2m^{1/2}} \geq c_3 2^{-m}\exp(-c_2 m\eta^4).$$

Therefore,

$$\Sigma \geq c_3 2^{-m}\exp(-c_2 m\eta^4) \sum_{|m/2-k|\leq m^{1/2}} C_m^k \geq 2c_1\exp(-c_2 m\eta^4). \qquad \square$$

We note that Theorem 4.91 is based on a probabilistic argument for $\rho(y|x)$ and reflects the fact that saturation of the error estimate at the level $m^{-1/2}$ is due to the probabilistic feature of the problem. We will show in the following theorem that the corresponding lower estimate in the case $r \in (0, 1/2]$ can be obtained for the Dirac measure $\rho(y|x)$. Thus, in this case ($r \in (0, 1/2]$) the lower estimate is entailed by the mixture of the deterministic (in $y$) and stochastic (in $x$) features of the problem.

**Theorem 4.92** *For any $r \in (0, 1/2]$ and for every $m \in \mathbb{N}$ there is $W \subset U(\mathcal{B}([0, 1])$ satisfying $\epsilon_n(W, \mathcal{B}) \leq (n/2)^{-r}$ for $n \in \mathbb{N}$ such that, for every estimator $f_{\mathbf{z}} \in W$, there is a $\rho$ such that $|y| \leq 1$ and*

$$\rho^m\{\mathbf{z} : \mathcal{E}(f_{\mathbf{z}}) - \mathcal{E}((f_\rho)w) \geq m^{-r/2}\} \geq 1/7.$$

*Proof* As above, let $X := [0, 1]$, $Y := [-1, 1]$ and let $\rho_X$ be the Lebesgue measure on $[0, 1]$. Define

$$\Gamma := \{\gamma = (\gamma_1, \ldots, \gamma_{2m}) : \gamma_i \in \{1, -1\} \, (i = 1, \ldots, 2m)\}.$$

For $\gamma \in \Gamma, x \in [0, 1)$ let

$$g_\gamma(x) := \gamma_{[2mx+1]},$$

$$f_\gamma(x) := g_\gamma(x)m^{-r},$$

where $[u]$ is the greatest integer not exceeding $u$. Let $W := \{f_\gamma : \gamma \in \Gamma\}$. It is clear that $\epsilon_n(W, \mathcal{B}) \leq (n/2)^{-r}$ as required.

We will consider a set $\mathcal{M}$ of probability measures $\rho$ such that $f_\rho = g_\gamma$, $\gamma \in \Gamma$ and $\rho(y|x)$ defined as the Dirac measure: $y_i = f_\rho(x_i)$. Clearly, if $f_\rho = g_\gamma$ then $(f_\rho)w = f_\gamma$. For every estimator $f_{\mathbf{z}} = f_{\gamma(\mathbf{z})}$ we have for $x$ satisfying $[2mx+1] = i$ and the $i$ such that $\gamma_i(\mathbf{z}) \neq \gamma_i$,

$$(f_{\mathbf{z}}(x) - f_\rho(x))^2 - ((f_\rho)w(x) - f_\rho(x))^2$$
$$= (1 + m^{-r})^2 - (1 - m^{-r})^2 = 4m^{-r}.$$

Therefore,

$$\mathcal{E}(f_\mathbf{z}) - \mathcal{E}((f_\rho)w) = \mathcal{E}_\rho(f_\mathbf{z}) - \mathcal{E}_\rho((f_\rho)w) = \| f_\mathbf{z} - f_\rho \|_2^2 - \|(f_\rho)w - f_\rho\|_2^2$$

$$= 2m^{-1-r} |\{ i : \gamma_i(\mathbf{z}) \neq \gamma_i \}| = m^{-1-r} \sum_{i=1}^{2m} |\gamma_i(\mathbf{z}) - \gamma_i|.$$

(4.130)

It is easy to conclude from here that always

$$|\mathcal{E}(f_\mathbf{z}) - \mathcal{E}((f_\rho)w)| \leq 4m^{-r}.$$  (4.131)

We will estimate

$$E_{\rho^m}(\mathcal{E}_\rho(f_\mathbf{z}) - \mathcal{E}_\rho((f_\rho)w)) = m^{-1-r} \sum_{i=1}^{2m} \int_{[0,1]^m} |\gamma_i((x, f_\rho(x)) - \gamma_i | dx_1 \cdots dx_m.$$

Note that $f_\rho = g_\gamma$. We now average over $\rho \in \mathcal{M}$ (over $\gamma \in \Gamma$). Consider

$$\Sigma_i := \sum_{\gamma \in \Gamma} |\gamma_i((x, g_\gamma(x)) - \gamma_i|.$$

For a given vector $\epsilon = (\epsilon_1, \ldots, \epsilon_m)$, $\epsilon_j = \pm 1$, $j = 1, \ldots, m$, we define

$$\Gamma_\epsilon(x) := \{ \gamma \in \Gamma : g_\gamma(x_j) = \epsilon_j, \quad j = 1, \ldots, m \}.$$

Then

$$\Sigma_i = \sum_\epsilon \sum_{\gamma \in \Gamma_\epsilon(x)} |\gamma_i(x, \epsilon) - \gamma_i|.$$

Define

$$I(x) := \{ [2mx_j + 1] : j = 1, \ldots, m \}.$$

For $i \notin I(x)$ we have the following property: If $\gamma \in \Gamma_\epsilon(x)$ then $(\gamma_1, \ldots, -\gamma_i, \ldots, \gamma_m) \in \Gamma_\epsilon(x)$. Therefore, for $i \notin I(x)$,

$$\sum_{\gamma \in \Gamma_\epsilon(x)} |\gamma_i(x, \epsilon) - \gamma_i| = |\Gamma_\epsilon(x)|$$

and

$$\Sigma_i = |\Gamma| = 2^{2m}.$$

We now fix $i \in [1, 2m]$ and consider

$$G_i := \{ x : i \notin I(x) \}.$$

It is clear that $\text{mes}(G_i) \geq (1 - 1/(2m))^m \geq 1/2$. Thus we obtain

$$\sum_{\rho \in \mathcal{M}} E_{\rho^m}(\mathcal{E}_\rho(f_{\mathbf{z}}) - \mathcal{E}_\rho((f_\rho)w) \geq m^{-r} 2^{2m}.$$

Let $\rho \in \mathcal{M}$ be such that

$$E_{\rho^m}(\mathcal{E}_\rho(f_{\mathbf{z}}) - \mathcal{E}_\rho((f_\rho)w) \geq m^{-r}. \tag{4.132}$$

Combining (4.131) and (4.132) we get

$$\rho^m\{\mathbf{z} : \mathcal{E}_\rho(f_{\mathbf{z}}) - \mathcal{E}_\rho((f_\rho)w) \geq m^{-r/2}\} \geq \frac{1}{7}. \qquad \square$$

### 4.5.3 Lower estimates for the Bernoulli scheme

We consider in this subsection the following estimation problem. Let $y$ be a random variable such that

$$\text{Prob}\{y = 1\} = \text{Prob}\{y = 0\} = 1/2.$$

Then $E(y) = 1/2$. We begin our discussion with the standard estimator $f_m := m^{-1} \sum_{i=1}^m y_i$. Then it is well known that

$$\text{Prob}\{|f_m - 1/2| \geq \epsilon\} = 2^{-m} \left( \sum_{|k-m/2| \geq m\epsilon} C_m^k \right),$$

where $C_m^k$ are the binomial coefficients. It is easy to check that

$$C_1 e^{-c_1 m \epsilon^2} \leq 2^{-m} \sum_{|k-m/2| \geq m\epsilon} C_m^k \leq C_2 e^{-c_2 m \epsilon^2}$$

with positive absolute constants $C_1, C_2, c_1, c_2$.

The main goal of this subsection is to prove that $f_m$ is optimal in a certain sense among all linear estimators. We will prove the following theorem.

**Theorem 4.93** *For any $\epsilon \in [0, 1/2]$, $m \geq 2$ and $w = (w_1, \ldots, w_m)$ we have*

$$\text{Prob}\left\{ |\sum_{i=1}^m w_i y_i - 1/2| \geq \epsilon \right\} \geq \exp\left( -cm\epsilon^2 - \frac{c}{4} \sum_{k=1}^{m-1} \frac{1}{k} \right)$$

*with $c = 25$.*

We present a proof of this theorem from Temlyakov (2006d), in which it is noted (see Temlyakov's remark 5.1) that one can deduce a slightly better bound

$$\text{Prob}\left\{ |\sum_{i=1}^m w_i y_i - 1/2| \geq \epsilon \right\} \geq \exp(-128m\epsilon^2 - 6 - \ln 8)$$

from Hitczenko and Kwapien (1994).

We begin with a technical lemma.

**Lemma 4.94** *Let* $\epsilon \in (0, \beta]$, $9n \geq \epsilon^{-2}$, $w_n \in [0, 1/n]$. *Then for* $\epsilon_1 := (\epsilon - w_n/2)(1 - w_n)^{-1}$, $\epsilon_2 := (\epsilon + w_n/2)(1 - w_n)^{-1}$ *one has, for* $c = 25$, $\beta = (\ln 2)^{1/2}/5$,

$$\exp\left(-c(n-1)\epsilon_1^2\right) + \exp(-c(n-1)\epsilon_2^2) \geq 2\exp\left(-cn\epsilon^2 - \frac{c}{4(n-1)}\right).$$
(4.133)

*Proof* We consider two cases separately: (i) $w_n \in [0, 1/(2n)]$ and (ii) $w_n \in (1/(2n), 1/n]$.

**Case (i)** Using the convexity of the function $e^{-x}$ we obtain, for any $C > 0$,

$$\exp(-C(n-1)\epsilon_1^2) + \exp(-C(n-1)\epsilon_2^2) \geq 2\exp(-C(n-1)(\epsilon_1^2 + \epsilon_2^2)/2).$$
(4.134)

Next,

$$\epsilon_1^2 + \epsilon_2^2 = (1-w_n)^{-2}((\epsilon - w_n/2)^2 + (\epsilon + w_n/2)^2) = (1-w_n)^{-2}(2\epsilon^2 + w_n^2/2).$$

Using the inequality

$$\frac{n-1}{(1-w_n)^2} \leq n \quad \text{for} \quad w_n \in [0, 1/(2n)],$$

we get

$$(n-1)(\epsilon_1^2 + \epsilon_2^2)/2 \leq n\epsilon^2 + 1/(16n).$$
(4.135)

Substituting (4.135) into (4.134), we obtain (4.133).

**Case (ii)** We rewrite

$$\begin{aligned} S &:= \exp(-c(n-1)\epsilon_1^2) + \exp(-c(n-1)\epsilon_2^2) \\ &= \exp(-c(n-1)\epsilon_1^2)\left(1 + \exp(-c(n-1)(\epsilon_2^2 - \epsilon_1^2))\right). \end{aligned}$$

We have an identity

$$\epsilon_2^2 - \epsilon_1^2 = 2w_n\epsilon(1 - w_n)^{-2}.$$

Denote $a_n := (n-1)(1-w_n)^{-2}$. We have

$$1 - 1/n \leq a_n/n \leq n/(n-1).$$
(4.136)

Let us estimate $\delta := n\epsilon^2 - (n-1)\epsilon_1^2$. We have

$$\delta = \epsilon^2\left(\frac{n}{n-1}(1-w_n)^2 - 1\right)a_n + a_nw_n\epsilon - a_nw_n^2/4.$$

Using

$$\frac{n}{n-1}(1-w_n)^2 - 1 = \frac{(1-w_n)^2}{1-1/n} - 1 \geq 1 - w_n - 1 = -w_n,$$

we get

$$\delta \geq a_n w_n \epsilon - a_n w_n \epsilon^2 - a_n w_n^2/4.$$

Therefore

$$S \geq \exp(-cn\epsilon^2 - ca_n w_n^2/4) 2 \cosh(ca_n w_n \epsilon) \exp(-ca_n w_n \epsilon^2).$$

We note that, by (4.136),

$$a_n w_n^2 \leq a_n n^{-2} \leq (n-1)^{-1}.$$

Thus we proceed to estimating $\cosh(A\epsilon) \exp(-A\epsilon^2)$ with $A := ca_n w_n$. By (4.136) and by our assumption $w_n > 1/(2n)$ we get

$$A \geq c(1 - 1/n)/2 \geq c/3, \quad n = 3, \ldots \tag{4.137}$$

It is easy to check that for the function $f(x) := \cosh(Ax) - \exp(Ax^2)$ we have $f(0) = 0$ and $f'(x) \geq 0$ for $x^2 \leq (\ln 4)/A$ in the case $A \geq 8$. The latter inequality $A \geq 8$ follows from (4.137). Therefore

$$\cosh(A\epsilon) \exp(-A\epsilon^2) \geq 1 \quad \text{if} \quad \epsilon^2 \leq \ln 4/A.$$

By (4.136) we have $A \leq cn/(n-1)$ and, hence, for $c = 25$ and $n \geq 2$ we have $\beta^2 = (1/5)^2 \ln 2 \leq \ln 4/A$ for all $A$ of the form $A = ca_n w_n$. This completes the proof of the lemma. $\qquad\square$

**Lemma 4.95** *For any $\epsilon \in [0, 1/2]$, $m \geq 2$ and $w_1 \geq w_2 \geq \cdots \geq w_m \geq 0$, $\sum_{i=1}^m w_i = 1$, we have*

$$\left| \left\{ \Lambda \subseteq [1, m] : \sum_{i \in \Lambda} w_i \geq 1/2 + \epsilon \right\} \right| \geq 2^m \exp\left( -cm\epsilon^2 - \frac{c}{4} \sum_{k=1}^{m-1} \frac{1}{k} \right) \tag{4.138}$$

*with $c = 25$.*

*Proof* Denote

$$\mathcal{L}(\epsilon, m, w) := \left\{ \Lambda \subseteq [1, m] : \sum_{i \in \Lambda} w_i \geq 1/2 + \epsilon \right\}.$$

Then for any $\epsilon \in [0, 1/2]$, $m$, $w$ we have $|\mathcal{L}(\epsilon, m, w))| \geq 1$. Therefore, (4.138) obviously holds for $m \leq 6$, $\epsilon \in [0, 1/2]$ and for any $m > 6$, $\epsilon \in [\beta, 1/2]$, $\beta = (\ln 2)^{1/2}/5$.

We first establish Lemma 4.95 for $\epsilon \in [0, (9m)^{-1/2}]$. We will use a simple property of the Rademacher functions $\{r_i(t)\}$.

**Lemma 4.96** *Let $\sum_{i=1}^{n} |c_i| = 1$. Then*

$$\text{mes}\left\{ t : |\sum_{i=1}^{n} c_i r_i(t)| \leq 2(9n)^{-1/2} \right\} \leq 1 - 5/(9n).$$

*Proof* Denote

$$g := \sum_{i=1}^{n} c_i r_i \quad \text{and} \quad E := \{t : |g(t)| \leq 2(9n)^{-1/2}\}.$$

Then we have, on the one hand,

$$\|g\|_2^2 = \sum_{i=1}^{n} c_i^2 \geq 1/n. \tag{4.139}$$

On the other hand,

$$\|g\|_2^2 \leq (4/(9n))|E| + (1 - |E|). \tag{4.140}$$

Comparing (4.139) and (4.140) we get

$$|E| \leq 1 - 5/(9n). \qquad \square$$

We continue the proof of Lemma 4.95 in the case $\epsilon \in [0, (9m)^{-1/2}]$. We observe that

$$2^{-m}|\mathcal{L}(\epsilon, m, w)| = \text{mes}\left\{ t : \sum_{i=1}^{m} w_i(r_i(t) + 1)/2 \geq 1/2 + \epsilon \right\}$$

$$= \text{mes}\left\{ t : \sum_{i=1}^{m} w_i r_i(t) \geq 2\epsilon \right\}. \tag{4.141}$$

Using Lemma 4.96 we obtain

$$2^{-m}|\mathcal{L}((9m)^{-1/2}, m, w)| \geq 5/(9m).$$

This inequality, combined with the following simple inequality

$$6 \sum_{k=1}^{m-1} \frac{1}{k} \geq \ln(2m), \quad m = 2, 3, \ldots,$$

gives us (4.138) in the case $\epsilon \in [0, (9m)^{-1/2}]$.

It remains to consider the case $\epsilon \in [(9m)^{-1/2}, \beta]$. The proof of this case goes by induction. As we have already mentioned, (4.138) holds for $m \leq 6$.

So, we assume that (4.138) holds for $m - 1$ and derive from it (4.138) for $m$. Denoting $w' := (w_1, \ldots, w_{m-1})$, $w^1 := w'(1 - w_m)^{-1}$ we get

$$\mathcal{L}(\epsilon, m, w) = \{\{m\} \cup \Lambda, \Lambda \in \mathcal{L}(\epsilon - w_m, m - 1, w')\} \cup \mathcal{L}(\epsilon, m - 1, w'). \quad (4.142)$$

Next,

$$\mathcal{L}(\epsilon - w_m, m - 1, w') = \mathcal{L}((\epsilon - w_m/2)(1 - w_m)^{-1}, m - 1, w^1),$$

$$\mathcal{L}(\epsilon, m - 1, w') = \mathcal{L}((\epsilon + w_m/2)(1 - w_m)^{-1}, m - 1, w^1).$$

Using the notations $\epsilon_1 := (\epsilon - w_m/2)(1 - w_m)^{-1}$, $\epsilon_2 := (\epsilon + w_m/2)(1 - w_m)^{-1}$, we obtain from (4.142)

$$|\mathcal{L}(\epsilon, m, w)| = |\mathcal{L}(\epsilon_1, m - 1, w^1)| + |\mathcal{L}(\epsilon_2, m - 1, w^1)|.$$

By the induction assumption we now obtain

$$|\mathcal{L}(\epsilon, m, w)| \geq 2^{m-1} \exp\left(-\frac{c}{4} \sum_{k=1}^{m-2} \frac{1}{k}\right)\left(\exp(-c(m-1)\epsilon_1^2) + \exp(-c(m-1)\epsilon_2^2)\right).$$

We want to apply Lemma 4.94 with $n = m$. The assumptions of Lemma 4.94, $\epsilon \in (0, \beta]$, $m \geq (3\epsilon)^{-2}$, follow from $\epsilon \in [(9m)^{-1/2}, \beta]$. Therefore, by Lemma 4.94 we obtain

$$|\mathcal{L}(\epsilon, m, w)| \geq 2^m \exp\left(-cm\epsilon^2 - \frac{c}{4} \sum_{k=1}^{m-1} \frac{1}{k}\right).$$

This completes the proof of Lemma 4.95. $\qquad\qquad\qquad\qquad\qquad\qquad\square$

**Theorem 4.97** *For any $\epsilon \in [0, 1/2]$, $m \geq 2$ and $w = (w_1, w_2, \ldots, w_m)$ we have*

$$\left|\left\{\Lambda \subseteq [1, m] : |\sum_{i \in \Lambda} w_i - 1/2| \geq \epsilon\right\}\right| \geq 2^m \exp\left(-cm\epsilon^2 - \frac{c}{4} \sum_{k=1}^{m-1} \frac{1}{k}\right)$$

*with $c = 25$.*

*Proof* Denote

$$\mathcal{L}'(\epsilon, m, w) := \left\{\Lambda \subseteq [1, m] : |\sum_{i \in \Lambda} w_i - 1/2| \geq \epsilon\right\}.$$

Similarly to (4.141) we have

$$2^{-m} |\mathcal{L}'(\epsilon, m, w)| = \text{mes}\left\{t : |\sum_{i=1}^{m} w_i(r_i(t) + 1)/2 - 1/2| \geq \epsilon\right\}. \quad (4.143)$$

Denoting $s := \sum_{i=1}^{m} w_i$ we continue (4.143):

$$= \operatorname{mes}\left\{t : \sum_{i=1}^{m} w_i r_i(t) \geq 1 - s + 2\epsilon\right\} + \operatorname{mes}\left\{t : \sum_{i=1}^{m} w_i r_i(t) \leq 1 - s - 2\epsilon\right\}$$

$$= \operatorname{mes}\left\{t : \sum_{i=1}^{m} |w_i| r_i(t) \geq 1 - s + 2\epsilon\right\}$$

$$+ \operatorname{mes}\left\{t : \sum_{i=1}^{m} |w_i| r_i(t) \leq 1 - s - 2\epsilon\right\} =: M_1 + M_2.$$

Denote $a := \sum_{i=1}^{m} |w_i|$ and $u_i := |w_i|/a$. In the case $a \geq 1$, $s \geq 1$ we have

$$M_1 = \operatorname{mes}\left\{t : \sum_{i=1}^{m} u_i r_i(t) \geq (1-s)/a + 2\epsilon/a\right\} \geq \operatorname{mes}\left\{t : \sum_{i=1}^{m} u_i r_i(t) \geq 2\epsilon\right\}.$$

We get the required estimate by Lemma 4.95. In the case $a \geq 1$, $s \leq 1$ we get in the same way as above

$$M_2 \geq \operatorname{mes}\left\{t : \sum_{i=1}^{m} u_i r_i(t) \leq -2\epsilon\right\}. \tag{4.144}$$

By Lemma 4.95 we complete the case.

Let $0 < a < 1$. Then using $s \leq a$ we get

$$(1-s)/a - 2\epsilon/a \geq -2\epsilon$$

and, therefore, (4.144) holds also in this case. It remains to use Lemma 4.95. Theorem 4.97 is now proved.                                             □

Theorem 4.93 is an immediate corollary of Theorem 4.97.

### 4.5.4 The proper function learning

We begin with the lower estimate of the accuracy confidence function from DeVore *et al.* (2006). We shall establish lower bounds in terms of a certain variant of the Kolmogorov entropy of $\Theta$, which we shall call *tight entropy*. This type of entropy has been used to prove lower bounds in approximation theory. Also, a similar type of entropy was used in Yang and Barron (1999) in statistical estimation. The entropy measure that we shall use is in general different from the Kolmogorov entropy, but, for classical smoothness sets $\Theta$, it is equivalent to the Kolmogorov entropy, and therefore our lower bounds will apply in these classical settings.

For a compact $\Theta$ in a Banach space $B$, we define the *packing numbers* as

$$P(\Theta, \delta) := P(\Theta, \delta, B) := \sup\{N : \exists f_1, ..., f_N \in \Theta,$$

with

$$\delta \leq \|f_i - f_j\|_B, \ \forall i \neq j\}. \tag{4.145}$$

It is well known, Pisier (1989), and easy to check (see Chapter 3) that $N(\Theta, \delta, B) \leq P(\Theta, \delta, B)$. The *tight packing numbers* are defined as follows. Let $1 \leq c_1 < \infty$ be a fixed real number. We define the tight packing numbers as

$$\bar{P}(\Theta, \delta) := \bar{P}(\Theta, \delta, c_1, B) := \sup\{N : \exists f_1, ..., f_N \in \Theta,$$

with

$$\delta \leq \|f_i - f_j\|_B \leq c_1\delta, \ \forall i \neq j\}. \tag{4.146}$$

It is clear that $\bar{P}(\Theta, \delta, c_1, B) \leq P(\Theta, \delta, B)$.

We let $\mu$ be any Borel probability measure defined on $X$ and let $\mathcal{M}(\Theta, \mu)$ denote the set of all $\rho \in \mathcal{M}(\Theta)$ such that $\rho_X = \mu, |y| \leq 1$. As above, $\mathcal{M}(\Theta) = \{\rho : f_\rho \in \Theta\}$. We specify $B = L_2(\mu)$ and assume that $\Theta \subset L_2(\mu)$. We will use the abbreviated notation $\bar{P}(\delta) := \bar{P}(\Theta, \delta, c_1, L_2(\mu))$.

Let us fix any set $\Theta$ and any Borel probability measure $\mu$ defined on $X$. We set $\mathcal{M} := \mathcal{M}(\Theta, \mu)$ as defined above. We also take $1 < c_1$ in an arbitrary way but then fix this constant. For any fixed $\delta > 0$, we let $\{f_i\}_{i=1}^{\bar{P}}$, with $\bar{P} := \bar{P}(\delta)$, be a net of functions satisfying (4.146). To each $f_i$ we shall associate the measure

$$d\rho_i(x, y) := (a_i(x)d\delta_1(y) + b_i(x)d\delta_{-1}(y))d\mu(x), \tag{4.147}$$

where $a_i(x) := (1 + f_i(x))/2$, $b_i(x) := (1 - f_i(x))/2$ and $d\delta_\xi$ denotes the Dirac delta with unit mass at $\xi$. Note that $(\rho_i)_X = \mu$ and $f_{\rho_i} = f_i$ and hence each $\rho_i$ is in $\mathcal{M}(\Theta, \mu)$.

The following theorem is from DeVore *et al.* (2006).

**Theorem 4.98** *Let $1 < c_1$ be a fixed constant. Suppose that $\Theta$ is a subset of $L_2(\mu)$ with tight packing numbers $\bar{P} := \bar{P}(\delta)$. In addition, suppose that, for $\delta = 2\eta > 0$, the net of functions $\{f_i\}_{i=0}^{\bar{P}}$ in (4.146) satisfies $\|f_i\|_{C(X)} \leq 1/2$, $i = 1, \ldots, \bar{P}$. Then for any estimator $f_\mathbf{z}$ we have, for some $i \in \{1, \ldots, \bar{P}\}$,*

$$\rho_i^m\{\mathbf{z} : \|f_\mathbf{z} - f_i\|_{L_2(\mu)} \geq \eta\} \geq \min(1/2, (\bar{P}(2\eta) - 1)^{1/2}e^{-6c_1^2 m\eta^2 - 3/e}),$$

*for all $\eta > 0$, $m = 1, 2, \ldots.$*

We begin with a lemma that relates the Kulback–Leibler information (defined in Section 4.2.3) of a pair $\rho_i$, $\rho_j$ and the $L_2(\mu)$ distance between the corresponding $f_i$ and $f_j$.

**Lemma 4.99** *For any Borel measure $\mu$ and the measures $\rho_i$ defined by (4.147), we have*

$$\mathcal{K}(\rho_i, \rho_j) \leq \frac{4}{3} \| f_i - f_j \|^2_{L_2(\mu)}, \quad i, j = 1, \ldots, \bar{P}. \tag{4.148}$$

*Proof* We fix $i$ and $j$. We have $d\rho_i(x, y) = g(x, y)d\rho_j(x, y)$, where

$$g(x, y) = \frac{1 + (\text{sign} y) f_i(x)}{1 + (\text{sign} y) f_j(x)} = 1 + \frac{(\text{sign} y)(f_i(x) - f_j(x))}{1 + (\text{sign} y) f_j(x)}.$$

Thus, by (4.30),

$$2\mathcal{K}(\rho_i, \rho_j) = \int_X F_{i,j}(x) d\mu(x), \tag{4.149}$$

where

$$F_{i,j}(x) := (1 + f_i(x)) \ln \left( 1 + \frac{f_i(x) - f_j(x)}{1 + f_j(x)} \right)$$
$$+ (1 - f_i(x)) \ln \left( 1 - \frac{f_i(x) - f_j(x)}{1 - f_j(x)} \right).$$

Using the inequality $\ln(1 + u) \leq u$, we obtain

$$F_{i,j}(x) \leq (f_i(x) - f_j(x)) \left\{ \frac{1 + f_i(x)}{1 + f_j(x)} - \frac{1 - f_i(x)}{1 - f_j(x)} \right\}$$
$$= \frac{2(f_i(x) - f_j(x))^2}{1 - f_j(x)^2} \leq (8/3)(f_i(x) - f_j(x))^2.$$

Putting this into (4.149), we deduce (4.148). $\qquad\square$

*Proof of Theorem 4.98* We define

$$A_i := \{ \mathbf{z} : \| f_{\mathbf{z}} - f_i \|_{L_2(\mu)} < \eta \}, \quad i = 1, \ldots, \bar{P}.$$

Then the sets $A_i$ are disjoint because of (4.146). We apply Fano's inequality (Lemma 4.29) with our measures $\rho_i^m$ and find that (see Corollary 4.30) for some $i$

$$\rho_i^m(Z^m \setminus A_i)$$

$$\geq \min \left( 1/2, e^{-3/e} \exp \left( - \inf_{1 \leq j \leq \bar{P}} (\bar{P} - 1)^{-1} \sum_{i: i \neq j} \mathcal{K}(\rho_i^m, \rho_j^m) + \frac{1}{2} \ln(\bar{P} - 1) \right) \right).$$

$$\tag{4.150}$$

Next, by Lemma 4.99 we get

$$\sum_{i \neq j} \mathcal{K}(\rho_i^m, \rho_j^m) \leq m \sum_{i:i \neq j} \frac{4}{3} \|f_i - f_j\|_{L_2(\mu)}^2.$$

Using (4.146) we continue:

$$\leq \frac{4m}{3}(\bar{P} - 1)c_1^2 \delta^2 \leq (\bar{P} - 1)6c_1^2 m \eta^2.$$

Substituting this estimate into (4.150), we complete the proof of Theorem 4.98.

$\square$

Theorem 4.98 provides lower estimates for classes $\Theta$ with known lower estimates for the tight packing numbers $\bar{P}(\Theta, \delta)$. We now show how this theorem can be used in a situation when we know the behavior of the packing numbers $P(\Theta, \delta)$.

**Lemma 4.100** *Let $\Theta$ be a compact subset of $B$. Assume that*

$$C_1 \varphi(\delta) \leq \ln P(\Theta, \delta) \leq C_2 \varphi(\delta), \quad \delta \in (0, \delta_1],$$

*with a function $\varphi(\delta)$ satisfying the following condition. For any $\gamma > 0$ there is $A_\gamma$ such that, for any $\delta > 0$,*

$$\varphi(A_\gamma \delta) \leq \gamma \varphi(\delta). \tag{4.151}$$

*Then there exists $c_1 \geq 1$ and $\delta_2 > 0$ such that*

$$\ln \bar{P}(\Theta, \delta, c_1, B) \geq C_3 \ln P(\Theta, \delta), \quad \delta \in (0, \delta_2].$$

*Proof* For $\delta > 0$ we take the set $F := \{f_i\}_{i=1}^{P(\Theta, \delta)} \subset \Theta$ satisfying (4.145). Considering a $l\delta$-net with $l \geq 1$ for covering $\Theta$ we obtain that one of the balls of radius $l\delta$ contains at least $P(\Theta, \delta)/P(\Theta, l\delta)$ points of the set $F$. Denote this set of points by $F_l = \{f_i\}_{i \in \Lambda(l)}$. Then, obviously, for any $i \neq j \in \Lambda(l)$ we have

$$\delta \leq \|f_i - f_j\| \leq 2l\delta.$$

Therefore

$$\ln \bar{P}(\Theta, \delta, 2l, B) \geq \ln P(\Theta, \delta) - \ln P(\Theta, l\delta) \geq C_1 \varphi(\delta) - C_2 \varphi(l\delta).$$

Specifying $\gamma = C_1/(2C_2)$, $l = A_\gamma$ and $\delta_2 := \delta_1/l$, we continue:

$$\geq C_1 \varphi(\delta)/2 \geq \frac{C_1}{2C_2} \ln P(\Theta, \delta), \quad \delta \in (0, \delta_2]. \qquad \square$$

As a corollary of Theorem 4.98 and Lemma 4.100 we obtain the following theorem.

**Theorem 4.101** *Assume* $\Theta$ *is a compact subset of* $L_2(\mu)$ *such that* $\Theta \subset$ $(1/4)U(\mathcal{C}(X))$ *and*

$$\epsilon_n(\Theta, L_2(\mu)) \asymp n^{-r}. \tag{4.152}$$

*Then there exist* $\delta_0 > 0$ *and* $\eta_m := \eta_m(r) \asymp m^{-r/(1+2r)}$ *such that*

$$\mathbf{AC}_m(\mathcal{M}(\Theta, \mu), \eta) \geq \delta_0 \quad \text{for} \quad \eta \leq \eta_m \tag{4.153}$$

*and*

$$\mathbf{AC}_m(\mathcal{M}(\Theta, \mu), \eta) \geq Ce^{-c(r)m\eta^2} \quad \text{for} \quad \eta \geq \eta_m. \tag{4.154}$$

*Proof* Condition (4.152) implies

$$C_1(r)\delta^{-1/r} \leq \ln P(\Theta, \delta) \leq C_2(r)\delta^{-1/r}, \quad \delta \in (0, \delta_1].$$

Clearly, the function $\varphi(\delta) = \delta^{-1/r}$ satisfies condition (4.151) from Lemma 4.100. Therefore by Lemma 4.100 we obtain

$$\ln \bar{P}(\Theta, \eta, c_1(r), L_2(\mu)) \geq C_3(r)\eta^{-1/r}, \quad \eta \in (0, \delta_2(r)],$$

with some $c_1(r) \geq 1$. It remains to use Theorem 4.98 with $\eta_m$ a solution of the equation

$$\frac{C_3(r)}{2}(2\eta)^{-1/r} - 6c_1(r)^2 m\eta^2 = 0.$$

It is clear that

$$\eta_m \asymp m^{-\frac{r}{1+2r}}. \qquad \square$$

**Remark 4.102** Theorem 4.101 holds in the case $\Theta \subset (M/4)U(\mathcal{C}(X))$, $|y| \leq M$, with constants allowed to depend on $M$.

We note that we do not impose direct restrictions on the measure $\mu$ in Theorem 4.101. However, the assumption (4.152) imposes an indirect restriction. For instance, if $\mu$ is a Dirac measure then we always have $\epsilon_n(\Theta, L_2(\mu)) \ll 2^{-n}$. Therefore, Theorem 4.101 does not apply in this case.

Let us make some comments on Theorem 4.101. It is clear that the parameter $r$ controls the size of the compact $\Theta$. The bigger the $r$, the smaller the compact $\Theta$. In the statement of Theorem 4.101 the parameter $r$ affects the rate of decay of $\eta_m$. The quantity $\eta_m$ is an important characteristic of the estimation process. The inequality (4.153) says that there is no way to estimate $f_\rho$ from $\Theta$ with accuracy $\leq \eta_m$ with high confidence ($>1-\delta_0$). It seems natural that this critical accuracy $\eta_m$ depends on the size of $\Theta$ (on the parameter $r$). The inequalities (4.153) and (4.154) give

$$\mathbf{AC}_m(\mathcal{M}(\Theta, \mu), \eta) \geq \delta_0 Ce^{-c(r)m\eta^2} \tag{4.155}$$

for all $\eta$. The exponent $m\eta^2$ in this inequality does not depend on the size of $\Theta$. This may indicate that the form of this exponent is related not to the size of $\Theta$ but rather to the stochastic nature of the problem. Another argument in support of the above observation is provided by an inequality from Section 4.5.3. We will use that inequality to show that in the case of a compact $\Theta$ consisting of only one function we have an analog of (4.155) in the case of linear estimators. Let $\Theta = \{1/2\}$. Suppose that we are looking for a linear estimator

$$f_\mathbf{z} = \sum_{i=1}^{m} w_i(x_1, \ldots, x_m, x) y_i \tag{4.156}$$

of the regression function $f_\rho$. Consider the following special case of the measure $\rho$. Let $\rho_X = \mu$ be any probabilistic measure on $X$. We define $\rho(y|x)$ as the Bernoulli measure:

$$\rho(1|x) = \rho(0|x) = 1/2, \quad x \in X.$$

Then for the above measure $\rho$ we have $f_\rho(x) \equiv 1/2 \in \Theta$. Then

$$\|f_\mathbf{z} - f_\rho\|_{L_2(\mu)} \geq \int_X |f_\mathbf{z} - f_\rho| d\mu$$

$$\geq |\int_X (f_\mathbf{z} - f_\rho) d\mu| = |\sum_{i=1}^{m} w_i(x_1, \ldots, x_m) y_i - 1/2|,$$

where

$$w_i(x_1, \ldots, x_m) := \int_X w_i(x_1, \ldots, x_m, x) d\mu.$$

Using Theorem 4.93 we get

$$\rho^m \{\mathbf{z} : \|f_\mathbf{z} - f_\rho\|_{L_2(\mu)} \geq \eta\} \geq \text{Prob}_{\mathbf{z}\in Z^m} \left\{ |\sum_{i=1}^{m} w_i(x_1, \ldots, x_m) y_i - 1/2| \geq \eta \right\}$$

$$\geq \exp\left(-25m\eta^2 - 6.25 \sum_{k=1}^{m-1} 1/k\right)$$

$$\geq m^{-6.25} \exp(-25m\eta^2 - 1).$$

Therefore, in the case where $\mathbb{E}(m)$ is the set of estimators of the form (4.156) we have for $\mathcal{M}_\mu := \{\rho : f_\rho = 1/2, \rho_X = \mu\}$

$$\mathbf{AC}_m(\mathcal{M}_\mu, \mathbb{E}, \eta) \geq m^{-6.25} \exp(-25m\eta^2 - 1).$$

### 4.6 Application of greedy algorithms in learning theory

The fundamental question of learning theory is: How should we build a good estimator? It is well known in statistics that the following way of building $f_{\mathbf{z}}$ provides a near optimal estimator in many cases. First, choose a right hypothesis space $\mathcal{H}$. Second, construct $f_{\mathbf{z}, \mathcal{H}} \in \mathcal{H}$ as the least squares estimator:

$$f_{\mathbf{z}, \mathcal{H}} = \arg \min_{f \in \mathcal{H}} \mathcal{E}_{\mathbf{z}}(f),$$

where

$$\mathcal{E}_{\mathbf{z}}(f) := \frac{1}{m} \sum_{i=1}^{m} (f(x_i) - y_i)^2.$$

Clearly, a crucial role in this approach is played by the choice of the hypothesis space $\mathcal{H}$. In other words, we need to begin our construction of an estimator with a decision of what should be the form of the estimator. In this section we discuss only the case oriented to the use of nonlinear approximation, in particular greedy approximation in such a construction. We want to construct a good estimator that will provide high accuracy and that will be practically implementable. We will discuss a realization of this plan in several stages. We begin with results on accuracy. We will give a presentation in a rather general form of nonlinear approximation.

Let $\mathcal{D}(n, q) := \{g_l^n\}_{l=1}^{N_n}$, $n \in \mathbb{N}$, $N_n \leq n^q$, $q \geq 1$, be a system of bounded functions defined on $X$. We will consider a sequence $\{\mathcal{D}(n, q)\}_{n=1}^{\infty}$ of such systems. In building an estimator, based on $\mathcal{D}(n, q)$, we are going to use $n$-term approximations with regard to $\mathcal{D}(n, q)$:

$$G_\Lambda := \sum_{l \in \Lambda} c_l g_l^n, \quad |\Lambda| = n. \tag{4.157}$$

A standard assumption that we make in supervised learning theory is that $|y| \leq M$ almost surely. This implies that we always assume that $|f_\rho| \leq M$. Denoting $\|f\|_{\mathcal{B}(X)} := \sup_{x \in X} |f(x)|$, we rewrite the above assumption in the form $\|f_\rho\|_{\mathcal{B}(X)} \leq M$. It is clear that with such an assumption it is natural to restrict our search to estimators $f_{\mathbf{z}}$ satisfying the same inequality $\|f_{\mathbf{z}}\|_{\mathcal{B}(X)} \leq M$. Now, there are two standard ways to go in learning theory. In the first way (i) we are looking for an estimator of the form (4.157) with an extra condition

$$\|G_\Lambda\|_{\mathcal{B}(X)} \leq M. \tag{4.158}$$

In the second way (ii) we take an approximant $G_\Lambda$ of the form (4.157) and truncate it, i.e. consider $T_M(G_\Lambda)$, where $T_M$ is a truncation operator:

$T_M(u) = u$ if $|u| \leq M$ and $T_M(u) = M$ sign $u$ if $|u| \geq M$. Then automatically $\|T_M(G_\Lambda)\|_{\mathcal{B}(X)} \leq M$.

Let us look in more detail at the hypothesis spaces generated in the above two cases. In case (i) we use the following compacts in $\mathcal{B}(X)$ as a source of estimators:

$$F_n(q) := \left\{ f : \exists \Lambda \subset [1, N_n], |\Lambda| = n, f = \sum_{l \in \Lambda} c_l g_l^n, \|f\|_{\mathcal{B}(X)} \leq M \right\}.$$

An important good feature of $F_n(q)$ is that it is a collection of sparse (at most $n$ terms) estimators. An important drawback is that it may not be easy to check if (4.158) is satisfied for a particular $G_\Lambda$ of the form (4.157).

In case (ii) we use the following sets in $\mathcal{B}(X)$ as a source of estimators:

$$F_n^T(q) := \left\{ f : \exists \Lambda \subset [1, N_n], |\Lambda| = n, f = T_M\left(\sum_{l \in \Lambda} c_l g_l^n\right) \right\}.$$

An obvious good feature of $F_n^T(q)$ is that, by definition, we have $\|f\|_{\mathcal{B}(X)} \leq M$ for any $f$ from $F_n^T(q)$. An important drawback of it is that $F_n^T(q)$ has (in general) a rather complex structure. In particular, applying the truncation operator $T_M$ to $G_\Lambda$ we lose (in general) the sparseness property of $G_\Lambda$.

Once we have specified our hypothesis spaces, we can look for an existing theory that provides the corresponding error bounds. The general theory is well developed in case (i). We will use a variant of such a general theory from Section 4.4 developed in Temlyakov (2008a). This theory is based on the following property of compacts $F_n(q)$, formulated in terms of covering numbers:

$$N(F_n(q), \epsilon, \mathcal{B}(X)) \leq (1 + 2M/\epsilon)^n n^{qn}. \tag{4.159}$$

For convenience, we now formulate the corresponding results from Section 4.4. For a compact $\Theta$ in a Banach space $B$ we denote by $N(\Theta, \epsilon, B)$ the covering number that is the minimal number of balls of radius $\epsilon$ with centers in $\Theta$ needed for covering $\Theta$. Let $a, b$ be two positive numbers. Consider a collection $\mathcal{K}(a, b)$ of compacts $K_n$ in $\mathcal{B}(X)$ that are contained in the $M$-ball of $\mathcal{B}(X)$ and satisfy the following covering numbers condition:

$$N(K_n, \epsilon, \mathcal{B}(X)) \leq (a(1 + 1/\epsilon))^n n^{bn}, \quad n = 1, 2, \ldots \tag{4.160}$$

We begin with the definition of our estimator. As above, let $\mathcal{K} := \mathcal{K}(a, b)$ be a collection of compacts $K_n$ in $\mathcal{B}(X)$ satisfying (4.160).

We take a parameter $A \geq 1$ and consider the following penalized least squares estimator (PLSE):

$$f_{\mathbf{z}}^A := f_{\mathbf{z}}^A(\mathcal{K}) := f_{\mathbf{z}, K_{n(\mathbf{z})}}$$

with

$$n(\mathbf{z}) := \arg \min_{1 \leq j \leq m} \left( \mathcal{E}_{\mathbf{z}}(f_{\mathbf{z}, K_j}) + \frac{A j \ln m}{m} \right).$$

Denote for a set $L$ of a Banach space $B$

$$d(\Theta, L)_B := \sup_{f \in \Theta} \inf_{g \in L} \| f - g \|_B.$$

**Theorem 4.103** *For* $\mathcal{K} := \{K_n\}_{n=1}^{\infty}$ *satisfying (4.160) and* $M > 0$ *there exists* $A_0 := A_0(a, b, M)$ *such that, for any* $A \geq A_0$ *and any* $\rho$ *such that* $|y| \leq M$ *a.s., we have*

$$\| f_{\mathbf{z}}^A - f_\rho \|_{L_2(\rho_X)}^2 \leq \min_{1 \leq j \leq m} \left( 3d(f_\rho, K_j)_{L_2(\rho_X)}^2 + \frac{4 A j \ln m}{m} \right)$$

*with probability* $\geq 1 - m^{-c(M)A}$.

It is clear from (4.159) and from the definition of $F_n(q)$ that we can apply Theorem 4.103 to the sequence of compacts $\{F_n(q)\}$ and obtain the following error bound with probability $\geq 1 - m^{-c(M)A}$:

$$\| f_{\mathbf{z}}^A - f_\rho \|_{L_2(\rho_X)}^2 \leq \min_{1 \leq j \leq m} \left( 3d(f_\rho, F_j(q))_{L_2(\rho_X)}^2 + \frac{4 A j \ln m}{m} \right). \quad (4.161)$$

We note that the inequality (4.161) is the Lebesgue-type inequality (see Section 2.6). Indeed, on the left-hand side of (4.161) we have an error of a particular estimator $f_{\mathbf{z}}^A$ built as the PLSE and on the right-hand side of (4.161) we have $d(f_\rho, F_j(q))_{L_2(\rho_X)}$ – the best error that we can get using estimators from $F_j(q)$, $j = 1, 2, \ldots$. We recall that, by construction, $f_{\mathbf{z}}^A \in F_{n(\mathbf{z})}(q)$.

Let us now discuss an application of the theory from Section 4.4 in case (ii). We cannot apply that theory directly to the sequence of sets $\{F_n^T(q)\}$ because we do not know if these sets satisfy the covering number condition (4.160). However, we can modify the sets $F_n^T(q)$ to make them satisfy the condition (4.160). Let $c \geq 0$ and define

$$F_n^T(q, c) := \left\{ f : \exists G_\Lambda := \sum_{l \in \Lambda} c_l g_l^n, \ \Lambda \subset [1, N_n], \ |\Lambda| = n, \right.$$

$$\left. \| G_\Lambda \|_{B(X)} \leq C_2 n^c, \ f = T_M(G_\Lambda) \right\}$$

with some fixed $C_2 \geq 1$. Then, using the inequality $|T_M(f_1(x)) - T_M(f_2(x))| \leq |f_1(x) - f_2(x)|$, $x \in X$, it is easy to get that

$$N(F_n^T(q, c), \epsilon, \mathcal{B}(X)) \leq (2C_2(1 + 1/\epsilon)^n n^{(q+c)n}.$$

Therefore, (4.160) is satisfied with $a = 2C_2$ and $b = q + c$. We note that, from a practical point of view, an extra restriction $\|G_\Lambda\|_{\mathcal{B}(X)} \leq C_2 n^c$ is not a big constraint.

The above estimators (built as the PLSE) are very good from the theoretical point of view. Their error bounds satisfy the Lebesgue-type inequalities. However, they are not good from the point of view of implementation. For example, there is no simple algorithm to find $f_{\mathbf{z}, F_n(q)}$ because $F_n(q)$ is a union of $\binom{Nn}{n}$ $M$-balls of $n$-dimensional subspaces. Thus, finding an exact LSE $f_{\mathbf{z}, F_n(q)}$ is practically impossible. We now use a remark from Temlyakov (2008a) that allows us to build an approximate LSE with a good approximation error. We proceed to the definition of the penalized approximate least squares estimator (PALSE) (see Temlyakov (2008a)). Let $\delta := \{\delta_{j,m}\}_{j=1}^m$ be a sequence of non-negative numbers. We define $f_{\mathbf{z}, \delta, K_j}$ as an estimator satisfying the relation

$$\mathcal{E}_{\mathbf{z}}(f_{\mathbf{z}, \delta, K_j}) \leq \mathcal{E}_{\mathbf{z}}(f_{\mathbf{z}, K_j}) + \delta_{j,m}. \tag{4.162}$$

In other words, $f_{\mathbf{z}, \delta, K_j}$ is an approximation to the least squares estimator $f_{\mathbf{z}, K_j}$.

Next, we take a parameter $A \geq 1$ and define the penalized approximate least squares estimator (PALSE)

$$f_{\mathbf{z}, \delta}^A := f_{\mathbf{z}, \delta}^A(\mathcal{K}) := f_{\mathbf{z}, \delta, K_{n(\mathbf{z})}}$$

with

$$n(\mathbf{z}) := \arg \min_{1 \leq j \leq m} \left( \mathcal{E}_{\mathbf{z}}(f_{\mathbf{z}, \delta, K_j}) + \frac{Aj \ln m}{m} \right).$$

The theory developed in Temlyakov (2008a) gives the following control of the error.

**Theorem 4.104** *Under the assumptions of Theorem 4.103 we have*

$$\|f_{\mathbf{z}, \delta}^A - f_\rho\|_{L_2(\rho_X)}^2 \leq \min_{1 \leq j \leq m} \left( 3d(f_\rho, K_j)_{L_2(\rho_X)}^2 + \frac{4Aj \ln m}{m} + 2\delta_{j,m} \right)$$

*with probability* $\geq 1 - m^{-c(M)A}$.

We point out here that the approximate least squares estimator $f_{\mathbf{z}, \delta, K_j}$ approximates the least squares estimator $f_{\mathbf{z}, K_j}$ in the sense that $\mathcal{E}_{\mathbf{z}}(f_{\mathbf{z}, \delta, K_j}) - \mathcal{E}_{\mathbf{z}}(f_{\mathbf{z}, K_j})$ is small and not in the sense that $\|f_{\mathbf{z}, \delta, K_j} - f_{\mathbf{z}, K_j}\|$ is small. Theorem 4.104 guarantees a good error bound for any penalized estimator built from $\{f_{\mathbf{z}, \delta, K_j}\}$ satisfying (4.162). We will use greedy algorithms in building an approximate estimator. We now present results from Temlyakov (2006e). We will need more specific compacts $F(n, q)$ and will impose some restrictions on $g_l^n$. We assume that $\|g_l^n\|_{B(X)} \leq C_1$ for all $n$ and $l$. We consider the following compacts instead of $F_n(q)$:

$$F(n, q) := \left\{ f : \exists \Lambda \subset [1, N_n], |\Lambda| = n, f = \sum_{l \in \Lambda} c_l g_l^n, \sum_{l \in \Lambda} |c_l| \leq 1 \right\}.$$

Then we have $\|f\|_{B(X)} \leq C_1$ for any $f \in F(n, q)$ and $\|f\|_{B(X)} \leq M$ if $M \geq C_1$. Let $\mathbf{z} = (z_1, \ldots, z_m)$, $z_i = (x_i, y_i)$, be given. Consider the following system of vectors in $\mathbb{R}^m$:

$$v^{j,l} := (g_l^j(x_1), \ldots, g_l^j(x_m)), \quad l \in [1, N_j].$$

We equip the $\mathbb{R}^m$ with the norm $\|v\| := (m^{-1} \sum_{i=1}^m v_i^2)^{1/2}$. Then

$$\|v^{j,l}\| \leq \|g_l^j\|_{B(X)} \leq C_1.$$

Consider the following system in $H = \mathbb{R}^m$ with the defined above norm $\|\cdot\|$:

$$\mathcal{G} := \{v^{j,l}\}_{l=1}^{N_j}.$$

Finding the estimator

$$f_{\mathbf{z}, F(j,q)} = \sum_{l \in \Lambda} c_l g_l^j, \quad \sum_{l \in \Lambda} |c_l| \leq 1, \quad |\Lambda| = j, \quad \Lambda \subset [1, N_j],$$

is equivalent to finding the best $j$-term approximant of $y \in \mathbb{R}^m$ from the $A_1(\mathcal{G})$ in the space $H$. We apply the RGA($\theta$) from Section 2.4 with $\theta = 2$ with respect to $\mathcal{G}$ to $y$ and find, after $j$ steps, an approximant

$$v^j := \sum_{l \in \Lambda'} a_l v^{j,l}, \quad \sum_{l \in \Lambda'} |a_l| \leq 1, \quad |\Lambda'| = j, \quad \Lambda' \subset [1, N_j],$$

such that

$$\|y - v^j\|^2 \leq d(y, A_1(\mathcal{G}))^2 + Cj^{-1}, \quad C = C(M, C_1).$$

We define an estimator

$$\hat{f}_{\mathbf{z}} := \hat{f}_{\mathbf{z}, F(j,q)} := \sum_{l \in \Lambda'} a_l g_l^j.$$

Then $\hat{f}_{\mathbf{z}} \in F(j, q)$ and

$$\mathcal{E}_{\mathbf{z}}(\hat{f}_{\mathbf{z}, F(j,q)}) \leq \mathcal{E}_{\mathbf{z}}(f_{\mathbf{z}, F(j,q)}) + Cj^{-1}.$$

We denote $\delta := \{Cj^{-1}\}_{j=1}^m$ and define, for $A \geq 1$,

$$f_{\mathbf{z}, \delta}^A := \hat{f}_{\mathbf{z}, F(n(\mathbf{z}), q)}$$

with

$$n(\mathbf{z}) := \arg \min_{1 \le j \le m} \left( \mathcal{E}_{\mathbf{z}}(\hat{f}_{\mathbf{z},F(j,q)}) + \frac{Aj \ln m}{m} \right).$$

By Theorem 4.104 we have for $A \ge A_0(M)$

$$\|f_{\mathbf{z},\delta}^A - f_\rho\|_{L_2(\rho_X)}^2 \le \min_{1 \le j \le m} \left( 3d(f_\rho, F(j,q))^2 + \frac{4Aj \ln m}{m} + 2Cj^{-1} \right)$$

(4.163)

with probability $\ge 1 - m^{-c(M)A}$.

In particular, (4.163) means that the estimator $f_{\mathbf{z},\delta}^A$ is an estimator that provides the error

$$\|f_{\mathbf{z},\delta}^A - f_\rho\|_{L_2(\rho_X)}^2 \ll \left( \frac{\ln m}{m} \right)^{2r/1+2r}$$

for $f_\rho$ such that $d(f_\rho, F(j,q))_{L_2(\rho_X)} \ll j^{-r}$, $r \le 1/2$. We note that the estimator $f_{\mathbf{z},\delta}^A$ is based on the greedy algorithm and it can easily be implemented.

We now describe an application of the greedy algorithm in learning theory from Barron *et al.* (2008). In this application one can use the Orthogonal Greedy Algorithm or the following variant of the Relaxed Greedy Algorithm.

Let $\alpha_1 := 0$ and $\alpha_m := 1 - 2/m$, $m \ge 2$. We set $f_0 := f$, $G_0 := 0$ and inductively define two sequences $\{\beta_m\}_{m=1}^\infty$, $\{\varphi_m\}_{m=1}^\infty$ as follows:

$$(\beta_m, \varphi_m) := \arg \min_{\beta \in \mathbb{R}, g \in \mathcal{D}} \|f - (\alpha_m G_{m-1} + \beta g)\|.$$

Then we set

$$f_m := f_{m-1} - \beta_m \varphi_m, \quad G_m := G_{m-1} + \beta_m \varphi_m.$$

For systems $\mathcal{D}(n, q)$ the following estimator is considered in Barron *et al.* (2008). As above, let $\mathbf{z} = (z_1, \dots, z_m)$, $z_i = (x_i, y_i)$, be given. Consider the following system of vectors in $\mathbb{R}^m$:

$$v^{j,l} := (g_l^j(x_1), \dots, g_l^j(x_m)), \quad l \in [1, N_j].$$

We equip the $\mathbb{R}^m$ with the norm $\|v\| := (m^{-1} \sum_{i=1}^m v_i^2)^{1/2}$ and normalize the above system of vectors. Denote the new system of vectors by $\mathcal{G}_j$. Now we apply either the OGA or the above defined version of the RGA to the vector $y \in \mathbb{R}$ with respect to the system $\mathcal{G}_j$. Similar to the above discussed case of the system $\mathcal{G}$, we obtain an estimator $\hat{f}_j$. Next, we look for the penalized estimator built from the estimators $\{\hat{f}_j\}$ in the following way. Let

$$n(\mathbf{z}) := \arg \min_{1 \le j \le m} \left( \mathcal{E}_{\mathbf{z}}(T_M(\hat{f}_j)) + \frac{Aj \log m}{m} \right).$$

Define

$$\hat{f} := T_M(\hat{f}_{n(\mathbf{z})}).$$

Assuming that the systems $\mathcal{D}(n, q)$ are normalized in $L_2(\rho_X)$, Barron *et al.* (2008) proved the following error estimate.

**Theorem 4.105** *There exists $A_0(M)$ such that for $A \geq A_0$ one has the following bound for the expectation of the error:*

$$E\left(\|f_\rho - \hat{f}\|_{L_2(\rho_X)}^2\right) \leq \min_{1 \leq j \leq m} (C(A, M, q) j \log m / m$$

$$+ \inf_{h \in \operatorname{span} \mathcal{D}(j,q)} (2\|f_\rho - h\|_{L_2(\rho_X)}^2 + 8\|h\|_{\mathcal{A}_1(\mathcal{D}(j,q))}^2 / j)).$$

$$(4.164)$$

Let us make a comparison of (4.164) with (4.163). First of all, (4.164) gives an error bound for the expectation and (4.163) gives an error bound with high probability. In this sense (4.163) is better than (4.164). However, the condition $\|g_l^n\|_{\mathcal{B}(X)} \leq C_1$ imposed on the systems $\mathcal{D}(n, q)$ in order to obtain (4.163) is more restrictive than the corresponding assumption for (4.164).

# 5

# Approximation in compressed sensing

## 5.1 Introduction

This chapter aims to provide the theoretical foundations of processing large data sets. The main technique used to achieve this goal is based on nonlinear sparse representations. Sparse representations of a function are not only a powerful theoretical tool, but also they are utilized in many applications in image/signal processing and numerical computation. Sparsity is a very important concept in applied mathematics.

We begin with a general setting of mathematical signal processing problems related to sparse representations. This setting consists of three components that we describe below: assumptions on a signal, tasks and methods of recovery of a signal. We only give a brief and schematic description of this setting in order to illustrate the idea.

First, assumptions are imposed on a signal.

(A1) **Strict sparsity** Usually, *sparse representation* of a signal (function) $f$ means that there is a given system of functions (dictionary) $\mathcal{D}$ such that $f$ has a representation as a linear combination of no greater than (or exactly) $k$ elements from the dictionary ($f$ is $k$-sparse with respect to $\mathcal{D}$):

$$f = \sum_{j=1}^{k} c_j g_j, \quad g_j \in \mathcal{D}, \quad j = 1, \ldots, k.$$

It is clear that in the case of a finite dictionary, assumption (A1) is equivalent to the assumption that $\sigma_k(f, \mathcal{D}) = 0$. A weaker form of the sparsity assumption (A1) is the following.

(A2) **Approximate sparsity** We describe sparsity properties of $f$ by the rate of decay of the sequence $\{\sigma_m(f, \mathcal{D})\}_{m=1}^{\infty}$ of best $m$-term approximations of $f$ with respect to $\mathcal{D}$. In other words, this assumption means that $f$ belongs to some approximation class.

Replacing an approximation class by a smoothness class, we obtain another type of assumption.

**(A3) Compressibility** Assume that $f$ has a representation

$$f = \sum_{j=1}^{\infty} c_j g_j, \quad g_j \in \mathcal{D}, \quad j = 1, \ldots, \quad \sum_{j=1}^{\infty} |c_j|^\beta \le R^\beta, \quad \beta \in (0, 1].$$

Second, we discuss two variants of our main task of recovery of the signal $f$. By *recovery* of a signal $f$ we mean finding its representation with respect to the dictionary $\mathcal{D}$. There are two forms of recovery widely used in signal processing:

**(T1) exact recovery**;
**(T2) approximate recovery**.

Usually, we apply task **(T1)** only in the case of sparsity assumption **(A1)**. In other cases we apply task **(T2)**. In the case of approximate recovery we need to specify how we measure the error of approximate recovery. There are two ways to define error, which lead to different settings. Suppose

$$f = \sum_{j=1}^{\infty} c_j g_j, \quad g_j \in \mathcal{D}, \quad j = 1, \ldots,$$

and its $m$-term approximant has the form

$$A_m(f) = \sum_{i=1}^{m} a_i \phi_i, \quad \phi_i \in \mathcal{D}, \quad i = 1, \ldots, m.$$

In the first method we measure the error between $f$ and $A_m(f)$ in the norm $\|\cdot\|$ of the space to which the function $f$ belongs (data space). Then the corresponding error is $\|f - A_m(f)\|$. In the second method we measure the error between $f$ and $A_m(f)$ in the norm of the space to which the coefficients of function $f$ belong (coefficient space). For instance, if $\phi_i = g_{j(i)}, i = 1, \ldots, m$, then the error in the $\ell_2$ norm of the coefficient space will be given by

$$\|f - A_m(f)\|_{\ell_2}^2 = \sum_{i=1}^{m} |a_i - c_{j(i)}|^2 + \sum_{j \ne j(i), i=1,\ldots,m} |c_j|^2.$$

Third, we choose a method of recovery and study its suitability (efficiency) for a task that we are interested in under one of the above assumptions. We only discuss theoretical results that measure the efficiency of a given method in terms of the magnitude of its error. It is clear that the quality of the method (rate of decay of the error) is determined by the representation system $\mathcal{D}$. The two typical and most popular methods of recovery are:

**(M1) optimization procedure;**
**(M2) greedy algorithm.**

The next step in the theoretical study of efficiency of a given recovery method is to find necessary and sufficient conditions on the dictionary $\mathcal{D}$ that guarantee that the chosen task is fulfilled under the chosen assumptions. This problem is solved only in some special (albeit important) cases. We discuss it at the end of Section 5.2. We now proceed to the compressed sensing setting, which is a particular case of the above discussed general setting of the recovery of sparse signals.

Recently, compressed sensing (compressive sampling) has attracted a lot of attention from both mathematicians and computer scientists. Compressed sensing refers to a problem of *economical* recovery of an unknown vector $u \in \mathbb{R}^m$ from the information provided by linear measurements $\langle u, \phi_j \rangle$, $\phi_j \in \mathbb{R}^m$, $j = 1, \ldots, n$. The goal is to design an algorithm that finds (approximates) $u$ from the information $y = (\langle u, \phi_1 \rangle, \ldots, \langle u, \phi_n \rangle)^T \in \mathbb{R}^n$. The crucial step here is to build a *sensing* set of vectors $\phi_j \in \mathbb{R}^m$, $j = 1, \ldots, n$, that is *good* for all vectors $u \in \mathbb{R}^m$ with a special property. Clearly, the terms *economical* and *good* should be clarified in a mathematical setting of the problem. A natural variant of this setting uses the concept of *sparsity*. We call a vector $u \in \mathbb{R}^m$ $k$-sparse if it has at most $k$ non-zero coordinates. Now, for a given pair $(m, n)$ we want to understand what is the biggest sparsity $k(m, n)$ such that there exists a set of vectors $\phi_j \in \mathbb{R}^m$, $j = 1, \ldots, n$, and economical algorithm $A$ mapping $y$ into $\mathbb{R}^m$ in such a way that, for any $u$ of sparsity $k(m, n)$, one would have an exact recovery $A(u) = u$. In other words, we want to describe matrices $\Phi$ with rows $\phi_j^T$, $j = 1, \ldots, n$, such that there exists an economical algorithm to solve the sparse recovery problem.

The sparse recovery problem is the problem of finding the vector $u^0 := u_\Phi^0(y) \in \mathbb{R}^m$ which solves the optimization problem

$$\min \|v\|_0 \quad \text{subject to} \quad \Phi v = y, \qquad (P_0)$$

where $\|v\|_0 := |\operatorname{supp}(v)|$. D. Donoho and coauthors (see, for example, Chen, Donoho and Saunders (2001) and Donoho, Elad and Temlyakov (2006) and historical remarks and references therein) have suggested an economical algorithm (the Basis Pursuit) and have started a systematic study of the following question: For which measurement matrices $\Phi$ should the highly non-convex combinatorial optimization problem $(P_0)$ be equivalent to its convex relaxation problem

$$\min \|v\|_1 \quad \text{subject to} \quad \Phi v = y, \tag{$P_1$}$$

where $\|v\|_1$ denotes the $\ell_1$-norm of the vector $v \in \mathbb{R}^m$? Denote the solution to $(P_1)$ by $A_\Phi(y)$. It is known that the problem $(P_1)$ can be solved using a linear programming technique. The $\ell_1$ minimization algorithm $A_\Phi$ from $(P_1)$ is an economical algorithm that we will inspect. It is known (see, for example, Donoho, Elad and Temlyakov (2006) and Theorem 5.40 below) that for $M$-coherent matrices $\Phi$ one has $u_\Phi^0(\Phi u) = A_\Phi(\Phi u) = u$ provided $u$ is $k$-sparse with $k < (1 + 1/M)/2$. This allows us to build rather simple deterministic matrices $\Phi$ with $k(m, n) \asymp n^{1/2}$ and recover sparse signals with the $\ell_1$ minimization algorithm $A_\Phi$ from $(P_1)$.

Recent progress (see Candes (2006) and DeVore (2006)) in compressed sensing has resulted in proving the existence of matrices $\Phi$ with $k(m, n) \asymp n/\log(m/n)$, which is substantially larger than $n^{1/2}$. We proceed to a detailed discussion of results on the recovery properties of the $\ell_1$ minimization algorithm $A_\Phi$.

Donoho (2006) formulated three properties of matrices $\Phi$ which have columns normalized in $\ell_2$ and proved the existence of such matrices. Let $T$ be a subset of indices from $[1, m]$. Denote by $\Phi_T$ a matrix consisting of columns of $\Phi$ with indices from $T$. Donoho's three properties are as follows.

**(CS1)** The minimal singular value of $\Phi_T$ is $\geq \eta_1 > 0$ uniformly in $T$, satisfying $|T| \leq \rho n/\log m$.

**(CS2)** Let $W_T$ denote the range of $\Phi_T$. Assume that for any $T$ satisfying $|T| \leq \rho n/\log m$ one has

$$\|w\|_1 \geq \eta_2 n^{1/2} \|w\|_2, \quad \forall w \in W_T, \quad \eta_2 > 0.$$

**(CS3)** Denote $T^c := \{j\}_{j=1}^m \setminus T$. For any $T$, $|T| \leq \rho n/\log m$, and for any $w \in W_{T^c}$ one has, for any $v$ satisfying $\Phi_{T^c} v = w$,

$$\|v\|_{\ell_1(T^c)} \geq \eta_3 (\log(m/n))^{-1/2} \|w\|_1, \quad \eta_3 > 0.$$

It is proved in Donoho (2006) that if $\Phi$ satisfies **(CS1)**–**(CS3)** then there exists $\rho_0 > 0$ such that $u_\Phi^0(\Phi u) = A_\Phi(\Phi u) = u$ provided $|\operatorname{supp} u| \leq \rho_0 n/\log m$. Analyses in Donoho (2006) relate the compressed sensing problem to the problem of estimating the Kolmogorov widths and their dual, the Gel'fand widths.

For a compact $F \subset \mathbb{R}^m$, the Kolmogorov width is given by

$$d_n(F, \ell_p) := \inf_{L_n : \dim L_n \le n} \sup_{f \in F} \inf_{a \in L_n} \|f - a\|_p,$$

where $L_n$ is a linear subspace of $\mathbb{R}^m$ and $\|\cdot\|_p$ denotes the $\ell_p$ norm. The Gel'fand width is defined as

$$d^n(F, \ell_p) := \inf_{V_n} \sup_{f \in F \cap V_n} \|f\|_p,$$

where the infimum is taken over linear subspaces $V_n$ with dimension $\ge m - n$. It is well known that the Kolmogorov and the Gel'fand widths are related by a duality formula. For instance, in the case when $F = B_p^m$ is a unit $\ell_p$-ball in $\mathbb{R}^m$ and $1 \le q, p \le \infty$ we have (see Theorem 5.13 below)

$$d_n(B_p^m, \ell_q) = d^n(B_{q'}^m, \ell_{p'}), \quad p' := p/(p - 1). \tag{5.1}$$

In a particular case when $p = 2$, $q = \infty$, (5.1) yields

$$d_n(B_2^m, \ell_\infty) = d^n(B_1^m, \ell_2). \tag{5.2}$$

It has been established in approximation theory (see Garnaev and Gluskin (1984) and Kashin (1977a)) that

$$d_n(B_2^m, \ell_\infty) \le C((1 + \log(m/n))/n)^{1/2}. \tag{5.3}$$

In other words, it was proved (see (5.3) and (5.2)) that for any pair $(m, n)$ there exists a subspace $V_n$, $\dim V_n \ge m - n$, such that for any $x \in V_n$ one has

$$\|x\|_2 \le C((1 + \log(m/n))/n)^{1/2} \|x\|_1. \tag{5.4}$$

It has been understood in Donoho (2006) that properties of the null space $\mathcal{N}(\Phi) := \{x : \Phi x = 0\}$ of a measurement matrix $\Phi$ play an important role in the compressed sensing problem. Donoho (2006) introduced the following two characteristics of $\Phi$ formulated in terms of $\mathcal{N}(\Phi)$:

$$w(\Phi, F) := \sup_{x \in F \cap \mathcal{N}(\Phi)} \|x\|_2$$

and

$$v(\Phi, T) := \sup_{x \in \mathcal{N}(\Phi)} \|x_T\|_1 / \|x\|_1,$$

where $x_T$ is a restriction of $x$ onto $T$: $(x_T)_j = x_j$ for $j \in T$ and $(x_T)_j = 0$ otherwise. He proved that if $\Phi$ obeys the following two conditions:

$$\nu(\Phi, T) \le \eta_1, \quad |T| \le \rho_1 n / \log m; \tag{A1}$$

$$w(\Phi, B_1^m) \le \eta_2 ((\log m)/n)^{1/2}, \tag{A2}$$

then, for any $u \in B_1^m$, we have

$$\|u - A_\Phi(\Phi u)\|_2 \le C((\log m)/n)^{1/2}.$$

We now proceed to the contributions of E. Candes, J. Romberg and T. Tao published in a series of papers. These authors (see Candes and Tao (2005)) introduced the *restricted isometry property* (RIP) of a sensing matrix $\Phi$: a number $\delta_S < 1$ is the $S$-restricted isometry constant of $\Phi$ if it is the smallest quantity such that

$$(1 - \delta_S)\|c\|_2^2 \le \|\Phi_T c\|_2^2 \le (1 + \delta_S)\|c\|_2^2$$

for all subsets $T$ with $|T| \le S$ and all coefficient sequences $\{c_j\}_{j \in T}$. A matrix $\Phi$ is said to have then RIP if this constant exists. Candes and Tao (2005) proved that if $\delta_{2S} + \delta_{3S} < 1$ then for $S$-sparse $u$ one has $A_\Phi(\Phi u) = u$ (recovery by $\ell_1$ minimization is exact). They also proved the existence of sensing matrices $\Phi$ obeying the condition $\delta_{2S} + \delta_{3S} < 1$ for large values of sparsity $S \asymp n / \log(m/n)$.

For a positive number $a$ denote

$$\sigma_a(v)_1 := \min_{w \in \mathbb{R}^m : |\operatorname{supp}(w)| \le a} \|v - w\|_1.$$

Candes, Romberg and Tao (2006) proved that, if $\delta_{3S} + 3\delta_{4S} < 2$, then

$$\|u - A_\Phi(\Phi u)\|_2 \le C S^{-1/2} \sigma_S(u)_1. \tag{5.5}$$

We note that properties of the RIP-type matrices have already been imployed in Kashin (1977a) (see Kashin and Temlyakov (2007) for a further discussion) for the widths estimation. The inequality (5.3) with an extra factor $1 + \log(m/n)$ has been established in Kashin (1977a). The proof in Kashin (1977a) is based on properties of a random matrix $\Phi$ with elements $\pm 1/\sqrt{n}$.

Further investigation of the compressed sensing problem has been conducted by Cohen, Dahmen and DeVore (2009). They proved that if $\Phi$ satisfies the RIP of order $2k$ with $\delta_{2k} < \delta < 1/3$ then one has

$$\|u - A_\Phi(\Phi u)\|_1 \le \frac{2 + 2\delta}{1 - 3\delta} \sigma_k(u)_1. \tag{5.6}$$

In Cohen, Dahmen and DeVore (2009) the inequality (5.6) has been called *instance optimality*. In the proof of (5.6) the authors used the following property (null space property) of matrices $\Phi$ satisfying the RIP of order $3k/2$: for any $x \in \mathcal{N}(\Phi)$ and any $T$ with $|T| \leq k$ we have

$$\|x\|_1 \leq C\|x_{T^c}\|_1. \tag{5.7}$$

The null space property (5.7) is closely related to the property $(A1)$ from Donoho (2006). The proof of (5.6) from Cohen, Dahmen and DeVore (2009) gives an inequality similar to (5.6) under the assumption that $\Phi$ has the null space property (5.7) with $C < 2$.

We now discuss the results of Kashin and Temlyakov (2007). We say that a measurement matrix $\Phi$ has a *strong compressed sensing property* with parameters $k$ and $C$ (SCSP$(k, C)$) if, for any $u \in \mathbb{R}^m$, we have

$$\|u - A_\Phi(\Phi u)\|_2 \leq Ck^{-1/2}\sigma_k(u)_1. \tag{5.8}$$

We define a *weak compressed sensing property* with parameters $k$ and $C$ (WCSP$(k, C)$) by replacing (5.8) by the weaker inequality

$$\|u - A_\Phi(\Phi u)\|_2 \leq Ck^{-1/2}\|u\|_1. \tag{5.9}$$

We say that $\Phi$ satisfies the *width property* with parameter $S$ (WP$(S)$) if the following version of (5.4) holds for the null space $\mathcal{N}(\Phi)$:

$$\|x\|_2 \leq S^{-1/2}\|x\|_1.$$

The main result of Kashin and Temlyakov (2007) states that the above three properties of $\Phi$ are equivalent (see Theorem 5.7 below).

We remark on the notations used in this chapter. Above we used $n$ and $m$ for the number of measurements and the size of a signal, respectively. These notations are motivated by their use in the theory of widths, and we continue to use these notations in Sections 5.2–5.4. In Sections 5.5–5.8 we use the letter $N$ for the size of a signal instead of $m$. This notation is more common in compressed sensing.

## 5.2 Equivalence of three approximation properties of the compressed sensing matrix

In this section we present a detailed discussion of results from Kashin and Temlyakov (2007). We mentioned in Section 5.1 that it is known that for any pair $(m, n)$, $n < m$, there exists a subspace $\Gamma \subset \mathbb{R}^m$ with $\dim \Gamma \geq m - n$ such that

$$\|x\|_2 \leq Cn^{-1/2}(\ln(em/n))^{1/2}\|x\|_1, \quad \forall x \in \Gamma.$$

We will study some properties of subspaces $\Gamma$, satisfying the following inequality (5.10), that are useful in compressed sensing:

$$\|x\|_2 \leq S^{-1/2}\|x\|_1, \quad \forall x \in \Gamma. \tag{5.10}$$

For $x = (x_1, \ldots, x_m) \in \mathbb{R}^m$ denote $\operatorname{supp}(x) := \{j : x_j \neq 0\}$.

**Lemma 5.1** *Let* $\Gamma$ *satisfy (5.10) and let* $x \in \Gamma$. *Then either* $x = 0$ *or* $|\operatorname{supp}(x)| \geq S$.

*Proof* Assume $x \neq 0$. Then $\|x\|_1 > 0$. Denote $\Lambda := \operatorname{supp}(x)$. We have

$$\|x\|_1 = \sum_{j \in \Lambda} |x_j| \leq |\Lambda|^{1/2} \left(\sum_{j \in \Lambda} |x_j|^2\right)^{1/2} \leq |\Lambda|^{1/2}\|x\|_2. \tag{5.11}$$

Using (5.10) we get from (5.11) that

$$\|x\|_1 \leq |\Lambda|^{1/2} S^{-1/2}\|x\|_1.$$

Thus

$$|\Lambda| \geq S.$$

$\square$

**Lemma 5.2** *Let* $\Gamma$ *satisfy (5.10) and let* $x \neq 0$, $x \in \Gamma$. *Then for any* $\Lambda$ *such that* $|\Lambda| < S/4$ *we have*

$$\sum_{j \in \Lambda} |x_j| < \|x\|_1/2.$$

*Proof* As in (5.11),

$$\sum_{j \in \Lambda} |x_j| \leq |\Lambda|^{1/2} S^{-1/2}\|x\|_1 < \|x\|_1/2.$$

$\square$

**Lemma 5.3** *Let* $\Gamma$ *satisfy (5.10). Suppose* $u \in \mathbb{R}^m$ *is sparse with* $|\operatorname{supp}(u)| < S/4$. *Then for any* $v = u + x$, $x \in \Gamma$, $x \neq 0$, *we have*

$$\|v\|_1 > \|u\|_1.$$

*Proof* Let $\Lambda := \operatorname{supp}(u)$. Then

$$\|v\|_1 = \sum_{j \in [1,m]} |v_j| = \sum_{j \in \Lambda} |u_j + x_j| + \sum_{j \notin \Lambda} |x_j|$$

$$\geq \sum_{j \in \Lambda} |u_j| - \sum_{j \in \Lambda} |x_j| + \sum_{j \notin \Lambda} |x_j| = \|u\|_1 + \|x\|_1 - 2\sum_{j \in \Lambda} |x_j|.$$

By Lemma 5.2,

$$\|x\|_1 - 2\sum_{j \in \Lambda} |x_j| > 0.$$

$\square$

Lemma 5.3 guarantees that the following algorithm, known as the Basis Pursuit (see $A_\Phi$ from Section 5.1), will find a sparse $u$ exactly, provided $|\operatorname{supp}(u)| < S/4$:

$$u_\Gamma := u + \arg\min_{x \in \Gamma} \|u + x\|_1.$$

**Theorem 5.4** *Let $\Gamma$ satisfy (5.10). Then for any $u \in \mathbb{R}^m$ and $u'$ such that $\|u'\|_1 \le \|u\|_1$, $u - u' \in \Gamma$, one has*

$$\|u - u'\|_1 \le 4\sigma_{S/16}(u)_1, \tag{5.12}$$

$$\|u - u'\|_2 \le (S/16)^{-1/2}\sigma_{S/16}(u)_1. \tag{5.13}$$

*Proof* It is given that $u - u' \in \Gamma$. Thus, (5.13) follows from (5.12) and (5.10). We now prove (5.12). Let $\Lambda$, $|\Lambda| = [S/16]$, be the set of indices with the largest absolute value coordinates of $u$. Denote by $u_\Lambda$ a restriction of $u$ onto this set, i.e $(u_\Lambda)_j = u_j$ for $j \in \Lambda$ and $(u_\Lambda)_j = 0$ for $j \notin \Lambda$. Also denote $u^\Lambda := u - u_\Lambda$. Then

$$\sigma_{S/16}(u)_1 = \sigma_{|\Lambda|}(u)_1 = \|u - u_\Lambda\|_1 = \|u^\Lambda\|_1. \tag{5.14}$$

We have

$$\|u - u'\|_1 = \|(u - u')_\Lambda\|_1 + \|(u - u')^\Lambda\|_1.$$

Next,

$$\|(u - u')^\Lambda\|_1 \le \|u^\Lambda\|_1 + \|(u')^\Lambda\|_1.$$

Using $\|u'\|_1 \le \|u\|_1$, we obtain

$$\|(u')^\Lambda\|_1 - \|u^\Lambda\|_1 = \|u'\|_1 - \|u\|_1 - \|u'_\Lambda\|_1 + \|u_\Lambda\|_1 \le \|(u - u')_\Lambda\|_1.$$

Therefore,

$$\|(u')^\Lambda\|_1 \le \|u^\Lambda\|_1 + \|(u - u')_\Lambda\|_1 \tag{5.15}$$

and

$$\|u - u'\|_1 \le 2\|(u - u')_\Lambda\|_1 + 2\|u^\Lambda\|_1. \tag{5.16}$$

Using the fact $u - u' \in \Gamma$ we estimate

$$\|(u - u')_\Lambda\|_1 \leq |\Lambda|^{1/2} \|(u - u')_\Lambda\|_2$$
$$\leq |\Lambda|^{1/2} \|u - u'\|_2 \leq |\Lambda|^{1/2} S^{-1/2} \|u - u'\|_1. \qquad (5.17)$$

Our assumption on $|\Lambda|$ guarantees that $|\Lambda|^{1/2} S^{-1/2} \leq 1/4$. Using this and substituting (5.17) into (5.16) we obtain

$$\|u - u'\|_1 \leq \|u - u'\|_1/2 + 2\|u^\Lambda\|_1,$$

which gives (5.12):

$$\|u - u'\|_1 \leq 4\|u^\Lambda\|_1.$$

$\square$

**Corollary 5.5** *Let $\Gamma$ satisfy (5.10). Then for any $u \in \mathbb{R}^m$ one has*

$$\|u - u_\Gamma\|_1 \leq 4\sigma_{S/16}(u)_1, \qquad (5.18)$$
$$\|u - u_\Gamma\|_2 \leq (S/16)^{-1/2}\sigma_{S/16}(u)_1. \qquad (5.19)$$

**Proposition 5.6** *Let $\Gamma$ be such that (5.9) holds with $u_\Gamma$ instead of $A_\Phi(\Phi u)$. Then $\Gamma$ satisfies (5.10) with $S = kC^{-2}$.*

*Proof* Let $u \in \Gamma$. Then $u_\Gamma = 0$ and we get, from (5.9),

$$\|u\|_2 \leq Ck^{-1/2}\|u\|_1.$$

$\square$

The following theorem states that the three properties – SCSP, WCSP, WP – are equivalent. In particular, it implies that the WP is a necessary and sufficient condition for either SCSP or WCSP. Thus, if we are looking for matrices $\Phi$ that are good for $\ell_1$ minimization operator $A_\Phi$, say, such that (5.8) holds, then we should study matrices $\Phi$ such that the corresponding null space $\mathcal{N}(\Phi)$ satisfies (5.10) with appropriate $S$. The RIP property is merely a sufficient condition for either the SCSP or WCSP.

**Theorem 5.7** *The following implications hold:*

$$\text{SCSP}(k, C) \Rightarrow \text{WCSP}(k, C);$$
$$\text{WP}(S) \Rightarrow \text{SCSP}(S/16, 1);$$
$$\text{WCSP}(k, C) \Rightarrow \text{WP}(C^{-2}k).$$

*Proof* It is obvious that $\text{SCSP}(k, C) \Rightarrow \text{WCSP}(k, C)$. Corollary 5.5 with $\Gamma = \mathcal{N}(\Phi)$ implies that $\text{WP}(S) \Rightarrow \text{SCSP}(S/16, 1)$. Proposition 5.6 with $\Gamma = \mathcal{N}(\Phi)$ implies that $\text{WCSP} \Rightarrow \text{WP}$. Thus the three properties are equivalent. $\square$

Let us discuss the application of this theorem to the setting discussed in Section 5.1. We study the $\ell_1$ minimization method $A_\Phi$. We impose the assumption (A2) of approximate sparsity of a signal. As a task let us take the following version of (T2):

$$\|u - A_\Phi(\Phi u)\|_2 \leq 4Ck^{-1/2}\sigma_k(u)_1. \tag{5.20}$$

This means that we measure the error of approximate recovery in the $\ell_2$ norm of the coefficient space. Then Theorem 5.7 states that the necessary condition on $\Phi$ to guarantee successful performance of the above task is that the null space $\mathcal{N}(\Phi)$ satisfies the following property: for any $x \in \mathcal{N}(\Phi)$ we have

$$\|x\|_2 \leq Ck^{-1/2}\|x\|_1. \tag{5.21}$$

On the other hand, if the null space $\mathcal{N}(\Phi)$ satisfies (5.21) then by Theorem 5.7 we get

$$\|u - A_\Phi(\Phi u)\|_2 \leq Ck^{-1/2}\sigma_{k/(16C^2)}(u)_1. \tag{5.22}$$

Thus the property (5.21) of the null space $\mathcal{N}(\Phi)$ is a necessary and sufficient condition (with different constants $C$) for fulfilling (T2) under assumptions (A2). We note that the necessary and sufficient condition on $k$ for the existence of matrices $\Phi$ satisfying (5.21) is the inequality $k \ll n/\log(m/n)$.

The result (5.5) of Candes, Romberg and Tao (2006) states that the RIP with $S \asymp n/\log(m/n)$ implies the SCSP. Therefore, by Theorem 5.7 it implies the WP with $S \asymp n/\log(m/n)$. One can find a direct proof of the above statement in Kashin and Temlyakov (2007).

## 5.3 Construction of a good matrix

The results of the preceding section justify the importance of finding subspaces $\Gamma$, $\dim \Gamma \geq m - n$, satisfying (5.10) with large $S$. In this section we prove the existence of such $\Gamma$ with $S \asymp n/\ln(em/n)$. A good matrix $\Phi$ is the one with $\mathcal{N}(\Phi) = \Gamma$. We present the proof suggested by Yu. Makovoz. This proof is based on the averaging argument, and it does not give an explicit construction of $\Gamma$. It is a very important open problem to give an explicit deterministic construction of $\Gamma$, $\dim \Gamma \geq m - n$, satisfying (5.10) with $S \asymp n/\ln(em/n)$.

### 5.3.1 Construction of a good subspace

**Theorem 5.8** *For any natural $m, n$, $n < m$, there exists a subspace $\Gamma \subset \mathbb{R}^m$, $\dim \Gamma \geq m - n$, such that, for $x \in \Gamma$,*

$$\|x\|_2 \leq Cn^{-1/2}\big(\ln(em/n)\big)^{1/2}\|x\|_1.$$

*Proof* Let

$$S := S^{m-1} := \{x \in \mathbb{R}^m : \|x\|_2 = 1\}$$

be the unit sphere in $\mathbb{R}^m$ and let $\mu$ be the normalized Lebesgue measure on $S$, i.e. $\mu(S) = 1$. Denote by $P$ the measure on the product set $Y := (S^{m-1})^n$ of $n$ unit spheres corresponding to the product of measures $\mu$. We denote

$$\bar{v} = (v_1, \ldots, v_n) \in Y, \qquad v_j \in S, \qquad j = 1, \ldots, n.$$

We first prove the following auxiliary statement.

**Lemma 5.9** *Let, for $x \in S$, $\bar{v} \in Y$,*

$$F(x, \bar{v}) := n^{-1} \sum_{j=1}^n |\langle x, v_j \rangle|.$$

*Then, for any $x \in S$,*

$$P\{\bar{v} \in Y : (0.01)m^{-1/2} \le F(x, \bar{v}) \le 3m^{-1/2}\} > 1 - e^{-n/2}.$$

*Proof* Let

$$E(n, t) = \int_Y e^{t F(x, \bar{v})} \, dP.$$

We make some simple remarks. The quantity $E(n, t)$ does not depend on $x$. The following inequality holds: for $t > 0$, $b > 0$,

$$P\{\bar{v} \in Y : F(x, \bar{v}) > b\} \le E(n, t)e^{-bt} \tag{5.23}$$

and, for $t < 0$, $a > 0$,

$$P\{\bar{v} \in Y : F(x, \bar{v}) < a\} \le E(n, t)e^{-at}. \tag{5.24}$$

Since

$$E(n, t) = E(1, t/n)^n, \tag{5.25}$$

it suffices to consider the case $n = 1$. We set $E(1, t) =: E(t)$. Taking $x = (1, 0, \ldots, 0)$, we have

$$E(t) = \int_S e^{t|y_1|} \, d\mu. \tag{5.26}$$

Further, let $m \ge 3$. Considering that the $(m - 2)$-dimensional volume of the sphere $S^{m-2}(r)$ of radius $r$ is proportional to $r^{m-2}$, we obtain

$$\mu\{v \in S : \alpha \le y_1 \le \beta\} = \int_\alpha^\beta |S^{m-2}((1 - r^2)^{1/2})|(1 - r^2)^{-1/2} \, dr$$

$$= C(m) \int_\alpha^\beta (1 - r^2)^{(m-3)/2} \, dr.$$

From the normalization condition of the measure $\mu$ we get

$$C(m) \int_0^1 (1 - r^2)^{(m-3)/2} \, dr = 1/2. \tag{5.27}$$

Thus,

$$E(t) = 2C(m) \int_0^1 e^{rt} (1 - r^2)^{(m-3)/2} \, dr. \tag{5.28}$$

To get the upper estimate for $E(t)$ we shall prove two estimates. First we prove the lower one:

$$\int_0^1 (1 - r^2)^{(m-3)/2} \, dr > \int_0^{m^{-1/2}} (1 - r^2)^{(m-3)/2} \, dr$$

$$> (1 - 1/m)^{(m-1)/2} m^{-1/2} > (em)^{-1/2}. \tag{5.29}$$

Then, using the inequality $1 - z \le e^{-z}$, $z \ge 0$, we prove the upper estimate:

$$\int_0^1 e^{tr} (1 - r^2)^{(m-3)/2} \, dr \le \int_0^1 e^{tr - mr^2/2 + 3r^2/2} \, dr$$

$$\le e^{3/2} \int_0^1 e^{tr - mr^2/2} \, dr$$

$$\le e^{3/2 + t^2/(2m)} m^{-1/2} \int_{-tm^{-1/2}}^{\infty} e^{-v^2/2} \, dv, \tag{5.30}$$

where $v = m^{1/2}(r - t/m)$.

From the relations (5.27)–(5.30), using (5.25), we get

$$E(n, t) < e^{2n + t^2/(2mn)} \left( \int_{-t/(nm^{1/2})}^{\infty} e^{-v^2/2} \, dv \right)^n. \tag{5.31}$$

From (5.31) and the inequality (5.23) with $b = 3m^{-1/2}$, $t = 3m^{1/2}n$, we get

$$P\{\bar{v} \in Y : F(x, \bar{v}) > 3m^{-1/2}\} < \left( e^{-5/2} (2\pi)^{1/2} \right)^n. \tag{5.32}$$

From (5.31) and (5.24) with $a = (0.01)m^{-1/2}$, $t = -100m^{1/2}n$, taking into account the inequality

$$\int_z^{\infty} e^{-v^2/2} \, dv < e^{-z^2/2}/z, \qquad z > 0,$$

we obtain

$$P\{\bar{v} \in Y : F(x, \bar{v}) < 0.01m^{-1/2}\} < (0.01e^3)^n. \tag{5.33}$$

The conclusion of Lemma 5.9 follows from (5.32) and (5.33). $\qquad\square$

We continue the proof of Theorem 5.8. Clearly, we can assume that $n$ is a sufficiently large number. For natural numbers $1 \le l \le m$ let $B^{m,l}$ be the set of all vectors from $B_1^m$ with coordinates of the form $k/l$, $k \in \mathbb{Z}$. The cardinality of $B^{m,l}$ does not exceed $2^l$ times the number of non-negative integer solutions of the inequality

$$l_1 + \cdots + l_m \le l.$$

Consequently ($\binom{m}{k}$ are the binomial coefficients),

$$|B^{m,l}| \le 2^l \sum_{j=0}^{l} \binom{m-1+j}{m-1} = 2^l \binom{m+l}{l}. \tag{5.34}$$

We set

$$l = \left[ An / \ln(em/n) \right], \tag{5.35}$$

where $A$ is a sufficiently small number not depending on $n, m$, such that there exists $\bar{v}^* \in Y$ with the property that, for all $x \in B^{m,l}$, we have

$$0.01 m^{-1/2} \|x\|_2 \le F(x, \bar{v}^*) \le 3m^{-1/2} \|x\|_2. \tag{5.36}$$

Indeed, due to Lemma 5.9 the $P$-measure of those $\bar{v}$ for which (5.36) does not hold is not greater than

$$|B^{m,l}| e^{-n/2} \le 2^l (2em/l)^l e^{-n/2}. \tag{5.37}$$

The number $A$ is chosen so that the right-hand side of (5.37) is less than 1. Let

$$\Gamma = \left\{ x \in \mathbb{R}^m : F(x, \bar{v}^*) = 0 \right\}.$$

It is clear that $\dim \Gamma \ge m - n$. We prove that, for $x \in \Gamma \cap B_1^m$,

$$\|x\|_2 \le 301 l^{-1/2}. \tag{5.38}$$

Let $x \in B_1^m$ and $x' \in B^{m,l}$ be such that $x_j$ and $x'_j$ have the same sign and $|x'_j| \le |x_j|$, $|x_j - x'_j| \le 1/l$, $j = 1, \ldots, m$. We consider $x'' := x - x'$. Then $x'' \in B_1^m \cap (1/l) B_\infty^m = \Pi$, and, consequently,

$$\|x''\|_2 \le \|x''\|_1^{1/2} \|x''\|_\infty^{1/2} \le l^{-1/2}. \tag{5.39}$$

Let us estimate $F(x'', \bar{v}^*)$. To do this we prove that $\Pi = \operatorname{conv}(V)$, where $V$ is the set of all vectors having exactly $l$ coordinates different from zero and equal to $\pm 1/l$. The set $\Pi$ as an intersection of two convex polytopes is a convex polytope. Clearly, the set $V$ belongs to the set of extreme points of $\Pi$. We prove that $\Pi$ has no other extreme points. Indeed, let $z \in \Pi \backslash V$ be a boundary point of $\Pi$ such that $\|z\|_1 = 1$, $\|z\|_\infty = 1/l$. Since $z \notin V$, there are $1 \leq j_1 < j_2 \leq m$ such that $0 < |z_{j_i}| < 1/l$, $i = 1, 2$. Then there is a $\delta > 0$ such that the vectors

$$z^1 = z + (0, \ldots, 0, \delta \operatorname{sing} z_{j_1}, 0, \ldots, 0, -\delta \operatorname{sing} z_{j_2}, 0, \ldots, 0),$$
$$z^2 = z + (0, \ldots, 0, -\delta \operatorname{sing} z_{j_1}, 0, \ldots, 0, \delta \operatorname{sing} z_{j_2}, 0, \ldots, 0)$$

belong to $\Pi$ and, obviously, $z = (z^1 + z^2)/2$, which shows that $z$ is not an extreme point of $\Pi$. Thus it is proved that the set of extreme points of $\Pi$ coincides with $V$. Consequently, $\Pi = \operatorname{conv}(V)$ and, for $x'' \in \Pi$,

$$F(x'', \overline{v}^*) \leq \max_{z \in V} F(z, \overline{v}^*)$$
$$\leq \max_{z \in B^{m,l}} \|z\|_2^{-1} F(z, \overline{v}^*) \max_{z \in V} \|z\|_2 \leq 3(ml)^{-1/2}. \tag{5.40}$$

If we now suppose that (5.38) does not hold, we get

$$\|x'\|_2 \geq \|x\|_2 - \|x''\|_2 > 300l^{-1/2},$$

which implies, by (5.36) for $x' \in B^{m,l}$,

$$F(x', \overline{v}^*) > 3(ml)^{-1/2}. \tag{5.41}$$

From (5.40) and (5.41) we find

$$F(x, \overline{v}^*) \geq F(x', \overline{v}^*) - F(x'', \overline{v}^*) > 0,$$

which contradicts the condition $x \in \Gamma$.

Relation (5.38) and Theorem 5.8 are proved. □

## 5.3.2 Duality principle

We now discuss the duality principle mentioned in Section 5.1. We will prove Nikol'skii's duality theorem. Let $X$ be a linear normed space (real or complex) and let $X^*$ be the conjugate (dual) space to $X$; that is, elements of $X^*$ are linear functionals $\varphi$ defined on $X$ with the norm

$$\|\varphi\| = \sup_{f \in X, \|f\| \leq 1} |\varphi(f)|.$$

Let $\Phi = \{\varphi_k\}_{k=1}^n$ be a set of functionals from $X^*$. Denote

$$X_\Phi := \{f \in X : \varphi_k(f) = 0, \quad k = 1, \ldots, n\}.$$

**Theorem 5.10 (Nikol'skii duality theorem)** *Let* $\Phi = \{\varphi_k\}_{k=1}^n$ *be a fixed system of functionals from* $X^*$. *Then, for any* $\varphi \in X^*$,

$$\inf_{\{c_k\}_{k=1}^n} \left\| \varphi - \sum_{k=1}^n c_k \varphi_k \right\| = \sup_{f \in X_\Phi, \|f\| \leq 1} |\varphi(f)|. \tag{5.42}$$

*Proof* Let us denote the left-hand side of (5.42) by $a$ and the right-hand side of (5.42) by $b$. From the relation

$$|\varphi(f)| = \left| \left( \varphi - \sum_{k=1}^n c_k \varphi_k \right)(f) \right| \leq \left\| \varphi - \sum_{k=1}^n c_k \varphi_k \right\|,$$

which is valid for any $f \in X_\Phi$, $\|f\| \leq 1$, it follows that $b \leq a$. We prove the inverse inequality. Clearly, we can assume that the system of functionals $\varphi_1, \ldots, \varphi_n$ is linearly independent.

**Lemma 5.11** *Let* $\varphi_1, \ldots, \varphi_n \in X^*$ *be linearly independent. There exists a set of elements* $f_1, \ldots, f_n \in X$ *which is biorthogonal to* $\varphi_1, \ldots, \varphi_n$, *i.e.* $\varphi_i(f_j) = 0$ *for* $1 \leq i \neq j \leq n$ *and* $\varphi_i(f_i) = 1$, $i = 1, \ldots, n$.

*Proof* The proof will be carried out by induction. The case $n = 1$ is evident. Let us assume that a biorthogonal system can be constructed if the number of functionals is less than $n$. Clearly, it suffices to prove the existence of $f_1 \in X$ such that

$$\varphi_1(f_1) = 1, \qquad \varphi_k(f_1) = 0, \qquad k = 2, \ldots, n.$$

Let $\Phi_1 = \{\varphi_k\}_{k=2}^n$ and $\{g_k\}_{k=2}^n$ be a biorthogonal system to $\Phi_1$. It is sufficient to prove the existence of $f_1 \in X_{\Phi_1}$ such that $\varphi_1(f_1) \neq 0$. Let us assume the contrary; that is, for any $f \in X_{\Phi_1}$ we have $\varphi_1(f) = 0$. We shall show that this contradicts the linear independence of the functionals $\varphi_1, \ldots, \varphi_n$. Let $f \in X$; then

$$f - \sum_{k=2}^n \varphi_k(f) g_k \in X_{\Phi_1}$$

and

$$\varphi_1 \left( f - \sum_{k=2}^n \varphi_k(f) g_k \right) = 0,$$

which implies

$$\varphi_1(f) = \sum_{k=2}^n \varphi_1(g_k) \varphi_k(f).$$

Consequently,

$$\varphi_1 = \sum_{k=2}^{n} \varphi_1(g_k)\varphi_k,$$

which is in contradiction with the linear independence of $\varphi_1, \dots, \varphi_n$.

The lemma is proved. $\square$

We continue the proof of the theorem. Let $\varphi \in X^*$. Along with $\varphi$ we consider a contraction $\varphi_\Phi$ of $\varphi$ to the subspace $X_\Phi$, i.e. a linear bounded functional $\varphi_\Phi$ defined on $X_\Phi$ such that $\varphi_\Phi(f) = \varphi(f)$ for all $f \in X_\Phi$. Any functional

$$\psi = \varphi - \sum_{k=1}^{n} c_k \varphi_k \qquad (5.43)$$

is an extension of $\varphi_\Phi$ to $X$. We prove that each extension of a functional $\varphi_\Phi$ from $X_\Phi$ to $X$ has the form (5.43). We use Lemma 5.11. Let the system $f_1, \dots, f_n$ be biorthogonal to $\Phi$; then, for any $f \in X$,

$$f - \sum_{k=1}^{n} \varphi_k(f) f_k \in X_\Phi.$$

Consequently, for any extension $\psi$ of a functional $\varphi_\Phi$ we have

$$\psi\left(f - \sum_{k=1}^{n} \varphi_k(f) f_k\right) = \varphi\left(f - \sum_{k=1}^{n} \varphi_k(f) f_k\right),$$

which implies

$$\psi(f) = \varphi(f) + \sum_{k=1}^{n} \left(\psi(f_k) - \varphi(f_k)\right)\varphi_k(f).$$

Thus, the representation (5.43) is valid for $\psi$.

Let $\psi$ be an extension of a functional $\varphi_\Phi$ such that $\|\psi\| = \|\varphi_\Phi\|$. The existence of such an extension follows from the Hahn–Banach theorem. Then

$$\|\psi\| = \left\|\varphi - \sum_{k=1}^{n} c_k \varphi_k\right\| = \|\varphi_\Phi\| = \sup_{f \in X_\Phi, \|f\| \le 1} |\varphi(f)|,$$

i.e. $a \le b$, which concludes the proof of the theorem. $\square$

**Corollary 5.12** *Let* $\varphi, \varphi_1, \dots, \varphi_n \in \ell_p^m$, $1 \le p \le \infty$; *then*

$$\inf_{c_k, k=1,\dots,n} \left\|\varphi - \sum_{k=1}^{n} c_k \varphi_k\right\|_p = \sup_{\|g\|_{p'} \le 1, \langle \varphi_k, g \rangle = 0, \ k=1,\dots,n} |\langle \varphi, g \rangle|.$$

*Proof* This theorem follows from Theorem 5.10. Indeed, let us consider an element $\varphi \in \ell_p^m$ as a functional $\varphi$ acting on $\ell_{p'}^m$ by the formula $\varphi(f) = \langle f, \varphi \rangle$. Then we have $\|\varphi\| = \|\varphi\|_p$. It remains to apply Theorem 5.10. □

We now prove (5.1) from Section 5.1.

**Theorem 5.13** *Let $B_p^m$ be a unit $\ell_p$-ball in $\mathbb{R}^m$ and $1 \leq q, p \leq \infty$. Then one has*

$$d_n(B_p^m, \ell_q^m) = d^n(B_{q'}^m, \ell_{p'}^m), \quad p' := p/(p-1).$$

*Proof* By the definition of the Kolmogorov width, we have

$$d_n(B_p^m, \ell_q^m) := \inf_{L_n: \dim L_n \leq n} \sup_{f \in B_p^m} \inf_{a \in L_n} \|f - a\|_q$$

$$= \inf_{\{\varphi_1, \ldots, \varphi_n\} \subset \mathbb{R}^m} \sup_{f \in B_p^m} \inf_{\{c_k\}} \left\| f - \sum_{k=1}^n c_k \varphi_k \right\|_q.$$

Applying Corollary 5.12 we continue:

$$= \inf_{\{\varphi_1, \ldots, \varphi_n\}} \sup_{f \in B_p^m} \sup_{g: \langle \varphi_k, g \rangle = 0, j = 1, \ldots, n, \|g\|_{q'} \leq 1} |\langle f, g \rangle|.$$

Changing the order of the two supremums in the above formula we obtain

$$d_n(B_p^m, \ell_q^m) = \inf_{\{\varphi_1, \ldots, \varphi_n\}} \sup_{g: \langle \varphi_k, g \rangle = 0, j = 1, \ldots, n, \|g\|_{q'} \leq 1} \|g\|_{p'}.$$

It follows from the definition of the Gel'fand width that

$$\inf_{\{\varphi_1, \ldots, \varphi_n\}} \sup_{g: \langle \varphi_k, g \rangle = 0, j = 1, \ldots, n, \|g\|_{q'} \leq 1} \|g\|_{p'} = d^n(B_{q'}^m, \ell_{p'}^m).$$

□

## 5.4 Dealing with noisy data

In Sections 5.1 and 5.2 we discussed the recovery of a signal $x \in \mathbb{R}^m$ from the exact data $y = \Phi x \in \mathbb{R}^n$. We will continue this study in Sections 5.5–5.7. In this section we show that the $\ell_1$ minimization algorithm is robust with respect to an additive noise. We assume in this section that the data vector $\Phi x \in \mathbb{R}^n$ is known with an error given by

$$y = \Phi x + z, \quad \|z\|_2 \leq \epsilon.$$

In such a situation we employ the following modification of the $\ell_1$ minimization algorithm $A_\Phi$:

$$\min \|v\|_1 \quad \text{subject to} \quad \|y - \Phi v\|_2 \le \epsilon. \tag{$P_\epsilon$}$$

We denote a solution to the above problem by $A_{\Phi,\epsilon}(y)$. As in Section 5.2 we consider a more general setting. We assume that there exists a subspace $\Gamma \subset \mathbb{R}^m$ and a semi-norm $\| \cdot \|$ on $\mathbb{R}^m$, generated by a semi-inner product $\langle \cdot, \cdot \rangle$, that have the following properties:

(1) for any $u \in \Gamma$ we have $\|u\| = 0$;
(2) for any $s \le S$ and any $s$-sparse vector $u$ one has

$$(1 - \delta_s)\|u\|_2^2 \le \|u\|^2 \le (1 + \delta_s)\|u\|_2^2, \quad \delta_s \in (0, 1). \tag{5.44}$$

As an example of the above semi-inner product we can take $\langle u, v \rangle :=\langle \Phi u, \Phi v \rangle$. Then $\Gamma = \mathcal{N}(\Phi)$, and (5.44) is equivalent to the RIP of $\Phi$.

We begin with a lemma that is a simple corollary of (5.44).

**Lemma 5.14** *Assume that $u$ and $v$ are vectors from $\mathbb{R}^m$ that are, respectively, $\mu$-sparse and $v$-sparse with $\mu + v \le S$. If $u$ and $v$ have disjoint supports then*

$$|\langle u, v \rangle| \le \delta_{\mu+v}\|u\|_2\|v\|_2.$$

*Proof* We carry out the proof for normalized vectors $\|u\|_2 = \|v\|_2 = 1$. By (5.44) applied to $u + v$ and $u - v$ we obtain

$$2(1 - \delta_{\mu+v}) \le \|u \pm v\|^2 \le 2(1 + \delta_{\mu+v}).$$

The parallelogram identity and the above inequalities yield

$$|\langle u, v \rangle| = \frac{1}{4}|\|u + v\|^2 - \|u - v\|^2| \le \delta_{\mu+v}.$$

$\square$

The presentation of the following theorem follows ideas from Candes (2008). We use the notations from Section 5.2.

**Theorem 5.15** *Assume $\delta_{2s} \le \sqrt{2} - 1$. Then, for any $u, u' \in \mathbb{R}^m$, such that $\|u'\|_1 \le \|u\|_1$ and $\|u' - u\| \le \epsilon$, we have*

$$\|u' - u\|_2 \le C_0 s^{-1/2}\sigma_s(u)_1 + C_1\epsilon.$$

*Proof* Set $h := u' - u$. Let $T_0$ be the set of $s$ largest coefficients of $u$. Let $h^{T_0} = (h_1, \ldots, h_m)$. Reorder the coordinates in a decreasing way:

$$|h_{i_1}| \ge |h_{i_2}| \ge \cdots \ge |h_{i_m}|.$$

For $j = 1, 2, \ldots$ define

$$T_j := \{i_{s(j-1)+1}, \ldots, i_{sj}\} \cap [1, m].$$

Then for cardinality of these sets we have $|T_j| = s$ provided $sj \leq m$. Clearly

$$h = \sum_{j \geq 0} h_{T_j}.$$

We have the following simple inequality:

$$\|h_{T_j}\|_2 \leq s^{-1/2} \|h_{T_{j-1}}\|_1, \quad j = 2, \ldots \tag{5.45}$$

Denote $T := T_0 \cup T_1$. Then the above inequality implies

$$\|h^T\|_2 \leq \sum_{j \geq 2} \|h_{T_j}\|_2 \leq \sum_{j \geq 2} s^{-1/2} \|h_{T_{j-1}}\|_1 \leq s^{-1/2} \|h^{T_0}\|_1. \tag{5.46}$$

We will use the above inequality (5.46) in estimating

$$\|h\|_2 \leq \|h_T\|_2 + \|h^T\|_2. \tag{5.47}$$

We now use the assumption $\|u'\|_1 \leq \|u\|_1$ to estimate $\|h^{T_0}\|_1$. We begin with a simple lemma.

**Lemma 5.16** *Let* $\|u'\|_1 \leq \|u\|_1$. *Then for any set of indexes* $\Lambda$ *we have*

$$\|(u')^\Lambda\|_1 \leq \|u^\Lambda\|_1 + \|(u - u')_\Lambda\|_1.$$

*Proof* Using $\|u'\|_1 \leq \|u\|_1$, we obtain

$$\|(u')^\Lambda\|_1 - \|u^\Lambda\|_1 = \|u'\|_1 - \|u\|_1 - \|u'_\Lambda\|_1 + \|u_\Lambda\|_1 \leq \|(u - u')_\Lambda\|_1.$$

Therefore,

$$\|(u')^\Lambda\|_1 \leq \|u^\Lambda\|_1 + \|(u - u')_\Lambda\|_1.$$

$\square$

We apply Lemma 5.16 with $\Lambda = T_0$ and get

$$\|h^{T_0}\|_1 \leq \|u^{T_0}\|_1 + \|(u')^{T_0}\|_1 \leq \|h_{T_0}\|_1 + 2\|u^{T_0}\|_1. \tag{5.48}$$

Bounds (5.46) and (5.48) imply

$$\|h^T\|_2 \leq s^{-1/2}(\|h_{T_0}\|_1 + 2\|u^{T_0}\|_1) \leq \|h_{T_0}\|_2 + 2s^{-1/2}\sigma_s(u)_1. \tag{5.49}$$

We now proceed to $\|h_T\|_2$. We have

$$h_T = h - \sum_{j \geq 2} h_{T_j}$$

and

$$\|h_T\|^2 = \langle h_T, h_T \rangle = \langle h_T, h - \sum_{j \geq 2} h_{T_j} \rangle = \langle h_T, h \rangle - \sum_{j \geq 2} \langle h_T, h_{T_j} \rangle.$$

By our assumption, $\|h\| \le \epsilon$. Therefore,

$$|\langle h_T, h \rangle| \le \|h_T\| \|h\| \le \epsilon \|h_T\|. \tag{5.50}$$

By (5.44) with $2s$ we get

$$\|h_T\| \le (1 + \delta_{2s})^{1/2} \|h_T\|_2. \tag{5.51}$$

Relations (5.50) and (5.51) imply

$$|\langle h_T, h \rangle| \le \epsilon(1 + \delta_{2s})^{1/2} \|h_T\|_2. \tag{5.52}$$

For $j \ge 2$ we have

$$|\langle h_T, h_{T_j} \rangle| \le |\langle h_{T_0}, h_{T_j} \rangle| + |\langle h_{T_1}, h_{T_j} \rangle|$$

and by Lemma 5.14 we continue:

$$\le \delta_{2s} \|h_{T_0}\|_2 \|h_{T_j}\|_2 + \delta_{2s} \|h_{T_1}\|_2 \|h_{T_j}\|_2$$
$$\le \sqrt{2}\delta_{2s} \|h_T\|_2 \|h_{T_j}\|_2.$$

By (5.44) we obtain

$$(1 - \delta_{2s})\|h_T\|_2^2 \le \|h_T\|^2 \le \|h_T\|_2 \left( \epsilon(1 + \delta_{2s})^{1/2} + \sqrt{2}\delta_{2s} \sum_{j\ge 2} \|h_{T_j}\|_2 \right).$$

This implies

$$\|h_T\|_2 \le \frac{\epsilon(1 + \delta_{2s})^{1/2}}{(1 - \delta_{2s})} + \frac{\sqrt{2}\delta_{2s}}{(1 - \delta_{2s})} \sum_{j\ge 2} \|h_{T_j}\|_2. \tag{5.53}$$

By (5.46) and (5.48) we get

$$\sum_{j\ge 2} \|h_{T_j}\|_2 \le s^{-1/2}\|h^{T_0}\|_1 \le s^{-1/2}(\|h_{T_0}\|_1 + 2\sigma_s(u)_1)$$

$$\le \|h_{T_0}\|_2 + 2s^{-1/2}\sigma_s(u)_1 \le \|h_T\|_2 + 2s^{-1/2}\sigma_s(u)_1. \tag{5.54}$$

Our assumption on $\delta_{2s}$ guarantees that

$$\frac{\sqrt{2}\delta_{2s}}{(1 - \delta_{2s})} < 1.$$

Substituting (5.54) into (5.53) and solving it for $\|h_T\|_2$ we get

$$\|h_T\|_2 \le C_2 s^{-1/2}\sigma_s(u)_1 + C_3\epsilon.$$

This bound and inequalities (5.47) and (5.49) complete the proof. $\qquad\square$

## 5.5 First results on exact recovery of sparse signals; the Orthogonal Greedy Algorithm

Let $\Phi$ be a matrix with columns $\varphi_i \in \mathbb{R}^n$, $i = 1, \ldots, N$. We will also denote by $\Phi$ the dictionary consisting of $\varphi_i$, $i = 1, \ldots, N$. Then the following two statements: (1) $u$ is $k$-sparse and (2) $\Phi u$ is $k$-sparse with respect to the dictionary $\Phi$ are equivalent. Results from Section 5.2 (see Theorem 5.7) imply that if the matrix $\Phi$ has the width property with parameter $S$ then the $\ell_1$ minimization algorithm $A_\Phi$ recovers exactly $S/16$-sparse signals. We begin this section with a discussion of the following question: Which property of $\Phi$ is necessary and sufficient for the existence of a recovery method that recovers exactly all $k$-sparse signals?

Let $\Sigma_k$ and $\Sigma_k(\Phi)$ denote all $k$-sparse vectors $u \in \mathbb{R}^N$ and all $k$-sparse with respect to the dictionary $\Phi$ vectors $y \in \mathbb{R}^n$, respectively. The following simple known theorem answers the above question.

**Theorem 5.17** *The following two conditions are equivalent:*

(i) *there is a recovery method $A$ such that $A(\Phi u) = u$ for any $u \in \Sigma_k$;*
(ii) $\Sigma_{2k} \cap \mathcal{N}(\Phi) = \{0\}$.

*Proof* We begin with the implication $(i) \Rightarrow (ii)$. The proof is by contradiction. Assume that there is a non-zero vector $v \in \Sigma_{2k} \cap \mathcal{N}(\Phi)$. Clearly, $v$ can be written in the form

$$v = v^1 - v^2, \quad v^1, v^2 \in \Sigma_k.$$

By our assumption $v \in \mathcal{N}(\Phi)$, we obtain $\Phi v^1 = \Phi v^2$. Therefore, by $(i)$ we get $v^1 = v^2$ and $v = 0$. The obtained contradiction proves that $(i) \Rightarrow (ii)$.

We now prove that $(ii) \Rightarrow (i)$. Define a recovery method in the following way. For $y \in \mathbb{R}^n$, $A(y)$ is a solution of problem $(P_0)$ from Section 5.1:

$$A(y) := \arg \min_{c:\Phi c=y} \|c\|_0. \tag{5.55}$$

Consider $v := A(\Phi u)$, $u \in \Sigma_k$. It follows from the definition of $A(y)$ that $v \in \Sigma_k$ and that $\Phi v = \Phi u$. Therefore, $v - u \in \Sigma_{2k}$ and $\Phi(v - u) = 0$. Assumption $(ii)$ implies that $v - u = 0$. $\square$

**Remark 5.18** Property $(ii)$ is equivalent to the following property of the dictionary $\Phi$:

(iii) any $2k$ distinct elements of the dictionary $\Phi$ are linearly independent.

As an example of a dictionary (matrix) $\Phi$ satisfying $(iii)$ we can take the following dictionary:

$$\varphi_j = (1, t_j, \dots, t_j^{2k-1})^T, \quad j = 1, \dots, N, \quad 0 < t_1 < \cdots < t_N.$$

Then by the property of the Vandermonde determinant this dictionary satisfies $(iii)$.

Theorem 5.17 and the above discussion mean that in order to be able to recover exactly all $k$-sparse signals it is sufficient to make $2k$ measurements with respect to a matrix satisfying $(iii)$. However, the method of recovery (5.55) is unfeasible numerically. Thus, in principle, we can recover $k$-sparse signals with $2k$ measurements and $2k$ is the best we can achieve. As we mentioned above, the $\ell_1$ minimization algorithm allows us to recover $k$-sparse signals with the number of measurements $n$ of order $k \log(N/k)$. This number of measurements is achieved on matrices $\Phi$ with $\mathcal{N}(\Phi)$ satisfying the WP($16k$). It is known (see Garnaev and Gluskin (1984) and Kashin (1977a)) and Section 5.3) that such matrices $\Phi$ exist when $n \asymp k \log(N/k)$. Their existence was proved by probabilistic methods. There are no deterministic constructions of such matrices. In this section we discuss the performance of the Weak Orthogonal Greedy Algorithm with parameter $t$ (WOGA($t$)) with respect to $M$-coherent dictionaries discussed in Section 2.6. We recall that, as in Section 2.6, we define the coherence parameter of a dictionary $\mathcal{D}$ by

$$M(\mathcal{D}) := \sup_{g \neq h; g,h \in \mathcal{D}} |\langle g, h \rangle|.$$

We proceed to the exact recovery of sparse signals by the WOGA($t$). We present our results in a general setting where we do not assume that $H = \mathbb{R}^n$ and the dictionary $\mathcal{D}$ is finite.

**Theorem 5.19** *Let $\mathcal{D}$ be an $M$-coherent dictionary. The WOGA($t$) recovers exactly any $f \in \Sigma_m(\mathcal{D})$ with $m < (t/(1+t))(1 + M^{-1})$.*

*Proof* Let

$$f = \sum_{i=1}^{m} c_i \varphi_i, \quad \varphi_i \in \mathcal{D}, \quad i = 1, \dots, m.$$

We prove that the WOGA($t$) picks one of $\varphi_i$ at each iteration $j = 1, \dots, m$. In particular, the argument that follows below proves Proposition 5.20.

**Proposition 5.20** *Let $\mathcal{D}$ be an $M$-coherent dictionary. Then for any $f \in \Sigma_m(\mathcal{D})$ with $m < (1/2)(1 + M^{-1})$ there exists $\varphi \in \mathcal{D}$ such that*

$$|\langle f, \varphi \rangle| = \sup_{g \in \mathcal{D}} |\langle f, g \rangle|.$$

Define $A := \max_{1 \leq i \leq m} |c_i|$. Then

$$\max_{1 \leq i \leq m} |\langle f, \varphi_i \rangle| \geq A - AM(m - 1).$$

For any $g \in \mathcal{D}$ distinct from $\varphi_i$, $i = 1, \ldots, m$, we have

$$|\langle f, g \rangle| \leq AMm.$$

The inequality

$$A(1 - M(m - 1))t > AMm \tag{5.56}$$

is equivalent to the inequality $m < (t/(1 + t))(1 + M^{-1})$ that we assumed to hold. Inequality (5.56) guarantees that at the first iteration the WOGA($t$) picks one of the $\varphi_i$, $i = 1, \ldots, m$, say $\varphi_{i_1}$. Then

$$f_1 := f_1^{o,t} = \sum_{i=1}^m c_i^1 \varphi_i, \quad \langle f_1, \varphi_{i_1} \rangle = 0.$$

Arguing as above, we prove that, at the second iteration, the WOGA($t$) picks one of $\varphi_i$ again, say $\varphi_{i_2}$. It is clear that $i_2 \neq i_1$. We continue this process till the $m$th iteration when all the $\varphi_i$, $i = 1, \ldots, m$, will be picked up. Taking the orthogonal projection, we obtain that $f_m = 0$. This completes the proof. $\square$

**Theorem 5.21** *Let $\mathcal{D}$ be an $M$-coherent dictionary. If, for $m < (1/2)$ $(1 + M^{-1})$, $\sigma_m(f) = 0$, then $f \in \Sigma_m(\mathcal{D})$.*

*Proof* The proof is by induction. First, consider the case $m = 1$. Let $\sigma_1(f) = 0$. If $f = 0$ then $f \in \Sigma_1(\mathcal{D})$. Suppose that $f \neq 0$. Take $\epsilon > 0$ and find $\varphi_1 \in \mathcal{D}$ and a coefficient $b_1$ such that

$$\|f - b_1 \varphi_1\| \leq \epsilon.$$

We prove that for sufficiently small $\epsilon$ the inequality

$$\|f - b_2 \varphi_2\| \leq \epsilon, \quad \varphi_2 \in \mathcal{D},$$

implies that $\varphi_2 = \varphi_1$. Indeed, assuming the contrary, $\varphi_2 \neq \varphi_1$, we obtain by Lemma 2.57 that

$$(b_1^2 + b_2^2)(1 - M) \leq \|b_1 \varphi_1 - b_2 \varphi_2\|^2 \leq 4\epsilon^2. \tag{5.57}$$

On the other hand,

$$|b_1| \geq \|f\| - \epsilon, \quad |b_2| \geq \|f\| - \epsilon. \tag{5.58}$$

It is clear that (5.57) and (5.58) contradict each other if $\epsilon$ is small enough compared to $\|f\|$. Thus, for sufficiently small $\epsilon$, only one dictionary element,

$\varphi_1$, provides a good approximation. This implies $\|f - \langle f, \varphi_1 \rangle \varphi_1\| = 0$, which in turn implies $f = \langle f, \varphi_1 \rangle \varphi_1$.

Consider the case $m > 1$. Following the induction argument assume that if $\sigma_{m-1}(f) = 0$ then $f \in \Sigma_{m-1}$. We now have $\sigma_m(f) = 0$. If $\sigma_{m-1}(f) = 0$ then, by the induction assumption, $f \in \Sigma_{m-1}$. So, assume that $\sigma_{m-1}(f) > 0$. Let

$$\|f - \sum_{i=1}^{m} b_i \varphi_i\| \le \epsilon, \quad \varphi_i \in \mathcal{D}, \quad i = 1, \ldots, m. \tag{5.59}$$

As in the case $m = 1$ suppose that there are $\psi_1, \ldots, \psi_m$ from the dictionary such that at least one of $\varphi_i$, say $\varphi_m$, is distinct from them and

$$\|f - \sum_{i=1}^{m} c_j \psi_j\| \le \epsilon, \quad \psi_i \in \mathcal{D}, \quad i = 1, \ldots, m. \tag{5.60}$$

Inequalities (5.59) and (5.60) imply

$$\| \sum_{i=1}^{m} b_i \varphi_i - \sum_{i=1}^{m} c_j \psi_j\| \le 2\epsilon.$$

By Lemma 2.57 we obtain from here

$$|b_m|^2 (1 - M(2m - 1)) \le 4\epsilon^2. \tag{5.61}$$

Therefore, for small enough $\epsilon$ we get

$$\|f - \sum_{i=1}^{m-1} b_i \varphi_i\| \le \epsilon + |b_m| < \sigma_{m-1}(f).$$

The obtained contradiction implies that, for small enough $\epsilon$, inequalities (5.59) and (5.60) imply that $\{\psi_1, \ldots, \psi_m\} = \{\varphi_1, \ldots, \varphi_m\}$. This reduces the problem to a finite dimensional case, where existence of best approximation is well known. In our case it is an orthogonal projection onto $\mathrm{span}(\varphi_1, \ldots, \varphi_m)$. The assumption $\sigma_m(f) = 0$ implies that $f$ is equal to that projection. $\square$

We now prove an inequality similar to the inequality (2.75) from Chapter 2.

**Theorem 5.22** *Let $\mathcal{D}$ be an $M$-coherent dictionary. For any $f \in H$ and any $m \le (4M)^{-1}$ we have for the OGA*

$$\|f_m^o\| \le (1 + (5m)^{1/2}) \sigma_m(f).$$

*Proof* In the case $\sigma_m(f) = 0$, Theorem 5.21 implies that $f \in \Sigma_m$. Then by Theorem 5.19 the OGA recovers it exactly after $m$ iterations. Assume that $\sigma_m(f) > 0$. Suppose

$$\left\| f - \sum_{i=1}^{m} c_i \varphi_i \right\| \leq (1 + \epsilon)\sigma_m(f), \quad \varphi_i \in \mathcal{D}, \quad i = 1, \ldots, m, \quad (5.62)$$

with some fixed $\epsilon > 0$. We will find a constant $C(m) := C(m, \epsilon)$ with the following property. If

$$\|f\| > C(m)\sigma_m(f)$$

then the OGA picks one of the $\varphi_i$, $i = 1, \ldots, m$, at the first iteration. If the OGA has picked elements from $\{\varphi_1, \ldots, \varphi_m\}$ at the first $k$ iterations and

$$\|f_k^o\| > C(m)\sigma_m(f), \quad (5.63)$$

then the OGA picks one of the $\varphi_i$, $i = 1, \ldots, m$, at the $(k + 1)$th iteration.

Define

$$a_m := \sum_{i=1}^{m} c_i \varphi_i, \quad G_k := G_k^o(f, \mathcal{D}) = \sum_{i \in \Lambda_k} b_i \varphi_i,$$

$$\Lambda_k \subset \{1, \ldots, m\}, \quad |\Lambda_k| = k.$$

Without loss of generality we assume that

$$a_m = P_{H_m}(f), \quad H_m := \text{span}(\varphi_1, \ldots, \varphi_m).$$

We write

$$a_m - G_k = \sum_{i=1}^{m} d_i \varphi_i$$

and define $A := \max_{1 \leq i \leq m} |d_i|$. Then

$$\max_{1 \leq i \leq m} |\langle f_k^o, \varphi_i \rangle| = \max_{1 \leq i \leq m} |\langle a_m - G_k, \varphi_i \rangle|$$

$$\geq A(1 - M(m - 1)) > A(1 - Mm).$$

For any $g \in \mathcal{D}$ distinct from $\{\varphi_1, \ldots, \varphi_m\}$ we obtain

$$|\langle f_k^o, g \rangle| = |\langle f - a_m + a_m - G_k, g \rangle| \leq (1 + \epsilon)\sigma_m(f) + AMm.$$

In order to prove our claim, it is sufficient to have

$$A(1 - Mm) > (1 + \epsilon)\sigma_m(f) + AMm$$

or

$$A > (1 + \epsilon)\sigma_m(f)(1 - 2Mm)^{-1} \geq 2(1 + \epsilon)\sigma_m(f).$$

From (5.62) and (5.63) we deduce

$$\|a_m - G_k\| = \|a_m - f + f - G_k\|$$
$$\geq \|f_k^o\| - \|f - a_m\| > (C(m) - 1 - \epsilon)\sigma_m(f).$$

By Lemma 2.57 the $M$-coherence assumption implies

$$A^2 m(1 + Mm) \geq \|a_m - G_k\|^2.$$

Therefore,

$$A \geq (C(m) - 1 - \epsilon)\sigma_m(f)m^{-1/2}(1 + Mm)^{-1/2}.$$

Thus, it is sufficient to choose $C(m)$ satisfying

$$(C(m) - 1 - \epsilon) \geq 2(1 + \epsilon)m^{1/2}(5/4)^{1/2} = (1 + \epsilon)m^{1/2}\sqrt{5},$$
$$C(m) \geq (1 + \epsilon)(1 + (5m)^{1/2}).$$

We complete the proof of Theorem 5.22 in the following way. Fix arbitrarily small $\epsilon > 0$ and choose $C(m) = (1 + \epsilon)(1 + (5m)^{1/2})$. If $\|f_k^o\| > C(m)\sigma_m(f)$ for $k = 0, \ldots, m - 1$ then, as we proved above, the OGA picks all $\varphi_1, \ldots, \varphi_m$, and therefore

$$\|f_m^o\| \leq (1 + \epsilon)\sigma_m(f).$$

If $\|f_k^o\| \leq C(m)\sigma_m(f)$ for some $k \leq m - 1$ then, by monotonicity,

$$\|f_m^o\| \leq \|f_k^o\| \leq C(m)\sigma_m(f) = (1 + \epsilon)(1 + (5m)^{1/2})\sigma_m(f).$$

This completes the proof of Theorem 5.22. $\qquad\square$

A slight modification of the above proof gives the following result for the WOGA($t$).

**Theorem 5.23** *Let $\mathcal{D}$ be an $M$-coherent dictionary. For the WOGA($t$) we have for any $f \in H$ and any $m \leq (t/2(1 + t))M^{-1}$*

$$\|f_m^{o,t}\| \leq (t^{-1} + 1)(1 + (5m)^{1/2})\sigma_m(f).$$

We now give an example from Temlyakov and Zheltov (2010) of a dictionary with small coherence that is difficult for the OGA. From Theorem 5.19 we know that the OGA recovers an $m$-sparse signal over an $M$-coherent dictionary $\mathcal{D}$ exactly in $m$ steps if

$$m < \frac{1}{2}\left(\frac{1}{M}+1\right).$$

We will show that the above bound is sharp.

**Theorem 5.24** *For any* $0 < M < 1$ *such that* $(1/2)(1/M + 1) \in \mathbb{N}$ *there exists an $M$-coherent dictionary $\mathcal{D}$ and an $m$-sparse signal $f$ such that* $m = (1/2)(1/M + 1)$, *but some realization of the OGA will never recover $f$ exactly.*

*Proof* Let $\{e_j\}_{j=1}^{\infty}$ be the standard basis for $H = \ell_2$ and a signal $f = \sum_{i=1}^{m} e_i$ with norm $\|f\| = \sqrt{m}$. Let the dictionary $\mathcal{D}$ consist of the following two kinds of elements:

$$\mathcal{D}_g := \mathcal{D}_{\text{good}} := \{\phi_i = \alpha e_i - \beta f, \quad i = 1, \dots, m\},$$
$$\mathcal{D}_b := \mathcal{D}_{\text{bad}} := \{\phi_j = \eta e_j + \gamma f, \quad j = m+1, \dots\}.$$

It is enough to consider $\alpha, \beta, \gamma > 0$. All elements $\phi_i$ are normalized. In particular, $\alpha, \beta$ satisfy the following equation:

$$(\alpha - \beta)^2 + (m-1)\beta^2 = 1. \tag{5.64}$$

The following are the scalar products of $f$ with the dictionary elements:

$$\langle f, \phi_i \rangle = \langle \alpha e_i - \beta f, f \rangle = \alpha - m\beta, \quad \phi_i \in \mathcal{D}_g,$$
$$\langle f, \phi_j \rangle = \langle \eta e_j + \gamma f, f \rangle = m\gamma, \quad \phi_j \in \mathcal{D}_b.$$

We equalize the above scalar products and introduce the notation $R := m\gamma = \alpha - m\beta$. This will allow some realization of the OGA to select $\phi_{m+1}$ at the first iteration. Now the scalar products of the elements in $\mathcal{D}$ are as follows: for distinct pairs $i, i'$ and $j, j'$

$$\langle \phi_i, \phi_{i'} \rangle = \langle \alpha e_i - \beta f, \alpha e_{i'} - \beta f \rangle = m\beta^2 - 2\alpha\beta = -m\beta(2\gamma + \beta), \ i, i' \le m,$$

$$\langle \phi_j, \phi_{j'} \rangle = \langle \eta e_j + \gamma f, \eta e_{j'} + \gamma f \rangle = m\gamma^2, \ j, j' > m,$$

$$\langle \phi_i, \phi_j \rangle = \langle \alpha e_i - \beta f, \eta e_j + \gamma f \rangle = \gamma(-m\beta + \alpha) = \gamma R = m\gamma^2, \ i \le m < j.$$

The coherence of such a dictionary is given by

$$M := \max(m\gamma^2, m\beta(2\gamma + \beta)).$$

We equalize $\gamma^2 = \beta(2\gamma + \beta)$. Solving this quadratic equation, we get $\gamma = (1 + \sqrt{2})\beta$. We find $\alpha = m\gamma + m\beta = m(2 + \sqrt{2})\beta$, and, plugging it into (5.64), we find $\beta$:

$$(m(2 + \sqrt{2}) - 1)^2 \beta^2 + (m - 1)\beta^2 = 1$$

$$\beta^2 = \frac{1}{m^2(2 + \sqrt{2})^2 - m(3 + 2\sqrt{2})}.$$

This gives the following value for the coherence:

$$M = m\gamma^2 = m(1 + \sqrt{2})^2 \beta^2 = \frac{(1 + \sqrt{2})^2}{m(2 + \sqrt{2})^2 - (3 + 2\sqrt{2})}.$$

Denote $A := 1 + \sqrt{2}$ and note that $2 + \sqrt{2} = \sqrt{2}A$, $\quad 3 + 2\sqrt{2} = A^2$. Then

$$M = \frac{A^2}{2A^2 \cdot m - A^2} = \frac{1}{2m - 1}, \quad \text{or} \quad m = \frac{1}{2}\left(\frac{1}{M} + 1\right).$$

We assume that our realization of the OGA picked an element $\psi_1$ from $\mathcal{D}_b$ at the first iteration. Then $f_1^o = f - \langle f, \psi_1 \rangle \psi_1$. We will prove by induction that the OGA may select elements from $\mathcal{D}_b$ at all iterations and will never select a correct element from $\mathcal{D}$. Suppose that by the $n$th iteration the OGA has selected $n$ elements $\Psi_n := \{\psi_i\}_{i=1}^n$ from $\mathcal{D}_b$. Due to orthogonal projection the OGA does not select an element from the dictionary twice. For $\phi \in \mathcal{D} \setminus \Psi_n$ we have

$$\left\langle f - \sum_{j=1}^n c_j \psi_j, \phi \right\rangle = \langle f, \phi \rangle - \sum_{j=1}^n c_j \langle \psi_j, \phi \rangle = R - M \sum_{j=1}^n c_j.$$

Since all the scalar products are still the same (they do not depend on $\phi$), some realization of the OGA will select another element from $\mathcal{D}_b$. This completes the induction argument. $\qquad\square$

## 5.6 Exact recovery of sparse signals; the Subspace Pursuit Algorithm

In this section we present results from Dai and Milenkovich (2009) on the Subspace Pursuit Algorithm. As above, let $\Phi = \{\varphi_i\}_{i=1}^N$ denote a finite dictionary from $\mathbb{R}^n$.

**Subspace Pursuit (SP)** Let $K$ be a given natural number and let $y \in \mathbb{R}^n$. At the beginning we take an arbitrary set of indices $T_0$ of cardinality $|T_0| = K$. At the $j$th iteration of the algorithm we construct a set of indices $T_j$ of cardinality $|T_j| = K$ performing the following steps.

(1) Take the $T_{j-1}$ from the previous iteration $j - 1$ and find

$$y^j := y - P_{T_{j-1}}(y),$$

where $P_A(y)$ denotes the orthogonal projection of $y$ onto the subspace span$(\varphi_i, i \in A)$.

(2) Find the set of indices $L_j$ such that $|L_j| = K$ and

$$\min_{i \in L_j} |\langle y^j, \varphi_i \rangle| \geq \max_{i \notin L_j} |\langle y^j, \varphi_i \rangle|.$$

(3) Consider

$$P_{T_{j-1} \cup L_j}(y) = \sum_{i \in T_{j-1} \cup L_j} c_i \varphi_i.$$

Define $T_j$ to be the set of $K$ indices $i$ that correspond to the largest values of $|c_i|$.

We study performance of the SP with respect to a special class of dictionaries. This class of dictionaries is a generalization of the class of classical Riesz bases. We give a definition in a general Hilbert space.

**Definition 5.25** A dictionary $\mathcal{D}$ is called the Riesz dictionary with depth $D$ and parameter $\delta \in (0, 1)$ if, for any $D$ distinct elements $g_1, \ldots, g_D$ of the dictionary and any coefficients $a = (a_1, \ldots, a_D)$, we have

$$(1 - \delta)\|a\|_2^2 \leq \|\sum_{i=1}^{D} a_i g_i\|^2 \leq (1 + \delta)\|a\|_2^2. \tag{5.65}$$

We denote the class of Riesz dictionaries with depth $D$ and parameter $\delta \in (0, 1)$ by $R(D, \delta)$.

It is clear that the term Riesz dictionary with depth $D$ and parameter $\delta \in (0, 1)$ is another name for a dictionary satisfying the restricted isometry property with parameters $D$ and $\delta$.

**Theorem 5.26** *Let* $\Phi \in R(D, \delta)$ *with* $\delta \leq 0.04$. *For any $K$-sparse vector $y$ with $3K \leq D$, the SP recovers $y$ exactly after $CK$ iterations.*

This theorem follows from Theorem 5.29 and Lemma 5.30 proved below. We begin with a simple lemma.

**Lemma 5.27** *Let* $\mathcal{D} \in R(D, \delta)$ *and let* $g_j \in \mathcal{D}$, $j = 1, \ldots, s$. *For* $f = \sum_{i=1}^{s} a_i g_i$ *and* $\Lambda \subset \{1, \ldots, s\}$ *denote*

$$S_\Lambda(f) := \sum_{i \in \Lambda} a_i g_i.$$

*If $s \leq D$ then*

$$\|S_\Lambda(f)\|^2 \leq (1 + \delta)(1 - \delta)^{-1}\|f\|^2.$$

*Proof* By (5.65) we obtain

$$\|S_\Lambda(f)\|^2 \leq (1+\delta)\left(\sum_{i\in\Lambda} |a_i|^2\right) \leq (1+\delta)\left(\sum_{i=1}^{s} |a_i|^2\right) \leq (1+\delta)(1-\delta)^{-1}\|f\|^2.$$

□

We now proceed to the evaluation of the $j$th iteration of the SP. First, we prove a lemma in a style of Lemma 5.14.

**Lemma 5.28** *Let* $\{g_i\}_{i=1}^k$ *and* $\{\psi_j\}_{j=1}^l$ *be disjoint sets of the dictionary elements. Consider*

$$u := \sum_{i=1}^{k} b_i g_i, \quad b := (b_1, \ldots, b_k), \qquad v := \sum_{j=1}^{l} c_j \psi_j \quad c := (c_1, \ldots, c_l).$$

*Assume that* $\mathcal{D} \in R(D, \delta)$ *and that* $k + l \leq D$. *Then*

$$|\langle u, v\rangle| \leq \delta\|b\|_2\|c\|_2 \tag{5.66}$$

*and*

$$\left(\sum_{j=1}^{l} |\langle u, \psi_j\rangle|^2\right)^{1/2} \leq \delta\|b\|_2 \leq \delta(1-\delta)^{-1/2}\|u\|. \tag{5.67}$$

*Proof* Without loss of generality, assume that $u$ and $v$ are non-zero and consider their normalized versions $u' := u/\|b\|_2$ and $v' := v/\|c\|_2$. We have by the parallelogram identity and (5.65)

$$|\langle u', v'\rangle| = \frac{1}{4}|\|u' + v'\|^2 - \|u' - v'\|^2| \leq \delta.$$

This proves (5.66). The inequality (5.67) follows from (5.66) by the chain of relations:

$$\left(\sum_{j=1}^{l} |\langle u, \psi_j\rangle|^2\right)^{1/2} = \max_{c:\|c\|_2\leq 1} |\sum_{j=1}^{l} c_j\langle u, \psi_j\rangle| \leq \delta\|b\|_2.$$

□

We introduce some notation. Let $T \subset \{1, \ldots, N\}$, $|T| = K$, be such that

$$y = \sum_{i\in T} x_i\varphi_i, \quad \varphi_i \in \Phi.$$

Define

$$T^j := T \setminus T_j, \quad y_{T^j} := S_{T^j}(y),$$

$$x_{T^j}: \quad (x_{T^j})_i = x_i, i \in T^j, \quad (x_{T^j})_i = 0, i \notin T^j,$$

where $T_j$ are from the SP. We prove the following theorem.

**Theorem 5.29** *Let* $\Phi \in R(D, \delta)$ *and let* $y$ *be a* $K$-*sparse vector with respect to* $\Phi$. *If* $3K \leq D$ *then*

$$\|y_{T^j}\| \leq 4\delta^{1/2}(1 + \delta)^{1/2}(1 - \delta)^{-3/2}\|y_{T^{j-1}}\|$$

*and*

$$\|x_{T^j}\|_2 \leq 4\delta^{1/2}(1 + \delta)(1 - \delta)^{-2}\|x_{T^{j-1}}\|_2.$$

*Proof* The second inequality follows from the first inequality and (5.65). We prove the first inequality. From the definition of $y^j$ and $T^{j-1}$ it follows that

$$y^j = y - P_{T_{j-1}}(y) = y_{T^{j-1}} - P_{T_{j-1}}(y_{T^{j-1}}). \tag{5.68}$$

Therefore

$$\|y^j\| \leq \|y_{T^{j-1}}\|. \tag{5.69}$$

Our goal is to estimate $\|y_{T^j}\|$. Before doing that we bound the quantity $\|y - P_{T_{j-1} \cup L_j}(y)\|$. We have

$$\|y - P_{T_{j-1} \cup L_j}(y)\|^2 = \|y^j - P_{T_{j-1} \cup L_j}(y^j)\|^2$$
$$\leq \|y^j - P_{L_j}(y^j)\|^2 = \|y^j\|^2 - \|P_{L_j}(y^j)\|^2. \tag{5.70}$$

We now estimate $\|P_{L_j}(y^j)\|$ from below. We have by the duality argument

$$\|P_{L_j}(y^j)\| = \max_{\{c_i, i \in L_j\}: \|\sum_{i \in L_j} c_i \varphi_i\| \leq 1} |\langle P_{L_j}(y^j), \sum_{i \in L_j} c_i \varphi_i \rangle|$$

$$= \max_{\{c_i, i \in L_j\}: \|\sum_{i \in L_j} c_i \varphi_i\| \leq 1} |\langle y^j, \sum_{i \in L_j} c_i \varphi_i \rangle|$$

$$\geq \max_{\{c_i, i \in L_j\}: \sum_{i \in L_j} |c_i|^2 \leq (1+\delta)^{-1}} |\sum_{i \in L_j} c_i \langle y^j, \varphi_i \rangle|$$

$$= (1 + \delta)^{-1/2} \left( \sum_{i \in L_j} |\langle y^j, \varphi_i \rangle|^2 \right)^{1/2}$$

$$\geq (1 + \delta)^{-1/2} \left( \sum_{i \in T^{j-1}} |\langle y^j, \varphi_i \rangle|^2 \right)^{1/2}. \tag{5.71}$$

Using (5.68) we obtain

$$
\left( \sum_{i \in T^{j-1}} |\langle y^j, \varphi_i \rangle|^2 \right)^{1/2}
$$

$$
\geq \left( \sum_{i \in T^{j-1}} |\langle y_{T^{j-1}}, \varphi_i \rangle|^2 \right)^{1/2} - \left( \sum_{i \in T^{j-1}} |\langle P_{T_{j-1}}(y_{T^{j-1}}), \varphi_i \rangle|^2 \right)^{1/2}.
$$

$$(5.72)$$

Further, using the duality argument again, we get

$$
\left( \sum_{i \in T^{j-1}} |\langle y_{T^{j-1}}, \varphi_i \rangle|^2 \right)^{1/2} \geq (1 - \delta)^{1/2} \| y_{T^{j-1}} \|. \tag{5.73}
$$

By Lemma 5.28,

$$
\left( \sum_{i \in T^{j-1}} |\langle P_{T_{j-1}}(y_{T^{j-1}}), \varphi_i \rangle|^2 \right)^{1/2} \leq \delta (1 - \delta)^{-1/2} \| P_{T_{j-1}}(y_{T^{j-1}}) \|
$$

$$
\leq \delta (1 - \delta)^{-1/2} \| y_{T^{j-1}} \|. \tag{5.74}
$$

Combining (5.72)–(5.74) we obtain

$$
\left( \sum_{i \in T^{j-1}} |\langle y^j, \varphi_i \rangle|^2 \right)^{1/2} \geq \| y_{T^{j-1}} \| ((1 - \delta)^{1/2} - \delta (1 - \delta)^{-1/2}). \tag{5.75}
$$

Substitution of (5.75) into (5.71) gives

$$
\| P_{L_j}(y^j) \| \geq (1 + \delta)^{-1/2} ((1 - \delta)^{1/2} - \delta (1 - \delta)^{-1/2}) \| y_{T^{j-1}} \|. \tag{5.76}
$$

Plugging (5.76) and (5.69) into (5.70) we get

$$
\begin{aligned}
\| y - P_{T_{j-1} \cup L_j}(y) \|^2 \\
\leq (1 - ((1 + \delta)^{-1/2} ((1 - \delta)^{1/2} - \delta (1 - \delta)^{-1/2}))^2) \| y_{T^{j-1}} \|^2 \\
\leq 4\delta \| y_{T^{j-1}} \|^2.
\end{aligned} \tag{5.77}
$$

We now proceed to the bound of $\| y_{T^j} \|$. By construction of $L_j$ we have that $T_{j-1} \cap L_j = \emptyset$ and $|T_{j-1}| = |L_j| = K$. Therefore

$$
|T_{j-1} \cup L_j| = 2K, \qquad |T_{j-1} \cup L_j \setminus T| \geq K.
$$

Define $R_j := T_{j-1} \cup L_j \setminus T_j$ and take any $\Lambda$ with the following properties:

$$\Lambda \subset T_{j-1} \cup L_j \setminus T, \qquad |\Lambda| = K.$$

Then by the definition of $T_j$ we have

$$\sum_{i \in R_j} |c_i|^2 \leq \sum_{i \in \Lambda} |c_i|^2. \tag{5.78}$$

Next

$$\|\sum_{i \in R_j} c_i \varphi_i\|^2 \leq (1+\delta) \sum_{i \in R_j} |c_i|^2 \leq (1+\delta) \sum_{i \in \Lambda} |c_i|^2$$

$$\leq (1+\delta)(1-\delta)^{-1} \|\sum_{i \in \Lambda} c_i \varphi_i\|^2.$$

From the definition of $\Lambda$ we find

$$\sum_{i \in \Lambda} c_i \varphi_i = S_\Lambda(P_{T_{j-1} \cup L_j}(y)) = S_\Lambda(P_{T_{j-1} \cup L_j}(y) - y).$$

By Lemma 5.27 and (5.77) we obtain from here that

$$\|\sum_{i \in \Lambda} c_i \varphi_i\| \leq (1+\delta)^{1/2}(1-\delta)^{-1/2} \|P_{T_{j-1} \cup L_j}(y) - y\|$$

$$\leq (1+\delta)^{1/2}(1-\delta)^{-1/2} 2\delta^{1/2} \|y_{T^{j-1}}\|.$$

By the definition of $T^j$ we find

$$y_{T^j} = S_{T^j}(y) = S_{T^j}\left(y - \sum_{i \in T_j} c_i \varphi_i\right).$$

Using Lemma 5.27 we get

$$\|y_{T^j}\| \leq (1+\delta)^{1/2}(1-\delta)^{-1/2} \|y - \sum_{i \in T_j} c_i \varphi_i\|$$

$$\leq (1+\delta)^{1/2}(1-\delta)^{-1/2}\left(\|y - \sum_{i \in T_{j-1} \cup L_j} c_i \varphi_i\| + \|\sum_{i \in R_j} c_i \varphi_i\|\right)$$

$$\leq (1+\delta)^{1/2}(1-\delta)^{-1/2} 2\delta^{1/2}(1 + (1+\delta)(1-\delta)^{-1})\|y_{T^{j-1}}\|$$

$$= (1+\delta)^{1/2}(1-\delta)^{-3/2} 4\delta^{1/2} \|y_{T^{j-1}}\|. \tag{5.79}$$

$\square$

We now need the following interesting lemma from Dai and Milenkovich (2009). For a vector $x = (x_1, \ldots, x_N)$ and a subset $A \subset \{1, 2, \ldots, N\}$ we define $x_A$ to be a vector such that $(x_A)_i = x_i$ for $i \in A$ and $(x_A)_i = 0$ for $i \notin A$. In the particular case $A = \{n + 1, \ldots, N\}$ we write $x^n := x_A$.

**Lemma 5.30** *Let $x_1 \geq x_2 \geq \cdots \geq x_S > x_{S+1} = \cdots = x_N = 0$. Suppose that for a given $q \in (0, 1)$ the non-empty sets $A_j \subset \{1, 2, \ldots, S\}$, $j = 1, \ldots, J$, satisfy the following property:*

$$\|x_{A_j}\|_2 \leq q\|x_{A_{j-1}}\|_2, \quad j = 2, \ldots, J. \tag{5.80}$$

*Then $J \leq C(q)S$.*

*Proof* It suffices to prove the lemma for $q = 1/2$. Indeed, for the sets $A_{lj+1}$, $j = 0, \ldots, [(J - 1)/l]$, with $l := [(\log(1/q))^{-1}] + 1$ we have

$$\|x_{A_{lj+1}}\|_2 \leq \frac{1}{2}\|x_{A_{l(j-1)+1}}\|_2.$$

Thus, assume that we have (5.80) with $q = 1/2$. Define $y_n := \|x^n\|_2/x_n$, $n = 1, \ldots, S$. Consider two cases:

$$(I) \quad y_1 < 1/2; \qquad (II) \quad y_1 \geq 1/2.$$

We claim that in case $(I)$ we have $1 \notin A_j$ for all $j = 2, \ldots, J$. Indeed, the inequality $y_1 < 1/2$ implies for $j = 2, \ldots, J$

$$\|x_{A_j}\|_2 \leq \frac{1}{2}\|x_{A_1}\|_2 \leq \frac{1}{2}\|x\|_2 \leq \frac{1}{2}(x_1 + \|x^1\|_2) \leq \frac{3}{4}x_1.$$

This implies that $1 \notin A_j$. Therefore, in case $(I)$ the problem is reduced to the same problem with $S$ and $J$ replaced by $S - 1$ and $J - 1$, respectively.

We proceed to case $(II)$. Let $y_1 \geq 1/2, \ldots, y_{s-1} \geq 1/2$ and $y_s < 1/2$. We note that $y_S = 0$. We prove that for all $k = 1, \ldots, s - 1$

$$x_{s-k} \leq 3^k x_s \quad \text{and} \quad \|x\|_2 \leq \frac{3^s}{2}x_s. \tag{5.81}$$

The proof of the first inequality in (5.81) is by induction. For $k = 1$ we have

$$y_s < 1/2 \quad \Rightarrow \quad \frac{1}{2}x_s > \|x^s\|_2,$$

$$y_{s-1} \geq 1/2 \quad \Rightarrow \quad \frac{1}{2}x_{s-1} \leq \|x^{s-1}\|_2 \leq x_s + \|x^s\|_2 \leq \frac{3}{2}x_s.$$

This proves the first inequality in (5.81) for $k = 1$. Let $x_{s-t} \leq 3^t x_s$ hold for $t = 1, 2, \ldots, k - 1, 1 < k < s$. Then

$$\frac{1}{2} x_{s-k} \leq \|x^{s-k}\|_2 \leq x_{s-k+1} + \cdots + x_s + \|x^s\|_2$$

$$\leq x_s(3^{k-1} + \cdots + 1 + 1/2) = \frac{3^k}{2} x_s.$$

This completes the proof of the first inequality in (5.81).

Next

$$\|x\|_2 \leq x_1 + \cdots + x_s + \|x^s\|_2 \leq (3^{s-1} + \cdots + 1 + 1/2)x_s = \frac{3^s}{2} x_s.$$

For $j > s \log 3 + 1$ we have

$$\|x_{A_j}\|_2 \leq (1/2)^{j-1} \|x\|_2 \leq 3^{-s} \|x\|_2 \leq \frac{1}{2} x_s.$$

Thus we have $A_j \subset \{s + 1, \ldots, S\}$ for all $j > s \log 3 + 1$. Therefore, in case $(II)$ the problem is reduced to the same problem with $S$ and $J$ replaced by $S - s$ and $J - [s \log 3] - 2$, respectively. Combination of cases $(I)$ and $(II)$ implies that $J \leq CS$. This completes the proof of the lemma. □

We define a variant of the SP for an arbitrary countable dictionary $\Phi = \{\varphi_i\}_{i=1}^{\infty}$ of a separable Hilbert space $H$. The following modification of the SP is in a spirit of weak greedy algorithms.

**Subspace Pursuit with weakness parameters $w_1$, $w_2$ (SP($w_1$, $w_2$))** Let $K$ be a given natural number and let $y \in H$. To start we take an arbitrary set of indices $T_0$ of cardinality $|T_0| = K$. At the $j$th iteration of the algorithm we construct a set of indices $T_j$ of cardinality $|T_j| = K$ performing the following steps.

(1) Take the $T_{j-1}$ from the previous iteration $j - 1$ and find

$$y^j := y - P_{T_{j-1}}(y),$$

where $P_A(y)$ denotes the orthogonal projection of $y$ onto the subspace span($\varphi_i, i \in A$).

(2) Find the set of indices $L_j$ such that $|L_j| = K$ and

$$\min_{i \in L_j} |\langle y_j, \varphi_i \rangle| \geq w_1 \sup_{i \notin L_j} |\langle y^j, \varphi_i \rangle|.$$

(3) Consider

$$P_{T_{j-1} \cup L_j}(y) = \sum_{i \in T_{j-1} \cup L_j} c_i \varphi_i.$$

Define $T_j$ to be the set of $K$ indices $i$ such that

$$\min_{i \in T_j} |c_i| \geq w_2 \max_{i \notin T_j} |c_i|.$$

The following theorem is an analog of Theorem 5.29.

**Theorem 5.31** *Let $\Phi \in R(D, \delta)$ and let $y$ be a $K$-sparse vector with respect to $\Phi$. Suppose that $3K \leq D$ and $w_1^2 \geq 1 - 4\delta$. Then for sufficiently small $\delta$ ($\delta \leq 0.02$) and large $w_2 \in (0, 1)$ ($w_2^2 \geq 7\delta^{1/2}$) there exists $q(\delta) \in (0, 1)$ such that for the iterations of the $SP(w_1, w_2)$ we have*

$$\|y_{T^j}\| \leq q(\delta)\|y_{T^{j-1}}\|, \quad \|x_{T^j}\|_2 \leq q(\delta)\|x_{T^{j-1}}\|_2.$$

*Proof* The proof goes along the lines of the proof of Theorem 5.29. We only point out the places which require modification. Instead of (5.71) we obtain

$$\|P_{L_j}(y^j)\| \geq (1 + \delta)^{-1/2} w_1 \left( \sum_{i \in T^{j-1}} |\langle y^j, \varphi_i \rangle|^2 \right)^{1/2}, \tag{5.82}$$

which results in the following change in (5.77):

$$\|y - P_{T_{j-1} \cup L_j}(y)\|^2 \leq (1 - w_1^2(1 - \delta^2)^{-1}(1 - 2\delta)^2)\|y_{T^{j-1}}\|^2.$$

Under assumption $w_1^2 \geq 1 - 4\delta$ we get

$$w_1^2(1 - \delta^2)^{-1}(1 - 2\delta)^2 \geq 1 - 8\delta$$

and

$$\|y - P_{T_{j-1} \cup L_j}(y)\|^2 \leq 8\delta\|y_{T^{j-1}}\|^2. \tag{5.83}$$

Instead of (5.78) we get

$$\sum_{i \in R_j} |c_i|^2 \leq w_2^{-2} \sum_{i \in \Lambda} |c_i|^2. \tag{5.84}$$

This inequality and (5.83) give us the following variant of (5.79):

$$\|y_{T^j}\| \leq (1 + \delta)^{1/2}(1 - \delta)^{-1/2}(8\delta)^{1/2}(1 + (1 + \delta)(1 - \delta)^{-1}w_2^{-2})\|y_{T^{j-1}}\|.$$

$\square$

Theorem 5.31 and Lemma 5.30 imply the following analog of Theorem 5.26.

**Theorem 5.32** *Let* $\Phi \in R(D, \delta)$ *with sufficiently small* $\delta$ $(\delta \leq 0.02)$. *For any* $K$-*sparse vector* $y$ *with* $3K \leq D$, *the* $SP(w_1, w_2)$, *with* $w_1^2 \geq 1 - 4\delta$ *and* $w_2^2 \geq 7\delta^{1/2}$, *recovers* $y$ *exactly after* $C(\delta)K$ *iterations.*

In this section we have only studied the problem of exact recovery of sparse signals by the SP. It is shown in Dai and Milenkovich (2009) that the SP is stable under signal and measurement perturbations in the spirit of Theorem 5.15.

### 5.7 On the size of incoherent systems

#### 5.7.1 Introduction

In this section we study special redundant systems, namely incoherent systems or systems with a small coherence parameter. Systems with a small coherence parameter are useful in signal processing because simple greedy-type algorithms perform well on such systems. For instance, known results (see Section 2.6 of Chapter 2 and Section 5.5 of this chapter) show that the smaller the coherence parameter the better performance of the Orthogonal Greedy Algorithm. Therefore, it is very desirable to build dictionaries with small coherence parameter. In this section we discuss the following problem for both $\mathbb{R}^n$ and $\mathbb{C}^n$: How large can a system with a coherence parameter not exceeding a fixed number $\mu$ be? We obtain upper and lower bounds for the maximal cardinality of such systems. Our presentation here follows Nelson and Temlyakov (2008). The main point of this section is to demonstrate how fundamental results from different areas of mathematics – linear algebra, probability, number theory – can be used in studying the above important problem from approximation theory.

Let $\mathcal{D} = \{g^k\}_{k=1}^N$ be a normalized ($\|g^k\| = 1$, $k = 1, \ldots, N$) system of vectors in $\mathbb{R}^n$ or $\mathbb{C}^n$ equipped with the Euclidean norm. As above we define the coherence parameter of the dictionary $\mathcal{D}$ as follows:

$$M(\mathcal{D}) := \sup_{k \neq l} |\langle g^k, g^l \rangle|.$$

In this section we discuss the following characteristics:

$$N(n, \mu) := \sup\{N : \exists \mathcal{D} \quad \text{such that} \quad \#\mathcal{D} \geq N, M(\mathcal{D}) \leq \mu\}.$$

The problem of studying $N(n, \mu)$ is equivalent to a fundamental problem of information theory. It is a problem on optimal spherical codes. A spherical code $\mathcal{S}(n, N, \mu)$ is a set of $N$ points (code words) on the $n$-dimensional unit sphere such that the inner products between any two code words is not greater than $\mu$. The problem is to find the largest $N^*$ such that the spherical code

$\mathcal{S}(n, N^*, \mu)$ exists. It is clear that $N^* = N(n, \mu)$. This problem is related to a known problem on Grassmannian frames (see, for example, Stromberg and Heath (2003)). A rigorous setting in this regard is the following. Let the cardinality of $\mathcal{D}$ be equal to $N$ ($\#\mathcal{D} = N$). Find

$$c(N, n) := \inf_{\mathcal{D}, \#\mathcal{D}=N} M(\mathcal{D})$$

and describe those dictionaries (Grassmannian dictionaries) that have $M(\mathcal{D}) = c(N, n)$, $\#\mathcal{D} = N$.

In a special case when $\mathcal{D}$ is assumed to be a frame, the above described dictionaries (frames) are known under the name Grassmannian frames. The theory of Grassmannian frames is a beautiful mathematical theory that has connections to areas such as spherical codes, algebraic geometry, graph theory and sphere packings. Some fundamental problems of this theory are still open. For instance, it is known that in the case of frames we have

$$c^{frame}(N, d) \geq \left(\frac{N-d}{d(N-1)}\right)^{1/2}. \tag{5.85}$$

However, it is not known for which pairs $(N, d)$ we have the equality in (5.85).

In this section we discuss the above problem for the case of general dictionaries (we do not limit ourselves to frames). We only present results on the growth (in the sense of order) of $N(n, \mu)$ with $n \to \infty$. We note that the problem is still open even in such a weaker form (we ask for the right bounds of the growth of $N(n, \mu)$ instead of a description of extremal dictionaries).

It is known (see below) that for a system $\mathcal{D}$ with $\#\mathcal{D} \geq 2n$ we have $M(\mathcal{D}) \geq (2n)^{-1/2}$. Thus, a natural range for $\mu$ is $[(2n)^{-1/2}, 1]$. We establish in this section that $N(n, Cn^{-1/2})$ has growth that is polynomial in $n$. In Section 5.7.2 we derive from N. Alon's result the bound

$$N(n, \mu) \leq \exp(C_1 n \mu^2 \ln(2/\mu)), \quad \mu \leq 1/2. \tag{5.86}$$

It is a fundamental result in information theory. The first and rather difficult proof of (5.86) was given by Levenshtein (1982, 1983). Gluskin (1986) gave a simpler proof that is based on lower bounds for the Kolmogorov width of an octahedron in the uniform norm. The technique used to derive (5.86) from Alon's result is simple.

In Section 5.7.3, using a probabilistic technique, we complement (5.86) by the following lower bound (for $\mathbb{R}^n$):

$$N(n, \mu) \geq C_2 \exp(n\mu^2/2). \tag{5.87}$$

The above bounds (5.86) and (5.87) combine into the following theorem.

**Theorem 5.33** *There exist two positive constants $C_1$ and $C_2$ such that for $\mu \in [(2n)^{-1/2}, 1/2]$ we have*

$$C_2 \exp(n\mu^2/2) \leq N(n, \mu) \leq \exp(C_1 n\mu^2 \ln(2/\mu)).$$

A very interesting and difficult problem is to provide an explicit (deterministic) construction of a large system with small coherence. We address this problem in Section 5.7.4, where we use Weil's sums to construct a system $\mathcal{D}$ in $\mathbb{C}^n$ with $M(\mathcal{D}) \leq \mu$ of cardinality of order $\exp(\mu n^{1/2} \ln n)$. We note that Kashin (1975) constructed a system $\mathcal{D}$ in $\mathbb{R}^n$ with $M(\mathcal{D}) \leq \mu$ of cardinality of order $\exp(C\mu n^{1/2} \ln n)$ using Legendre symbols. Similar results have been obtained by Gilbert, Muthukrishnan and Strauss (2003) with combinatorial designs and by DeVore (2007) with the finite fields technique.

The problem of finding $N(n, \mu)$ is equivalent to the Grassmannian packing problem for subspaces in the case of one-dimensional subspaces (see Conway, Hardin and Sloane (1996) and Stromberg and Heath (2003)). We refer the reader to Conway and Sloane (1998) and Stromberg and Heath (2003) for further discussions of related problems.

In Section 5.7.5 we give bounds on the maximal size of dictionaries that have RIP$(S, \delta)$. We derive from known results the fact that it behaves in the sense of order as $n \exp(Cn/S)$.

### 5.7.2 Upper bounds

We begin this subsection with an argument that establishes that, for big dictionaries $\mathcal{D}$ ($\#\mathcal{D} \geq 2n$), the coherence parameter $M(\mathcal{D})$ is always bounded from below by $c_0 n^{-1/2}$, where $c_0 > 0$ is an absolute constant. This is an elementary linear algebra argument that works for both $\mathbb{R}^n$ and $\mathbb{C}^n$.

Let $\mathcal{D} := \{g^j\}_{j=1}^N$ be a normalized system of vectors in $\mathbb{R}^n$, $g^j = (g_1^j, \ldots, g_n^j)^T$. Denote by

$$\Phi := [g^1, \ldots, g^N]$$

an $n \times N$ matrix formed by column vectors $\{g^j\}$. Consider the transposed matrix $\Phi^T$ that is formed by the row vectors $(g_1^j, \ldots, g_n^j)$, $j = 1, \ldots, N$, or by the column vectors $h_i := (g_i^1, \ldots, g_i^N)^T$, $i = 1, \ldots, n$. Then the Gram matrix $G$ of the system $\{g^j\}_{j=1}^N$ can be written as

$$G = \Phi^T \Phi.$$

It is well known and easy to understand that $\operatorname{rank} G \leq n$ (indeed, the columns of $G$ are linear combinations of $n$ columns $h_i$, $i = 1, \ldots, n$). Therefore, the positive (non-negative) definite symmetric matrix $G$ has at most $n$ non-zero eigenvalues $\lambda_k > 0$. By the normalization assumption $\|g^j\| = 1$, $j = 1, \ldots, N$, we obtain from the property of traces of matrices that

$$\sum_k \lambda_k = N.$$

By the Hilbert–Schmidt theory for the singular value decompositions we get

$$\sum_{i,j=1}^{N} (\langle g^i, g^j \rangle)^2 = \sum_k \lambda_k^2. \tag{5.88}$$

By Cauchy's inequality,

$$\sum_k \lambda_k^2 \geq n^{-1} \left( \sum_k \lambda_k \right)^2 = N^2/n.$$

Therefore, (5.88) yields

$$N + (N^2 - N)M(\mathcal{D})^2 \geq \sum_{i,j=1}^{N} (\langle g^i, g^j \rangle)^2 \geq N^2/n$$

and

$$M(\mathcal{D}) \geq \left( \frac{N - n}{n(N - 1)} \right)^{1/2}. \tag{5.89}$$

In particular, (5.89) implies for $N \geq 2n$ that $M(\mathcal{D}) \geq (2n)^{-1/2}$.

The lower bound (5.89) has been derived from the property $\operatorname{rank} G \leq n$ of the Gram matrix of the system $\mathcal{D}$. We now use a fundamental result of Alon (2003) to derive an upper bound for $N(n, \mu)$ from the property $\operatorname{rank} G \leq n$.

**Theorem 5.34** *Let $A := \|a_{i,j}\|_{i,j=1}^{N}$ be a square matrix of the form $a_{i,i} = 1$, $i = 1, \ldots, N$; $|a_{i,j}| \leq \epsilon < 1/2$, $i \neq j$. Then*

$$\min(N, (\ln N)(\epsilon^2 \ln(2/\epsilon))^{-1}) \leq C_1 \operatorname{rank} A \tag{5.90}$$

*with an absolute constant $C_1$.*

We apply this theorem with $A = G$ and $\epsilon = \mu$. For $N \geq C_2 n$ with large enough $C_2$ the inequality $\mu \geq (2n)^{-1/2}$ implies that

$$N \geq (\ln N)(\epsilon^2 \ln(2/\epsilon))^{-1}.$$

Therefore, (5.90) gives the inequality

$$(\ln N)(\mu^2 \ln(2/\mu))^{-1} \leq C_1 n$$

and

$$N \leq \exp(C_1 n \mu^2 \ln(2/\mu)). \tag{5.91}$$

In particular, in the case $\mu = C_3 n^{-1/2}$, (5.91) gives the polynomial bound $N \leq n^{C_4}$.

*Proof of Theorem 5.34*   In the notations of Theorem 5.34 an analog of (5.89) will read as follows:

$$\epsilon \geq \left(\frac{N-r}{r(N-1)}\right)^{1/2} \tag{5.92}$$

with $r := \text{rank } A$. This implies the bound

$$r \geq \min(N/2, (2\epsilon^2)^{-1}). \tag{5.93}$$

Now we improve (5.93) by using the Hadamard multiplication of matrices. For two vectors $u, v \in \mathbb{R}^N$, denote the Hadamard product as follows:

$$uHv := (u_1 v_1, \ldots, u_N v_N)^T, \qquad u = (u_1, \ldots, u_N)^T, \qquad v = (v_1, \ldots, v_N)^T.$$

The $m$th Hadamard power of a matrix $A := \|a_{i,j}\|_{i,j=1}^N$ is defined by $A^{Hm} := \|a_{i,j}^m\|_{i,j=1}^N$. We are interested in an upper bound for the rank $A^{Hm}$ in terms of rank $A$. An elementary bound is rank $A^{Hm} \leq (\text{rank } A)^m$. We need a more accurate bound. Let $A_j$ denote the $j$th column vector of $A$ and let $A_j^{Hm}$ denote the $j$th column vector of $A^{Hm}$. Since $A_j \in \text{span}\{v_k\}_{k=1}^r$; then

$$A_j^{Hm} \in S := \text{span}\{v_{k_1} H v_{k_2} H \cdots H v_{k_m}\}_{1 \leq k_1 \leq k_2 \leq \cdots \leq k_m \leq r}.$$

Therefore,

$$\text{rank } A^{Hm} \leq \dim S \leq \binom{r+m-1}{m} \leq (e(r+m)/m)^m. \tag{5.94}$$

We choose $m$ as the largest integer such that $\epsilon^{-2m} \leq N$ or $m := [\ln N/(2\ln(1/\epsilon))]$. Applying inequality (5.93) to the matrix $A^{Hm}$ with $\epsilon^m$ instead of $\epsilon$, we obtain

$$\text{rank } A^{Hm} \geq (\epsilon^{-2m})/2. \tag{5.95}$$

Combining (5.94) and (5.95) we get the required lower bound for $r$.  □

### 5.7.3 Lower bounds; probabilistic approach

In this subsection we prove the existence of large systems with small coherence. We demonstrate how two classical systems – the trigonometric system and the Walsh system – can be used in such constructions. We use the trigonometric system $\{e^{ikx}\}$ for a construction in $\mathbb{C}^n$ and the Walsh system for a construction in $\mathbb{R}^n$. We note that the lower bound for $\mathbb{R}^n$ implies the corresponding lower bound for $\mathbb{C}^n$. The proof is based on the following Hoeffding inequality (see Section 4.2).

**Theorem 5.35** *Let $\xi_i$ be real random variables on $(X, \Sigma, \rho)$ such that $|\xi_i - E\xi_i| \leq b_i$, $i = 1, \ldots, m$, almost surely. Consider a new random variable $\zeta$ on $(X^m, \Sigma^m, \rho^m)$ defined as*

$$\zeta(\omega) := \sum_{i=1}^{m} \xi_i(\omega_i), \quad \omega = (\omega_1, \ldots, \omega_m).$$

*Then for $t > 0$*

$$\rho^m \{\omega : |\zeta(\omega) - E\zeta| \geq mt\} \leq 2 \exp\left(-\frac{m^2 t^2}{2\|b\|_2^2}\right). \tag{5.96}$$

Theorem 5.35 implies the following inequality in the case of complex random variables. If $z = a + ib$ then $|z| \geq t$ implies that either $|a| \geq 2^{-1/2} t$ or $|b| \geq 2^{-1/2} t$. Therefore, in the complex case we have the following inequality instead of (5.96):

$$\rho^m \{\omega : |\zeta(\omega) - E\zeta| \geq mt\} \leq 4 \exp\left(-\frac{m^2 t^2}{4\|b\|_2^2}\right). \tag{5.97}$$

Let us begin with a construction in the $\mathbb{C}^n$. Consider random variables $\xi_k = e^{i2\pi kx}$, $x \in [0, 1]$, $k = -N, \ldots, -1, 1, \ldots, N$. Let $\rho$ be the Lebesgue measure on $[0, 1]$. Then by (5.97) with $t = \mu$ we get, for each $k$,

$$\rho^n \left\{ (x_1, \ldots, x_n) : |n^{-1} \sum_{j=1}^{n} e^{i2\pi kx_j}| > \mu \right\} \leq 4 \exp(-n\mu^2/4). \tag{5.98}$$

Therefore, for any $N$ satisfying

$$N < 8^{-1} \exp(n\mu^2/4)$$

there exists a set of points $y_1, \ldots, y_n$ such that, for all $|k| \leq N, k \neq 0$, we have

$$|n^{-1} \sum_{j=1}^{n} e^{i2\pi ky_j}| \leq \mu.$$

We now define a system $\mathcal{D}_N = \{g^l\}_{l=1}^N$ by

$$g^l := n^{-1/2}(e^{i2\pi ly_1}, \ldots, e^{i2\pi ly_n})^T.$$

It is a normalized system with the property

$$|\langle g^l, g^m \rangle| = |n^{-1} \sum_{j=1}^n e^{i2\pi(l-m)y_j}| \le \mu$$

provided $l \ne m$. Thus, we have built a normalized system of $N$ vectors with coherence $\le \mu$ with $N$ of the order of $\exp(n\mu^2/4)$ that is very close to the corresponding upper bound (see (5.91)) $N \le \exp(C_1 n\mu^2 \ln(1/\mu))$.

We proceed to a construction in $\mathbb{R}^n$. This construction is similar to the above one with the exponential functions $e^{i2\pi kx}$ replaced by the Walsh functions $w_k(x)$. We recall the definition of the Walsh system of functions (see, for example, Kashin and Saakyan (1989)). Let

$$r_k(x) := \text{sign}\sin(2^k \pi x), \quad x \in [0,1], \quad k = 1, 2, \ldots,$$

be the Rademacher system. We define $w_m(x)$ for $m = 1, 2, \ldots$ in the following way. Let

$$m = \sum_{j=0}^l a_j 2^j, \quad a_j = 0, 1, \quad j = 0, 1, \ldots, l.$$

Denote $J_m := \{j : a_j = 1\}$. Define

$$w_m(x) := \prod_{j \in J_m} r_{j+1}(x).$$

Then $\{w_m(x)\}_{m=1}^\infty$ forms an orthonormal system on $[0,1]$. It has the following property convenient for us. Let

$$m = \sum_{j=0}^l a_j 2^j, \quad k = \sum_{j=0}^l b_j 2^j, \quad s(m,k) := \sum_{j=0}^l |a_j - b_j| 2^j.$$

Then for all $x$, except maybe dyadic rationals, we have

$$w_m(x)w_k(x) = w_{s(m,k)}(x).$$

It is clear that $s(m,k) \le m + k$.

Consider real random variables $\xi_k = w_k(x)$, $x \in [0,1]$, $k = 1, \ldots, 2N$. Let $\rho$ be the Lebesgue measure on $[0,1]$. Then by Hoeffding's inequality (5.96) with $t = \mu$ we get, for each $k = 1, \ldots, 2N$,

$$\rho^n\{(x_1, \ldots, x_n) : |n^{-1} \sum_{i=1}^n w_k(x_i)| > \mu\} \le 2\exp(-n\mu^2/2).$$

Therefore, for any $N < 4^{-1} \exp(n\mu^2/2)$ there exists a set of points $y_1, \ldots, y_n$ that are not dyadic rationals such that, for all $k \in [1, 2N]$, we have

$$|n^{-1} \sum_{i=1}^{n} w_k(y_i)| \leq \mu.$$

Consider the following system $W_N = \{g^l\}_{l=1}^{N}$ in $\mathbb{R}^n$:

$$g^l := n^{-1/2}(w_l(y_1), \ldots, w_l(y_n))^T.$$

It is a normalized system satisfying, for $l \neq m$,

$$|\langle g^l, g^m \rangle| = |n^{-1} \sum_{i=1}^{n} w_l(y_i)w_m(y_i)| = |n^{-1} \sum_{i=1}^{n} w_{s(l,m)}(y_i)| \leq \mu.$$

Thus, the above system $W_N$ with $N = [4^{-1} \exp(n\mu^2/2)] - 1$ provides an example of a large system with coherence parameter $\leq \mu$.


### 5.7.4 Lower bounds; deterministic construction

In this subsection we present a deterministic construction of large systems with small coherence. The construction is based on the following variant of the A. Weil theorem (see Carlitz and Uchiyama (1957)).

**Theorem 5.36** *Let $r \geq 2$ be a natural number and let $p > r$ be a prime number. Denote for $a := (a_1, \ldots, a_r)$, where $a_j$ are integers,*

$$F(a, u) := a_r u^r + \cdots + a_1 u.$$

*Then for $a \neq (0, \ldots, 0)$ mod $p$ we have*

$$|S(a)| \leq (r - 1)p^{1/2}, \quad S(a) := \sum_{u=1}^{p} e^{2\pi i F(a,u)/p}. \tag{5.99}$$

Clearly, in a particular case $r = 2$, the inequality (5.99) gives the classical result for the magnitude of the Gaussian sums. Consider the following set $W(r, p)$ of vectors in $\mathbb{C}^p$:

$$v^a := p^{-1/2}(e^{2\pi i F(a,1)/p}, \ldots, e^{2\pi i F(a,p)/p})^T$$

for $a_j \in [1, p]$, $j = 1, \ldots, r$. It is clear that this is a set of normalized vectors. The size of this set is $p^r$. Now, it remains to find the magnitude of the coherence parameter. If we consider $v^a$ and $v^{a'}$, where $a \neq a'$, we may use the above Theorem 5.36 to bound the required inner product directly:

$$
|\langle v^a, v^{a'} \rangle| = p^{-1} | \sum_{u=1}^{p} e^{2\pi i (F(a,u) - F(a',u))/p} |
$$

$$
= p^{-1} | \sum_{u=1}^{p} e^{2\pi i F(a-a',u)/p} | \leq (r-1) p^{-1/2}. \tag{5.100}
$$

For given $n$ and $\mu \geq (2/n)^{1/2}$ we set $p$ to be the biggest prime not exceeding $n$. Then $n/2 \leq p \leq n$. We specify $r$ to be the biggest natural number such that $(r-1) p^{-1/2} \leq \mu$. Then by (5.100) $M(W(r,p)) \leq \mu$. For the cardinality of the $W(r,p)$ we have

$$
\#W(r,p) = p^r = e^{r \ln p} \geq e^{\mu p^{1/2} \ln p}.
$$

It is clear that for our purposes of construction of deterministic dictionaries with small coherence it would be sufficient to have Theorem 5.36 for a slowly growing sequence $\{p_k\}$ instead of a sequence of primes. We give here a proof of an analog of Theorem 5.36 for $p$ replaced by $p^2$.

**Theorem 5.37** *Let $r \geq 2$ be a natural number and let $p > r$ be a prime number. Denote for $a := (a_1, \ldots, a_r)$, where $a_j$ are integers,*

$$
F(a, u) := a_r u^r + \cdots + a_1 u.
$$

*Then for $a \neq (0, \ldots, 0) \mod p$ we have*

$$
|S(a, 2)| \leq (r-1) p, \quad S(a, 2) := \sum_{u=1}^{p^2} e^{2\pi i F(a,u)/p^2}. \tag{5.101}
$$

*Proof* For a fixed $a$ denote for brevity $f(u) := F(a, u)$. Let $y$ and $z$ run over the complete system of residues mod $p$. Then the sum $y + pz$ runs over the complete system of residues mod $p^2$. It is clear from Taylor's expansion formula that

$$
f(y + pz) = f(y) + f'(y)pz \mod p^2.
$$

Therefore

$$
\sum_{u=1}^{p^2} e^{2\pi i f(u)/p^2} = \sum_{y=1}^{p} \sum_{z=1}^{p} e^{2\pi i f(y+pz)/p^2} = \sum_{y=1}^{p} e^{2\pi i f(y)/p^2} \sum_{z=1}^{p} e^{2\pi i f'(y)z/p}.
$$

$$
\tag{5.102}
$$

Introducing the notation $T := \{y \in [1, p] : f'(y) = 0 \mod p\}$ we continue:

$$\sum_{u=1}^{p^2} e^{2\pi i f(u)/p^2} = p \sum_{y \in T} e^{2\pi i f(y)/p^2}.$$

Thus

$$|\sum_{u=1}^{p^2} e^{2\pi i f(u)/p^2}| \le p|T|. \tag{5.103}$$

We now need to estimate $|T|$. By the assumptions $a \ne (0, \ldots, 0) \mod p$ and $p > r$ we obtain that at least one of the coefficients of the polynomial

$$f'(y) = a_1 + 2a_2 y + \cdots + r a_r y^{r-1}$$

is mutually prime with $p$. This implies that $|T| \le r - 1$. Substituting this bound into (5.103) we complete the proof. $\square$

## 5.7.5 Dictionaries with the RIP

In this subsection we give bounds on the maximal size of dictionaries in $\mathbb{R}^n$ and $\mathbb{C}^n$ satisfying the restricted isometry property. We are interested here in the dependence of the maximal size of a dictionary that has RIP$(S, \delta)$ on the parameters $n$ and $S$. It is known (see, for example, Candes (2006) and Section 5.8 below) that one can use random matrices to build (prove the existence) of dictionaries with RIP$(S, \delta)$ of size $N$ such that

$$S \ge Cn/\ln(N/n).$$

This means that the lower bound on the maximal size of the RIP$(S, \delta)$ dictionaries is given by

$$N \ge n \exp(Cn/S). \tag{5.104}$$

We now discuss the corresponding upper bound. For the problem $\Phi v = y$ consider the following $\ell_1$ minimization problem (see Section 5.1 above)

$$\min \|v\|_1 \quad \text{subject to} \quad \Phi v = y, \tag{$P_1$}$$

where $\|v\|_1$ denotes the $\ell_1$ norm of the vector $v \in \mathbb{R}^N$. Denote as above the solution to $(P_1)$ by $A_\Phi(y)$.

Candes, Romberg and Tao (2006) proved that if $\Phi$ has RIP$(4S, \delta)$ with $\delta$ small enough (say $\delta < 1/2$), then for any $u \in \mathbb{R}^N$

$$\|u - A_\Phi(\Phi u)\|_2 \le CS^{-1/2}\sigma_S(u)_1, \tag{5.105}$$

where

$$\sigma_S(u)_1 := \min_{w \in \mathbb{R}^N : |\operatorname{supp}(w)| \le S} \|u - w\|_1.$$

By Theorem 5.7 the bound (5.105) implies that the null space $\mathcal{N}(\Phi) := \{u : \Phi u = 0\}$ has the following width property. For any $u \in \mathcal{N}(\Phi)$ we have

$$\|u\|_2 \le CS^{-1/2}\|u\|_1. \tag{5.106}$$

We note that it was proved in Section 5.2 that the width property (5.106) of the $\mathcal{N}(\Phi)$ implies (5.105) with $S$ replaced by $S/16$.

The inequality (5.106) is related to the concept of Gelfand's width. We recall that the Gel'fand width of a compact $F$ in the $\ell_2^N$ norm is defined as follows:

$$d^k(F, \ell_2^N) := \inf_{V_k} \sup_{f \in F \cap V_k} \|f\|_2,$$

where the infimum is taken over linear subspaces $V_k$ with dimension $\ge N - k$. The inequality (5.106) says that

$$d^n(B_1^N, \ell_2^N) \le CS^{-1/2}, \tag{5.107}$$

where $B_1^N := \{u : \|u\|_1 \le 1\}$.

We now use the following Gluskin lower bound for $d^n(B_1^N, \ell_2^N)$ (see Garnaev and Gluskin (1984)):

$$d^n(B_1^N, \ell_2^N) \ge \frac{1}{4}\min(1, (C_1 \ln(1 + N/n))/n)^{1/2}. \tag{5.108}$$

Combining (5.107) and (5.108) we obtain

$$N \le n \exp(Cn/S). \tag{5.109}$$

The bounds (5.104) and (5.109) describe the behavior of maximal size of dictionaries with RIP$(S, \delta)$.

### 5.7.6 Incoherence and the Kolmogorov widths

We begin this subsection with a discussion of a remark from Gluskin (1986) about the relation between $c(N, n)$, defined at the beginning of this section, and the Kolmogorov width $d_n(B_1^N, \ell_\infty^N)$. We begin with a simple observation.

**Proposition 5.38** *Let $\Phi$ be a dictionary in $\mathbb{R}^n$ of cardinality $N$. Then*

$$d_n(B_1^N, \ell_\infty^N) \le \frac{M(\Phi)}{1 + M(\Phi)}. \tag{5.110}$$

*Proof* Let $\Phi = \{x^i\}_{i=1}^N$, $x^i = (x_1^i, \dots, x_n^i)^T$. The octahedron $B_1^N$ is a convex hull of its vertices $\pm e^j$, $e^j = (0, \dots, 0, 1, 0, \dots, 0)^T$ (with 1 at the $j$th coordinate), $j = 1, \dots, N$. Therefore, it is sufficient to construct an $n$-dimensional linear subspace that approximates $e^j$, $j = 1, \dots, N$, in the $\ell_\infty^N$ with the error from (5.110). Define $n$ vectors $w^k := (x_k^1, \dots, x_k^N)^T$, $k = 1, \dots, n$, in $\mathbb{R}^N$. As an approximation subspace we take $L_n := \text{span}(w^1, \dots, w^n)$. We approximate each $e^j$ by a vector

$$a^j := (1 + M)^{-1} \sum_{k=1}^n x_k^j w^k, \quad M := M(\Phi)$$

from $L_n$. For the $j$th coordinate of $e^j - a^j$ we have

$$|(e^j - a^j)_j| = |1 - (1 + M)^{-1} \sum_{k=1}^n x_k^j x_k^j| = M(1 + M)^{-1}.$$

For the $i$th coordinate of $e^j - a^j$ with $i \neq j$ we have

$$|(e^j - a^j)_i| = |(1 + M)^{-1} \sum_{k=1}^n x_k^j x_k^i| \leq M(1 + M)^{-1}.$$

The above inequalities prove Proposition 5.38. $\qquad\square$

Proposition 5.38 implies the following inequality:

$$d_n(B_1^N, \ell_\infty^N) \leq \frac{c(N, n)}{1 + c(N, n)}. \tag{5.111}$$

It implies that $d_n(B_1^N, \ell_\infty^N) \leq c(N, n)$. However, it is not known if the following relation holds: $d_n(B_1^N, \ell_\infty^N) \asymp c(N, n)$.

We discussed in Section 5.3 the duality property of the Kolmogorov and Gel'fand widths. In the spirit of that duality we prove the following analog of Proposition 5.38.

**Proposition 5.39** *Let $\Phi$ be an $M$-coherent dictionary in $\mathbb{R}^n$. Then the null space $\mathcal{N}(\Phi)$ has the following property. For $x \in \mathcal{N}(\Phi)$ we have*

$$\|x\|_\infty \leq \frac{M}{1 + M} \|x\|_1, \quad \|x\|_2 \leq \left(\frac{M}{1 + M}\right)^{1/2} \|x\|_1. \tag{5.112}$$

*Proof* We use the notations from the proof of Proposition 5.38. As in the above proof let $\Phi = \{x^i\}_{i=1}^N$, $x^i = (x_1^i, \dots, x_n^i)^T$. For $x \in \mathcal{N}(\Phi)$ we have for any $w \in L_n$

$$\|x\|_\infty = \max_{y:\|y\|_1 \leq 1} |\langle x, y \rangle| = \max_{y:\|y\|_1 \leq 1} |\langle x, y - w \rangle|$$

$$\leq \|x\|_1 \max_{y:\|y\|_1 \leq 1} \|y - w\|_\infty.$$

Minimizing $\|y - w\|_\infty$ over $w \in L_n$, and using the argument from the proof of Proposition 5.38 we get

$$\|x\|_\infty \leq \frac{M}{1 + M} \|x\|_1.$$

This proves the first inequality in (5.112). The second inequality in (5.112) follows from the first one and the inequality $\|x\|_2^2 \leq \|x\|_1 \|x\|_\infty$. $\square$

The second inequality in (5.112) means that the null space $\mathcal{N}(\Phi)$ of an $M$-coherent dictionary satisfies the WP($S$) with $S := (1 + M^{-1})$. In particular, this implies by Theorem 5.7 that $M$-coherent dictionaries are good for exact recovery of sparse signals by $\ell_1$ minimization. We now present a direct proof of a theorem on the exact recovery of sparse signals by $\ell_1$ minimization that gives the optimal bound on sparsity that allows exact recovery: $< \frac{1}{2}(1 + M^{-1})$.

**Theorem 5.40** *Let* $\Phi$ *be an* $M$-*coherent dictionary. Then for any* $S$-*sparse signal* $u$ *with* $S < (1/2)(1 + M^{-1})$ *we have*

$$A_\Phi(u) = u.$$

*Proof* We begin with a lemma that is an analog of Lemma 5.2.

**Lemma 5.41** *Let* $\Phi$ *be an* $M$-*coherent dictionary and let* $x \in \mathcal{N}(\Phi)$, $x \neq 0$. *Then for any set of indices* $\Lambda$, $|\Lambda| < (1/2)(1 + M^{-1})$, *we have*

$$\sum_{k \in \Lambda} |x_k| < \|x\|_1/2.$$

*Proof* Let $\Phi = \{\varphi_i\}_{i=1}^N$. By our assumption $x \in \mathcal{N}(\Phi)$, we have

$$\sum_{i=1}^N x_i \varphi_i = 0,$$

and, therefore, for all $k$,

$$|x_k| = |\sum_{i \neq k} x_i \langle \varphi_i, \varphi_k \rangle| \leq M(\|x\|_1 - |x_k|).$$

This implies

$$|x_k| \leq \frac{M}{1+M} \|x\|_1$$

and

$$\sum_{k \in \Lambda} |x_k| \leq |\Lambda| \frac{M}{1+M} \|x\|_1 < \frac{1}{2} \|x\|_1.$$

□

The following lemma is an analog of Lemma 5.3 and is derived from Lemma 5.41 in the same way as Lemma 5.3 was derived from Lemma 5.2.

**Lemma 5.42** *Let* $\Phi$ *be an M-coherent dictionary. Suppose that* $u \in \mathbb{R}^N$ *is S-sparse with* $S < (1/2)(1 + M^{-1})$. *Then for any* $v = u + x$, $x \in \mathcal{N}(\Phi)$, $x \neq 0$, *we have*

$$\|v\|_1 > \|u\|_1.$$

Theorem 5.40 is a direct corollary of Lemma 5.42. □

## 5.8 Restricted Isometry Property for random matrices

In this section we show how one can prove the existence of the RIP matrices; its presentation is based on Baraniuk *et al.* (2008). We consider a special case of Bernoulli matrices and remark that this technique works for more general settings. We use the probabilistic approach to prove the existence of the RIP matrices. Let $\Phi(w) = [\varphi_1(w), \dots, \varphi_N(w)]$, $w = (w_1, \dots, w_n)$, $\varphi_i(w) = n^{-1/2}(r_i(w_1), \dots, r_i(w_n))^T$, where $r_i$ are the Rademacher functions, $w_j \in [0, 1]$, $j = 1, \dots, n$. Consider the random variables

$$\xi_j := |\sum_{i=1}^{N} u_i r_i(w_j)|^2 \|u\|_{\ell_2^N}^{-2}, \quad j = 1, \dots, n.$$

Then

$$E\xi_j = \left(\sum_{i=1}^{N} u_i^2\right) \|u\|_{\ell_2^N}^{-2} = 1.$$

For $k = 2, 3, \dots$, by Khinchin's inequality (see Lindenstrauss and Tzafriri (1977), p. 66) we get

$$E\xi_j^k \leq k^k.$$

Therefore,

$$E|\xi_j - E\xi_j|^k \leq 1 + \sum_{\nu=1}^{k} \binom{n}{\nu} E\xi_j^\nu \leq 1 + \sum_{\nu=1}^{k} \binom{n}{\nu} \nu^\nu$$

$$\leq 1 + \sum_{\nu=1}^{k} \binom{n}{\nu} k^\nu \leq (k+1)^k \leq ek^k.$$

By Theorem 4.24 we get, for $t \in (0, 1)$,

$$\rho^n \left\{ w : |\sum_{k=1}^{n} \xi_j(w_j) - n| \geq nt \right\} \leq 2\exp(-c_0 nt^2), \tag{5.113}$$

with $c_0 = e^{-3}/8$.

We use the above inequality (5.113) to prove the following lemma. For a set $T \subset \{1, \ldots, N\}$, define $X_T := \{x : x_i = 0, i \notin T\}$.

**Lemma 5.43** *Let* $\Phi(w)$ *be as defined above. Then, for any set of indices* $T$, $|T| = s \leq n$, *and any* $\delta \in (0, 1)$, *we have, for all* $x \in X_T$,

$$(1 - \delta)\|x\|_{\ell_2^N} \leq \|\Phi(w)x\|_{\ell_2^n} \leq (1 + \delta)\|x\|_{\ell_2^N}$$

*with probability*

$$\geq 1 - 2(12/\delta)^s \exp(-c_0 n\delta^2/4).$$

*Proof* It is clear that it suffices to prove Lemma 5.43 in the case $\|x\|_{\ell_2^N} = 1$. We deduce it from (5.113) by building the corresponding $\delta/4$-nets. It follows from the proof of Corollary 3.4 that there exists a finite set of points $Q_T$ such that $Q_T \subset X_T$, $\|q\|_{\ell_2^N} = 1$, for all $q \in Q_T$, $|Q_T| \leq (12/\delta)^s$, and for all $x \in X_T$ with $\|x\|_{\ell_2^N} = 1$ we have

$$\min_{q \in Q_T} \|x - q\|_{\ell_2^N} \leq \delta/4. \tag{5.114}$$

By inequality (5.113) with $t = \delta/2$ we get

$$(1 - \delta/2)\|q\|_{\ell_2^N}^2 \leq \|\Phi(w)q\|_{\ell_2^n}^2 \leq (1 + \delta/2)\|q\|_{\ell_2^N}^2, \quad q \in Q_T, \tag{5.115}$$

with probability $\geq 1 - 2(12/\delta)^s \exp(-c_0 n\delta^2/4)$. The above inequalities imply

$$(1 - \delta/2)\|q\|_{\ell_2^N} \leq \|\Phi(w)q\|_{\ell_2^n} \leq (1 + \delta/2)\|q\|_{\ell_2^N}, \quad q \in Q_T. \tag{5.116}$$

We now define $A$ as the smallest number such that

$$\|\Phi(w)x\|_{\ell_2^n} \leq (1+A)\|x\|_{\ell_2^N}, \quad x \in X_T, \quad \|x\|_{\ell_2^N} = 1. \qquad (5.117)$$

We show that $A \leq \delta$. Indeed, for any $x \in X_T$ with $\|x\|_{\ell_2^N} = 1$ we pick a $q \in Q_T$ such that $\|x - q\|_{\ell_2^N} \leq \delta/4$. Then we have

$$\|\Phi(w)x\|_{\ell_2^n} \leq \|\Phi(w)q\|_{\ell_2^n} + \|\Phi(w)(x-q)\|_{\ell_2^n} \leq 1 + \delta/2 + (1+A)\delta/4. \qquad (5.118)$$

Since by the definition $A$ is the smallest number for which (5.117) holds, we obtain $A \leq \delta/2 + (1+A)\delta/4$. This implies $A \leq \delta$ and proves the upper bound in Lemma 5.43. The corresponding lower bound follows from

$$\|\Phi(w)x\|_{\ell_2^n} \geq \|\Phi(w)q\|_{\ell_2^n} - \|\Phi(w)(x-q)\|_{\ell_2^n}$$
$$\geq 1 - \delta/2 - (1+\delta)\delta/4 \geq 1 - \delta,$$

which completes the proof. $\qquad\square$

Lemma 5.43 implies the following result on the RIP matrices.

**Theorem 5.44** *Suppose that $n$, $N$ and $\delta \in (0,1)$ are given. Let $\Phi(w)$ be the Bernoulli matrix defined above. Then there exist two positive constants $C_1(\delta)$ and $C_2(\delta)$ such that, for any $s$-sparse signal $x$ with $s \leq C_1(\delta)n/\ln(eN/n)$, we have*

$$(1-\delta)\|x\|_{\ell_2^N} \leq \|\Phi(w)x\|_{\ell_2^n} \leq (1+\delta)\|x\|_{\ell_2^N} \qquad (5.119)$$

*with probability*

$$\geq 1 - \exp(-C_2(\delta)n).$$

*Proof* Lemma 5.43 guarantees that for each of the $s$-dimensional subspaces $X_T$, the matrix $\Phi(w)$ fails to satisfy (5.119) with probability

$$\leq 2(12/\delta)^s \exp(-c_0 n\delta^2/4).$$

There are $\binom{N}{s} \leq (eN/s)^s$ such subspaces. Therefore (5.119) fails to hold with probablity

$$\leq (eN/s)^s 2(12/\delta)^s \exp(-c_0 n\delta^2/4).$$

The claim of the theorem follows from here by simple calculations. $\qquad\square$

### 5.9 Some further remarks

We have already explained in Sections 5.1 and 5.2 a connection between the results on the widths that were obtained in the 1970s and the current results in compressed sensing. The early theoretical results on the widths did not consider the question of practical recovery methods. The celebrated contribution of the work by Candes, Tao and Donoho was to show that the recovery can be done by the $\ell_1$ minimization. While the $\ell_1$ minimization technique plays an important role in designing computationally tractable recovery methods, its complexity is still impractical for many applications. An attractive alternative to $\ell_1$ minimization is a family of greedy algorithms. They include the Orthogonal Greedy Algorithm (called the Orthogonal Matching Pursuit (OMP) in signal processing) discussed in Section 5.5, the Regularized Orthogonal Matching Pursuit (see Needell and Vershynin (2009)) and the Subspace Pursuit discussed in Section 5.6. The compressed sensing problem discussed in this chapter can be cast as a coding–decoding problem. For a given signal $u \in \mathbb{R}^m$, we encode it by $n$ linear measurements $\langle u, \phi_j \rangle$, $\phi_j \in \mathbb{R}^m$, $j = 1, \ldots, n$. Then we decode the information $y = (\langle u, \phi_1 \rangle, \ldots, \langle u, \phi_n \rangle)^T \in \mathbb{R}^n$ to obtain an approximant of $u$. The matrix $\Phi$ (also called the CS matrix) provides the encoding procedure. Usually in compressed sensing the encoding procedure is assumed to be linear. The above discussed $\ell_1$ minimization and greedy algorithms are examples of decoders used to map the information vector $y$ into the signal (coefficients) space $\mathbb{R}^m$. Let us use the notation $\Delta$ for a general decoder which is a mapping from $\mathbb{R}^n$ to $\mathbb{R}^m$. We emphasize that $\Delta$ is not assumed to be linear. Then, for a class $K \subset \mathbb{R}^m$ of signals, we consider the following optimization problem in a spirit of widths:

$$E_n(K)_X := \inf_{(\Phi, \Delta)} \sup_{u \in K} \|u - \Delta(\Phi u)\|_X,$$

where the infimum is taken over all $n \times m$ matrices $\Phi$ and all decoders $\Delta : \mathbb{R}^n \to \mathbb{R}^m$. It is known (see, for example, Cohen, Dahmen and DeVore (2009)) that the above optimization problem is closely related to the Gelfand width of $K$. Under mild assumptions on $K$ we have

$$d^n(K, X) \le E_n(K)_X \le Cd^n(K, X). \tag{5.120}$$

In the particular case $K = B_1^m$, $X = \ell_2^m$, we know from Garnaev and Gluskin (1984) that

$$d^n(B_1^m, \ell_2^m) \ge C_1((\ln(em/n))/n)^{1/2}. \tag{5.121}$$

Theorem 5.7 guarantees that for a matrix $\Phi$ such that

$$\|x\|_2 \le C_2((\ln(em/n))/n)^{1/2}\|x\|_1, \quad x \in \mathcal{N}(\Phi), \tag{5.122}$$

the $\ell_1$ minimization provides the optimal rate of recovery:

$$\sup_{u \in B_1^m} \|u - A_\Phi(\Phi u)\|_2 \leq C_3((\ln(em/n))/n)^{1/2}.$$

In this chapter we have mostly discussed the deterministic setting, and we have proved results of the following type. Assume that a matrix $\Phi$ has a certain property. Then we guarantee that a specific method of recovery ($\ell_1$ minimization, OGA, etc.) is good (recovers exactly sparse signals or provides a good error of approximation for general signals). There are many interesting results in the probabilistic setting. Here we mention only some of them in order to illustrate an idea. For comparison we begin with the deterministic results on exact recovery. Theorem 5.7 guarantees that for a matrix $\Phi$ satisfying (5.122) the $\ell_1$ minimization recovers exactly $s$-sparse signals provided $s \leq C_4 n/ \ln(em/n)$. There is no such recovery result for the OMP. It is known that random matrices (such as Gaussian and Bernoulli) satisfy (5.122) with high probability. The following result was proved by Tropp and Gilbert (2007). Let $\Phi$ be a random matrix (say Gaussian or Bernoulli) and let $s \leq cn/\ln m$ with sufficiently small $c$. Then the OMP recovers exactly, after $s$ iterations, any $s$-sparse signal with high probability.

We now proceed to a discussion of the concept of *instance optimality* introduced and studied in Cohen, Dahmen and DeVore (2009). We mentioned this important concept in Section 5.1. A pair $(\Phi, \Delta)$ of encoder $\Phi : \mathbb{R}^m \to \mathbb{R}^n$ and decoder $\Delta : \mathbb{R}^n \to \mathbb{R}^m$ is called *instance optimal* of order $k$ with constant $C_0$ for the norm $X$ if

$$\|x - \Delta(\Phi x)\|_X \leq C_0 \sigma_k(x)_X \tag{5.123}$$

for all $x \in \mathbb{R}^m$. This concept is related to the concept of Lebesgue-type inequalities studied in Section 2.6. In the case when $\Delta$ provides a $k$-sparse approximant of $x$, inequality (5.123) is the Lebesgue-type inequality. In particular, when the decoder $\Delta$ is the OGA at the $k$th iteration, (5.123) is the Lebesgue-type inequality for the OGA in the coefficients norm. In Section 2.6 and in Section 5.5 we proved some non-trivial Lebesgue-type inequalities for the OGA in the case of the $\ell_2$ norm in the data space $\mathbb{R}^n$. It is proved in Cohen, Dahmen and DeVore (2009) that there are no non-trivial Lebesgue-type inequalities (more generally, no non-trivial inequalities (5.123)) in the case of the $\ell_2$ norm in the coefficients space $\mathbb{R}^m$. This indicates an advantage of measuring the error in the data space. We note that it was established in Cohen, Dahmen and DeVore (2007) that there is a Lebesgue-type inequality for the OGA in the case of the $\ell_2$ norm in the coefficients space $\mathbb{R}^m$ in the probabilistic setting.

We remark further on a relation between the approximation theory setting and the compressed sensing (CS) setting. In the CS setting, $H = \mathbb{R}^n$ equipped with the Euclidean norm $\|x\| := \langle x, x \rangle^{1/2}$ and $\mathcal{D} = \Phi = \{\varphi_i\}_{i=1}^m$ is a finite set of elements (column vectors) of $\mathbb{R}^n$. Then the dictionary $\mathcal{D}$ is associated with an $n \times m$ matrix $\Phi = [\varphi_1 \ldots \varphi_m]$. The condition $y \in A_1(\mathcal{D})$ is equivalent to the existence of $x \in \mathbb{R}^m$ such that $y = \Phi x$ and

$$\|x\|_1 := |x_1| + \cdots + |x_m| \leq 1. \tag{5.124}$$

As a direct corollary of Theorem 2.19, we get for any $y \in A_1(\mathcal{D})$ that the Orthogonal Greedy Algorithm guarantees the following upper bound for the error:

$$\|y - G_k^o(y, \mathcal{D})\| \leq k^{-1/2}. \tag{5.125}$$

The bound (5.125) holds for any $\mathcal{D}$ (any $\Phi$).

In compressed sensing the relation $y = \Phi x$ has the following interpretation. Let $\phi_1, \ldots, \phi_n$ be the rows of the matrix $\Phi$. Then the corresponding column vectors $\phi_i^T$ belong to $\mathbb{R}^m$. The relation $y = \Phi x$ is equivalent to $y_i = \langle x, \phi_i^T \rangle$, $i = 1, \ldots, n$. The number $y_i = \langle x, \phi_i^T \rangle$ is understood as a linear measurement of an unknown vector $x$. The goal is to recover (or approximately recover) the unknown vector $x$ from its measurements $y$.

The following error bound that was discussed in Sections 5.1 and 5.2 is one of the fundamental results of CS. Under some conditions on the matrix $\Phi$ we have for $x$ satisfying (5.124)

$$\|x - A_\Phi(\Phi x)\|_{\ell_2^m} \leq C k^{-1/2}. \tag{5.126}$$

The inequalities (5.125) and (5.126) look alike. However, they provide the error bounds in different spaces: (5.125) in $\mathbb{R}^n$ (the data space) and (5.126) in $\mathbb{R}^m$ (the coefficients space).

## 5.10 Open problems

We formulate here a few open problems on the theoretical aspects of compressed sensing discussed in this chapter.

**5.1.** Give an explicit deterministic construction of a subspace $\Gamma$ from Theorem 5.8.

The following two problems are about the performance of the OGA. Let $y = \Phi x$ and let $y_k^o = \Phi x_k^o$ denote the $k$th residual of the OGA applied to $y$ with respect to $\Phi$.

**5.2.** Find necessary and sufficient conditions on $\Phi$ such that for a given $k$ we have

$$\|x_k^o\|_{\ell_2^N} \leq Ck^{-1/2}\|x\|_{\ell_1^N}$$

for all $x \in \mathbb{R}^N$.

**5.3.** Find necessary and sufficient conditions on $\Phi$ such that for a given $k$ we have

$$\|x_k^o\|_{\ell_2^N} \leq Ck^{-1/2}\sigma_k(x)_{\ell_1^N}$$

for all $x \in \mathbb{R}^N$.

**5.4.** Find the right order of the $N(n, \mu)$ from Theorem 5.33.

**5.5.** Give an explicit deterministic construction of a system $\mathcal{D}$ in $\mathbb{R}^n$, $M(\mathcal{D}) \leq \mu$, with the lower bound on the cardinality better than the existing bound $\exp(C\mu n^{1/2} \ln n)$.

**5.6.** Does the relation $d_n(B_1^N, \ell_\infty^N) \geq Cc(N, n)$ hold with a positive absolute constant $C$?

# 6

# Greedy approximation with respect to dictionaries: Banach spaces

## 6.1 Introduction

In this chapter we move from Hilbert spaces to more general Banach spaces. Let $X$ be a Banach space with norm $\|\cdot\|$. We say that a set of elements (functions) $\mathcal{D}$ from $X$ is a dictionary, respectively, symmetric dictionary, if each $g \in \mathcal{D}$ has a norm bounded by 1 ($\|g\| \leq 1$),

$$g \in \mathcal{D} \quad \text{implies} \quad -g \in \mathcal{D},$$

and the closure of span $\mathcal{D}$ is $X$. We denote the closure (in $X$) of the convex hull of $\mathcal{D}$ by $A_1(\mathcal{D})$. We introduce a new norm, associated with a dictionary $\mathcal{D}$, in the dual space $X^*$ by the formula

$$\|F\|_{\mathcal{D}} := \sup_{g \in \mathcal{D}} F(g), \quad F \in X^*.$$

In this chapter we will study greedy algorithms with regard to $\mathcal{D}$. For a non-zero element $f \in X$ we let $F_f$ denote a norming (peak) functional for $f$:

$$\|F_f\| = 1, \qquad F_f(f) = \|f\|.$$

The existence of such a functional is guaranteed by the Hahn–Banach theorem.

We begin with a generalization of the Pure Greedy Algorithm. The greedy step of the PGA can be interpreted in two ways. First, we look at the $m$th step for an element $\varphi_m \in \mathcal{D}$ and a number $\lambda_m$ satisfying

$$\|f_{m-1} - \lambda_m \varphi_m\|_H = \inf_{g \in \mathcal{D}, \lambda} \|f_{m-1} - \lambda g\|_H. \tag{6.1}$$

Second, we look for an element $\varphi_m \in \mathcal{D}$ such that

$$\langle f_{m-1}, \varphi_m \rangle = \sup_{g \in \mathcal{D}} \langle f_{m-1}, g \rangle. \tag{6.2}$$

In a Hilbert space both versions (6.1) and (6.2) result in the same PGA. In a general Banach space the corresponding versions of (6.1) and (6.2) lead to different greedy algorithms. The Banach space version of (6.1) is straightforward: instead of the Hilbert norm $\|\cdot\|_H$ in (6.1) we use the Banach norm $\|\cdot\|_X$. This results in the following greedy algorithm (see Temlyakov (2003a)).

**$X$-Greedy Algorithm (XGA)** We define $f_0 := f$, $G_0 := 0$. Then, for each $m \geq 1$, we have the following inductive definition.

(1) $\varphi_m \in \mathcal{D}$, $\lambda_m \in \mathbb{R}$ are such that (we assume existence)

$$\|f_{m-1} - \lambda_m \varphi_m\|_X = \inf_{g \in \mathcal{D}, \lambda} \|f_{m-1} - \lambda g\|_X. \tag{6.3}$$

(2) Denote

$$f_m := f_{m-1} - \lambda_m \varphi_m, \qquad G_m := G_{m-1} + \lambda_m \varphi_m.$$

The second version of the PGA in a Banach space is based on the concept of a norming (peak) functional. We note that in a Hilbert space a norming functional $F_f$ acts as follows:

$$F_f(g) = \langle f/\|f\|, g \rangle.$$

Therefore, (6.2) can be rewritten in terms of the norming functional $F_{f_{m-1}}$ as follows:

$$F_{f_{m-1}}(\varphi_m) = \sup_{g \in \mathcal{D}} F_{f_{m-1}}(g). \tag{6.4}$$

This observation leads to the class of dual greedy algorithms. We define the Weak Dual Greedy Algorithm with weakness $\tau$ (WDGA($\tau$)) (see Dilworth, Kutzarova and Temlyakov (2002) and Temlyakov (2003a)) that is a generalization of the Weak Greedy Algorithm.

**Weak Dual Greedy Algorithm (WDGA($\tau$))** Let $\tau := \{t_m\}_{m=1}^{\infty}$, $t_m \in [0, 1]$, be a weakness sequence. We define $f_0 := f$. Then, for each $m \geq 1$, we have the following inductive definition.

(1) $\varphi_m \in \mathcal{D}$ is any element satisfying

$$F_{f_{m-1}}(\varphi_m) \geq t_m \|F_{f_{m-1}}\|_{\mathcal{D}}. \tag{6.5}$$

(2) Define $a_m$ as

$$\|f_{m-1} - a_m \varphi_m\| = \min_{a \in \mathbb{R}} \|f_{m-1} - a \varphi_m\|.$$

(3) Let

$$f_m := f_{m-1} - a_m \varphi_m.$$

Let us make a remark that justifies the idea of the dual greedy algorithms in terms of real analysis. We consider here approximation in uniformly smooth Banach spaces. For a Banach space $X$ we define the modulus of smoothness:

$$\rho(u) := \sup_{\|x\|=\|y\|=1} \left( \frac{1}{2}(\|x + uy\| + \|x - uy\|) - 1 \right).$$

The uniformly smooth Banach space is the one with the property

$$\lim_{u \to 0} \rho(u)/u = 0.$$

It is easy to see that for any Banach space $X$ its modulus of smoothness $\rho(u)$ is an even convex function satisfying the inequalities

$$\max(0, u - 1) \le \rho(u) \le u, \quad u \in (0, \infty).$$

We note that from the definition of modulus of smoothness we get the following inequality (6.6).

**Lemma 6.1** *Let $x \ne 0$. Then*

$$0 \le \|x + uy\| - \|x\| - uF_x(y) \le 2\|x\|\rho(u\|y\|/\|x\|). \tag{6.6}$$

*Proof* We have

$$\|x + uy\| \ge F_x(x + uy) = \|x\| + uF_x(y).$$

This proves the first inequality. Next, from the definition of modulus of smoothness it follows that

$$\|x + uy\| + \|x - uy\| \le 2\|x\|(1 + \rho(u\|y\|/\|x\|)). \tag{6.7}$$

Also,

$$\|x - uy\| \ge F_x(x - uy) = \|x\| - uF_x(y). \tag{6.8}$$

Combining (6.7) and (6.8), we obtain

$$\|x + uy\| \le \|x\| + uF_x(y) + 2\|x\|\rho(u\|y\|/\|x\|).$$

This proves the second inequality. □

**Proposition 6.2** *Let $X$ be a uniformly smooth Banach space. Then, for any $x \ne 0$ and $y$ we have*

$$F_x(y) = \left( \frac{d}{du} \|x + uy\| \right)(0) = \lim_{u \to 0} (\|x + uy\| - \|x\|)/u. \tag{6.9}$$

*Proof* The equality (6.9) follows from (6.6) and the property that, for a uniformly smooth Banach space, $\lim_{u \to 0} \rho(u)/u = 0$. $\qquad\square$

Proposition 6.2 shows that in the WDGA we are looking for an element $\varphi_m \in \mathcal{D}$ that provides a large derivative of the quantity $\|f_{m-1} + ug\|$. Thus, we have two classes of greedy algorithms in Banach spaces. The first one is based on a greedy step of the form (6.3). We call this class the class of $X$-greedy algorithms. The second one is based on a greedy step of the form (6.5). We call this class the class of dual greedy algorithms. A very important feature of the dual greedy algorithms is that they can be modified into a weak form. The term "weak" in the definition of the WDGA means that at the greedy step (6.5) we do not aim for the optimal element of the dictionary which realizes the corresponding supremum, but are satisfied with a weaker property than being optimal. The obvious reason for this is that we do not know, in general, that the optimal one exists. Another, practical reason is that the weaker the assumption, the easier it is satisfied and, therefore, the easier it is to realize in practice.

The greedy algorithms defined above (XGA, WDGA) are the generalizations of the PGA and the WGA, studied in Chapter 2, to the case of Banach spaces. The results of Chapter 2 show that the PGA is not the most efficient greedy algorithm for the approximation of elements of $A_1(\mathcal{D})$. It was mentioned in Chapter 2 (see Livshitz and Temlyakov (2003) for the proof) that there exist a dictionary $\mathcal{D}$, a positive constant $C$ and an element $f \in A_1(\mathcal{D})$ such that, for the PGA,

$$\|f_m\| \geq Cm^{-0.27}. \tag{6.10}$$

We note that, even before the lower estimate (6.10) was proved, researchers began looking for other greedy algorithms that provide good rates of approximation of functions from $A_1(\mathcal{D})$. Two different ideas have been used at this step. The first idea was that of relaxation: see Barron (1993), DeVore and Temlyakov (1996), Jones (1992) and Temlyakov (2000b). The corresponding algorithms (for example, the WRGA, studied in Chapter 2) were designed for approximation of functions from $A_1(\mathcal{D})$. These algorithms do not provide an expansion into a series but they have other good features. It was established (see Theorem 2.21) for the WRGA with $\tau = \{1\}$ in a Hilbert space that for $f \in A_1(\mathcal{D})$

$$\|f_m\| \leq Cm^{-1/2}.$$

Also, for the WRGA we always have $G_m \in A_1(\mathcal{D})$. The latter property clearly limits the applicability of the WRGA to the $A_1(\mathcal{D})$.

The second idea was that of building the best approximant from the span($\varphi_1, \ldots, \varphi_m$) instead of the use of only one element $\varphi_m$ for an update of the approximant. This idea was realized in the Weak Orthogonal Greedy Algorithm (see Chapter 2) in the case of a Hilbert space and in the Weak Chebyshev Greedy Algorithm (WCGA) (see Temlyakov (2001b)) in the case of a Banach space.

The realization of both ideas resulted in the construction of algorithms (WRGA and WCGA) that are good for the approximation of functions from $A_1(\mathcal{D})$. We present results on the WCGA in Section 6.2 and results on the WRGA in Section 6.3. The WCGA has the following advantage over the WRGA. It will be proved in Section 6.2 that the WCGA (under some assumptions on the weakness sequence $\tau$) converges for each $f \in X$ in any uniformly smooth Banach space. The WRGA is simpler than the WCGA in the sense of computational complexity. However, the WRGA has limited applicability. It converges only for elements of the closure of the convex hull of a dictionary. In Sections 6.4 and 6.5 we study algorithms that combine good features of both algorithms the WRGA and the WCGA. In the construction of such algorithms we use different forms of relaxation.

The Weak Greedy Algorithm with Free Relaxation (WGAFR) (Temlyakov (2008c)), studied in Section 6.4, is the most powerful of the versions considered here. We prove convergence of the WGAFR in Theorem 6.22. This theorem is the same as the corresponding convergence result for the WCGA (see Theorem 6.6). The results on the rate of convergence for the WGAFR and the WCGA are also the same (see Theorems 6.23 and Theorem 6.14). Thus, the WGAFR performs in the same way as the WCGA from the point of view of convergence and rate of convergence, and outperforms the WCGA in terms of computational complexity.

In the WGAFR we are optimizing over two parameters $w$ and $\lambda$ at each step of the algorithm. In other words we are looking for the best approximation from a two-dimensional linear subspace at each step. In the other version of the weak relaxed greedy algorithms (see the GAWR), considered in Section 6.5, we approximate from a one-dimensional linear subspace at each step of the algorithm. This makes computational complexity of these algorithms very close to that of the PGA. The analysis of the GAWR version turns out to be more complicated than the analysis of the WGAFR. Also, the results obtained for the GAWR are not as general as in the case of the WGAFR. For instance, we present results on the GAWR only in the case $\tau = \{t\}$, when the weakness parameter $t$ is the same for all steps.

The XGA and WDGA have a good feature that distinguishes them from all relaxed greedy algorithms, and also from the WCGA. For an element $f \in X$ they provide an expansion into a series,

$$f \sim \sum_{j=1}^{\infty} c_j(f) g_j(f), \quad g_j(f) \in \mathcal{D}, \quad c_j(f) > 0, \quad j = 1, 2, \ldots \quad (6.11)$$

such that

$$G_m = \sum_{j=1}^{m} c_j(f) g_j(f), \quad f_m = f - G_m.$$

In Section 6.7 we discuss other greedy algorithms that provide the expansion (6.11).

All the algorithms studied in Sections 6.2–6.7 belong to the class of dual greedy algorithms. Results obtained in Sections 6.2–6.7 confirm that dual greedy algorithms provide powerful methods of nonlinear approximation. In Section 6.8 we present some results on the $X$-greedy algorithms. These results are similar to those for the dual greedy algorithms.

The algorithms studied in Sections 6.2–6.8 are very general approximation methods that work well in an arbitrary uniformly smooth Banach space $X$ for any dictionary $\mathcal{D}$. This motivates an attempt, made in Section 6.10, to modify these theoretical approximation methods in the direction of practical applicability. In Section 6.10 we illustrate this idea by modifying the WCGA. We note that Section 6.6 is also devoted to modification of greedy algorithms in order to make them more practically feasible. The main idea of Section 6.6 is to replace the most difficult (expensive) step of an algorithm, namely the greedy step, by a thresholding step.

In Section 6.11 we give an example of how the greedy algorithms can be used in constructing deterministic cubature formulas with error estimates similar to those for the Monte Carlo method.

As a typical example of a uniformly smooth Banach space we will use a space $L_p$, $1 < p < \infty$. It is well known (see, for example, Donahue *et al.* (1997), lemma B.1) that in the case $X = L_p$, $1 \leq p < \infty$, we have

$$\rho(u) \leq u^p/p \quad \text{if} \quad 1 \leq p \leq 2 \quad \text{and} \quad \rho(u) \leq (p-1)u^2/2 \quad \text{if} \quad 2 \leq p < \infty. \tag{6.12}$$

It is also known (see Lindenstrauss and Tzafriri (1977), p. 63) that, for any $X$ with dim $X = \infty$, one has

$$\rho(u) \geq (1 + u^2)^{1/2} - 1$$

and for every $X$, dim $X \geq 2$,

$$\rho(u) \geq Cu^2, \quad C > 0.$$

This limits the power type modulus of smoothness of non-trivial Banach spaces to the case $u^q$, $1 \leq q \leq 2$.

## 6.2 The Weak Chebyshev Greedy Algorithm

Let $\tau := \{t_k\}_{k=1}^{\infty}$ be a given weakness sequence of non-negative numbers $t_k \leq 1, k = 1, \ldots$. We define first the Weak Chebyshev Greedy Algorithm (WCGA) (see Temlyakov (2001b)), which is a generalization for Banach spaces of the Weak Orthogonal Greedy Algorithm.

**Weak Chebyshev Greedy Algorithm (WCGA)** We define $f_0^c := f_0^{c,\tau} := f$. Then for each $m \geq 1$ we have the following inductive definition.

(1) $\varphi_m^c := \varphi_m^{c,\tau} \in \mathcal{D}$ is any element satisfying

$$F_{f_{m-1}^c}(\varphi_m^c) \geq t_m \| F_{f_{m-1}^c} \|_{\mathcal{D}}.$$

(2) Define

$$\Phi_m := \Phi_m^\tau := \text{span}\{\varphi_j^c\}_{j=1}^m,$$

and define $G_m^c := G_m^{c,\tau}$ to be the best approximant to $f$ from $\Phi_m$.

(3) Let

$$f_m^c := f_m^{c,\tau} := f - G_m^c.$$

**Remark 6.3** It follows from the definition of the WCGA that the sequence $\{\| f_m^c \|\}$ is a non-increasing sequence.

We proceed to a theorem on the convergence of the WCGA. In the formulation of this theorem we need a special sequence which is defined for a given modulus of smoothness $\rho(u)$ and a given $\tau = \{t_k\}_{k=1}^{\infty}$.

**Definition 6.4** Let $\rho(u)$ be an even convex function on $(-\infty, \infty)$ with the property $\rho(2) \geq 1$ and

$$\lim_{u \to 0} \rho(u)/u = 0.$$

For any $\tau = \{t_k\}_{k=1}^{\infty}$, $0 < t_k \leq 1$, and $0 < \theta \leq 1/2$ we define $\xi_m := \xi_m(\rho, \tau, \theta)$ as a number $u$ satisfying the equation

$$\rho(u) = \theta t_m u. \tag{6.13}$$

**Remark 6.5** Assumptions on $\rho(u)$ imply that the function

$$s(u) := \rho(u)/u, \quad u \neq 0, \quad s(0) = 0,$$

is a continuous increasing function on $[0, \infty)$ with $s(2) \geq 1/2$. Thus (6.13) has a unique solution $\xi_m = s^{-1}(\theta t_m)$ such that $0 < \xi_m \leq 2$.

The following theorem from Temlyakov (2001b) gives a sufficient condition for convergence of the WCGA.

**Theorem 6.6** *Let $X$ be a uniformly smooth Banach space with modulus of smoothness $\rho(u)$. Assume that a sequence $\tau := \{t_k\}_{k=1}^{\infty}$ satisfies the following condition: for any $\theta > 0$ we have*

$$\sum_{m=1}^{\infty} t_m \xi_m(\rho, \tau, \theta) = \infty.$$

*Then, for any $f \in X$ we have*

$$\lim_{m \to \infty} \| f_m^{c,\tau} \| = 0.$$

**Corollary 6.7** *Let a Banach space $X$ have modulus of smoothness $\rho(u)$ of power type $1 < q \leq 2$, i.e. $\rho(u) \leq \gamma u^q$. Assume that*

$$\sum_{m=1}^{\infty} t_m^p = \infty, \quad p = \frac{q}{q-1}. \tag{6.14}$$

*Then the WCGA converges for any $f \in X$.*

*Proof* Denote $\rho^q(u) := \gamma u^q$. Then

$$\rho(u)/u \leq \rho^q(u)/u,$$

and therefore for any $\theta > 0$ we have

$$\xi_m(\rho, \tau, \theta) \geq \xi_m(\rho^q, \tau, \theta).$$

For $\rho^q$ we get from the definition of $\xi_m$ that

$$\xi_m(\rho^q, \tau, \theta) = (\theta t_m/\gamma)^{1/(q-1)}.$$

Thus (6.14) implies that

$$\sum_{m=1}^{\infty} t_m \xi_m(\rho, \tau, \theta) \geq \sum_{m=1}^{\infty} t_m \xi_m(\rho^q, \tau, \theta) \asymp \sum_{m=1}^{\infty} t_m^p = \infty.$$

It remains to apply Theorem 6.6. $\qquad\square$

The following theorem from Temlyakov (2001b) gives the rate of convergence of the WCGA for $f$ in $A_1(\mathcal{D})$.

**Theorem 6.8** *Let $X$ be a uniformly smooth Banach space with modulus of smoothness $\rho(u) \leq \gamma u^q$, $1 < q \leq 2$. Then, for a sequence $\tau := \{t_k\}_{k=1}^{\infty}$, $t_k \leq 1$, $k = 1, 2, \ldots$, we have for any $f \in A_1(\mathcal{D})$ that*

$$\|f_m^{c,\tau}\| \leq C(q, \gamma)\left(1 + \sum_{k=1}^{m} t_k^p\right)^{-1/p}, \quad p := \frac{q}{q-1},$$

*with a constant $C(q, \gamma)$ which may depend only on $q$ and $\gamma$.*

We will use the following two simple and well known lemmas in the proof of the above two theorems.

**Lemma 6.9** *Let $X$ be a uniformly smooth Banach space and let $L$ be a finite dimensional subspace of $X$. For any $f \in X \setminus L$ let $f_L$ denote the best approximant of $f$ from $L$. Then we have*

$$F_{f-f_L}(\phi) = 0$$

*for any $\phi \in L$.*

*Proof* Let us assume the contrary: there is a $\phi \in L$ such that $\|\phi\| = 1$ and

$$F_{f-f_L}(\phi) = \beta > 0.$$

For any $\lambda$ we have from the definition of $\rho(u)$ that

$$\|f - f_L - \lambda\phi\| + \|f - f_L + \lambda\phi\| \leq 2\|f - f_L\|\left(1 + \rho\left(\frac{\lambda}{\|f - f_L\|}\right)\right). \tag{6.15}$$

Next

$$\|f - f_L + \lambda\phi\| \geq F_{f-f_L}(f - f_L + \lambda\phi) = \|f - f_L\| + \lambda\beta. \tag{6.16}$$

Combining (6.15) and (6.16) we get

$$\|f - f_L - \lambda\phi\| \leq \|f - f_L\|\left(1 - \frac{\lambda\beta}{\|f - f_L\|} + 2\rho\left(\frac{\lambda}{\|f - f_L\|}\right)\right). \tag{6.17}$$

Taking into account that $\rho(u) = o(u)$, we find $\lambda' > 0$ such that

$$\left(1 - \frac{\lambda'\beta}{\|f - f_L\|} + 2\rho\left(\frac{\lambda'}{\|f - f_L\|}\right)\right) < 1.$$

Then (6.17) yields

$$\|f - f_L - \lambda'\phi\| < \|f - f_L\|,$$

which contradicts the assumption that $f_L \in L$ is the best approximant of $f$.

$\square$

**Lemma 6.10** *For any bounded linear functional $F$ and any dictionary $\mathcal{D}$, we have*

$$\|F\|_{\mathcal{D}} := \sup_{g \in \mathcal{D}} F(g) = \sup_{f \in A_1(\mathcal{D})} F(f).$$

*Proof* The inequality

$$\sup_{g \in \mathcal{D}} F(g) \leq \sup_{f \in A_1(\mathcal{D})} F(f)$$

is obvious. We prove the opposite inequality. Take any $f \in A_1(\mathcal{D})$. Then for any $\epsilon > 0$ there exist $g_1^\epsilon, \ldots, g_N^\epsilon \in \mathcal{D}$ and numbers $a_1^\epsilon, \ldots, a_N^\epsilon$ such that $a_i^\epsilon > 0, a_1^\epsilon + \cdots + a_N^\epsilon \leq 1$ and

$$\|f - \sum_{i=1}^{N} a_i^\epsilon g_i^\epsilon\| \leq \epsilon.$$

Thus

$$F(f) \leq \|F\|\epsilon + F\left(\sum_{i=1}^{N} a_i^\epsilon g_i^\epsilon\right) \leq \epsilon\|F\| + \sup_{g \in \mathcal{D}} F(g),$$

which proves Lemma 6.10.

$\square$

We will also need one more lemma from Temlyakov (2001b).

**Lemma 6.11** *Let $X$ be a uniformly smooth Banach space with modulus of smoothness $\rho(u)$. Take a number $\epsilon \geq 0$ and two elements $f$, $f^\epsilon$ from $X$ such that*

$$\|f - f^\epsilon\| \leq \epsilon, \quad f^\epsilon/A(\epsilon) \in A_1(\mathcal{D}),$$

*with some number $A(\epsilon) > 0$. Then we have*

$$\|f_m^{c,\tau}\| \leq \|f_{m-1}^{c,\tau}\| \inf_{\lambda \geq 0} \left(1 - \lambda t_m A(\epsilon)^{-1}\left(1 - \frac{\epsilon}{\|f_{m-1}^{c,\tau}\|}\right) + 2\rho\left(\frac{\lambda}{\|f_{m-1}^{c,\tau}\|}\right)\right),$$

*for $m = 1, 2, \ldots$.*

*Proof* We have for any $\lambda$

$$\|f_{m-1}^c - \lambda\varphi_m^c\| + \|f_{m-1}^c + \lambda\varphi_m^c\| \leq 2\|f_{m-1}^c\|\left(1 + \rho\left(\frac{\lambda}{\|f_{m-1}^c\|}\right)\right) \quad (6.18)$$

and by (1) from the definition of the WCGA and Lemma 6.10 we get

$$F_{f_{m-1}^c}(\varphi_m^c) \geq t_m \sup_{g \in \mathcal{D}} F_{f_{m-1}^c}(g)$$

$$= t_m \sup_{\phi \in A_1(\mathcal{D})} F_{f_{m-1}^c}(\phi) \geq t_m A(\epsilon)^{-1} F_{f_{m-1}^c}(f^\epsilon).$$

By Lemma 6.9 we obtain

$$F_{f_{m-1}^c}(f^\epsilon) = F_{f_{m-1}^c}(f + f^\epsilon - f) \geq F_{f_{m-1}^c}(f) - \epsilon$$

$$= F_{f_{m-1}^c}(f_{m-1}^c) - \epsilon = \|f_{m-1}^c\| - \epsilon.$$

Thus, as in (6.17) we get from (6.18)

$$\|f_m^c\| \leq \inf_{\lambda \geq 0} \|f_{m-1}^c - \lambda \varphi_m^c\|$$

$$\leq \|f_{m-1}^c\| \inf_{\lambda \geq 0} \left( 1 - \lambda t_m A(\epsilon)^{-1} \left( 1 - \frac{\epsilon}{\|f_{m-1}^c\|} \right) + 2\rho \left( \frac{\lambda}{\|f_{m-1}^c\|} \right) \right),$$

$$(6.19)$$

which proves the lemma. □

*Proof of Theorem 6.6* The definition of the WCGA implies that $\{\|f_m^c\|\}$ is a non-increasing sequence. Therefore we have

$$\lim_{m \to \infty} \|f_m^c\| = \alpha.$$

We prove that $\alpha = 0$ by contradiction. Assume to the contrary that $\alpha > 0$. Then, for any $m$ we have

$$\|f_m^c\| \geq \alpha.$$

We set $\epsilon = \alpha/2$ and find $f^\epsilon$ such that

$$\|f - f^\epsilon\| \leq \epsilon \quad \text{and} \quad f^\epsilon/A(\epsilon) \in A_1(\mathcal{D}),$$

with some $A(\epsilon)$. Then, by Lemma 6.11 we get

$$\|f_m^c\| \leq \|f_{m-1}^c\| \inf_\lambda (1 - \lambda t_m A(\epsilon)^{-1}/2 + 2\rho(\lambda/\alpha)).$$

Let us specify $\theta := \alpha/8A(\epsilon)$ and take $\lambda = \alpha\xi_m(\rho, \tau, \theta)$. Then we obtain

$$\|f_m^c\| \leq \|f_{m-1}^c\|(1 - 2\theta t_m \xi_m).$$

The assumption

$$\sum_{m=1}^{\infty} t_m \xi_m = \infty$$

implies that

$$\|f_m^c\| \to 0 \quad \text{as} \quad m \to \infty.$$

We have a contradiction, which proves the theorem. $\qquad\square$

*Proof of Theorem 6.8* By Lemma 6.11 with $\epsilon = 0$ and $A(\epsilon) = 1$ we have for $f \in A_1(\mathcal{D})$ that

$$\|f_m^c\| \leq \|f_{m-1}^c\| \inf_{\lambda \geq 0} \left( 1 - \lambda t_m + 2\gamma \left( \frac{\lambda}{\|f_{m-1}^c\|} \right)^q \right). \tag{6.20}$$

Choose $\lambda$ from the equation

$$\frac{1}{2}\lambda t_m = 2\gamma \left( \frac{\lambda}{\|f_{m-1}^c\|} \right)^q,$$

which implies that

$$\lambda = \|f_{m-1}^c\|^{q/(q-1)} (4\gamma)^{-1/(q-1)} t_m^{1/(q-1)}.$$

Let

$$A_q := 2(4\gamma)^{1/(q-1)}.$$

Using the notation $p := q/(q-1)$ we get from (6.20)

$$\|f_m^c\| \leq \|f_{m-1}^c\| \left( 1 - \frac{1}{2}\lambda t_m \right) = \|f_{m-1}^c\|(1 - t_m^p \|f_{m-1}^c\|^p / A_q).$$

Raising both sides of this inequality to the power $p$ and taking into account the inequality $x^r \leq x$ for $r \geq 1, 0 \leq x \leq 1$, we obtain

$$\|f_m^c\|^p \leq \|f_{m-1}^c\|^p (1 - t_m^p \|f_{m-1}^c\|^p / A_q).$$

By an analog of Lemma 2.16 (see Temlyakov (2000b), lemma 3.1), using the estimate $\|f\|^p \leq 1 < A_q$ we get

$$\|f_m^c\|^p \leq A_q \left( 1 + \sum_{n=1}^m t_n^p \right)^{-1},$$

which implies

$$\|f_m^c\| \leq C(q, \gamma) \left( 1 + \sum_{n=1}^m t_n^p \right)^{-1/p}.$$

Theorem 6.8 is now proved. $\qquad\square$

**Remark 6.12** Theorem 6.8 holds for a slightly modified version of the WCGA, the WCGA(1), for which at step (1) we require

$$F_{f_{m-1}^{c(1)}}(\varphi_m^{c(1)}) \geq t_m \|f_{m-1}^{c(1)}\|. \tag{6.21}$$

This statement follows from the fact that, in the proof of Theorem 6.8, the relation

$$F_{f_{m-1}^c}(\varphi_m^c) \geq t_m \sup_{g \in \mathcal{D}} F_{f_{m-1}^c}(g)$$

was used only to get (6.21).

**Proposition 6.13** *Condition (6.14) in Corollary 6.7 is sharp.*

*Proof* Let $1 < q \leq 2$. Consider $X = \ell_q$. It is known (Lindenstrauss and Tzafriri (1977), p. 67) that $\ell_q$, $1 < q \leq 2$, is a uniformly smooth Banach space with modulus of smoothness $\rho(u)$ of power type $q$. Denote $p := q/(q-1)$ and take any $\{t_k\}_{k=1}^\infty$, $0 < t_k \leq 1$, such that

$$\sum_{k=1}^\infty t_k^p < \infty. \tag{6.22}$$

Choose $\mathcal{D}$ as a standard basis $\{e_j\}_{j=1}^\infty$, $e_j := (0, \ldots, 0, 1, 0, \ldots)$, for $\ell_q$. Consider the following realization of the WCGA for

$$f := (1, t_1^{1/(q-1)}, t_2^{1/(q-1)}, \ldots).$$

First of all, (6.22) guarantees that $f \in \ell_q$. Next, it is well known that $F_f$ can be identified as

$$F_f = (1, t_1, t_2, \ldots)/\left(1 + \sum_{k=1}^\infty t_k^p\right)^{1/p} \in \ell_p.$$

At the first step of the WCGA we pick $\varphi_1 = e_2$ and get

$$f_1^c = (1, 0, t_2^{1/(q-1)}, \ldots).$$

We continue with $f$ replaced by $f_1$ and so on. After $m$ steps we get

$$f_m^c = (1, 0, \ldots, 0, t_{m+1}^{1/(q-1)}, \ldots).$$

It is clear that for all $m$ we have $\|f_m^c\|_{\ell_q} \geq 1$. □

The following variant of Theorem 6.8 (see Temlyakov (2008c)) follows from Lemma 6.11.

**Theorem 6.14** *Let $X$ be a uniformly smooth Banach space with modulus of smoothness $\rho(u) \leq \gamma u^q$, $1 < q \leq 2$. Take a number $\epsilon \geq 0$ and two elements $f$, $f^\epsilon$ from $X$ such that*

$$\|f - f^\epsilon\| \leq \epsilon, \quad f^\epsilon / A(\epsilon) \in A_1(\mathcal{D}),$$

*with some number $A(\epsilon) > 0$. Then we have ($p := q/(q-1)$)*

$$\|f_m^{c,\tau}\| \leq \max\left(2\epsilon, C(q, \gamma)(A(\epsilon) + \epsilon)\left(1 + \sum_{k=1}^{m} t_k^p\right)^{-1/p}\right). \tag{6.23}$$

## 6.3 Relaxation; co-convex approximation

In this section we study a generalization for Banach spaces of relaxed greedy algorithms considered in Chapter 2. We present here results from Temlyakov (2001b). Let $\tau := \{t_k\}_{k=1}^{\infty}$ be a given weakness sequence of numbers $t_k \in [0, 1]$, $k = 1, \dots$.

**Weak Relaxed Greedy Algorithm (WRGA)** We define $f_0^r := f_0^{r,\tau} := f$ and $G_0^r := G_0^{r,\tau} := 0$. Then, for each $m \geq 1$ we have the following inductive definition.

(1) $\varphi_m^r := \varphi_m^{r,\tau} \in \mathcal{D}$ is any element satisfying

$$F_{f_{m-1}^r}(\varphi_m^r - G_{m-1}^r) \geq t_m \sup_{g \in \mathcal{D}} F_{f_{m-1}^r}(g - G_{m-1}^r).$$

(2) Find $0 \leq \lambda_m \leq 1$ such that

$$\|f - ((1 - \lambda_m)G_{m-1}^r + \lambda_m \varphi_m^r)\| = \inf_{0 \leq \lambda \leq 1} \|f - ((1 - \lambda)G_{m-1}^r + \lambda \varphi_m^r)\|$$

and define

$$G_m^r := G_m^{r,\tau} := (1 - \lambda_m)G_{m-1}^r + \lambda_m \varphi_m^r.$$

(3) Let

$$f_m^r := f_m^{r,\tau} := f - G_m^r.$$

**Remark 6.15** It follows from the definition of the WRGA that the sequence $\{\|f_m^r\|\}$ is a non-increasing sequence.

We call the WRGA *relaxed* because at the $m$th step of the algorithm we use a linear combination (convex combination) of the previous approximant $G_{m-1}^r$ and a new element $\varphi_m^r$. The relaxation parameter $\lambda_m$ in the WRGA is chosen at the $m$th step depending on $f$. We prove here the analogs of Theorems 6.6 and 6.8 for the Weak Relaxed Greedy Algorithm.

**Theorem 6.16** *Let $X$ be a uniformly smooth Banach space with modulus of smoothness $\rho(u)$. Assume that a sequence $\tau := \{t_k\}_{k=1}^{\infty}$ satisfies the following condition: for any $\theta > 0$ we have*

$$\sum_{m=1}^{\infty} t_m \xi_m(\rho, \tau, \theta) = \infty.$$

*Then, for any $f \in A_1(\mathcal{D})$ we have*

$$\lim_{m \to \infty} \|f_m^{r,\tau}\| = 0.$$

**Theorem 6.17** *Let $X$ be a uniformly smooth Banach space with modulus of smoothness $\rho(u) \leq \gamma u^q$, $1 < q \leq 2$. Then, for a sequence $\tau := \{t_k\}_{k=1}^{\infty}$, $t_k \leq 1$, $k = 1, 2, \ldots$, we have for any $f \in A_1(\mathcal{D})$ that*

$$\|f_m^{r,\tau}\| \leq C_1(q, \gamma) \left( 1 + \sum_{k=1}^{m} t_k^p \right)^{-1/p}, \quad p := \frac{q}{q-1},$$

*with a constant $C_1(q, \gamma)$ which may depend only on $q$ and $\gamma$.*

*Proof of Theorems 6.16 and 6.17* This proof is similar to the proof of Theorems 6.6 and 6.8. Instead of Lemma 6.11 we use the following lemma.

**Lemma 6.18** *Let $X$ be a uniformly smooth Banach space with modulus of smoothness $\rho(u)$. Then, for any $f \in A_1(\mathcal{D})$ we have*

$$\|f_m^{r,\tau}\| \leq \|f_{m-1}^{r,\tau}\| \inf_{0 \leq \lambda \leq 1} \left( 1 - \lambda t_m + 2\rho \left( \frac{2\lambda}{\|f_{m-1}^{r,\tau}\|} \right) \right), \quad m = 1, 2, \ldots$$

*Proof* We have

$$f_m^r := f - ((1 - \lambda_m) G_{m-1}^r + \lambda_m \varphi_m^r) = f_{m-1}^r - \lambda_m (\varphi_m^r - G_{m-1}^r)$$

and

$$\|f_m^r\| = \inf_{0 \leq \lambda \leq 1} \|f_{m-1}^r - \lambda(\varphi_m^r - G_{m-1}^r)\|.$$

As for (6.18) we have, for any $\lambda$,

$$\|f_{m-1}^r - \lambda(\varphi_m^r - G_{m-1}^r)\| + \|f_{m-1}^r + \lambda(\varphi_m^r - G_{m-1}^r)\|$$

$$\leq 2\|f_{m-1}^r\| \left( 1 + \rho \left( \frac{\lambda\|\varphi_m^r - G_{m-1}^r\|}{\|f_{m-1}^r\|} \right) \right). \tag{6.24}$$

Next for $\lambda \geq 0$ we obtain

$$\|f_{m-1}^r + \lambda(\varphi_m^r - G_{m-1}^r)\| \geq F_{f_{m-1}^r}(f_{m-1}^r + \lambda(\varphi_m^r - G_{m-1}^r))$$

$$= \|f_{m-1}^r\| + \lambda F_{f_{m-1}^r}(\varphi_m^r - G_{m-1}^r)$$

$$\geq \|f_{m-1}^r\| + \lambda t_m \sup_{g \in \mathcal{D}} F_{f_{m-1}^r}(g - G_{m-1}^r).$$

Using Lemma 6.10 we continue:

$$= \|f_{m-1}^r\| + \lambda t_m \sup_{\phi \in A_1(\mathcal{D})} F_{f_{m-1}^r}(\phi - G_{m-1}^r)$$

$$\geq \|f_{m-1}^r\| + \lambda t_m \|f_{m-1}^r\|.$$

Using the trivial estimate $\|\varphi_m^r - G_{m-1}^r\| \leq 2$ we obtain

$$\|f_{m-1}^r - \lambda(\varphi_m^r - G_{m-1}^r)\| \leq \|f_{m-1}^r\| \left(1 - \lambda t_m + 2\rho\left(\frac{2\lambda}{\|f_{m-1}^r\|}\right)\right), \quad (6.25)$$

from (6.24), which proves Lemma 6.18. $\qquad\square$

The remaining part of the proof uses the inequality (6.25) in the same way (6.19) was used in the proof of Theorems 6.6 and 6.8. The only additional difficulty here is that we are optimizing over $0 \leq \lambda \leq 1$. However, it is easy to check that the corresponding $\lambda$ chosen in a similar way always satisfies the restriction $0 \leq \lambda \leq 1$. In the proof of Theorem 6.16 we choose $\theta = \alpha/8$ and $\lambda = \alpha \xi_m(\rho, \tau, \theta)/2$ and in the proof of Theorem 6.17 we choose $\lambda$ from the equation

$$\frac{1}{2}\lambda t_m = 2\gamma(2\lambda)^q \|f_{m-1}^r\|^{-q}.$$

$\qquad\square$

**Remark 6.19** Theorems 6.16 and 6.17 hold for a slightly modified version of the WRGA, the WRGA(1), for which at step (1) we require

$$F_{f_{m-1}^{r(1)}}(\varphi_m^{r(1)} - G_{m-1}^{r(1)}) \geq t_m \|f_{m-1}^{r(1)}\|. \quad (6.26)$$

This follows from the observation that in the proof of Lemma 6.18 we used the inequality from step (1) of the WRGA only to derive (6.26). It is clear from Lemma 6.10 that in the case of approximation of $f \in A_1(\mathcal{D})$, the requirement (6.26) is weaker and easier to check than (1) of the WRGA.

## 6.4 Free relaxation

Both of the above algorithms, the WCGA and the WRGA, use the functional $F_{f_{m-1}}$ in a search for the $m$th element $\varphi_m$ from the dictionary to be used in the approximation. The construction of the approximant in the WRGA is different from the construction in the WCGA. In the WCGA we build the approximant $G_m^c$ so as to use maximally the approximation power of the elements $\varphi_1, \ldots, \varphi_m$. The WRGA, by definition, is designed for the approximation of functions from $A_1(\mathcal{D})$. In building the approximant in the WRGA we keep the property $G_m^r \in A_1(\mathcal{D})$. As mentioned in Section 6.3 the relaxation parameter $\lambda_m$ in the WRGA is chosen at the $m$th step depending on $f$. The following modification of the above idea of relaxation in greedy approximation will be studied in this section (see Temlyakov (2008c)).

**Weak Greedy Algorithm with Free Relaxation (WGAFR)** Let $\tau := \{t_m\}_{m=1}^{\infty}, t_m \in [0, 1]$, be a weakness sequence. We define $f_0 := f$ and $G_0 := 0$. Then for each $m \geq 1$ we have the following inductive definition.

(1) $\varphi_m \in \mathcal{D}$ is any element satisfying

$$F_{f_{m-1}}(\varphi_m) \geq t_m \| F_{f_{m-1}} \|_{\mathcal{D}}.$$

(2) Find $w_m$ and $\lambda_m$ such that

$$\| f - ((1 - w_m)G_{m-1} + \lambda_m \varphi_m) \| = \inf_{\lambda, w} \| f - ((1 - w)G_{m-1} + \lambda \varphi_m) \|$$

and define

$$G_m := (1 - w_m)G_{m-1} + \lambda_m \varphi_m.$$

(3) Let

$$f_m := f - G_m.$$

We begin with an analog of Lemma 6.11.

**Lemma 6.20** *Let $X$ be a uniformly smooth Banach space with modulus of smoothness $\rho(u)$. Take a number $\epsilon \geq 0$ and two elements $f$, $f^\epsilon$ from $X$ such that*

$$\| f - f^\epsilon \| \leq \epsilon, \quad f^\epsilon / A(\epsilon) \in A_1(\mathcal{D}),$$

*with some number $A(\epsilon) \geq \epsilon$. Then we have for the WGAFR*

$$\| f_m \| \leq \| f_{m-1} \| \inf_{\lambda \geq 0} \left( 1 - \lambda t_m A(\epsilon)^{-1} \left( 1 - \frac{\epsilon}{\| f_{m-1} \|} \right) + 2\rho \left( \frac{5\lambda}{\| f_{m-1} \|} \right) \right),$$

*for*  $m = 1, 2, \ldots.$

*Proof* By the definition of $f_m$

$$\|f_m\| \leq \inf_{\lambda \geq 0, w} \|f_{m-1} + wG_{m-1} - \lambda\varphi_m\|.$$

As in the arguments in the proof of Lemma 6.11, we use the inequality

$$\|f_{m-1} + wG_{m-1} - \lambda\varphi_m\| + \|f_{m-1} - wG_{m-1} + \lambda\varphi_m\|$$
$$\leq 2\|f_{m-1}\|(1 + \rho(\|wG_{m-1} - \lambda\varphi_m\|/\|f_{m-1}\|)) \qquad (6.27)$$

and estimate for $\lambda \geq 0$

$$\|f_{m-1} - wG_{m-1} + \lambda\varphi_m\| \geq F_{f_{m-1}}(f_{m-1} - wG_{m-1} + \lambda\varphi_m)$$
$$\geq \|f_{m-1}\| - F_{f_{m-1}}(wG_{m-1}) + \lambda t_m \sup_{g \in \mathcal{D}} F_{f_{m-1}}(g).$$

By Lemma 6.10, we continue:

$$= \|f_{m-1}\| - F_{f_{m-1}}(wG_{m-1}) + \lambda t_m \sup_{\phi \in A_1(\mathcal{D})} F_{f_{m-1}}(\phi)$$
$$\geq \|f_{m-1}\| - F_{f_{m-1}}(wG_{m-1})$$
$$+ \lambda t_m A(\epsilon)^{-1} F_{f_{m-1}}(f^\epsilon)$$
$$\geq \|f_{m-1}\| - F_{f_{m-1}}(wG_{m-1})$$
$$+ \lambda t_m A(\epsilon)^{-1}(F_{f_{m-1}}(f) - \epsilon).$$

We set $w^* := \lambda t_m A(\epsilon)^{-1}$ and obtain

$$\|f_{m-1} - w^*G_{m-1} + \lambda\varphi_m\| \geq \|f_{m-1}\| + \lambda t_m A(\epsilon)^{-1}(\|f_{m-1}\| - \epsilon). \quad (6.28)$$

Combining (6.27) and (6.28) we get

$$\|f_m\| \leq \|f_{m-1}\| \inf_{\lambda \geq 0}(1 - \lambda t_m A(\epsilon)^{-1}(1 - \epsilon/\|f_{m-1}\|)$$
$$+ 2\rho(\|w^*G_{m-1} - \lambda\varphi_m\|/\|f_{m-1}\|)).$$

We now estimate

$$\|w^*G_{m-1} - \lambda\varphi_m\| \leq w^*\|G_{m-1}\| + \lambda.$$

Next,

$$\|G_{m-1}\| = \|f - f_{m-1}\| \leq 2\|f\| \leq 2(\|f^\epsilon\| + \epsilon) \leq 2(A(\epsilon) + \epsilon).$$

Thus, under assumption $A(\epsilon) \geq \epsilon$ we get

$$w^*\|G_{m-1}\| \leq 2\lambda t_m(A(\epsilon) + \epsilon)/A(\epsilon) \leq 4\lambda.$$

Finally,

$$\|w^*G_{m-1} - \lambda\varphi_m\| \leq 5\lambda.$$

This completes the proof of Lemma 6.20.                                    □

**Remark 6.21** It follows from the definition of the WGAFR that the sequence $\{\|f_m\|\}$ is a non-increasing sequence.

We now prove a convergence theorem for an arbitrary uniformly smooth Banach space. The modulus of smoothness $\rho(u)$ of a uniformly smooth Banach space is an even convex function such that $\rho(0) = 0$ and $\lim_{u\to 0} \rho(u)/u = 0$. The function $s(u) := \rho(u)/u$, $s(0) := 0$, associated with $\rho(u)$ is a continuous increasing function on $[0, \infty)$. Therefore, the inverse function $s^{-1}(\cdot)$ is well defined.

**Theorem 6.22** *Let $X$ be a uniformly smooth Banach space with modulus of smoothness $\rho(u)$. Assume that a sequence $\tau := \{t_k\}_{k=1}^{\infty}$ satisfies the following condition. For any $\theta > 0$ we have*

$$\sum_{m=1}^{\infty} t_m s^{-1}(\theta t_m) = \infty. \qquad (6.29)$$

*Then, for any $f \in X$ we have for the WGAFR*

$$\lim_{m\to\infty} \|f_m\| = 0.$$

*Proof* By Remark 6.21, $\{\|f_m\|\}$ is a non-increasing sequence. Therefore we have

$$\lim_{m\to\infty} \|f_m\| = \beta.$$

We prove that $\beta = 0$ by contradiction. Assume the contrary, that $\beta > 0$. Then, for any $m$ we have

$$\|f_m\| \geq \beta.$$

We set $\epsilon = \beta/2$ and find $f^\epsilon$ such that

$$\|f - f^\epsilon\| \leq \epsilon \quad \text{and} \quad f^\epsilon/A(\epsilon) \in A_1(\mathcal{D}),$$

with some $A(\epsilon) \geq \epsilon$. Then, by Lemma 6.20 we get

$$\|f_m\| \leq \|f_{m-1}\| \inf_{\lambda\geq 0}(1 - \lambda t_m A(\epsilon)^{-1}/2 + 2\rho(5\lambda/\beta)).$$

Let us specify $\theta := \beta/(40A(\epsilon))$ and take $\lambda = \beta s^{-1}(\theta t_m)/5$. Then we obtain

$$\|f_m\| \le \|f_{m-1}\|(1 - 2\theta t_m s^{-1}(\theta t_m)).$$

The assumption

$$\sum_{m=1}^{\infty} t_m s^{-1}(\theta t_m) = \infty$$

implies that

$$\|f_m\| \to 0 \quad \text{as} \quad m \to \infty.$$

We have a contradiction, which proves the theorem. $\qquad\square$

**Theorem 6.23** *Let $X$ be a uniformly smooth Banach space with modulus of smoothness $\rho(u) \le \gamma u^q$, $1 < q \le 2$. Take a number $\epsilon \ge 0$ and two elements $f$, $f^\epsilon$ from $X$ such that*

$$\|f - f^\epsilon\| \le \epsilon, \quad f^\epsilon/A(\epsilon) \in A_1(\mathcal{D}),$$

*with some number $A(\epsilon) > 0$. Then we have for the WGAFR*

$$\|f_m\| \le \max\left(2\epsilon, C(q, \gamma)(A(\epsilon) + \epsilon)\left(1 + \sum_{k=1}^{m} t_k^p\right)^{-1/p}\right), \quad p := q/(q-1).$$

*Proof* It is clear that it suffices to consider the case $A(\epsilon) \ge \epsilon$. Otherwise, $\|f_m\| \le \|f\| \le \|f^\epsilon\| + \epsilon \le 2\epsilon$. Also, assume $\|f_m\| > 2\epsilon$ (otherwise Theorem 6.23 trivially holds). Then, by Remark 6.21 we have for all $k = 0, 1, \ldots, m$ that $\|f_k\| > 2\epsilon$. By Lemma 6.20 we obtain

$$\|f_k\| \le \|f_{k-1}\| \inf_{\lambda \ge 0}\left(1 - \lambda t_k A(\epsilon)^{-1}/2 + 2\gamma\left(\frac{5\lambda}{\|f_{k-1}\|}\right)^q\right). \qquad (6.30)$$

Choose $\lambda$ from the equation

$$\frac{\lambda t_k}{4A(\epsilon)} = 2\gamma\left(\frac{5\lambda}{\|f_{k-1}\|}\right)^q,$$

which implies that

$$\lambda = \|f_{k-1}\|^{q/(q-1)} 5^{-q/(q-1)}(8\gamma A(\epsilon))^{-1/(q-1)} t_k^{1/(q-1)}.$$

Define

$$A_q := 4(8\gamma)^{1/(q-1)} 5^{q/(q-1)}.$$

Using the notation $p := q/(q-1)$, we get from (6.30)

$$\|f_k\| \le \|f_{k-1}\| \left(1 - \frac{1}{4}\frac{\lambda t_k}{A(\epsilon)}\right) = \|f_{k-1}\| \left(1 - \frac{t_k^p \|f_{k-1}\|^p}{A_q A(\epsilon)^p}\right).$$

Raising both sides of this inequality to the power $p$ and taking into account the inequality $x^r \le x$ for $r \ge 1, 0 \le x \le 1$, we obtain

$$\|f_k\|^p \le \|f_{k-1}\|^p \left(1 - \frac{t_k^p \|f_{k-1}\|^p}{A_q A(\epsilon)^p}\right).$$

By an analog of Lemma 2.16 (see Temlyakov (2000b), lemma 3.1), using the estimates $\|f\| \le A(\epsilon) + \epsilon$ and $A_q > 1$, we get

$$\|f_m\|^p \le A_q (A(\epsilon) + \epsilon)^p \left(1 + \sum_{k=1}^m t_k^p\right)^{-1}$$

which implies

$$\|f_m\| \le C(q, \gamma)(A(\epsilon) + \epsilon) \left(1 + \sum_{k=1}^m t_k^p\right)^{-1/p}.$$

Theorem 6.23 is proved. □

## 6.5 Fixed relaxation

In this section we consider a relaxed greedy algorithm with relaxation prescribed in advance. Let a sequence $\mathbf{r} := \{r_k\}_{k=1}^\infty$, $r_k \in [0, 1)$, of relaxation parameters be given. Then at each step of our new algorithm we build the $m$th approximant of the form $G_m = (1 - r_m)G_{m-1} + \lambda\varphi_m$. With an approximant of this form we are not limited to the approximation of functions from $A_1(\mathcal{D})$ as in the WRGA. In this section we study the Greedy Algorithm with Weakness parameter $t$ and Relaxation $\mathbf{r}$ (GAWR$(t, \mathbf{r})$). In addition to the acronym GAWR$(t, \mathbf{r})$ we will use the abbreviated acronym GAWR. We give a general definition of the algorithm in the case of a weakness sequence $\tau$. We present in this section results from Temlyakov (2008c).

**GAWR$(\tau, \mathbf{r})$** Let $\tau := \{t_m\}_{m=1}^\infty$, $t_m \in (0, 1]$, be a weakness sequence and let $\mathbf{r} := \{r_m\}_{m=1}^\infty$, $r_m \in [0, 1)$, be a relaxation sequence. We define $f_0 := f$ and $G_0 := 0$. Then, for each $m \ge 1$ we have the following inductive definition.

(1) $\varphi_m \in \mathcal{D}$ is any element satisfying

$$F_{f_{m-1}}(\varphi_m) \ge t_m \|F_{f_{m-1}}\|_{\mathcal{D}}.$$

(2) Find $\lambda_m \geq 0$ such that

$$\|f - ((1 - r_m)G_{m-1} + \lambda_m \varphi_m)\| = \inf_{\lambda \geq 0} \|f - ((1 - r_m)G_{m-1} + \lambda \varphi_m)\|$$

and define

$$G_m := (1 - r_m)G_{m-1} + \lambda_m \varphi_m.$$

(3) Let

$$f_m := f - G_m.$$

In the case $\tau = \{t\}$ we write $t$ instead of $\tau$ in the notation. We note that in the case $r_k = 0, k = 1, \ldots,$ when there is no relaxation the GAWR($\tau, \mathbf{0}$) coincides with the Weak Dual Greedy Algorithm. We now proceed to the GAWR. We begin with an analog of Lemma 6.11.

**Lemma 6.24** *Let $X$ be a uniformly smooth Banach space with modulus of smoothness $\rho(u)$. Take a number $\epsilon \geq 0$ and two elements $f$, $f^\epsilon$ from $X$ such that*

$$\|f - f^\epsilon\| \leq \epsilon, \quad f^\epsilon / A(\epsilon) \in A_1(\mathcal{D}),$$

*with some number $A(\epsilon) > 0$. Then we have for the GAWR($t, \mathbf{r}$)*

$$\|f_m\| \leq \|f_{m-1}\|(1 - r_m(1 - \epsilon/\|f_{m-1}\|))$$
$$+ 2\rho((r_m(\|f\| + A(\epsilon)/t))/((1 - r_m)\|f_{m-1}\|)), \quad m = 1, 2, \ldots$$

**Theorem 6.25** *Let a sequence $\mathbf{r}$ satisfy the conditions*

$$\sum_{k=1}^{\infty} r_k = \infty, \quad r_k \to 0 \quad as \quad k \to \infty.$$

*Then the GAWR($t, \mathbf{r}$) converges in any uniformly smooth Banach space for each $f \in X$ and for all dictionaries $\mathcal{D}$.*

*Proof* We prove this theorem in two steps.

(I) First, we prove that $\liminf_{m \to \infty} \|f_m\| = 0$. The proof goes by contradiction. We want to prove that $\liminf_{m \to \infty} \|f_m\| = 0$. Assume the contrary. Then there exists $K$ and $\beta > 0$ such that we have for all $k \geq K$ that $\|f_k\| \geq \beta$. By Lemma 6.24 for $m > K$

$$\|f_m\| \leq \|f_{m-1}\| \left(1 - r_m \left(1 - \frac{\epsilon}{\beta}\right) + 2\rho \left(\frac{r_m(\|f\| + A(\epsilon)/t)}{(1 - r_m)\beta}\right)\right).$$

We choose $\epsilon := \beta/2$. Using the assumption that $X$ is uniformly smooth and the assumption $r_k \to 0$ as $k \to \infty$, we find $N \geq K$ such that for $m \geq N$ we have

$$2\rho \left( \frac{r_m(\|f\| + A(\epsilon)/t)}{(1 - r_m)\beta} \right) \leq r_m/4.$$

Then, for $m > N$,

$$\|f_m\| \leq \|f_{m-1}\|(1 - r_m/4).$$

The assumption $\sum_{m=1}^{\infty} r_m = \infty$ implies that $\|f_m\| \to 0$ as $m \to \infty$. The obtained contradiction to the assumption $\beta > 0$ completes the proof of part (I).

(II) Secondly, we prove that $\lim_{m \to \infty} \|f_m\| = 0$. Using the assumption $r_k \to 0$ as $k \to \infty$ we find $N_1$ such that for $k \geq N_1$ we have $r_k \leq 1/2$. For such $k$ we obtain from Lemma 6.24

$$\|f_k\| - \epsilon \leq (1 - r_k)(\|f_{k-1}\| - \epsilon) + 2\|f_{k-1}\|\rho \left( \frac{Br_k}{\|f_{k-1}\|} \right), \tag{6.31}$$

with $B := 2(\|f\| + A(\epsilon)/t)$. Denote $a_k := \|f_{k-1}\| - \epsilon$. We note that from the definition of $f_k$ it follows that

$$a_{k+1} \leq a_k + r_k\|f\|. \tag{6.32}$$

Using the fact that the function $\rho(u)/u$ is monotone increasing on $[0, \infty)$, we obtain from (6.31) for $a_k > 0$

$$a_{k+1} \leq a_k \left( 1 - r_k + 2\frac{\|f_{k-1}\|}{a_k}\rho \left( \frac{Br_k}{\|f_{k-1}\|} \right) \right) \leq a_k \left( 1 - r_k + 2\rho \left( \frac{Br_k}{a_k} \right) \right). \tag{6.33}$$

We now introduce an auxiliary sequence $\{b_k\}$ of positive numbers that is defined by the equation

$$2\rho(Br_k/b_k) = r_k.$$

The property $\rho(u)/u \to 0$ as $u \to 0$ implies $b_k \to 0$ as $k \to \infty$. Inequality (6.33) guarantees that for $k \geq N_1$ such that $a_k \geq b_k$ we have $a_{k+1} \leq a_k$.

Let

$$U := \{k : \quad k \geq N_1, \quad a_k \geq b_k\}.$$

If the set $U$ is finite then we get

$$\limsup_{k \to \infty} a_k \leq \lim_{k \to \infty} b_k = 0.$$

This implies

$$\limsup_{m \to \infty} \| f_m \| \le \epsilon.$$

Consider the case when $U$ is infinite. We note that part (I) of the proof implies that there is a subsequence $\{k_j\}$ such that $a_{k_j} \le 0$, $j = 1, 2, \ldots$. This means that

$$U = \cup_{j=1}^{\infty} [l_j, n_j]$$

with the property $n_{j-1} < l_j - 1$. For $k \notin U, k \ge N_1$ we have

$$a_k < b_k. \tag{6.34}$$

For $k \in [l_j, n_j]$, we have by (6.32) and the monotonicity property of $a_k$, when $k \in [l_j, n_j]$, that

$$a_k \le a_{l_j} \le a_{l_j-1} + r_{l_j-1} \| f \| \le b_{l_j-1} + r_{l_j-1} \| f \|. \tag{6.35}$$

By (6.34) and (6.35) we obtain

$$\limsup_{k \to \infty} a_k \le 0 \Rightarrow \limsup_{m \to \infty} \| f_m \| \le \epsilon.$$

Taking into account that $\epsilon > 0$ is arbitrary, we complete the proof. $\square$

We now proceed to results on the rate of approximation. We will need the following technical lemma (see Temlyakov (1999, 2008c) and Lemma 2.34).

**Lemma 6.26** *Let a sequence* $\{a_n\}_{n=1}^{\infty}$ *have the following property. For given positive numbers* $\alpha < \gamma \le 1$, $A > a_1$, *we have for all* $n \ge 2$

$$a_n \le a_{n-1} + A(n - 1)^{-\alpha}. \tag{6.36}$$

*If for some* $v \ge 2$ *we have*

$$a_v \ge A v^{-\alpha}$$

*then*

$$a_{v+1} \le a_v(1 - \gamma/v). \tag{6.37}$$

*Then there exists a constant* $C(\alpha, \gamma)$ *such that for all* $n = 1, 2, \ldots$ *we have*

$$a_n \le C(\alpha, \gamma) A n^{-\alpha}.$$

**Theorem 6.27** *Let* $X$ *be a uniformly smooth Banach space with modulus of smoothness* $\rho(u) \le \gamma u^q$, $1 < q \le 2$. *Let* $\mathbf{r} := \{2/(k + 2)\}_{k=1}^{\infty}$. *Consider the* GAWR$(t, \mathbf{r})$. *For a pair of functions* $f, f^\epsilon$ *satisfying*

$$\| f - f^\epsilon \| \le \epsilon, \quad f^\epsilon / A(\epsilon) \in A_1(\mathcal{D}),$$

*we have*

$$\|f_m\| \le \epsilon + C(q, \gamma)(\|f\| + A(\epsilon)/t)m^{-1+1/q}.$$

*Proof* By Lemma 6.24 we obtain

$$\|f_k\| - \epsilon \le (1 - r_k)(\|f_{k-1}\| - \epsilon) + C\gamma\|f_{k-1}\| \left( \frac{r_k(\|f\| + A(\epsilon)/t)}{\|f_{k-1}\|} \right)^q.$$
$$(6.38)$$

Consider, as in the proof of Theorem 6.25, the sequence $a_n := \|f_{n-1}\| - \epsilon$. We plan to apply Lemma 6.26 to the sequence $\{a_n\}$. We set $\alpha := 1 - 1/q \le 1/2$. The parameters $\gamma \in (\alpha, 1]$ and $A$ will be chosen later. We note that

$$\|f_m\| \le \|f_{m-1}\| + r_m\|f\|.$$

Therefore, the condition (6.36) of Lemma 6.26 is satisfied with $A \ge 2\|f\|$. Let $a_k \ge Ak^{-\alpha}$. Then by (6.38) we get

$$a_{k+1} \le a_k(1 - r_k + C\gamma(r_k(\|f\| + A(\epsilon)/t)/a_k)^q)$$
$$\le a_k \left( 1 - \frac{2}{k+2} + \frac{C\gamma(\|f\| + A(\epsilon)/t)^q 2^q}{A^q} \frac{k^{\alpha q}}{(k+2)^q} \right).$$

Setting $A := \max(2\|f\|, 2(2C\gamma)^{1/q}(\|f\| + A(\epsilon)/t))$, we obtain

$$a_{k+1} \le a_k \left( 1 - \frac{3}{2(k+2)} \right).$$

Thus condition (6.37) of Lemma 6.26 is satisfied with $\gamma = 3/4$. Applying Lemma 6.26 we obtain

$$\|f_m\| \le \epsilon + C(q, \gamma)(\|f\| + A(\epsilon)/t)m^{-1+1/q}.$$

$\square$

We conclude this section with the following remark. The algorithms GAWR and WGAFR are both dual type greedy algorithms. The first steps are similar for both algorithms: we use the norming functional $F_{f_{m-1}}$ in the search for an element $\varphi_m$. The WGAFR provides more freedom than the GAWR in choosing good coefficients $w_m$ and $\lambda_m$. This results in more flexibility in choosing the weakness sequence $\tau = \{t_m\}$. For instance, condition (6.29) of Theorem 6.22 is satisfied if $\tau = \{t\}, t \in (0, 1]$, for any uniformly smooth Banach space. In the case $\rho(u) \le \gamma u^q, 1 < q \le 2$, condition (6.29) is satisfied if

$$\sum_{m=1}^{\infty} t_m^p = \infty, \quad p := q/(q - 1).$$

## 6.6 Thresholding algorithms

We begin with a remark on the computational complexity of greedy algorithms. The main point of Section 6.4 is in proving that relaxation allows us to build greedy algorithms (see the WGAFR) that are computationally simpler than the WCGA and perform as well as the WCGA. We note that the WCGA and the WGAFR differ in the second step of the algorithm. However, the most computationally involved step of all greedy algorithms is the greedy step (the first step of the algorithm). One of the goals of relaxation was to get rid of the assumption $f \in A_1(\mathcal{D})$ (as in the WRGA). All relaxed greedy algorithms from Sections 6.4 and 6.5 are applicable to (and converge for) any $f \in X$. We want to point out that the information $f \in A_1(\mathcal{D})$ allows us to simplify substantially the greedy step of the algorithm. It is remarked in Section 6.2 (see Remark 6.12) that we can replace the first step of the WCGA by the following search criterion:

$$F_{f_{m-1}}(\varphi_m) \geq t_m \| f_{m-1} \|. \tag{6.39}$$

A similar remark (see Section 6.3, Remark 6.19) holds for the WRGA. The requirement (6.39) is weaker than the requirement of the greedy step of the WCGA. However, Theorem 6.8 holds for this modification of the WCGA. The relation (6.39) is a threshold-type inequality and can be checked easier than the greedy inequality.

We now consider two algorithms defined and studied in Temlyakov (2008c) with a different type of thresholding. These algorithms work for any $f \in X$. We begin with the Dual Greedy Algorithm with Relaxation and Thresholding (DGART).

**DGART** We define $f_0 := f$ and $G_0 := 0$. Then for a given parameter $\delta \in (0, 1/2]$ we have the following inductive definition for $m \geq 1$.

(1) $\varphi_m \in \mathcal{D}$ is any element satisfying

$$F_{f_{m-1}}(\varphi_m) \geq \delta. \tag{6.40}$$

If there is no $\varphi_m \in \mathcal{D}$ satisfying (6.40) then we stop.

(2) Find $w_m$ and $\lambda_m$ such that

$$\| f - ((1 - w_m)G_{m-1} + \lambda_m \varphi_m) \| = \inf_{\lambda, w} \| f - ((1 - w)G_{m-1} + \lambda \varphi_m) \|$$

and define

$$G_m := (1 - w_m)G_{m-1} + \lambda_m \varphi_m.$$

(3) Let

$$f_m := f - G_m.$$

If $\|f_m\| \leq \delta \|f\|$ then we stop, otherwise we proceed to the $(m+1)$th iteration.

The following algorithm is a thresholding-type modification of the WCGA. This modification can be applied to any $f \in X$.

**Chebyshev Greedy Algorithm with Thresholding (CGAT)** For a given parameter $\delta \in (0, 1/2]$, we conduct instead of the greedy step of the WCGA the following thresholding step: find $\varphi_m \in \mathcal{D}$ such that $F_{f_{m-1}}(\varphi_m) \geq \delta$. Choosing such a $\varphi_m$, if one exists, we apply steps (2) and (3) of the WCGA. If such $\varphi_m$ does not exist, then we stop. We also stop if $\|f_m\| \leq \delta \|f\|$.

**Theorem 6.28** *Let $X$ be a uniformly smooth Banach space with modulus of smoothness $\rho(u) \leq \gamma u^q$, $1 < q \leq 2$. Take a number $\epsilon \geq 0$ and two elements $f$, $f^\epsilon$ from $X$ such that*

$$\|f - f^\epsilon\| \leq \epsilon, \quad f^\epsilon / A(\epsilon) \in A_1(\mathcal{D}),$$

*with some number $A(\epsilon) > 0$. Then the DGART (CGAT) will stop after $m \leq C(\gamma) \delta^{-p} \ln(1/\delta)$, $p := q/(q-1)$, iterations with*

$$\|f_m\| \leq \epsilon + \delta A(\epsilon).$$

*Proof* We begin with the error bound. For both algorithms, the DGART and the CGAT, our stopping criterion guarantees that either $\|F_{f_m}\|_{\mathcal{D}} \leq \delta$ or $\|f_m\| \leq \delta \|f\|$. In the latter case the required bound follows from simple inequalities

$$\|f\| \leq \epsilon + \|f^\epsilon\| \leq \epsilon + A(\epsilon).$$

Thus, assume that $\|F_{f_m}\|_{\mathcal{D}} \leq \delta$ holds. In the case of CGAT we apply Lemma 6.9 with $L = \text{span}(\varphi_1, \ldots, \varphi_m)$ and obtain

$$\|f_m\| = F_{f_m}(f_m) = F_{f_m}(f) \leq \epsilon + F_{f_m}(f^\epsilon) \leq \epsilon + \|F_{f_m}\|_{\mathcal{D}} A(\epsilon) \leq \epsilon + \delta A(\epsilon).$$

For the DGART we apply Lemma 6.9 with $f_{m-1}$ and $L = \text{span}(G_{m-1}, \varphi_m)$, and get

$$\|f_m\| = F_{f_m}(f_m) = F_{f_m}(f_{m-1}) = F_{f_m}(f)$$
$$\leq \epsilon + F_{f_m}(f^\epsilon) \leq \epsilon + \|F_{f_m}\|_{\mathcal{D}} A(\epsilon) \leq \epsilon + \delta A(\epsilon).$$

This proves the required bound.

We now proceed to the bound of $m$. We prove the bound for both algorithms simultaneously. We note that for the DGART

$$\|f_k\| = \inf_{\lambda, w} \|f_{k-1} + wG_{k-1} - \lambda \varphi_k\| \leq \inf_{\lambda \geq 0} \|f_{k-1} - \lambda \varphi_k\|.$$

We write for all $k \leq m, \lambda \geq 0$

$$\|f_{k-1} - \lambda \varphi_k\| + \|f_{k-1} + \lambda \varphi_k\| \leq 2\|f_{k-1}\|(1 + \rho(\lambda/\|f_{k-1}\|)). \qquad (6.41)$$

Next,

$$\|f_{k-1} + \lambda \varphi_k\| \geq F_{f_{k-1}}(f_{k-1} + \lambda \varphi_k) \geq \|f_{k-1}\| + \lambda \delta. \qquad (6.42)$$

Combining (6.41) with (6.42), we obtain

$$\|f_k\| \leq \inf_{\lambda \geq 0} \|f_{k-1} - \lambda \varphi_k\| \leq \inf_{\lambda \geq 0} \left( \|f_{k-1}\| - \lambda \delta + 2\|f_{k-1}\| \gamma (\lambda/\|f_{k-1}\|)^q \right).$$
$$(6.43)$$

Solving the equation $\delta x/2 = 2\gamma x^q$, we get $x_1 = (\delta/(4\gamma))^{1/(q-1)}$. Setting $\lambda := x_1 \|f_{k-1}\|$ we obtain

$$\|f_k\| \leq \|f_{k-1}\|(1 - \delta x_1/2) = \|f_{k-1}\|(1 - c(\gamma)\delta^p).$$

Thus,

$$\|f_k\| \leq \|f\|(1 - c(\gamma)\delta^p)^k.$$

By the stopping condition $\|f_m\| \leq \delta \|f\|$, we deduce that $m \leq n$, where $n$ is the smallest integer for which

$$(1 - c(\gamma)\delta^p)^n \leq \delta.$$

This implies

$$m \leq C(\gamma)\delta^{-p} \ln(1/\delta).$$

$\square$

We proceed to one more thresholding-type algorithm (see Temlyakov (2005a)). Keeping in mind possible applications of this algorithm, we do not assume that a dictionary $\mathcal{D}$ is symmetric: $g \in \mathcal{D}$ implies $-g \in \mathcal{D}$. To indicate this we use the notation $\mathcal{D}^+$ for such a dictionary. We do not assume that elements of a dictionary $\mathcal{D}^+$ are normalized ($\|g\| = 1$ if $g \in \mathcal{D}^+$) and assume only that $\|g\| \leq 1$ if $g \in \mathcal{D}^+$. By $A_1(\mathcal{D}^+)$ we denote the closure of the convex hull of $\mathcal{D}^+$. Let $\epsilon = \{\epsilon_n\}_{n=1}^{\infty}, \epsilon_n > 0, n = 1, 2, \dots$ .

**Incremental Algorithm with schedule $\epsilon$ (IA($\epsilon$))** Let $f \in A_1(\mathcal{D}^+)$. Denote $f_0^{i,\epsilon} := f$ and $G_0^{i,\epsilon} := 0$. Then, for each $m \geq 1$ we have the following inductive definition.

(1) $\varphi_m^{i,\epsilon} \in \mathcal{D}^+$ is any element satisfying

$$F_{f_{m-1}^{i,\epsilon}} (\varphi_m^{i,\epsilon} - f) \geq -\epsilon_m.$$

(2) Define

$$G_m^{i,\epsilon} := (1 - 1/m) G_{m-1}^{i,\epsilon} + \varphi_m^{i,\epsilon} / m.$$

(3) Let

$$f_m^{i,\epsilon} := f - G_m^{i,\epsilon}.$$

We note that, as in Lemma 6.10, we have for any bounded linear functional $F$ and any $\mathcal{D}^+$

$$\sup_{g \in \mathcal{D}^+} F(g) = \sup_{f \in A_1(\mathcal{D}^+)} F(f).$$

Therefore, for any $F$ and any $f \in A_1(\mathcal{D}^+)$

$$\sup_{g \in \mathcal{D}^+} F(g) \geq F(f).$$

This guarantees the existence of $\varphi_m^{i,\epsilon}$.

**Theorem 6.29** *Let $X$ be a uniformly smooth Banach space with modulus of smoothness $\rho(u) \leq \gamma u^q$, $1 < q \leq 2$. Define*

$$\epsilon_n := K_1 \gamma^{1/q} n^{-1/p}, \qquad p = \frac{q}{q-1}, \quad n = 1, 2, \ldots$$

*Then, for any $f \in A_1(\mathcal{D}^+)$ we have*

$$\| f_m^{i,\epsilon} \| \leq C(K_1) \gamma^{1/q} m^{-1/p}, \qquad m = 1, 2 \ldots$$

*Proof* We will use the abbreviated notation $f_m := f_m^{i,\epsilon}$, $\varphi_m := \varphi_m^{i,\epsilon}$, $G_m := G_m^{i,\epsilon}$. Writing

$$f_m = f_{m-1} - (\varphi_m - G_{m-1})/m$$

we immediately obtain the trivial estimate

$$\| f_m \| \leq \| f_{m-1} \| + 2/m. \tag{6.44}$$

Since

$$f_m = (1 - 1/m) f_{m-1} - (\varphi_m - f)/m$$
$$= (1 - 1/m)(f_{m-1} - (\varphi_m - f)/(m-1)) \tag{6.45}$$

we obtain

$$\| f_{m-1} - (\varphi_m - f)/(m-1) \|$$
$$\leq \| f_{m-1} \| (1 + 2\rho(2((m-1)\| f_{m-1} \|)^{-1})) + \epsilon_m (m-1)^{-1}, \qquad (6.46)$$

in a similar way to (6.43). Using the definition of $\epsilon_m$ and the assumption $\rho(u) \leq \gamma u^q$, we make the following observation. There exists a constant $C(K_1)$ such that, if

$$\| f_{m-1} \| \geq C(K_1)\gamma^{1/q}(m-1)^{-1/p} \qquad (6.47)$$

then

$$2\rho(2((m-1)\| f_{m-1} \|)^{-1}) + \epsilon_m((m-1)\| f_{m-1} \|)^{-1} \leq 1/(4m), \qquad (6.48)$$

and therefore, by (6.45) and (6.46)

$$\| f_m \| \leq (1 - 3/(4m))\| f_{m-1} \|. \qquad (6.49)$$

Taking into account (6.44) we apply Lemma 6.26 to the sequence $a_n = \| f_n \|$, $n = 1, 2, \ldots$, with $\alpha = 1/p$, $\beta = 3/4$ and complete the proof of Theorem 6.29. $\qquad \square$

## 6.7 Greedy expansions

### 6.7.1 Introduction

From the definition of a dictionary it follows that any element $f \in X$ can be approximated arbitrarily well by finite linear combinations of the dictionary elements. The primary goal of this section is to study representations of an element $f \in X$ by a series

$$f \sim \sum_{j=1}^{\infty} c_j(f)g_j(f), \quad g_j(f) \in \mathcal{D}, \quad c_j(f) > 0, \quad j = 1, 2, \ldots \quad (6.50)$$

In building the representation (6.50) we should construct two sequences: $\{g_j(f)\}_{j=1}^{\infty}$ and $\{c_j(f)\}_{j=1}^{\infty}$. In this section the construction of $\{g_j(f)\}_{j=1}^{\infty}$ will be based on ideas used in greedy type nonlinear approximation (greedy type algorithms). This justifies the use of the term *greedy expansion* for (6.50) considered in this section. The construction of $\{g_j(f)\}_{j=1}^{\infty}$ is, clearly, the most important and difficult part in building the representation (6.50). On the basis of the contemporary theory of nonlinear approximation with respect to redundant dictionaries, we may conclude that the method of using a norming functional in greedy steps of an algorithm is the most productive in approximation in Banach spaces. This method was utilized in the Weak Chebyshev

Greedy Algorithm and in the Weak Dual Greedy Algorithm. We use this same method in new algorithms considered in this section. A new qualitative result of this section establishes that we have a lot of flexibility in constructing a sequence of coefficients $\{c_j(f)\}_{j=1}^{\infty}$.

Denote

$$r_{\mathcal{D}}(f) := \sup_{F_f} \|F_f\|_{\mathcal{D}} := \sup_{F_f} \sup_{g \in \mathcal{D}} F_f(g).$$

We note that, in general, a norming functional $F_f$ is not unique. This is why we take $\sup_{F_f}$ over all norming functionals of $f$ in the definition of $r_{\mathcal{D}}(f)$. It is known that in the case of uniformly smooth Banach spaces (our primary object here) the norming functional $F_f$ is unique. In such a case we do not need $\sup_{F_f}$ in the definition of $r_{\mathcal{D}}(f)$; we have $r_{\mathcal{D}}(f) = \|F_f\|_{\mathcal{D}}$.

We begin with a description of a general scheme that provides an expansion for a given element $f$. Later, specifying this general scheme, we will obtain different methods of expansion.

**Dual-Based Expansion (DBE)** Let $t \in (0, 1]$ and $f \neq 0$. Denote $f_0 := f$. Assume that $\{f_j\}_{j=0}^{m-1} \subset X$, $\{\varphi_j\}_{j=1}^{m-1} \subset \mathcal{D}$ and a set of coefficients $\{c_j\}_{j=1}^{m-1}$ of expansion have already been constructed. If $f_{m-1} = 0$ then we stop (set $c_j = 0$, $j = m, m+1, \ldots$ in the expansion) and get $f = \sum_{j=1}^{m-1} c_j \varphi_j$. If $f_{m-1} \neq 0$ then we conduct the following two steps.

(1) Choose $\varphi_m \in \mathcal{D}$ such that

$$\sup_{F_{f_{m-1}}} F_{f_{m-1}}(\varphi_m) \geq t r_{\mathcal{D}}(f_{m-1}).$$

(2) Define

$$f_m := f_{m-1} - c_m \varphi_m,$$

where $c_m > 0$ is a coefficient either prescribed in advance or chosen from a concrete approximation procedure.

We call the series

$$f \sim \sum_{j=1}^{\infty} c_j \varphi_j \tag{6.51}$$

the Dual-Based Expansion of $f$ with coefficients $c_j(f) := c_j$, $j = 1, 2, \ldots$, with respect to $\mathcal{D}$.

Denote

$$S_m(f, \mathcal{D}) := \sum_{j=1}^{m} c_j \varphi_j.$$

Then it is clear that

$$f_m = f - S_m(f, \mathcal{D}).$$

We prove some convergence results for the DBE in Sections 6.7.2 and 6.7.3. In Section 6.7.3 we consider a variant of the Dual-Based Expansion with coefficients chosen by a certain simple rule. The rule depends on two numerical parameters, $t \in (0, 1]$ (the weakness parameter from the definition of the DBE) and $b \in (0, 1)$ (the tuning parameter of the approximation method). The rule also depends on a majorant $\mu$ of the modulus of smoothness of the Banach space $X$.

**Dual Greedy Algorithm with parameters** $(t, b, \mu)$ **(DGA$(t, b, \mu)$)** Let $X$ be a uniformly smooth Banach space with modulus of smoothness $\rho(u)$, and let $\mu(u)$ be a continuous majorant of $\rho(u)$: $\rho(u) \le \mu(u)$, $u \in [0, \infty)$. For parameters $t \in (0, 1]$, $b \in (0, 1]$ we define sequences $\{f_m\}_{m=0}^{\infty}$, $\{\varphi_m\}_{m=1}^{\infty}$, $\{c_m\}_{m=1}^{\infty}$ inductively. Let $f_0 := f$. If for $m \ge 1$, $f_{m-1} = 0$ then we set $f_j = 0$ for $j \ge m$ and stop. If $f_{m-1} \ne 0$ then we conduct the following three steps.

(1) Take any $\varphi_m \in \mathcal{D}$ such that

$$F_{f_{m-1}}(\varphi_m) \ge t r_{\mathcal{D}}(f_{m-1}). \tag{6.52}$$

(2) Choose $c_m > 0$ from the equation

$$\|f_{m-1}\| \mu(c_m / \|f_{m-1}\|) = \frac{tb}{2} c_m r_{\mathcal{D}}(f_{m-1}). \tag{6.53}$$

(3) Define

$$f_m := f_{m-1} - c_m \varphi_m. \tag{6.54}$$

In Section 6.7.3 we prove the following convergence result.

**Theorem 6.30** *Let $X$ be a uniformly smooth Banach space with the modulus of smoothness $\rho(u)$ and let $\mu(u)$ be a continuous majorant of $\rho(u)$ with the property $\mu(u)/u \downarrow 0$ as $u \to +0$. Then, for any $t \in (0, 1]$ and $b \in (0, 1)$ the DGA$(t, b, \mu)$ converges for each dictionary $\mathcal{D}$ and all $f \in X$.*

The following result from Section 6.7.3 gives the rate of convergence.

**Theorem 6.31** *Assume $X$ has a modulus of smoothness $\rho(u) \le \gamma u^q$, $q \in (1, 2]$ and $b \in (0, 1)$. Denote $\mu(u) = \gamma u^q$. Then, for any dictionary $\mathcal{D}$ and any $f \in A_1(\mathcal{D})$, the rate of convergence of the DGA$(t, b, \mu)$ is given by*

$$\|f_m\| \le C(t, b, \gamma, q) m^{-t(1-b)/(p(1+t(1-b)))}, \qquad p := \frac{q}{q-1}.$$

### 6.7.2 Convergence of the Dual-Based Expansion

We begin with the following lemma.

**Lemma 6.32** *Let $f \in X$. Assume that the coefficients $\{c_j\}_{j=1}^{\infty}$ of the expansion*

$$f \sim \sum_{j=1}^{\infty} c_j \varphi_j, \qquad f_m := f - \sum_{j=1}^{m} c_j \varphi_j$$

*are non-negative and satisfy the following two conditions:*

$$\sum_{j=1}^{\infty} c_j r_D(f_{j-1}) < \infty, \tag{6.55}$$

$$\sum_{j=1}^{\infty} c_j = \infty. \tag{6.56}$$

*Then*

$$\liminf_{m \to \infty} \| f_m \| = 0. \tag{6.57}$$

*Proof* The proof of this lemma is similar to the proof of lemma 1 from Ganichev and Kalton (2003). Denote $s_n := \sum_{j=1}^{n} c_j$. Then (6.56) implies (see Bary (1961), p. 904) that

$$\sum_{n=1}^{\infty} c_n / s_n = \infty. \tag{6.58}$$

Using (6.55) we get

$$\sum_{n=1}^{\infty} s_n r_D(f_{n-1}) c_n / s_n = \sum_{n=1}^{\infty} c_n r_D(f_{n-1}) < \infty.$$

Thus, by (6.58),

$$\liminf_{n \to \infty} s_n r_D(f_{n-1}) = 0$$

and also $(s_{n-1} \leq s_n)$

$$\liminf_{n \to \infty} s_n r_D(f_n) = 0.$$

Let

$$\lim_{k \to \infty} s_{n_k} r_D(f_{n_k}) = 0. \tag{6.59}$$

Consider $\{F_{f_{n_k}}\}$. The unit sphere in the dual $X^*$ is weakly* compact (see Habala, Hájek and Zizler (1996), p. 45). Let $\{F_i\}_{i=1}^\infty$, $F_i := F_{f_{n_{k_i}}}$ be a $w^*$-convergent subsequence. Denote

$$F := w^* - \lim_{i \to \infty} F_i.$$

We will complete the proof of Lemma 6.32 by contradiction. We assume that (6.57) does not hold; that is, there exist $\alpha > 0$ and $N \in \mathbb{N}$ such that

$$\|f_m\| \geq \alpha, \quad m \geq N, \tag{6.60}$$

and will then derive a contradiction.

We begin by deducing from (6.60) that $F \neq 0$. Indeed, we have

$$F(f) = \lim_{i \to \infty} F_i(f) \tag{6.61}$$

and

$$F_i(f) = F_i \left( f_{n_{k_i}} + \sum_{j=1}^{n_{k_i}} c_j \varphi_j \right) = \|f_{n_{k_i}}\| + \sum_{j=1}^{n_{k_i}} c_j F_i(\varphi_j) \geq \alpha - s_{n_{k_i}} r_{\mathcal{D}}(f_{n_{k_i}})$$

$$\tag{6.62}$$

for large $i$. Relations (6.61), (6.62) and (6.59) imply that $F(f) \geq \alpha$, and hence $F \neq 0$. This implies that there exists $g \in \mathcal{D}$ for which $F(g) > 0$. However,

$$F(g) = \lim_{i \to \infty} F_i(g) \leq \lim_{i \to \infty} r_{\mathcal{D}}(f_{n_{k_i}}) = 0.$$

We have a contradiction, which completes the proof of Lemma 6.32. $\qquad\square$

In Temlyakov (2007b) we pushed to the extreme the flexibility choice of the coefficients $c_j(f)$ in (6.50). We made these coefficients independent of an element $f \in X$. Surprisingly, for properly chosen coefficients we obtained results for the corresponding dual greedy expansion similar to Theorems 6.30 and 6.31. Even more surprisingly, we obtained similar results for the corresponding $X$-greedy expansions. We proceed to the formulation of these results. Let $\mathcal{C} := \{c_m\}_{m=1}^\infty$ be a fixed sequence of positive numbers. We restrict ourselves to positive numbers because of the symmetry of the dictionary $\mathcal{D}$.

**$X$-Greedy Algorithm with coefficients $\mathcal{C}$ (XGA($\mathcal{C}$))** We define $f_0 := f$, $G_0 := 0$. Then, for each $m \geq 1$ we have the following inductive definition.

(1) $\varphi_m \in \mathcal{D}$ is such that (assuming existence)

$$\|f_{m-1} - c_m \varphi_m\|_X = \inf_{g \in \mathcal{D}} \|f_{m-1} - c_m g\|_X.$$

(2) Let

$$f_m := f_{m-1} - c_m \varphi_m, \qquad G_m := G_{m-1} + c_m \varphi_m.$$

**Dual Greedy Algorithm with weakness $\tau$ and coefficients $\mathcal{C}$ (DGA($\tau, \mathcal{C}$))**
Let $\tau := \{t_m\}_{m=1}^{\infty}$, $t_m \in [0, 1]$, be a weakness sequence. We define $f_0 := f$,
$G_0 := 0$. Then, for each $m \geq 1$ we have the following inductive definition.

(1) $\varphi_m \in \mathcal{D}$ is any element satisfying

$$F_{f_{m-1}}(\varphi_m) \geq t_m \|F_{f_{m-1}}\|_{\mathcal{D}}.$$

(2) Let

$$f_m := f_{m-1} - c_m \varphi_m, \qquad G_m := G_{m-1} + c_m \varphi_m.$$

In the case $\tau = \{t\}$, $t \in (0, 1]$, we write $t$ instead of $\tau$ in the notation.
The first result on convergence properties of the DGA($t, \mathcal{C}$) was obtained in
Temlyakov (2007a). We prove it here.

**Theorem 6.33** *Let $X$ be a uniformly smooth Banach space with the modulus
of smoothness $\rho(u)$. Assume $\mathcal{C} = \{c_j\}_{j=1}^{\infty}$ is such that $c_j \geq 0$, $j = 1, 2, \ldots$,*

$$\sum_{j=1}^{\infty} c_j = \infty,$$

*and for any $y > 0$*

$$\sum_{j=1}^{\infty} \rho(y c_j) < \infty. \tag{6.63}$$

*Then, for the DGA($t, \mathcal{C}$) we have*

$$\liminf_{m \to \infty} \|f_m\| = 0. \tag{6.64}$$

*Proof* The proof is by contradiction. Assume (6.64) does not hold. Then there
exist $\alpha > 0$ and $N \in \mathbb{N}$ such that, for all $m \geq N$,

$$\|f_m\| \geq \alpha > 0.$$

From the definition of the modulus of smoothness we have

$$\|f_{n-1} - c_n \varphi_n\| + \|f_{n-1} + c_n \varphi_n\| \leq 2\|f_{n-1}\|(1 + \rho(c_n / \|f_{n-1}\|)). \tag{6.65}$$

Using the definition of $\varphi_n$,

$$F_{f_{n-1}}(\varphi_n) \geq t r_{\mathcal{D}}(f_{n-1}), \tag{6.66}$$

we get

$$\|f_{n-1} + c_n\varphi_n\| \geq F_{f_{n-1}}(f_{n-1} + c_n\varphi_n) \tag{6.67}$$
$$= \|f_{n-1}\| + c_n F_{f_{n-1}}(\varphi_n) \geq \|f_{n-1}\| + c_n tr_{\mathcal{D}}(f_{n-1}).$$

Combining (6.65) and (6.67), we get

$$\|f_n\| = \|f_{n-1} - c_n\varphi_n\| \leq \|f_{n-1}\|(1 + 2\rho(c_n/\|f_{n-1}\|)) - c_n tr_{\mathcal{D}}(f_{n-1}). \tag{6.68}$$

We note that by Remark 6.5

$$\|f_{n-1}\|\rho(c_n/\|f_{n-1}\|) \leq \alpha\rho(c_n/\alpha), \quad n > N.$$

Therefore, by the assumption (6.63)

$$\sum_{n=1}^{\infty} \|f_{n-1}\|\rho(c_n/\|f_{n-1}\|) < \infty. \tag{6.69}$$

This and (6.68) imply

$$\sum_{n=1}^{\infty} c_n r_{\mathcal{D}}(f_{n-1}) \leq t^{-1}\left(\|f\| + 2\sum_{n=1}^{\infty} \|f_{n-1}\|\rho(c_n/\|f_{n-1}\|)\right) < \infty.$$

It remains to apply Lemma 6.32 to complete the proof. □

In Temlyakov (2007b) we proved an analog of Theorem 6.33 for the XGA($\mathcal{C}$) and improved upon the convergence in Theorem 6.33 in the case of uniformly smooth Banach spaces with power-type modulus of smoothness. Under an extra assumption on $\mathcal{C}$ we replaced lim inf by lim. Here is the corresponding result from Temlyakov (2007b).

**Theorem 6.34** *Let $\mathcal{C} \in \ell_q \setminus \ell_1$ be a monotone sequence. Then the DGA($t, \mathcal{C}$) and the XGA($\mathcal{C}$) converge for each dictionary and all $f \in X$ in any uniformly smooth Banach space $X$ with modulus of smoothness $\rho(u) \leq \gamma u^q$, $q \in (1, 2]$.*

In Temlyakov (2007b) we also addressed a question of rate of approximation for $f \in A_1(\mathcal{D})$. We proved the following theorem.

**Theorem 6.35** *Let $X$ be a uniformly smooth Banach space with modulus of smoothness $\rho(u) \leq \gamma u^q$, $q \in (1, 2]$. We set $s := (1 + 1/q)/2$ and $\mathcal{C}_s := \{k^{-s}\}_{k=1}^{\infty}$. Then the DGA($t, \mathcal{C}_s$) and XGA($\mathcal{C}_s$) (for this algorithm $t = 1$) converge for $f \in A_1(\mathcal{D})$ with the following rate: for any $r \in (0, t(1-s))$*

$$\|f_m\| \leq C(r, t, q, \gamma)m^{-r}.$$

In the case $t = 1$, Theorem 6.35 provides the rate of convergence $m^{-r}$ for $f \in A_1(\mathcal{D})$ with $r$ arbitrarily close to $(1 - 1/q)/2$. Theorem 6.31 provides a similar rate of convergence. It would be interesting to know if the rate $m^{-(1-1/q)/2}$ is the best that can be achieved in greedy expansions (for each $\mathcal{D}$, any $f \in A_1(\mathcal{D})$ and any $X$ with $\rho(u) \leq \gamma u^q$, $q \in (1, 2]$). We note that there are greedy approximation methods that provide error bounds of the order $m^{1/q-1}$ for $f \in A_1(\mathcal{D})$ (see Temlyakov (2003a, 2008c) for recent results). However, these approximation methods do not provide an expansion.

### 6.7.3  Modification of the Weak Dual Greedy Algorithm

We begin this subsection with a proof of Theorem 6.30. Here we give a definition of the DGA$(\tau, b, \mu)$, $\tau = \{t_k\}_{k=1}^{\infty}$, $t_k \in (0, 1]$ that coincides with the definition of DGA$(t, b, \mu)$ from Section 6.7.1 in the case $\tau = \{t\}$.

**Dual Greedy Algorithm with parameters** $(\tau, b, \mu)$ **(DGA$(\tau, b, \mu)$)** Let $X$ be a uniformly smooth Banach space with modulus of smoothness $\rho(u)$ and let $\mu(u)$ be a continuous majorant of $\rho(u)$: $\rho(u) \leq \mu(u)$, $u \in [0, \infty)$. For a sequence $\tau = \{t_k\}_{k=1}^{\infty}$, $t_k \in (0, 1]$ and a parameter $b \in (0, 1]$ we define sequences $\{f_m\}_{m=0}^{\infty}$, $\{\varphi_m\}_{m=1}^{\infty}$, $\{c_m\}_{m=1}^{\infty}$ inductively. Let $f_0 := f$. If $f_{m-1} = 0$ for some $m \geq 1$, then we set $f_j = 0$ for $j \geq m$ and stop. If $f_{m-1} \neq 0$ then we conduct the following three steps.

(1) Take any $\varphi_m \in \mathcal{D}$ such that

$$F_{f_{m-1}}(\varphi_m) \geq t_m r_{\mathcal{D}}(f_{m-1}). \tag{6.70}$$

(2) Choose $c_m > 0$ from the equation

$$\|f_{m-1}\| \mu(c_m / \|f_{m-1}\|) = \frac{t_m b}{2} c_m r_{\mathcal{D}}(f_{m-1}). \tag{6.71}$$

(3) Define

$$f_m := f_{m-1} - c_m \varphi_m. \tag{6.72}$$

*Proof of Theorem 6.30* In this case $\tau = \{t\}$, $t \in (0, 1]$. We have by (6.68)

$$\|f_m\| = \|f_{m-1} - c_m \varphi_m\| \leq \|f_{m-1}\|(1 + 2\rho(c_m / \|f_{m-1}\|)) - c_m t r_{\mathcal{D}}(f_{m-1}). \tag{6.73}$$

Using the choice of $c_m$ we find

$$\|f_m\| \leq \|f_{m-1}\| - t(1 - b)c_m r_{\mathcal{D}}(f_{m-1}). \tag{6.74}$$

In particular, (6.74) implies that $\{\|f_m\|\}$ is a monotone decreasing sequence and

$$t(1-b)c_m r_{\mathcal{D}}(f_{m-1}) \leq \|f_{m-1}\| - \|f_m\|.$$

Thus

$$\sum_{m=1}^{\infty} c_m r_{\mathcal{D}}(f_{m-1}) < \infty. \qquad (6.75)$$

We have the following two cases:

$$\text{(I)} \quad \sum_{m=1}^{\infty} c_m = \infty, \qquad \text{(II)} \quad \sum_{m=1}^{\infty} c_m < \infty.$$

In case (I) by Lemma 6.32 we obtain

$$\liminf_{m \to \infty} \|f_m\| = 0 \Rightarrow \lim_{m \to \infty} \|f_m\| = 0.$$

It remains to consider case (II). We prove convergence in this case by contradiction. Assume

$$\lim_{m \to \infty} \|f_m\| = \alpha > 0. \qquad (6.76)$$

By (II) we have $f_m \to f_\infty \neq 0$ as $m \to \infty$. We note that by uniform smoothness of $X$ we get

$$\lim_{m \to \infty} \|F_{f_m} - F_{f_\infty}\| = 0.$$

We have $F_{f_\infty} \neq 0$, and therefore there is a $g \in \mathcal{D}$ such that $F_{f_\infty}(g) > 0$. However,

$$F_{f_\infty}(g) = \lim_{m \to \infty} F_{f_m}(g) \leq \lim_{m \to \infty} r_{\mathcal{D}}(f_m) = 0. \qquad (6.77)$$

Indeed, by (6.71) and (6.76) we get

$$r_{\mathcal{D}}(f_{m-1}) \leq \alpha c_m^{-1} \mu(c_m/\alpha) \frac{2}{tb} \to 0$$

as $m \to \infty$.

Theorem 6.30 is proved. $\qquad \square$

**Remark 6.36** It is clear from the above prove that Theorem 6.30 holds for an algorithm obtained from the DGA($\tau, b, \mu$), by replacing (6.71) by

$$\|f_{m-1}\| \mu(c_m/\|f_{m-1}\|) = \frac{b}{2} c_m F_{f_{m-1}}(\varphi_m). \qquad (6.78)$$

Also, a parameter $b$ in (6.71) and (6.78) can be replaced by varying parameters $b_m \in (a, b) \subset (0, 1)$.

We proceed to study the rate of convergence of the DGA$(\tau, b, \mu)$ in the uniformly smooth Banach spaces with the power-type majorant of modulus of smoothness: $\rho(u) \le \mu(u) = \gamma u^q$, $1 < q \le 2$. We now prove a statement more general than Theorem 6.31.

**Theorem 6.37** *Let $\tau := \{t_k\}_{k=1}^{\infty}$ be a non-increasing sequence $1 \ge t_1 \ge t_2 \ge \ldots > 0$ and $b \in (0, 1)$. Assume $X$ has a modulus of smoothness $\rho(u) \le \gamma u^q$, $q \in (1, 2]$. Denote $\mu(u) = \gamma u^q$. Then, for any dictionary $\mathcal{D}$ and any $f \in A_1(\mathcal{D})$, the rate of convergence of the DGA$(\tau, b, \mu)$ is given by*

$$\|f_m\| \le C(b, \gamma, q) \left(1 + \sum_{k=1}^{m} t_k^p\right)^{-\frac{t_m(1-b)}{p(1+t_m(1-b))}}, \quad p := \frac{q}{q-1}.$$

*Proof* As in (6.74), we get

$$\|f_m\| \le \|f_{m-1}\| - t_m(1 - b)c_m r_{\mathcal{D}}(f_{m-1}). \tag{6.79}$$

Thus we need to estimate $c_m r_{\mathcal{D}}(f_{m-1})$ from below. It is clear that

$$\|f_{m-1}\|_{A_1(\mathcal{D})} = \|f - \sum_{j=1}^{m-1} c_j \varphi_j\|_{A_1(\mathcal{D})} \le \|f\|_{A_1(\mathcal{D})} + \sum_{j=1}^{m-1} c_j. \tag{6.80}$$

Denote $b_n := 1 + \sum_{j=1}^{n} c_j$. Then, by (6.80) we get

$$\|f_{m-1}\|_{A_1(\mathcal{D})} \le b_{m-1}.$$

Next, by Lemma 6.10 we get

$$r_{\mathcal{D}}(f_{m-1}) = \sup_{g \in \mathcal{D}} F_{f_{m-1}}(g) = \sup_{\varphi \in A_1(\mathcal{D})} F_{f_{m-1}}(\varphi)$$

$$\ge \|f_{m-1}\|_{A_1(\mathcal{D})}^{-1} F_{f_{m-1}}(f_{m-1}) \ge \|f_{m-1}\| / b_{m-1}. \tag{6.81}$$

Substituting (6.81) into (6.79), we get

$$\|f_m\| \le \|f_{m-1}\|(1 - t_m(1 - b)c_m / b_{m-1}). \tag{6.82}$$

From the definition of $b_m$ we find

$$b_m = b_{m-1} + c_m = b_{m-1}(1 + c_m / b_{m-1}).$$

Using the inequality

$$(1 + x)^\alpha \le 1 + \alpha x, \quad 0 \le \alpha \le 1, \quad x \ge 0,$$

we obtain

$$b_m^{t_m(1-b)} \le b_{m-1}^{t_m(1-b)}(1 + t_m(1-b)c_m/b_{m-1}). \tag{6.83}$$

Multiplying (6.82) and (6.83), and using that $t_m \le t_{m-1}$, we get

$$\|f_m\| b_m^{t_m(1-b)} \le \|f_{m-1}\| b_{m-1}^{t_{m-1}(1-b)} \le \|f\| \le 1. \tag{6.84}$$

The function $\mu(u)/u = \gamma u^{q-1}$ is increasing on $[0, \infty)$. Therefore the $c_m$ from (6.71) is greater than or equal to $c_m'$ from (see (6.81))

$$\gamma \|f_{m-1}\| (c_m'/\|f_{m-1}\|)^q = \frac{t_m b}{2} c_m' \|f_{m-1}\|/b_{m-1}, \tag{6.85}$$

$$c_m' = \left(\frac{t_m b}{2\gamma}\right)^{1/(q-1)} \frac{\|f_{m-1}\|^{q/(q-1)}}{b_{m-1}^{1/(q-1)}}. \tag{6.86}$$

Using the notations

$$p := \frac{q}{q-1}, \qquad A^{-1} := (1-b)\left(\frac{b}{2\gamma}\right)^{1/(q-1)} \le 1/2,$$

we obtain

$$\|f_m\| \le \|f_{m-1}\|\left(1 - \frac{t_m^p}{A}\frac{\|f_{m-1}\|^p}{b_{m-1}^p}\right), \tag{6.87}$$

from (6.82) and (6.86). Noting that $b_m \ge b_{m-1}$, we infer from (6.87) that

$$(\|f_m\|/b_m)^p \le (\|f_{m-1}\|/b_{m-1})^p(1 - A^{-1}t_m^p(\|f_{m-1}\|/b_{m-1})^p). \tag{6.88}$$

Taking into account that $\|f\| \le 1 < A$, we obtain from (6.88) by an analog of Lemma 2.16 (see Temlyakov (2000b), lemma 3.1)

$$(\|f_m\|/b_m)^p \le A\left(1 + \sum_{k=1}^m t_k^p\right)^{-1}. \tag{6.89}$$

Combining (6.84) and (6.89), we get

$$\|f_m\| \le C(b, \gamma, q)\left(1 + \sum_{k=1}^m t_k^p\right)^{-\frac{t_m(1-b)}{p(1+t_m(1-b))}}, \qquad p := \frac{q}{q-1}.$$

This completes the proof of Theorem 6.37. $\qquad\square$

In the case $\tau = \{t\}$, $t \in (0, 1]$ we get Theorem 6.31 from Theorem 6.37.

**Remark 6.38** Theorem 6.39 holds for an algorithm obtained from the DGA$(\tau, b, \mu)$ by replacing (6.71) by (6.78).

It follows from the proof of Theorem 6.37 that it holds for a modification of the DGA$(\tau, b, \mu)$, where we replace the quantity $r_{\mathcal{D}}(f_{m-1})$ in the definition by its lower estimate (see (6.81)) $\|f_{m-1}\|/b_{m-1}$, with $b_{m-1} := 1 + \sum_{j=1}^{m-1} c_j$. Clearly, this modification is more ready for practical implementation than the DGA$(\tau, b, \mu)$. We formulate the above remark as a separate result.

**Modified Dual Greedy Algorithm** $(\tau, b, \mu)$ **(MDGA$(\tau, b, \mu)$)** Let $X$ be a uniformly smooth Banach space with modulus of smoothness $\rho(u)$ and let $\mu(u)$ be a continuous majorant of $\rho(u)$: $\rho(u) \leq \mu(u)$, $u \in [0, \infty)$. For a sequence $\tau = \{t_k\}_{k=1}^{\infty}$, $t_k \in (0, 1]$ and a parameter $b \in (0, 1)$, we define for $f \in A_1(\mathcal{D})$ sequences $\{f_m\}_{m=0}^{\infty}$, $\{\varphi_m\}_{m=1}^{\infty}$, $\{c_m\}_{m=1}^{\infty}$ inductively. Let $f_0 := f$. If $f_{m-1} = 0$ for some $m \geq 1$, then we set $f_j = 0$ for $j \geq m$ and stop. If $f_{m-1} \neq 0$ then we conduct the following three steps.

(1) Take any $\varphi_m \in \mathcal{D}$ such that

$$F_{f_{m-1}}(\varphi_m) \geq t_m \|f_{m-1}\| \left(1 + \sum_{j=1}^{m-1} c_j\right)^{-1}.$$

(2) Choose $c_m > 0$ from the equation

$$\mu(c_m/\|f_{m-1}\|) = \frac{t_m b}{2} c_m \left(1 + \sum_{j=1}^{m-1} c_j\right)^{-1}.$$

(3) Define

$$f_m := f_{m-1} - c_m \varphi_m.$$

**Theorem 6.39** *Let* $\tau := \{t_k\}_{k=1}^{\infty}$ *be a non-increasing sequence* $1 \geq t_1 \geq t_2 \geq \ldots > 0$ *and* $b \in (0, 1)$. *Assume* $X$ *has a modulus of smoothness* $\rho(u) \leq \gamma u^q$, $q \in (1, 2]$. *Denote* $\mu(u) = \gamma u^q$. *Then, for any dictionary* $\mathcal{D}$ *and any* $f \in A_1(\mathcal{D})$, *the rate of convergence of the MDGA$(\tau, b, \mu)$ is given by*

$$\|f_m\| \leq C(b, \gamma, q) \left(1 + \sum_{k=1}^{m} t_k^p\right)^{-\frac{t_m(1-b)}{p(1+t_m(1-b))}}, \qquad p := \frac{q}{q-1}.$$

Let us discuss an application of Theorem 6.31 in the case of a Hilbert space. It is well known and easy to check that, for a Hilbert space $H$, one has

$$\rho(u) \leq (1 + u^2)^{1/2} - 1 \leq u^2/2.$$

Therefore, by Theorem 6.31 with $\mu(u) = u^2/2$, the DGA$(t, b, \mu)$ provides the following error estimate:

$$\|f_m\| \leq C(t, b)m^{-\frac{t(1-b)}{2(1+t(1-b))}} \quad \text{for} \quad f \in A_1(\mathcal{D}). \tag{6.90}$$

The estimate (6.90) with $t = 1$ gives

$$\|f_m\| \leq C(b)m^{-(1-b)/(2(2-b))} \quad \text{for} \quad f \in A_1(\mathcal{D}). \tag{6.91}$$

The exponent $(1 - b)/(2(2 - b))$ in this estimate tends to $1/4$ when $b$ tends to zero. Comparing (6.91) with the upper estimate for the PGA (see Section 2.3), we observe that the DGA$(1, b, u^2/2)$ with small $b$ has a better upper estimate for the rate of convergence than the known estimates for the PGA. We note also that inequality (2.40) from Chapter 2 indicates that the exponent in the power rate of decay of error for the PGA is less than 0.1898.

Let us figure out how the DGA$(1, b, u^2/2)$ works in Hilbert space. Consider its $m$th step. Let $\varphi_m \in \mathcal{D}$ be from (6.52). Then it is clear that $\varphi_m$ maximizes the $\langle f_{m-1}, g \rangle$ over the dictionary $\mathcal{D}$ and

$$\langle f_{m-1}, \varphi_m \rangle = \|f_{m-1}\|r_{\mathcal{D}}(f_{m-1}).$$

The PGA would use $\varphi_m$ with the coefficient $\langle f_{m-1}, \varphi_m \rangle$ at this step. The DGA$(1, b, u^2/2)$ uses the same $\varphi_m$ and only a fraction of $\langle f_{m-1}, \varphi_m \rangle$:

$$c_m = b\|f_{m-1}\|r_{\mathcal{D}}(f_{m-1}). \tag{6.92}$$

Thus the choice $b = 1$ in (6.92) corresponds to the PGA. However, it is clear from the above considerations that our technique, designed for general Banach spaces, does not work in the case $b = 1$. The above discussion brings us the following surprising observation. The use of a small fraction ($c_m = b\langle f_{m-1}, g \rangle$) of an optimal coefficient results in an improvement of the upper estimate for the rate of convergence.

### 6.7.4 Convergence of the WDGA

We now study convergence of the Weak Dual Greedy Algorithm (WDGA) defined in Section 6.1. We present in this subsection results from Ganichev and Kalton (2003). We will prove the convergence result under an extra assumption on a Banach space $X$.

**Definition 6.40 (Property $\Gamma$)** A uniformly smooth Banach space has property $\Gamma$ if there is a constant $\beta > 0$ such that, for any $x, y \in X$ satisfying $F_x(y) = 0$, we have

$$\|x + y\| \geq \|x\| + \beta F_{x+y}(y).$$

Property $\Gamma$ in the above form was introduced in Ganichev and Kalton (2003). This condition (formulated somewhat differently) was considered previously in the context of greedy approximation in Livshitz (2003).

**Theorem 6.41** *Let $X$ be a uniformly smooth Banach space with property $\Gamma$. Then the WDGA($\tau$) with $\tau = \{t\}$, $t \in (0, 1]$, converges for each dictionary and all $f \in X$.*

*Proof* Let $\{f_m\}_{m=0}^{\infty}$ be a sequence generated by the WDGA($t$). Then

$$f_{m-1} = f_m + a_m \varphi_m, \quad F_{f_m}(\varphi_m) = 0. \tag{6.93}$$

We use property $\Gamma$ with $x := f_m$ and $y := a_m \varphi_m$ and obtain

$$\|f_{m-1}\| \geq \|f_m\| + \beta a_m F_{f_{m-1}}(\varphi_m). \tag{6.94}$$

This inequality, and monotonicity of the sequence $\{\|f_m\|\}$, imply that

$$\sum_{m=1}^{\infty} a_m F_{f_{m-1}}(\varphi_m) < \infty \Rightarrow \sum_{m=1}^{\infty} a_m r_{\mathcal{D}}(f_{m-1}) < \infty. \tag{6.95}$$

As in the proof of Theorem 6.30, we consider separately two cases:

$$\text{(I)} \quad \sum_{m=1}^{\infty} a_m = \infty, \qquad \text{(II)} \quad \sum_{m=1}^{\infty} a_m < \infty.$$

In case (I), by (6.95) and Lemma 6.32 we obtain

$$\liminf_{m \to \infty} \|f_m\| = 0 \Rightarrow \lim_{m \to \infty} \|f_m\| = 0.$$

In case (II) we argue by contradiction. Assume

$$\lim_{m \to \infty} \|f_m\| = \alpha > 0.$$

Then, by (II) we have $f_m \to f_\infty \neq 0$ as $m \to \infty$. By the uniform smoothness of $X$ we get

$$\lim_{m \to \infty} \|F_{f_m} - F_{f_\infty}\| = 0, \quad \lim_{m \to \infty} \|F_{f_m} - F_{f_{m-1}}\| = 0. \tag{6.96}$$

In particular, (6.93) and (6.96) imply that

$$\lim_{m \to \infty} F_{f_{m-1}}(\varphi_m) = 0 \Rightarrow \lim_{m \to \infty} r_{\mathcal{D}}(f_m) = 0. \tag{6.97}$$

We have $F_{f_\infty} \neq 0$, and therefore there is a $g \in \mathcal{D}$ such that $F_{f_\infty}(g) > 0$. However, by (6.96) and (6.97),

$$F_{f_\infty}(g) = \lim_{m \to \infty} F_{f_m}(g) \le \lim_{m \to \infty} r_{\mathcal{D}}(f_m) = 0.$$

The obtained contradiction completes the proof.

We now give a direct proof in case (I) that does not use Lemma 6.32. By property $\Gamma$ we get

$$\|f_m\| \le \|f_{m-1}\| - \beta a_m F_{f_{m-1}}(\varphi_m) \le \|f_{m-1}\| - t\beta a_m \|F_{f_{m-1}}\|_{\mathcal{D}}. \tag{6.98}$$

Let $\epsilon > 0$, $A(\epsilon) > 0$ and $f^\epsilon$ be such that

$$\|f - f^\epsilon\| \le \epsilon, \quad f^\epsilon / A(\epsilon) \in A_1(\mathcal{D}).$$

Then

$$\|f_{m-1}\| = F_{f_{m-1}}(f_{m-1}) = F_{f_{m-1}}(f - f^\epsilon + f^\epsilon - G_{m-1})$$
$$\le \epsilon + \|F_{f_{m-1}}\|_{\mathcal{D}}(A(\epsilon) + b_m),$$

where $b_m := \sum_{k=1}^{m-1} a_k$. Therefore,

$$\|F_{f_{m-1}}\|_{\mathcal{D}} \ge (\|f_{m-1}\| - \epsilon)/(A(\epsilon) + b_m). \tag{6.99}$$

We complete the proof by obtaining a contradiction. If $\lim_{m \to \infty} \|f_m\| = \alpha > 0$ and $\epsilon := \alpha/2$, then (6.98) and (6.99) imply

$$\|f_m\| \le \|f_{m-1}\| \left(1 - \frac{t\beta a_m}{2(A(\epsilon) + b_m)}\right).$$

Assumption (I) implies

$$\sum_{m=1}^\infty \frac{a_m}{A(\epsilon) + b_m} = \infty \Rightarrow \|f_m\| \to 0.$$

$\square$

We now turn to the $L_p$-spaces. The following results, Proposition 6.42 and Theorem 6.43, are from Ganichev and Kalton (2003).

**Proposition 6.42** *The $L_p$-space with $1 < p < \infty$ has property $\Gamma$.*

*Proof* Let $p \in (1, \infty)$. Consider the following function:

$$\phi_p(u) := \frac{u|1 + u|^{p-2}(1 + u) - u}{|1 + u|^p - pu - 1}, \quad u \neq 0, \quad \phi_p(0) := 2/p.$$

We note that $|1 + u|^p - pu - 1 > 0$ for $u \neq 0$. Indeed, it is sufficient to check the inequality for $u \geq -1/p$. In this case $|1 + u|^p = (1 + u)^p > 1 + pu$, $u \neq 0$. It is easy to check that

$$\lim_{u \to 0} \phi_p(u) = 2/p.$$

Thus, $\phi_p(u)$ is continuous on $(-\infty, \infty)$. This and

$$\lim_{u \to -\infty} \phi_p(u) = \lim_{u \to \infty} \phi_p(u) = 1$$

imply that $\phi_p(u) \leq C_p$.

We now proceed to property $\Gamma$. For any two real functions $x(s)$, $y(s)$ the inequality $\phi_p(u) \leq C_p$ implies

$$|x(s) + y(s)|^{p-2}(x(s) + y(s))y(s) - |x(s)|^{p-2}x(s)y(s)$$
$$\leq C_p(|x(s) + y(s)|^p - p|x(s)|^{p-2}x(s)y(s) - |x(s)|^p). \qquad (6.100)$$

Suppose that $F_x(y) = 0$. This means that

$$\int |x(s)|^{p-2}x(s)y(s)ds = 0. \qquad (6.101)$$

Integrating inequality (6.100) and taking into account (6.101), we get

$$\|x + y\|^{p-1}F_{x+y}(y) \leq C_p(\|x + y\|^p - \|x\|^p). \qquad (6.102)$$

Next,

$$\|x\| = F_x(x) = F_x(x + y) \leq \|x + y\|.$$

Therefore, (6.102) implies

$$F_{x+y}(y) \leq pC_p(\|x + y\| - \|x\|). \qquad (6.103)$$

It remains to note that (6.103) is equivalent to property $\Gamma$ with $\beta = (pC_p)^{-1}$.

□

Combining Theorem 6.41 with Proposition 6.42 we obtain the following result.

**Theorem 6.43** *Let* $p \in (1, \infty)$. *Then the WDGA($\tau$) with* $\tau = \{t\}$, $t \in (0, 1]$, *converges for each dictionary and all* $f \in L_p$.

## 6.8 Relaxation; $X$-greedy algorithms

In Sections 6.2–6.7 we studied dual greedy algorithms. In this section we define some generalizations of the $X$-Greedy Algorithm using the idea of relaxation. We begin with an analog of the WGAFR.

**X-Greedy Algorithm with Free Relaxation (XGAFR)** We define $f_0 := f$ and $G_0 := 0$. Then for each $m \geq 1$ we have the following inductive definition.

(1) $\varphi_m \in \mathcal{D}$ and $\lambda_m \geq 0$, $w_m$ are such that

$$\|f-((1-w_m)G_{m-1}+\lambda_m\varphi_m)\| = \inf_{g \in \mathcal{D}, \lambda \geq 0, w} \|f-((1-w)G_{m-1}+\lambda g)\|$$

and

$$G_m := (1 - w_m)G_{m-1} + \lambda_m\varphi_m.$$

(2) Let

$$f_m := f - G_m.$$

Using this definition, we obtain that, for any $t \in (0, 1]$,

$$\|f_m\| \leq \inf_{\lambda \geq 0, w} \|f - ((1 - w)G_{m-1} + \lambda\varphi_m^t)\|,$$

where the $\varphi_m^t \in \mathcal{D}$ is an element, satisfying

$$F_{f_{m-1}}(\varphi_m^t) \geq t\|F_{f_{m-1}}\|_{\mathcal{D}}.$$

Setting $t = 1$ we obtain a version of Lemma 6.20 for the XGAFR.

**Lemma 6.44** *Let $X$ be a uniformly smooth Banach space with modulus of smoothness $\rho(u)$. Take a number $\epsilon \geq 0$ and two elements $f$, $f^\epsilon$ from $X$ such that*

$$\|f - f^\epsilon\| \leq \epsilon, \quad f^\epsilon/A(\epsilon) \in A_1(\mathcal{D}),$$

*with some number $A(\epsilon) \geq \epsilon$. Then we have for the XGAFR*

$$\|f_m\| \leq \|f_{m-1}\| \inf_{\lambda \geq 0}\left(1 - \lambda A(\epsilon)^{-1}\left(1 - \frac{\epsilon}{\|f_{m-1}\|}\right) + 2\rho\left(\frac{5\lambda}{\|f_{m-1}\|}\right)\right),$$

*for $m = 1, 2, \ldots$.*

Theorems 6.22 and 6.23 were derived from Lemma 6.20. In the same way we derive from Lemma 6.44 the following analogs of Theorems 6.22 and 6.23 for the XGAFR.

**Theorem 6.45** *Let $X$ be a uniformly smooth Banach space with modulus of smoothness $\rho(u)$. Then, for any $f \in X$ we have for the XGAFR*

$$\lim_{m \to \infty} \|f_m\| = 0.$$

**Theorem 6.46** *Let $X$ be a uniformly smooth Banach space with modulus of smoothness $\rho(u) \le \gamma u^q$, $1 < q \le 2$. Take a number $\epsilon \ge 0$ and two elements $f$, $f^\epsilon$ from $X$ such that*

$$\|f - f^\epsilon\| \le \epsilon, \quad f^\epsilon / A(\epsilon) \in A_1(\mathcal{D}),$$

*with some number $A(\epsilon) > 0$. Then we have for the XGAFR*

$$\|f_m\| \le \max\left(2\epsilon, C(q,\gamma)(A(\epsilon) + \epsilon)(1 + m)^{-1/p}\right), \quad p := q/(q-1).$$

We now proceed to an analog of the GAWR.

**$X$-Greedy Algorithm with Relaxation r (XGAR(r))** Given a relaxation sequence $\mathbf{r} := \{r_m\}_{m=1}^\infty$, $r_m \in [0, 1)$, we define $f_0 := f$ and $G_0 := 0$. Then, for each $m \ge 1$ we have the following inductive definition.

(1) $\varphi_m \in \mathcal{D}$ and $\lambda_m \ge 0$ are such that

$$\|f - ((1-r_m)G_{m-1} + \lambda_m\varphi_m)\| = \inf_{g \in \mathcal{D}, \lambda \ge 0} \|f - ((1-r_m)G_{m-1} + \lambda g)\|$$

and

$$G_m := (1 - r_m)G_{m-1} + \lambda_m\varphi_m.$$

(2) Let

$$f_m := f - G_m.$$

We note that in the case $r_k = 0$, $k = 1, \ldots$, when there is no relaxation, the XGAR(0) coincides with the $X$-Greedy Algorithm. Practically nothing is known about the convergence and rate of convergence of the $X$-Greedy Algorithm. However, relaxation helps to prove convergence results for the XGAR(**r**). Here are analogs of the corresponding results for the GAWR.

**Lemma 6.47** *Let $X$ be a uniformly smooth Banach space with modulus of smoothness $\rho(u)$. Take a number $\epsilon \ge 0$ and two elements $f$, $f^\epsilon$ from $X$ such that*

$$\|f - f^\epsilon\| \le \epsilon, \quad f^\epsilon / A(\epsilon) \in A_1(\mathcal{D}),$$

*with some number $A(\epsilon) > 0$. Then we have for the XGAR(**r**)*

$$\|f_m\| \le \|f_{m-1}\| \left(1 - r_m\left(1 - \frac{\epsilon}{\|f_{m-1}\|}\right) + 2\rho\left(\frac{r_m(\|f\| + A(\epsilon))}{(1-r_m)\|f_{m-1}\|}\right)\right),$$

*for $m = 1, 2, \ldots$.*

**Theorem 6.48** *Let a sequence* $\mathbf{r} := \{r_k\}_{k=1}^{\infty}$, $r_k \in [0, 1)$, *satisfy the conditions*

$$\sum_{k=1}^{\infty} r_k = \infty, \quad r_k \to 0 \quad as \quad k \to \infty.$$

*Then the XGAR($\mathbf{r}$) converges in any uniformly smooth Banach space for each* $f \in X$ *and for all dictionaries* $\mathcal{D}$.

**Theorem 6.49** *Let $X$ be a uniformly smooth Banach space with modulus of smoothness* $\rho(u) \leq \gamma u^q$, $1 < q \leq 2$. *Let* $\mathbf{r} := \{2/(k+2)\}_{k=1}^{\infty}$. *Consider the XGAR($\mathbf{r}$). For a pair of functions $f$, $f^{\epsilon}$, satisfying*

$$\|f - f^{\epsilon}\| \leq \epsilon, \quad f^{\epsilon}/A(\epsilon) \in A_1(\mathcal{D})$$

*we have*

$$\|f_m\| \leq \epsilon + C(q, \gamma)(\|f\| + A(\epsilon))m^{-1+1/q}.$$

## 6.9 Incoherent dictionaries and exact recovery

We present here a generalization of the concept of an $M$-coherent dictionary to the case of Banach spaces.

Let $\mathcal{D}$ be a dictionary in a Banach space $X$. We define the coherence parameter of this dictionary in the following way:

$$M(\mathcal{D}) := \sup_{g \neq h; g, h \in \mathcal{D}} \sup_{F_g} |F_g(h)|.$$

We note that, in general, a norming functional $F_g$ is not unique. This is why we take $\sup_{F_g}$ over all norming functionals of $g$ in the definition of $M(\mathcal{D})$. We do not need $\sup_{F_g}$ in the definition of $M(\mathcal{D})$ if for each $g \in \mathcal{D}$ there is a unique norming functional $F_g \in X^*$. Then we define $\mathcal{D}^* := \{F_g, g \in \mathcal{D}\}$ and call $\mathcal{D}^*$ a *dual dictionary* to a dictionary $\mathcal{D}$. It is known that the uniqueness of the norming functional $F_g$ is equivalent to the property that $g$ is a point of Gateaux smoothness:

$$\lim_{u \to 0} (\|g + uy\| + \|g - uy\| - 2\|g\|)/u = 0$$

for any $y \in X$. In particular, if $X$ is uniformly smooth then $F_f$ is unique for any $f \neq 0$. We considered in Temlyakov (2006a) the following greedy algorithm.

**Weak Quasi-Orthogonal Greedy Algorithm (WQOGA)** Let $t \in (0, 1]$. Denote $f_0 := f_0^{q,t} := f$ (here and in the following index $q$ stands for *quasi-orthogonal*) and find $\varphi_1 := \varphi_1^{q,t} \in \mathcal{D}$ such that

$$|F_{\varphi_1}(f_0)| \geq t \sup_{g \in \mathcal{D}} |F_g(f_0)|.$$

Next we find $c_1$ satisfying

$$F_{\varphi_1}(f - c_1\varphi_1) = 0.$$

Denote $f_1 := f_1^{q,t} := f - c_1\varphi_1$.

We continue this construction in an inductive way. Assume that we have already constructed residuals $f_0, f_1, \ldots, f_{m-1}$ and dictionary elements $\varphi_1, \ldots, \varphi_{m-1}$. Now we pick an element $\varphi_m := \varphi_m^{q,t} \in \mathcal{D}$ such that

$$|F_{\varphi_m}(f_{m-1})| \geq t \sup_{g \in \mathcal{D}} |F_g(f_{m-1})|.$$

Next, we look for $c_1^m, \ldots, c_m^m$ satisfying

$$F_{\varphi_j}\left(f - \sum_{i=1}^m c_i^m \varphi_i\right) = 0, \quad j = 1, \ldots, m. \tag{6.104}$$

If there is no solution to (6.104) then we stop, otherwise we denote $f_m := f_m^{q,t} := f - \sum_{i=1}^m c_i^m \varphi_i$ with $c_1^m, \ldots, c_m^m$ satisfying (6.104).

**Remark 6.50** We note that (6.104) has a unique solution if $\det \|F_{\varphi_j}(\varphi_i)\|_{i,j=1}^m \neq 0$. We apply the WQOGA in the case of a dictionary with the coherence parameter $M := M(\mathcal{D})$. Then, by a simple well known argument on the linear independence of the rows of the matrix $\|F_{\varphi_j}(\varphi_i)\|_{i,j=1}^m$, we conclude that (6.104) has a unique solution for any $m < 1 + 1/M$. Thus, in the case of an $M$-coherent dictionary $\mathcal{D}$, we can run the WQOGA for at least $[1/M]$ iterations.

The following result was obtained in Temlyakov (2006a).

**Theorem 6.51** *Let $t \in (0, 1]$. Assume that $\mathcal{D}$ has coherence parameter $M$. Let $S < (t/(1 + t))(1 + 1/M)$. Then for any $f$ of the form*

$$f = \sum_{i=1}^S a_i \psi_i,$$

*where $\psi_i$ are distinct elements of $\mathcal{D}$, we have that $f_S^{q,t} = 0$.*

We will prove the above theorem in a more general setting. Instead of a pair $(\mathcal{D}, \mathcal{D}^*)$ of a dictionary $\mathcal{D}$ and its dual dictionary $\mathcal{D}^*$, we now consider a pair $(\mathcal{D}, \mathcal{W})$ of a dictionary $\mathcal{D}$ and a set $\mathcal{W}$ of normalized elements $w$ indexed by elements from $\mathcal{D}$. We define

$$\mathcal{W} := \{w_g \in X^*, \|w_g\|_{X^*} = 1, g \in \mathcal{D}\}$$

and define the coherence parameter of the pair $(\mathcal{D}, \mathcal{W})$ in the following way:

$$M(\mathcal{D}, \mathcal{W}) := \sup_{g \neq h; g, h \in \mathcal{D}} |w_g(h)|.$$

We assume that the pair $(\mathcal{D}, \mathcal{W})$ satisfies the condition

$$w_g(g) \geq 1 - \delta, \qquad g \in \mathcal{D}, \tag{6.105}$$

with some $\delta \in [0, 1)$. If $\delta = 0$ then $w_g$ is a norming functional of $g$. For a pair $(\mathcal{D}, \mathcal{W})$ we define an analog of the WQOGA in the following way.

**Weak Projective Greedy Algorithm (WPGA)** Let $t \in (0, 1]$. Denote $f_0 := f_0^{p,t} := f$ (here and below index $p$ stands for *projective*) and find $\varphi_1 := \varphi_1^{p,t} \in \mathcal{D}$ such that

$$|w_{\varphi_1}(f_0)| \geq t \sup_{g \in \mathcal{D}} |w_g(f_0)|.$$

Next, we find $c_1$ satisfying

$$w_{\varphi_1}(f - c_1 \varphi_1) = 0.$$

Denote $f_1 := f_1^{p,t} := f - c_1 \varphi_1$.

We continue this construction in an inductive way. Assume that we have already constructed residuals $f_0, f_1, \ldots, f_{m-1}$ and dictionary elements $\varphi_1, \ldots, \varphi_{m-1}$. Now we pick an element $\varphi_m := \varphi_m^{p,t} \in \mathcal{D}$ such that

$$|w_{\varphi_m}(f_{m-1})| \geq t \sup_{g \in \mathcal{D}} |w_g(f_{m-1})|.$$

Next, we look for $c_1^m, \ldots, c_m^m$ satisfying

$$w_{\varphi_j}\left(f - \sum_{i=1}^{m} c_i^m \varphi_i\right) = 0, \quad j = 1, \ldots, m. \tag{6.106}$$

If there is no solution to (6.106) then we stop, otherwise we denote $f_m := f_m^{p,t} := f - \sum_{i=1}^{m} c_i^m \varphi_i$ with $c_1^m, \ldots, c_m^m$ satisfying (6.106).

The following remark is an analog of Remark 6.50.

**Remark 6.52** The system (6.106) has a unique solution if $\det \|w_{\varphi_j}(\varphi_i)\|_{i,j=1}^{m} \neq 0$. We apply the WPGA in the case of a pair $(\mathcal{D}, \mathcal{W})$ with coherence parameter $M := M(\mathcal{D}, \mathcal{W})$. Then, by a simple well known argument on the linear independence of the rows of the matrix $\|w_{\varphi_j}(\varphi_i)\|_{i,j=1}^{m}$, we conclude that (6.106) has a unique solution for any $m < 1 + (1 - \delta)/M$. In this case we can run the WPGA for at least $[(1 - \delta)/M]$ iterations.

We begin with an auxiliary statement.

**Lemma 6.53** *Let* $t \in (0, 1]$. *Assume that the pair* $(\mathcal{D}, \mathcal{W})$ *has coherence parameter* $M := M(\mathcal{D}, \mathcal{W})$ *and satisfies (6.105). Let* $S < (t/(1 + t))$ $(1 + (1 - \delta)/M)$. *Then, for any* $f$ *of the form*

$$f = \sum_{i=1}^{S} a_i \psi_i,$$

*where* $\psi_i$ *are distinct elements of* $\mathcal{D}$, *we have that* $\varphi_1^{p,t} = \psi_j$ *with some* $j \in [1, S]$.

*Proof* Let $A := \max_i |a_i| = |a_p| > 0$. Then, on the one hand,

$$|w_{\psi_p}(f)| \geq |w_{\psi_p}(a_p \psi_p)| - \sum_{j \neq p} |a_j w_{\psi_p}(\psi_j)| \geq A(1 - \delta - M(S - 1)).$$

Therefore,

$$\max_i |w_{\psi_i}(f)| \geq A(1 - \delta - M(S - 1)). \tag{6.107}$$

On the other hand, for any $g \in \mathcal{D}$ different from $\psi_i, i = 1, \ldots, S$, we get from our assumptions

$$|w_g(f)| \leq \sum_{i=1}^{S} |a_i w_g(\psi_i)| \leq AMS < tA(1 - \delta - M(S - 1)). \tag{6.108}$$

Comparing (6.107) with (6.108) we conclude the proof.                    □

We now prove an analog of Theorem 6.51.

**Theorem 6.54** *Let* $t \in (0, 1]$. *Assume that the pair* $(\mathcal{D}, \mathcal{W})$ *has coherence parameter* $M := M(\mathcal{D}, \mathcal{W})$ *and satisfies (6.105). Let* $S < (t/(1 + t))$ $\times (1 + (1 - \delta)/M)$. *Then, for any* $f$ *of the form*

$$f = \sum_{i=1}^{S} a_i \psi_i,$$

*where* $\psi_i$ *are distinct elements of* $\mathcal{D}$, *we have that* $f_S^{p,t} = 0$.

*Proof* By Lemma 6.53 we obtain that each $f_m^{p,t}, m \leq S$, has a form

$$f_m^{p,t} = \sum_{i=1}^{S} a_i^m \psi_i.$$

We note that by (6.107) $|w_{\varphi_{m+1}^{p,t}}(f_m^{p,t})| > 0$ for $m < S$ provided $f_m^{p,t} \neq 0$. Therefore, $\varphi_{m+1}^{p,t}$ is different from all the previous $\varphi_1^{p,t}, \ldots, \varphi_m^{p,t}$. Thus, assuming, without loss of generality, that $f_{S-1}^{p,t} \neq 0$, we conclude that the set $\varphi_1^{p,t}, \ldots, \varphi_S^{p,t}$ coincides with the set $\psi_1, \ldots, \psi_S$.

The condition

$$w_{\psi_j}\left(\sum_{i=1}^{S}(a_i - a_i^S)\psi_i\right) = 0, \quad j = 1, \ldots, S,$$

implies $a_i = a_i^S$, $i = 1, \ldots, S$. Here we used the fact that $\det \|w_{\varphi_j}(\varphi_i)\|_{i,j=1}^S \neq 0$. □

## 6.10 Greedy algorithms with approximate evaluations and restricted search

In this section we study a modification of the WCGA that is motivated by numerical applications. In this modification, we allow steps of the WCGA to be performed approximately with some error control. We show that the modified version of the WCGA performs as well as the WCGA. We develop the theory of the Approximate Weak Chebyshev Greedy Algorithm in a general setting: $X$ is an arbitrary uniformly smooth Banach space and $\mathcal{D}$ is any dictionary. We begin with some remarks on the WCGA. It is clear that in the case of an infinite dictionary $\mathcal{D}$ there is no direct computationally feasible way to evaluate $\sup_{g\in\mathcal{D}} F_{f_{m-1}^c}(g)$. This makes the greedy step, even in a weak version, very difficult to realize in practice. At the second step of the WCGA we are looking for the best approximant of $f$ from $\Phi_m$. We know that such an approximant exists. However, in practice we cannot find it exactly: we can only find it approximately.

The above observations motivated us to consider a variant of the WCGA with an eye towards practically implementable algorithms.

In Temlyakov (2005a) we studied the following modification of the WCGA. Let three sequences $\tau = \{t_k\}_{k=1}^{\infty}$, $\delta = \{\delta_k\}_{k=0}^{\infty}$, $\eta = \{\eta_k\}_{k=1}^{\infty}$ of numbers from $[0, 1]$ be given.

**Approximate Weak Chebyshev Greedy Algorithm (AWCGA)** We define $f_0 := f_0^{\tau,\delta,\eta} := f$. Then, for each $m \geq 1$ we have the following inductive definition.

(1) $F_{m-1}$ is a functional with properties

$$\|F_{m-1}\| \leq 1, \qquad F_{m-1}(f_{m-1}) \geq \|f_{m-1}\|(1 - \delta_{m-1});$$

and $\varphi_m := \varphi_m^{\tau,\delta,\eta} \in \mathcal{D}$ is any element satisfying

$$F_{m-1}(\varphi_m) \ge t_m \sup_{g \in \mathcal{D}} F_{m-1}(g).$$

(2) Define

$$\Phi_m := \operatorname{span}\{\varphi_j\}_{j=1}^m,$$

and let

$$E_m(f) := \inf_{\varphi \in \Phi_m} \|f - \varphi\|.$$

Let $G_m \in \Phi_m$ be such that

$$\|f - G_m\| \le E_m(f)(1 + \eta_m).$$

(3) Let

$$f_m := f_m^{\tau,\delta,\eta} := f - G_m.$$

The term *approximate* in this definition means that we use a functional $F_{m-1}$ that is an approximation to the norming (peak) functional $F_{f_{m-1}}$ and also that we use an approximant $G_m \in \Phi_m$ which satisfies a weaker assumption than being a best approximant to $f$ from $\Phi_m$. Thus, in the *approximate* version of the WCGA, we have addressed the issue of non-exact evaluation of the norming functional and the best approximant. We did not address the issue of finding the $\sup_{g \in \mathcal{D}} F_{f_{m-1}^c}(g)$. In Temlyakov (2005b) we addressed this issue. We did it in two steps. First we considered the corresponding modification of the WCGA, and then the modification of the AWCGA. These modifications are done in the style of the concept of *depth search* from Donoho (2001).

We now consider a countable dictionary $\mathcal{D} = \{\pm\psi_j\}_{j=1}^\infty$. We denote $\mathcal{D}(N) := \{\pm\psi_j\}_{j=1}^N$. Let $\mathcal{N} := \{N_j\}_{j=1}^\infty$ be a sequence of natural numbers.

**Restricted Weak Chebyshev Greedy Algorithm (RWCGA)** We define $f_0 := f_0^{c,\tau,\mathcal{N}} := f$. Then, for each $m \ge 1$ we have the following inductive definition.

(1) $\varphi_m := \varphi_m^{c,\tau,\mathcal{N}} \in \mathcal{D}(N_m)$ is any element satisfying

$$F_{f_{m-1}}(\varphi_m) \ge t_m \sup_{g \in \mathcal{D}(N_m)} F_{f_{m-1}}(g).$$

(2) Define

$$\Phi_m := \Phi_m^{\tau,\mathcal{N}} := \operatorname{span}\{\varphi_j\}_{j=1}^m,$$

and define $G_m := G_m^{c,\tau,\mathcal{N}}$ to be the best approximant to $f$ from $\Phi_m$.

(3) Let

$$f_m := f_m^{c,\tau,\mathcal{N}} := f - G_m.$$

We formulate some results from Temlyakov (2005a) and Temlyakov (2005b) in a particular case of a uniformly smooth Banach space with modulus of smoothness of power type (see Temlyakov (2005a, b) for the general case). The following theorem was proved in Temlyakov (2005a).

**Theorem 6.55** *Let a Banach space $X$ have modulus of smoothness $\rho(u)$ of power type $1 < q \leq 2$, that is $\rho(u) \leq \gamma u^q$. Assume that*

$$\sum_{m=1}^{\infty} t_m^p = \infty, \quad p = \frac{q}{q-1},$$

*and*

$$\delta_m = o(t_m^p), \qquad \eta_m = o(t_m^p).$$

*Then the AWCGA converges for any $f \in X$.*

We now give two theorems from Temlyakov (2005b) on greedy algorithms with restricted search.

**Theorem 6.56** *Let a Banach space $X$ have modulus of smoothness $\rho(u)$ of power type $1 < q \leq 2$, that is $\rho(u) \leq \gamma u^q$. Assume that $\lim_{m \to \infty} N_m = \infty$ and*

$$\sum_{m=1}^{\infty} t_m^p = \infty, \quad p = \frac{q}{q-1}.$$

*Then the RWCGA converges for any $f \in X$.*

For $b > 0$, $K > 0$, we define the class

$$A_1^b(K, \mathcal{D}) := \{f : d(f, A_1(\mathcal{D}(n))) \leq Kn^{-b}, \quad n = 1, 2, \ldots\}.$$

Here, $A_1(\mathcal{D}(n))$ is a convex hull of $\{\pm \psi_j\}_{j=1}^n$ and, for a compact set $F$,

$$d(f, F) := \inf_{\phi \in F} \|f - \phi\|.$$

**Theorem 6.57** *Let $X$ be a uniformly smooth Banach space with modulus of smoothness $\rho(u) \leq \gamma u^q$, $1 < q \leq 2$. Then, for $t \in (0, 1]$ there exist $C_1(t, \gamma, q, K)$, $C_2(t, \gamma, q, K)$ such that, for $\mathcal{N}$ with $N_m \geq C_1(t, \gamma, q, K)m^{r/b}$, $m = 1, 2, \ldots$, we have, for any $f \in A_1^b(K, \mathcal{D})$,*

$$\|f_m^{c,\tau,\mathcal{N}}\| \leq C_2(t, \gamma, q, K)m^{-r}, \quad \tau = \{t\}, \quad r := 1 - 1/q.$$

We note that we can choose an algorithm from Theorem 6.57 that satisfies the *polynomial depth search* condition $N_m \leq Cm^a$ from Donoho (2001).

We proceed to an algorithm that combines approximate evaluations with restricted search. Let three sequences $\tau = \{t_k\}_{k=1}^{\infty}$, $\delta = \{\delta_k\}_{k=0}^{\infty}$, $\eta = \{\eta_k\}_{k=1}^{\infty}$ of numbers from $[0, 1]$ be given. Let $\mathcal{N} := \{N_j\}_{j=1}^{\infty}$ be a sequence of natural numbers.

**Restricted Approximate Weak Chebyshev Greedy Algorithm (RAWCGA)**
We define $f_0 := f_0^{\tau,\delta,\eta,\mathcal{N}} := f$. Then for each $m \geq 1$ we have the following inductive definition.

(1) $F_{m-1}$ is a functional with properties

$$\|F_{m-1}\| \leq 1, \qquad F_{m-1}(f_{m-1}) \geq \|f_{m-1}\|(1 - \delta_{m-1}),$$

and $\varphi_m := \varphi_m^{\tau,\delta,\eta,\mathcal{N}} \in \mathcal{D}(N_m)$ is any element satisfying

$$F_{m-1}(\varphi_m) \geq t_m \sup_{g \in \mathcal{D}(N_m)} F_{m-1}(g).$$

(2) Define

$$\Phi_m := \text{span}\{\varphi_j\}_{j=1}^{m},$$

and let

$$E_m(f) := \inf_{\varphi \in \Phi_m} \|f - \varphi\|.$$

Let $G_m \in \Phi_m$ be such that

$$\|f - G_m\| \leq E_m(f)(1 + \eta_m).$$

(3) Let

$$f_m := f_m^{\tau,\delta,\eta,\mathcal{N}} := f - G_m.$$

**Theorem 6.58** *Let a Banach space $X$ have modulus of smoothness $\rho(u)$ of power type $1 < q \leq 2$, that is $\rho(u) \leq \gamma u^q$. Assume that $\lim_{m \to \infty} N_m = \infty$,*

$$\sum_{m=1}^{\infty} t_m^p = \infty, \quad p = \frac{q}{q-1},$$

*and*

$$\delta_m = o(t_m^p), \qquad \eta_m = o(t_m^p).$$

*Then the RAWCGA converges for any $f \in X$.*

We now make some general remarks on $m$-term approximation with the depth search constraint. The depth search constraint means that for a given $m$ we restrict ourselves to systems of elements (subdictionaries) containing at most $N := N(m)$ elements. Let $X$ be a linear metric space and for a set $\mathcal{D} \subset X$ let $\mathcal{L}_m(\mathcal{D})$ denote the collection of all linear subspaces spanned by $m$ elements of $\mathcal{D}$. For a linear subspace $L \subset X$, the $\epsilon$-neighborhood $U_\epsilon(L)$ of $L$ is the set of all $x \in X$ which are at a distance not exceeding $\epsilon$ from $L$ (i.e. those $x \in X$ which can be approximated to an error not exceeding $\epsilon$ by the elements of $L$). For any compact set $F \subset X$ and any integers $N, m \geq 1$, we define the $(N, m)$-entropy numbers (see Temlyakov (2003a), p. 94)

$$\epsilon_{N,m}(F, X) := \inf_{\#\mathcal{D}=N} \inf\{\epsilon : F \subset \cup_{L \in \mathcal{L}_m(\mathcal{D})} U_\epsilon(L)\}.$$

We let $\Sigma_m(\mathcal{D})$ denote the collection of all functions (elements) in $X$ which can be expressed as a linear combination of at most $m$ elements of $\mathcal{D}$. Thus each function $s \in \Sigma_m(\mathcal{D})$ can be written in the form

$$s = \sum_{g \in \Lambda} c_g g, \quad \Lambda \subset \mathcal{D}, \quad \#\Lambda \leq m,$$

where the $c_g$ are real or complex numbers. For a function $f \in X$ we define its best $m$-term approximation error as follows:

$$\sigma_m(f) := \sigma_m(f, \mathcal{D}) := \inf_{s \in \Sigma_m(\mathcal{D})} \|f - s\|.$$

For a function class $F \subset X$ we define

$$\sigma_m(F) := \sigma_m(F, \mathcal{D}) := \sup_{f \in F} \sigma_m(f, \mathcal{D}).$$

We can express $\sigma_m(F, \mathcal{D})$ as

$$\sigma_m(F, \mathcal{D}) = \inf\{\epsilon : F \subset \cup_{L \in \mathcal{L}_m(\mathcal{D})} U_\epsilon(L)\}.$$

It follows therefore that

$$\inf_{\#\mathcal{D}=N} \sigma_m(F, \mathcal{D}) = \epsilon_{N,m}(F, X).$$

In other words, finding the best dictionaries consisting of $N$ elements for an $m$-term approximation of $F$ is the same as finding sets $\mathcal{D}$ which attain the $(N, m)$-entropy numbers $\epsilon_{N,m}(F, X)$. It is easy to see that $\epsilon_{m,m}(F, X) = d_m(F, X)$, where $d_m(F, X)$ is the Kolmogorov width of $F$ in $X$. This establishes a connection between $(N, m)$-entropy numbers and the Kolmogorov widths. One can find a further discussion on the nonlinear Kolmogorov $(N, m)$-widths and the entropy numbers in Temlyakov (2003a).

### 6.11 An application of greedy algorithms for the discrepancy estimates

We begin with a brief historical survey on discrepancy and numerical integration. We refer the reader for a complete survey on discrepancy to the following books: Beck and Chen (1987), Chazelle (2000), Kuipers and Niederreiter (1974) and Matoušek (1999). We formulate all results in the notation of this book and in a form convenient for us. Let $1 \leq q \leq \infty$. We recall the definition of the $L_q$ discrepancy (the $L_q$-star discrepancy) of points $\{\xi^1, \ldots, \xi^m\} \subset \Omega_d := [0, 1]^d$. Let $\chi_{[a,b]}(\cdot)$ be a characteristic function of the interval $[a, b]$. Denote for $x, y \in \Omega_d$

$$B(x, y) := \prod_{j=1}^{d} \chi_{[0,x_j]}(y_j).$$

Then the $L_q$ discrepancy of $\xi := \{\xi^1, \ldots, \xi^m\} \subset \Omega_d$ is defined by

$$D(\xi, m, d)_q := \| \int_{\Omega_d} B(x, y)dy - \frac{1}{m} \sum_{\mu=1}^{m} B(x, \xi^\mu) \|_{L_q(\Omega_d)}.$$

The first result in this area was the following conjecture of van der Corput (1935a, b). Let $\xi^j \in [0, 1]$, $j = 1, 2, \ldots$, then we have

$$\limsup_{m \to \infty} m D((\xi^1, \ldots, \xi^m), m, 1)_\infty = \infty.$$

This conjecture was proved by van Aardenne-Ehrenfest (1945):

$$\limsup_{m \to \infty} \frac{\log \log \log m}{\log \log m} m D((\xi^1, \ldots, \xi^m), m, 1)_\infty > 0.$$

Let us denote

$$D(m, d)_q := \inf_\xi D(\xi, m, d)_q, \qquad 1 \leq q \leq \infty.$$

Roth (1954) proved that

$$D(m, d)_2 \geq C(d) m^{-1} (\log m)^{(d-1)/2}. \tag{6.109}$$

Schmidt (1972) proved

$$D(m, 2)_\infty \geq C m^{-1} \log m. \tag{6.110}$$

Schmidt (1977) proved

$$D(m, d)_q \geq C(d, q) m^{-1} (\log m)^{(d-1)/2}, \qquad 1 < q \leq \infty. \tag{6.111}$$

Halász (1981) proved

$$D(m, d)_1 \geq C(d) m^{-1} (\log m)^{1/2}. \tag{6.112}$$

The following conjecture has been formulated in Beck and Chen (1987) as an excruciatingly difficult great open problem.

**Conjecture 6.59** *We have for $d \geq 3$*

$$D(m, d)_\infty \geq C(d)m^{-1}(\log m)^{d-1}.$$

This problem is still open. Recently, Bilyk and Lacey (2008) and Bilyk, Lacey and Vagharshakyan (2008) made substantial progress on this problem. The authors proved that, for some $\eta = \eta(d) > 0$,

$$D(m, d)_\infty \geq C(d)m^{-1}(\log m)^{((d-1)/2)+\eta}.$$

They also conjectured that the right order of $D(m, d)_\infty$ is $m^{-1}(\log m)^{d/2}$.

We now proceed to a brief discussion of numerical integration. Numerical integration seeks good ways of approximating an integral

$$\int_\Omega f(x)d\mu$$

by an expression of the form

$$\Lambda_m(f, \xi) := \sum_{j=1}^{m} \lambda_j f(\xi^j), \quad \xi = (\xi^1, \ldots, \xi^m), \quad \xi^j \in \Omega, \quad j = 1, \ldots, m.$$

$$(6.113)$$

It is clear that we must assume that $f$ is integrable and defined at the points $\xi^1, \ldots, \xi^m$. Equation (6.113) is called a cubature formula $(\Lambda, \xi)$ (if $\Omega \subset \mathbb{R}^d$, $d \geq 2$) or a quadrature formula $(\Lambda, \xi)$ (if $\Omega \subset \mathbb{R}$) with knots $\xi = (\xi^1, \ldots, \xi^m)$ and weights $\Lambda = (\lambda_1, \ldots, \lambda_m)$. For a function class $W$ we introduce a concept of error of the cubature formula $\Lambda_m(\cdot, \xi)$ by

$$\Lambda_m(W, \xi) := \sup_{f \in W} |\int_\Omega f \, d\mu - \Lambda_m(f, \xi)|. \quad (6.114)$$

In order to orient the reader we will begin with the case of univariate periodic functions. Let for $r > 0$

$$F_r(x) := 1 + 2\sum_{k=1}^{\infty} k^{-r} \cos(kx - r\pi/2) \quad (6.115)$$

and

$$W_p^r := \{f : f = \varphi * F_r, \quad \|\varphi\|_p \leq 1\}. \quad (6.116)$$

It is well known that for $r > 1/p$ the class $W_p^r$ is embedded into the space of continuous functions $\mathcal{C}(\mathbb{T})$. In a particular case of $W_1^1$ we also have embedding into $\mathcal{C}(\mathbb{T})$. From the definitions (6.113), (6.114) and (6.116) we see that, for the normalized measure $d\mu = (1/2\pi)dx$,

$$\Lambda_m(W_p^r, \xi) = \sup_{\|\varphi\|_p \le 1} |\frac{1}{2\pi} \int_{\mathbb{T}} \left( \int_{\mathbb{T}} F_r(x - y)d\mu - \sum_{j=1}^{m} \lambda_j F_r(\xi^j - y) \right) \varphi(y)dy|$$

$$= \|1 - \sum_{j=1}^{m} \lambda_j F_r(\xi^j - \cdot)\|_{p'}, \quad p' := \frac{p}{p-1}. \tag{6.117}$$

Thus the quality of the quadrature formula $\Lambda_m(\cdot, \xi)$ for the function class $W_p^r$ is controlled by the quality of $\Lambda_m(\cdot, \xi)$ for the representing kernel $F_r(x - y)$. In the particular case of $W_1^1$ we have

$$\Lambda_m(W_1^1, \xi) = \max_{y} |1 - \sum_{j=1}^{m} \lambda_j F_1(\xi^j - y)|. \tag{6.118}$$

In this case the function

$$F_1(x) = 1 + 2 \sum_{k=1}^{\infty} k^{-1} \sin kx = 1 + S(x)$$

has a simple form: $S(x) = 0$ for $x = l\pi$ and $S(x) = \pi - x$ for $x \in (0, 2\pi)$. This allows us to associate the quantity $\Lambda_m(W_1^1, \xi)$ with the one that has a simple geometrical interpretation. Denote by $\chi$ the class of all characteristic functions $\chi_{[0,a]}(x)$, $a \in [0, 2\pi)$. Then we have the following simple property (see Temlyakov (2003b) for the proof).

**Proposition 6.60** *There exist two positive absolute constants $C_1$ and $C_2$ such that, for any $\Lambda_m(\cdot, \xi)$ with a property $\sum_j \lambda_j = 1$, we have*

$$C_1 \Lambda_m(\chi, \xi) \le \Lambda_m(W_1^1, \xi) \le C_2 \Lambda_m(\chi, \xi). \tag{6.119}$$

We proceed to the multivariate case. For $x = (x_1, \ldots, x_d)$ denote

$$F_r(x) := \prod_{j=1}^{d} F_r(x_j)$$

and

$$MW_p^r := \{f : f = \varphi * F_r, \quad \|\varphi\|_p \le 1\}.$$

For $f \in MW_p^r$ we will denote $f^{(r)} := \varphi$, where $\varphi$ is such that $f = \varphi * F_r$. The letter $M$ in the notation $MW_p^r$ stands for "mixed," because in the case of integer $r$ the class $MW_p^r$ is very close to the class of functions $f$, satisfying

$\|f^{(r,...,r)}\|_p \leq 1$, where $f^{(r,...,r)}$ is the mixed derivative of $f$ of order $rd$. A multivariate analog of the class $\chi$ is the class

$$\chi^d := \left\{ \chi_{[0,a]}(x) := \prod_{j=1}^{d} \chi_{[0,a_j]}(x_j), \quad a_j \in [0, 2\pi), \quad j = 1, \ldots, d \right\}.$$

As in the univariate case one obtains analogs of (6.117), (6.118) and Proposition 6.60 (see Temlyakov (2003b)):

$$\Lambda_m(MW_p^r, \xi) = \left\| 1 - \sum_{j=1}^{m} \lambda_j F_r(\xi^j - \cdot) \right\|_{p'}; \tag{6.120}$$

$$\Lambda_m(MW_1^1, \xi) = \max_y \left| 1 - \sum_{j=1}^{m} \lambda_j F_1(\xi^j - y) \right|. \tag{6.121}$$

**Proposition 6.61** *There exist two positive constants $C_1(d)$ and $C_2(d)$ such that, for any $\Lambda_m(\cdot, \xi)$ with a property $\sum_j \lambda_j = 1$, we have*

$$C_1(d)\Lambda_m(\chi^d, \xi) \leq \Lambda_m(MW_1^1, \xi) \leq C_2(d)\Lambda_m(\chi^d, \xi). \tag{6.122}$$

It is clear from the definition of discrepancy that for a set $\xi$ of points $\xi^1, \ldots, \xi^m \subset [0, 1]^d$

$$D(\xi, m, d)_\infty = \Lambda_m(\chi^d, 2\pi\xi) \quad \text{with} \quad \lambda_1 = \cdots = \lambda_m = 1/m.$$

Thus the classical concept of discrepancy is directly related to the efficiency of the corresponding cubature formulas for a special function class $MW_1^1$. It is well known that the $W_1^1$ is very close to the class of functions of bounded variation and that the $MW_1^1$ is very close to the class of functions with bounded variation in the sense of Hardy–Vitali. At the beginning of the twentieth century D. Vitali and G. Hardy generalized the definition of variation to the multivariate case. Roughly speaking, in the one-dimensional case the condition that $f$ be of bounded variation is close to the condition $\|f'\|_1 < \infty$. In the multidimensional case the condition that a function has bounded variation in the sense of Hardy–Vitali is close to that of requiring $\|f^{(1,...,1)}\|_1 < \infty$, where $f^{(1,...,1)}$ is a mixed derivative of $f$.

In the 1930s, in connection with applications in mathematical physics, S. L. Sobolev introduced the classes of functions by imposing the following restrictions:

$$\|f^{(n_1,...,n_d)}\|_p \leq 1 \tag{6.123}$$

for all $n = (n_1, \ldots, n_d)$ such that $n_1 + \cdots + n_d \leq R$. These classes appeared as natural ways to measure smoothness in many multivariate problems including numerical integration. It was established that for Sobolev classes the optimal error of numerical integration by formulas with $m$ knots is of order $m^{-R/d}$. Assume now for the sake of simplicity (to avoid fractional differentiation) that $R = rd$, where $r$ is a natural number. At the end of the 1950s, N. M. Korobov discovered the following phenomenon. Let us consider the class of functions which satisfy (6.123) for all $n$ such that $n_j \leq r$, $j = 1, \ldots, d$ (compare to the above classes $MW_p^r$). It is clear that this new class (the class of functions with a dominating mixed derivative) is wider than the Sobolev class with $R = rd$. For example, all functions of the form

$$f(x) = \prod_{j=1}^{d} f_j(x_j), \quad \|f_j^{(r)}\|_p \leq 1,$$

are in this class, while not necessarily in the Sobolev class (it would require, roughly, $\|f_j^{(rd)}\|_p \leq 1$). Korobov constructed a cubature formula with $m$ knots which guaranteed the accuracy of numerical integration for this class of order $m^{-r}(\log m)^{rd}$, i.e. almost the same accuracy that we had for the Sobolev class. Korobov's discovery pointed out the importance of the classes of functions with dominating mixed derivative in fields such as approximation theory and numerical analysis. The (convenient for us) definition of these classes (classes of functions with bounded mixed derivative) is given above (see the definition of $MW_p^r$).

In addition to the classes of $2\pi$-periodic functions, it will be convenient for us to consider the classes of non-periodic functions defined on $\Omega_d = [0, 1]^d$.

Let $r$ be a natural number and let $MW_p^r(\Omega_d)$, $1 \leq p \leq \infty$, denote the closure in the uniform metric of the set of $rd$-times continuously differentiable functions $f(x)$ such that

$$\|f\|_{MW_p^r} := \sum_{0 \leq n_j \leq r; j=1,\ldots,d} \left\| \frac{\partial^{n_1 + \cdots + n_d} f}{\partial x_1^{n_1} \ldots \partial x_d^{n_d}} \right\|_p \leq 1, \qquad (6.124)$$

where

$$\|g\|_p = \left( \int_{\Omega_d} |g(x)|^p \, dx \right)^{1/p}.$$

It will be convenient for us to consider the subclass $M\dot{W}_p^r(\Omega_d)$ of the class $MW_p^r(\Omega_d)$ consisting of the functions $f(x)$ representable in the form

$$f(x) = \int_{\Omega_d} B_r(t, x)\varphi(t)dt, \qquad \|\varphi\|_p \leq 1,$$

where

$$B_r(t, x) := \prod_{j=1}^{d} \left((r-1)!\right)^{-1} (t_j - x_j)_+^{r-1}, \qquad t, x \in \Omega_d, \quad (a)_+ = \max(a, 0).$$

We note that $B_1(t, x)$ coincides with $B(t, x)$ defined above.

In connection with the definition of the class $M\dot{W}_p^r(\Omega_d)$, we remark here that for the error of the cubature formula $(\Lambda, \xi)$ with weights $\Lambda = (\lambda_1, \ldots, \lambda_m)$ and knots $\xi = (\xi^1, \ldots, \xi^m)$ the following relation holds. Let

$$\left| \Lambda_m(f, \xi) - \int_{\Omega_d} f(x)dx \right| =: R_m(\Lambda, \xi, f),$$

then, similarly to (6.117) and (6.120), one obtains $(p' := p/(p-1))$

$$\Lambda_m\left(M\dot{W}_p^r(\Omega_d), \xi\right) := \sup_{f \in M\dot{W}_p^r(\Omega_d)} R_m(\Lambda, \xi, f)$$

$$= \left\| \sum_{\mu=1}^{m} \lambda_\mu B_r(t, \xi^\mu) - \prod_{j=1}^{d} (t_j^r / r!) \right\|_{p'} =: D_r(\xi, \Lambda, m, d)_{p'}.$$

(6.125)

The quantity $D_r(\xi, \Lambda, m, d)_q$ in the case $r = 1$, $\Lambda = (1/m, \ldots, 1/m)$, is the classical discrepancy of the set of points $\{\xi^\mu\}$. The quantity $D_r(\xi, \Lambda, m, d)_q$ defined in (6.125) is a natural generalization of the concept of discrepancy. This generalization contains two ingredients: general weights $\Lambda$ instead of the special case of equal weights $(1/m, \ldots, 1/m)$ and any natural number $r$ instead of $r = 1$. We note that in approximation theory we usually study the whole scale of smoothness classes rather than an individual smoothness class. The above generalization of discrepancy for arbitrary positive integer $r$ allows us to study the question of how smoothness $r$ affects the rate of decay of generalized discrepancy.

We now present the results on the lower estimates for the $D_r(\xi, \Lambda, m, d)_q$. We denote

$$D_r^o(m, d)_q := \inf_{\xi, \Lambda} D_r(\xi, \Lambda, m, d)_q,$$

where $D_r(\xi, \Lambda, m, d)_q$ is defined in (6.125). The superscript $o$ stands here for *optimal* to emphasize that we are optimizing over the weights $\Lambda$. The first result in estimating the generalized discrepancy was obtained by Bykovskii (1985):

$$D_r^o(m, d)_2 \geq C(r, d) m^{-r} (\log m)^{(d-1)/2}.$$

(6.126)

This result is a generalization of Roth's result (6.109). The generalization of Schmidt's result (6.111) was obtained by Temlyakov (1990) (see theorem 3.5 of Temlyakov (2003b)):

$$D_r^o(m, d)_q \geq C(r, d, q)m^{-r}(\log m)^{(d-1)/2}, \qquad 1 < q \leq \infty. \qquad (6.127)$$

We formulate an open problem (see Temlyakov (2003b)) for the case $q = \infty$.

**Conjecture 6.62** *For all $d, r \in \mathbb{N}$ we have*

$$D_r^o(m, d)_\infty \geq C(r, d)m^{-r}(\log m)^{d-1}.$$

The above lower estimates for $D_1^o(m, d)_q$ are formally stronger than the corresponding estimates for $D(m, d)_q$ because in $D_1^o(m, d)_q$ we are optimizing over the weights $\Lambda$. However, the proofs for $D(m, d)_q$ could be adjusted to give the estimates for $D_1^o(m, d)_q$. The results (6.126) and (6.127) for the generalized discrepancy were obtained as a corollary of the corresponding results on cubature formulas. We do not know if existing methods for $D(m, d)_q$ could be modified to get the estimates for $D_r^o(m, d)_q, r \geq 2$.

**Remark 6.63** In the case of natural $r$ the class $MW_p^r$ turns into the subclass of the class $MW_p^r(\Omega_d)B := \{f : f/B \in MW_p^r(\Omega_d)\}$, after the linear change of variables

$$x_j = -\pi + 2\pi t_j, \qquad j = 1, \ldots, d.$$

We are interested in the dependence on $m$ of the quantities

$$\delta_m(W) = \inf_{\lambda_1, \ldots, \lambda_m; \xi^1, \ldots, \xi^m} \Lambda_m(W, \xi)$$

for the classes $W$ defined above. Remark 6.63 shows that

$$\delta_m(MW_p^r) \ll \delta_m(MW_p^r(\Omega_d)). \qquad (6.128)$$

Let $M\overset{0}{W}_p^r(\Omega_d)$ denote the subset of functions in $MW_p^r(\Omega_d)$, which is the closure in the uniform metric of the set of functions $f$ satisfying the condition (6.124) which have the following property: $f(x)$ is $rd$-times continuously differentiable and for all $0 \leq n_j \leq r, j = 1, \ldots, d, f^{(n_1, \ldots, n_d)}(x) = 0$ on the boundary of the cube $\Omega_d$. The following theorem (see Temlyakov (2003b)) shows that the above defined classes are the same from the point of view of optimal (in the sense of order) numerical integration.

**Theorem 6.64** *Let $1 \leq p \leq \infty$. Then*

$$\delta_m(M\overset{0}{W}_p^r(\Omega_d)) \asymp \delta_m(M\dot{W}_p^r(\Omega_d)) \asymp \delta_m(MW_p^r(\Omega_d)). \qquad (6.129)$$

We proceed to the lower estimates for cubature formulas. The following bound was obtained in Temlyakov (1990) (see also Temlyakov (2003b), theorem 3.2).

**Theorem 6.65** *The following lower estimate is valid for any cubature formula* $(\Lambda, \xi)$ *with $m$ knots* $(r > 1/p)$

$$\Lambda_m(MW_p^r, \xi) \geq C(r, d, p)m^{-r}(\log m)^{(d-1)/2}, \qquad 1 \leq p < \infty.$$

There are two big open problems in this area. They are formulated in Temlyakov (2003b) as conjectures.

**Conjecture 6.66** *For any $d \geq 2$ and any $r \geq 1$ we have*

$$\delta_m(MW_1^r) \geq C(r, d)m^{-r}(\log m)^{d-1}.$$

**Conjecture 6.67** *For any $d \geq 2$ and any $r > 0$ we have*

$$\delta_m(MW_\infty^r) \geq C(r, d)m^{-r}(\log m)^{(d-1)/2}.$$

We note that by Proposition 6.61, Theorem 6.64 and (6.125), Conjecture 6.66 implies Conjecture 6.62 and Conjecture 6.67 implies

$$D_r^o(m, d)_1 \geq C(r, d)m^{-r}(\log m)^{(d-1)/2}. \tag{6.130}$$

We turn to the upper estimates. We begin with the cubature formulas. We have already made a historical remark on classes with bounded mixed derivative. We will discuss only these classes here. For results on cubature formulas for the Sobolev-type classes, we refer the reader to books by Novak (1988), Sobolev (1974) and Temlyakov (1993a). The first result in this direction was obtained by Korobov (1959). He used the cubature formulas $P_m(f, a)$ defined as follows (see Temlyakov (2003b) for more details). Let $m \in \mathbb{N}$, $a = (a_1, \ldots, a_d) \in \mathbb{Z}^d$. Korobov considered the cubature formulas

$$P_m(f, a) = m^{-1} \sum_{\mu=1}^m f\left(2\pi\left\{\frac{\mu a_1}{m}\right\}, \ldots, 2\pi\left\{\frac{\mu a_d}{m}\right\}\right),$$

where $\{x\}$ is the fractional part of the number $x$, which are called the Korobov cubature formulas. His results lead to the following estimate:

$$\delta_m(MW_1^r) \leq C(r, d)m^{-r}(\log m)^{rd}, \qquad r > 1. \tag{6.131}$$

Bakhvalov (1959) improved (6.131) to

$$\delta_m(MW_1^r) \leq C(r, d)m^{-r}(\log m)^{r(d-1)}, \qquad r > 1.$$

The first best possible upper estimate for the classes $MW_p^r$ was obtained by Bakhvalov (1963). He proved in the case $d = 2$ that

$$\delta_m(MW_2^r) \leq C(r)m^{-r}(\log m)^{1/2}, \qquad r \in \mathbb{N}. \tag{6.132}$$

Bakhvalov used the Fibonacci cubature formulas defined as follows. For periodic functions of two variables we consider the Fibonacci cubature formulas

$$\Phi_n(f) = b_n^{-1} \sum_{\mu=1}^{b_n} f(2\pi\mu/b_n, \ 2\pi\{\mu b_{n-1}/b_n\}),$$

where $b_0 = b_1 = 1$ and $b_n = b_{n-1} + b_{n-2}$ are the Fibonacci numbers.

Frolov (1976) used other cubature formulas that we call the Frolov cubature formulas. We give a definition of these cubature formulas. The following lemma plays a fundamental role in the construction of such cubature formulas.

**Lemma 6.68** *There exists a matrix A such that the lattice $L(\mathbf{m}) = A\mathbf{m}$, where $\mathbf{m}$ is a (column) vector with integer coordinates, has the following properties:*

$$\left( L(\mathbf{m}) = \left( \begin{array}{c} L_1(\mathbf{m}) \\ \vdots \\ L_d(\mathbf{m}) \end{array} \right) \right),$$

*(1⁰)* $\qquad \left| \prod_{j=1}^d L_j(\mathbf{m}) \right| \geq 1$ *for all* $\mathbf{m} \neq \mathbf{0}$;

*(2⁰)* *each parallelepiped P with volume $|P|$ whose edges are parallel to the coordinate axes contains no more than $|P| + 1$ lattice points.*

Let $a > 1$ and $A$ be the matrix from Lemma 6.68. We consider the cubature formula

$$\Phi(a, A)(f) = \left( a^d |\det A| \right)^{-1} \sum_{\mathbf{m} \in \mathbb{Z}^d} f\left( \frac{(A^{-1})^T \mathbf{m}}{a} \right)$$

for $f$ continuous with support in $\Omega_d$. Clearly, the number $N$ of points of this cubature formula does not exceed $C(A)a^d |\det A|$.

Frolov used the above defined cubature formulas to extend (6.132) to the case $d > 2$ as follows:

$$\delta_m(MW_2^r) \leq C(r, d)m^{-r}(\log m)^{(d-1)/2}, \qquad r \in \mathbb{N}. \tag{6.133}$$

This estimate was further generalized by Bykovskii (1985) to $r \in \mathbb{R}, r \geq 1$. Bykovskii also used the Frolov cubature formulas. One can find these results in sect. 4 of Temlyakov (2003b). We note that there are no sharp results for $\delta_m(MW_p^r)$ in the case of small smoothness $1/p < r < 1$. It is an interesting open problem. The approach based on nonlinear $m$-term approximation (see sect. 2 of Temlyakov (2003b)) can be useful in this case.

The Frolov cubature formulas (see Frolov (1979)) give the following estimate

$$\delta_m(MW_1^r) \le C(r, d)m^{-r}(\log m)^{d-1}, \qquad r > 1. \tag{6.134}$$

Thus the lower estimate in Conjecture 6.66 is the best possible.

Skriganov (1994) proved the following estimate:

$$\delta_m(MW_p^r) \le C(r, d, p)m^{-r}(\log m)^{(d-1)/2}, \quad 1 < p \le \infty, \quad r \in \mathbb{N}. \tag{6.135}$$

This estimate combined with Theorem 6.65 implies

$$\delta_m(MW_p^r) \asymp m^{-r}(\log m)^{(d-1)/2}, \quad 1 < p < \infty, \quad r \in \mathbb{N}. \tag{6.136}$$

We now present the upper estimates for the discrepancy. Davenport (1956) proved that

$$D(m, 2)_2 \le Cm^{-1}(\log m)^{1/2}.$$

Other proofs of this estimate were later given by Vilenkin (1967), Halton and Zaremba (1969) and Roth (1976). Roth (1979) proved

$$D(m, 3)_2 \le Cm^{-1}\log m,$$

and Roth (1980) and Frolov (1980) proved

$$D(m, d)_2 \le C(d)m^{-1}(\log m)^{(d-1)/2}.$$

Chen (1980) proved

$$D(m, d)_q \le C(d)m^{-1}(\log m)^{(d-1)/2}, \qquad q < \infty. \tag{6.137}$$

The estimate (6.134) and Theorem 6.64 imply

$$D_r^o(m, d)_\infty \le C(r, d)m^{-r}(\log m)^{d-1}, \qquad r \ge 2.$$

We note that the upper estimates for $D(m, d)_q$ are stronger than the same upper estimates for $D_1^o(m, d)_q$.

The quantity $\delta_m(MW_p^r)$, discussed above, is a function of four variables $m, r, p, d$. We discussed above a traditional setting of the problem when we want to find the right order of decay of $\delta_m(MW_p^r)$ as a function of $m$ when $m$ increases to infinity and other parameters are fixed. As we have seen, there are important problems in this setting which remain open. Clearly, the problem of finding the right dependence of $\delta_m(MW_p^r)$ of all four parameters $m, r, p, d$ is a more difficult problem. Recently, driven by possible applications in finance, global optimization and path integrals, some researchers studied the problem of dependence of $\delta_m(F_d)$ on dimension $d$ for various function classes $F_d$. Today this topic is being thoroughly studied in information-based complexity, and a number of results have been recently obtained. We refer the reader to a nice survey of the topic written by Novak and Wozniakowski (2001).

We now proceed to applications of greedy approximation for the discrepancy estimates in high dimensions. It will be convenient for us to study a slight modification of $D(\xi, m, d)_q$. For $a, t \in [0, 1]$ denote

$$H(a, t) := \chi_{[0,a]}(t) - \chi_{[a,1]}(t),$$

and for $x, y \in \Omega_d$

$$H(x, y) := \prod_{j=1}^{d} H(x_j, y_j).$$

We define the symmetrized $L_q$ discrepancy by

$$D^s(\xi, m, d)_q := \| \int_{\Omega_d} H(x, y) dy - \frac{1}{m} \sum_{\mu=1}^{m} H(x, \xi^\mu) \|_{L_q(\Omega_d)}.$$

Using the identity

$$\chi_{[0,x_j]}(y_j) = \frac{1}{2}(H(1, y_j) + H(x_j, y_j))$$

we get a simple inequality

$$D(\xi, m, d)_\infty \le D^s(\xi, m, d)_\infty. \tag{6.138}$$

We are interested in $\xi$ with small discrepancy. Consider

$$D^s(m, d)_q := \inf_\xi D^s(\xi, m, d)_q.$$

As already mentioned above (see (6.111) and (6.137)) the following relation is known:

$$D(m, d)_q \asymp m^{-1}(\ln m)^{(d-1)/2}, \qquad 1 < q < \infty, \tag{6.139}$$

with constants in $\asymp$ depending on $q$ and $d$. The right order of $D(m, d)_q$, $q = 1, \infty$, for $d \ge 3$ is unknown. The following estimate has been obtained in Heinrich *et al.* (2001):

$$D(m, d)_\infty \le Cd^{1/2}m^{-1/2}. \tag{6.140}$$

It is pointed out in Heinrich *et al.* (2001) that (6.140) is only an existence theorem and that even a constant $C$ in (6.140) is unknown. Their proof is a probabilistic one. There are also some other estimates in Heinrich *et al.* (2001) with explicit constants. We mention one of them,

$$D(m, d)_\infty \leq C(d \ln d)^{1/2}((\ln m)/m)^{1/2}, \qquad (6.141)$$

with an explicit constant $C$. The proof of (6.141) is also probabilistic.

In this section we apply greedy type algorithms to obtain upper estimates of $D(m, d)_q$, $1 \leq q \leq \infty$, in the style of (6.140) and (6.141). The important feature of our proof is that it is deterministic and moreover it is constructive. Formally the optimization problem

$$D(m, d)_q = \inf_\xi D(\xi, m, d)_q$$

is deterministic: one needs to minimize over $\{\xi^1, \ldots, \xi^m\} \subset \Omega_d$. However, minimization by itself does not provide any upper estimate. It is known (see Davis, Mallat and Avellaneda (1997)) that simultaneous optimization over many parameters ($\{\xi^1, \ldots, \xi^m\}$ in our case) is a very difficult problem. We note that

$$D(m, d)_q = \sigma_m^e(J, \mathcal{B})_q := \inf_{g_1, \ldots, g_m \in \mathcal{B}} \left\| J(\cdot) - \frac{1}{m} \sum_{\mu=1}^m g_\mu \right\|_{L_q(\Omega_d)},$$

where

$$J(x) = \int_{\Omega_d} B(x, y) dy$$

and

$$\mathcal{B} = \{B(x, y), \quad y \in \Omega_d\}.$$

It has been proved in Davis, Mallat and Avellaneda (1997) that if an algorithm finds a best $m$-term approximation for each $f \in \mathbb{R}^N$ for every dictionary $\mathcal{D}$ with the number of elements of order $N^k$, $k \geq 1$, then this algorithm solves an $NP$-hard problem. Thus, in nonlinear $m$-term approximation we look for methods (algorithms) which provide approximation close to best $m$-term approximation and at each step solve an optimization problem over only one parameter ($\xi^\mu$ in our case). In this section we will provide such an algorithm for estimating $\sigma_m^e(J, \mathcal{B})_q$. We call this algorithm *constructive* because it provides an explicit construction with feasible one-parameter optimization steps.

We proceed to the construction. We will use in our construction the IA($\epsilon$) which has been studied in Section 6.6. We will use the following corollaries of Theorem 6.29.

**Corollary 6.69** *We apply Theorem 6.29 for* $X = L_q(\Omega_d)$, $q \in [2, \infty)$, $\mathcal{D}^+ = \{H(x, y), y \in \Omega_d\}$, $f = J^s(x)$, *where*

$$J^s(x) = \int_{\Omega_d} H(x, y) dy \in A_1(\mathcal{D}^+).$$

*Using (6.12) we get by Theorem 6.29 a constructive set* $\xi^1, \ldots, \xi^m$ *such that*

$$D^s(\xi, m, d)_q = \|(J^s)_m^{i,\epsilon}\|_{L_q(\Omega_d)} \leq Cq^{1/2}m^{-1/2}$$

*with absolute constant C.*

**Corollary 6.70** *We apply Theorem 6.29 for* $X = L_q(\Omega_d)$, $q \in [2, \infty)$, $\mathcal{D}^+ = \{B(x, y), y \in \Omega_d\}$, $f = J(x)$, *where*

$$J(x) = \int_{\Omega_d} B(x, y) dy \in A_1(\mathcal{D}^+).$$

*Using (6.12) we get by Theorem 6.29 a constructive set* $\xi^1, \ldots, \xi^m$ *such that*

$$D(\xi, m, d)_q = \|J_m^{i,\epsilon}\|_{L_q(\Omega_d)} \leq Cq^{1/2}m^{-1/2}$$

*with absolute constant C.*

**Corollary 6.71** *We apply Theorem 6.29 for* $X = L_q(\Omega_d)$, $q \in [2, \infty)$, $\mathcal{D}^+ = \{B(x, y)/\|B(\cdot, y)\|_{L_q(\Omega_d)}, \quad y \in \Omega_d\}$, $f = J(x)$. *Using (1.12) we get by Theorem 6.29 a constructive set* $\xi^1, \ldots, \xi^m$ *such that*

$$\left\| \int_{\Omega_d} B(x, y) dy - \frac{1}{m} \sum_{\mu=1}^m \left( \frac{q}{q+1} \right)^d \left( \prod_{j=1}^d (1 - \xi_j^\mu)^{-1/q} \right) B(x, \xi^\mu) \right\|_{L_q(\Omega_d)}$$

$$\leq C \left( \frac{q}{q+1} \right)^d q^{1/2} m^{-1/2}$$

*with absolute constant C.*

We note that in the case $X = L_q(\Omega_d)$, $q \in [2, \infty)$, $\mathcal{D}^+ = \{H(x, y), x \in \Omega_d\}$, $f = J^s(y)$, the implementation of the IA($\epsilon$) is a sequence of maximization steps when we maximize functions of $d$ variables. An important advantage of the $L_q$ spaces is a simple and explicit form of the norming functional $F_f$ of a function $f \in L_q(\Omega_d)$. The $F_f$ acts as (for real $L_q$ spaces)

$$F_f(g) = \int_{\Omega_d} \|f\|_q^{1-q} |f|^{q-2} fg\, dy.$$

Thus the IA($\epsilon$) should find at step $m$ an approximate solution to the following optimization problem (over $y \in \Omega_d$):

$$\int_{\Omega_d} |f_{m-1}^{i,\epsilon}(x)|^{q-2} f_{m-1}^{i,\epsilon}(x) H(x, y) dx \rightarrow \max.$$

Let us discuss possible applications of the WRGA instead of the IA($\epsilon$). An obvious change is that, instead of the cubature formula

$$\frac{1}{m} \sum_{\mu=1}^{m} H(x, \xi^\mu),$$

in the case of the IA($\epsilon$) we have a cubature formula

$$\sum_{\mu=1}^{m} w_\mu^m H(x, \xi^\mu), \quad \sum_{\mu=1}^{m} |w_\mu^m| \leq 1,$$

in the case of the WRGA; this is a disadvantage of the WRGA. An advantage of the WRGA is that we are more flexible in selecting an element $\varphi_m^r$:

$$F_{f_{m-1}^r}(\varphi_m^r - G_{m-1}^r) \geq t_m \sup_{g \in \mathcal{D}} F_{f_{m-1}^r}(g - G_{m-1}^r)$$

than an element $\varphi_m^{i,\epsilon}$:

$$F_{f_{m-1}^{i,\epsilon}}(\varphi_m^{i,\epsilon} - f) \geq -\epsilon_m.$$

We will now derive an estimate for $D(m, d)_\infty$ from Corollary 6.70.

**Proposition 6.72** *For any $m$ there exists a constructive set $\xi = \{\xi^1, \dots, \xi^m\} \subset \Omega_d$ such that*

$$D(\xi, m, d)_\infty \leq Cd^{3/2} (\max(\ln d, \ln m))^{1/2} m^{-1/2}, \quad d, m \geq 2 \quad (6.142)$$

*with effective absolute constant $C$.*

*Proof* We use the inequality from Niederreiter, Tichy and Turnwald (1990)

$$D(\xi, m, d)_\infty \leq c(d, q) d(3d + 4) D(\xi, m, d)_q^{q/(q+d)} \quad (6.143)$$

and the estimate for $c(d, q)$ from Heinrich *et al.* (2001)

$$c(d, q) \leq 3^{1/3} d^{-1+2/(1+q/d)}. \quad (6.144)$$

Specifying $q = d \max(\ln d, \ln m)$ and using Corollary 6.70 we get (6.142) from (6.143) and (6.144). □

## 6.12 Open problems

There are no results on the convergence of $X$-Greedy Algorithms in general uniformly smooth Banach spaces. Here are two open problems that concern the convergence of the XGA.

**6.1.** Does the XGA converge for all dictionaries $\mathcal{D}$ and each element $f \in X$ in uniformly smooth Banach spaces $X$ with modulus of smoothness of fixed power type $q$, $1 < q \leq 2$, $(\rho(u) \leq \gamma u^q)$?

**6.2.** Characterize Banach spaces $X$ such that the $X$-Greedy Algorithm converges for all dictionaries $\mathcal{D}$ and each element $f$.

A little more is known about convergence of the WDGA (see Section 6.7). Here are some open problems.

**6.3.** Characterize Banach spaces $X$ such that the WDGA($t$), $t \in (0, 1]$, converges for all dictionaries $\mathcal{D}$ and each element $f$.

**6.4.** (Conjecture.) Prove that the WDGA($t$), $t \in (0, 1]$, converges for all dictionaries $\mathcal{D}$ and each element $f \in X$ in uniformly smooth Banach spaces $X$ with modulus of smoothness of fixed power type $q$, $1 < q \leq 2$, $(\rho(u) \leq \gamma u^q)$.

**6.5.** Find the necessary and sufficient conditions on a weakness sequence $\tau$ to guarantee convergence of the Weak Dual Greedy Algorithm in uniformly smooth Banach spaces $X$ with modulus of smoothness of fixed power type $q$, $1 < q \leq 2$, $(\rho(u) \leq \gamma u^q)$ for all dictionaries $\mathcal{D}$ and each element $f \in X$.

**6.6.** Let $p \in (1, \infty)$. Find the necessary and sufficient conditions on a weakness sequence $\tau$ to guarantee convergence of the Weak Dual Greedy Algorithm in the $L_p$ space for all dictionaries $\mathcal{D}$ and each element $f \in L_p$.

**6.7.** Characterize Banach spaces $X$ such that the WCGA($t$), $t \in (0, 1]$, converges for every dictionary $\mathcal{D}$ and for every $f \in X$.

**6.8.** Characterize Banach spaces $X$ such that the WGAFR with the weakness sequence $\tau = \{t\}$, $t \in (0, 1]$, converges for every dictionary $\mathcal{D}$ and for every $f \in X$.

**6.9.** Characterize Banach spaces $X$ such that the XGAFR converges for every dictionary $\mathcal{D}$ and for every $f \in X$.

# References

van Aardenne-Ehrenfest, T. (1945), Proof of the impossibility of a just distribution of an infinite sequence of points over an interval, *Proc. Kon. Ned. Akad. v. Wetensch* **48**, 266–271.

Alon, N. (2003), Problems and results in extremal combinatorics, *Discrete Math.*, **273**, 31–53.

Baishanski, B. M. (1983), Approximation by polynomials of given length, *Illinois J. Math.*, **27**, 449–458.

Bakhvalov, N. S. (1959), On the approximate computation of multiple integrals, *Vestnik Moskov. Univ. Ser. Mat. Mekh. Astr. Fiz. Khim.*, **4**, 3–18.

Bakhvalov, N. S. (1963), Optimal convergence bounds for quadrature processes and integration methods of Monte Carlo type for classes of functions, *Zh. Vychisl. Mat. i Mat. Fiz. Suppl.*, **4**, 5–63.

Baraniuk, R., Davenport, M., DeVore, R. and Wakin, M. (2008), A simple proof of the Restricted Isometry Property for random matrices, *Construct. Approx.*, **28**, 253–263.

Barron, A. R. (1991), Complexity regularization with applications to artificial neural networks. In *Nonparametric Functional Estimation*, G. Roussas, ed. (Dordrecht: Kluwer), pp. 561–576.

Barron, A. R. (1993), Universal approximation bounds for superposition of $n$ sigmoidal functions, *IEEE Trans. Inf. Theory*, **39**, 930–945.

Barron, A., Birgé, L. and Massart, P. (1999), Risk bounds for model selection via penalization, *Prob. Theory Related Fields*, **113**, 301–413.

Barron, A., Cohen, A., Dahmen, W. and DeVore, R. (2008), Approximation and learning by greedy algorithms, *Ann. Stat.*, **36**, 64–94.

Bass, R. F. (1988), Probability estimates for multiparameter Brownian processes, *Ann. Prob.*, **16**, 251–264.

Bary, N. K. (1961), *Trigonometric Series*, (Moscow: Nauka) (in Russian); English translation (Oxford: Pergamon Press, 1964).

Beck, J. and Chen, W. (1987), *Irregularities of Distribution*, (Cambridge: Cambridge University Press).

Bednorz, W. (2008), Greedy bases are best for $m$-term approximation, *Construct. Approx.*, **28**, 265–275.

Belinsky, E. S. (1998), Estimates of entropy numbers and Gaussian measures for classes of functions with bounded mixed derivative, *J. Approx. Theory*, **93**, 114–127.

Bilyk, D. and Lacey, M. (2008), On the small ball inequality in three dimensions, *Duke Math J.*, **143**, 81–115.

Bilyk, D., Lacey, M. and Vagharshakyan, A. (2008), On the small ball inequality in all dimensions, *J. Func. Anal.*, **254**, 2470–2502.

Binev, P., Cohen, A., Dahmen, W., DeVore, R. and Temlyakov, V. (2005), Universal algorithms for learning theory. Part I: Piecewise constant functions, *J. Machine Learning Theory (JMLT)*, **6**, 1297–1321.

Bourgain, J. (1992), A remark on the behaviour of $L^p$-multipliers and the range of operators acting on $L^p$-spaces, *Israel J. Math.*, **79**, 193–206.

Bykovskii, V. A. (1985), On the correct order of the error of optimal cubature formulas in spaces with dominant derivative, and on quadratic deviations of grids, *Preprint, Computing Center Far-Eastern Scientific Center, Akad. Sci. USSR, Vladivostok.*

Candes, E. (2006), Compressive sampling, *ICM Proc., Madrid*, **3**, 1433–1452.

Candes, E. (2008), The restricted isometry property and its implications for compressed sensing, *C. R. Acad. Sci. Paris, Ser.* **I 346**, 589–592.

Candes, E., Romberg, J. and Tao, T. (2006), Stable signal recovery from incomplete and inaccurate measurements, *Commun. Pure Appl. Math.*, **59**, 1207–1223.

Candes, E. and Tao, T. (2005), Decoding by linear programming, *IEEE Trans. Inform. Theory*, **51**, 4203–4215.

Carl, B. (1981), Entropy numbers, $s$-numbers, and eigenvalue problems, *J. Func. Anal.*, **41**, 290–306.

Carlitz, L. and Uchiyama, S. (1957), Bounds for exponential sums, *Duke Math. J.*, **24**, 37–41.

Chazelle, B. (2000), *The Discrepancy Method*, (Cambridge: Cambridge University Press).

Chen, W. W. L. (1980), On irregularities of distribution, *Mathematika*, **27**, 153–170.

Chen, S. S., Donoho, D. L. and Saunders, M. A. (2001), Atomic decomposition by basis pursuit, *SIAM Rev.*, **43**, 129–159.

Cohen, A., Dahmen, W. and DeVore, R. (2007), A taste of compressed sensing, *Proc. SPIE*, Orlando, March 2007.

Cohen, A., Dahmen, W. and DeVore, R. (2009), Compressed sensing and $k$-term approximation, *J. Amer. Math. Soc.*, **22**, 211–231.

Cohen, A., DeVore, R. A. and Hochmuth, R. (2000), Restricted nonlinear approximation, *Construct. Approx.*, **16**, 85–113.

Coifman, R. R. and Wickerhauser, M. V. (1992), Entropy-based algorithms for best-basis selection, *IEEE Trans. Inform. Theory*, **38**, 713–718.

Conway, J. H., Hardin, R. H. and Sloane, N. J. A. (1996), Packing lines, planes, etc.: packing in Grassmannian spaces, *Experiment. Math.* **5**, 139–159.

Conway, J. H. and Sloane, N. J. A. (1998), *Sphere Packing, Lattices and Groups* (New York: Springer-Verlag).

Cordoba, A. and Fernandez, P. (1998), Convergence and divergence of decreasing rearranged Fourier series, *SIAM J. Math. Anal.*, **29**, 1129–1139.

van der Corput, J. G. (1935a), Verteilungsfunktionen. I, *Proc. Kon. Ned. Akad. v. Wetensch.*, **38**, 813–821.

van der Corput, J. G. (1935b), Verteilungsfunktionen. II, *Proc. Kon. Ned. Akad. v. Wetensch.*, **38**, 1058–1066.

Cucker, F. and Smale, S. (2001), On the mathematical foundations of learning, *Bull. AMS*, **39**, 1–49.

Dai, W. and Milenkovich, O. (2009), Subspace pursuit for compressive sensing signal reconstruction, *IEEE Trans. Inform. Theory*, **55**, 2230–2249.

Davenport, H. (1956), Note on irregularities of distribution, *Mathematika*, **3**, 131–135.

Davis, G., Mallat, S. and Avellaneda, M. (1997), Adaptive greedy approximations, *Construct. Approx.*, **13**, 57–98.

DeVore, R. A. (1998), Nonlinear approximation, *Acta Numerica*, **7**, 51–150.

DeVore, R. A. (2006), Optimal computation, *ICM Proc., Madrid*, **1**, 187–215.

DeVore, R. A. (2007), Deterministic constructions of compressed sensing matrices *J. Complex.*, **23**, 918–925.

DeVore, R. A., Jawerth, B. and Popov, V. (1992), Compression of wavelet decompositions, *Amer. J. Math.*, **114**, 737–785.

DeVore, R. A., Konyagin, S. V. and Temlyakov, V. N. (1998), Hyperbolic wavelet approximation, *Construct. Approx.*, **14**, 1–26.

DeVore, R. A. and Lorenz, G. G. (1993), *Constructive Approximation* (Berlin: Springer-Verlag).

DeVore, R. A., Petrova, G. and Temlyakov, V. N. (2003), Best basis selection for approximation in $L_p$, *Found. Comput. Math.*, **3**, 161–185.

DeVore, R. A. and Popov, V. A. (1988), *Interpolation Spaces and Non-linear Approximation*, Lecture Notes in Mathematics 1302 (Berlin: Springer), pp. 191–205.

DeVore, R. A. and Temlyakov, V. N. (1995), Nonlinear approximation by trigonometric sums, *J. Fourier Anal. Appl.*, **2**, 29–48.

DeVore, R. A. and Temlyakov, V. N. (1996), Some remarks on Greedy Algorithms, *Adv. Comp. Math.*, **5**, 173–187.

DeVore, R. A. and Temlyakov, V. N. (1997), Nonlinear approximation in finite-dimensional spaces, *J. Complexity*, **13**, 489–508.

DeVore, R. A., Kerkyacharian, G., Picard, D. and Temlyakov, V. (2004), On mathematical methods of learning, *IMI Preprints*, **10**, 1–24.

DeVore, R. A., Kerkyacharian, G., Picard, D. and Temlyakov, V. (2006), Mathematical methods for supervised learning, *Found. Comput. Math.*, **6**, 3–58.

Dilworth, S. J., Kalton, N. J. and Kutzarova, D. (2003), On the existence of almost greedy bases in Banach spaces, *Studia Math.*, **158**, 67–101.

Dilworth, S. J., Kutzarova, D. and Temlyakov, V. (2002), Convergence of some Greedy Algorithms in Banach spaces, *J. Fourier Anal. Applic.*, **8**, 489–505.

Dilworth, S. J., Kutzarova, D. and Wojtaszczyk, P. (2002), On approximate $\ell_1$ systems in Banach spaces, *J. Approx. Theory*, **114**, 214–241.

Dilworth, S. J., Kalton, N. J., Kutzarova, D. and Temlyakov, V. N. (2003), The Thresholding Greedy Algorithm, greedy bases, and duality, *Construct. Approx.*, **19**, 575–597.

Ding Dung (1985), Approximation of multivariate functions by means of harmonic analysis, Hab. Dissertation, Moscow, MGU.

Donahue, M., Gurvits, L., Darken, C. and Sontag, E. (1997), Rate of convex approximation in non-Hilbert spaces, *Construct. Approx.*, **13**, 187–220.

Donoho, D. L. (1993), Unconditional bases are optimal bases for data compression and for statistical estimation, *Appl. Comput. Harmon. Anal.*, **1**, 100–115.

Donoho, D. L. (1997), CART and Best-Ortho-Basis: a connection, *Ann. Stat.*, **25**, 1870–1911.

Donoho, D. L. (2001), Sparse components of images and optimal atomic decompositions, *Construct. Approx.*, **17**, 353–382.

Donoho, D. L. (2006), Compressed sensing, *IEEE Trans. Inform. Theory*, **52**, 1289–1306.

Donoho, D. L., Elad, M. and Temlyakov, V. N. (2006), Stable recovery of sparse overcomplete representations in the presence of noise, *IEEE Trans. Inf. Theory*, **52**, 6–18.

Donoho, D. L., Elad, M. and Temlyakov, V. N. (2007), On the Lebesgue type inequalities for greedy approximation, *J. Approx. Theory*, **147**, 185–195.

Donoho, D. L. and Johnstone, I. (1994), Ideal spatial adaptation via wavelet shrinkage, *Biometrica*, **81**, 425–455.

Dubinin, V. V. (1997), Greedy algorithms and applications, Ph.D. Thesis, University of South Carolina.

Dudley, R. M. (1967), The sizes of compact subsets of Hilbert space and continuity of Gaussian processes, *J. Func. Anal.*, **1**, 290–330.

Fefferman, C. and Stein, E. (1972), $H^p$ spaces of several variables, *Acta Math.*, **129**, 137–193.

Figiel, T., Johnson, W. B. and Schechtman, G. (1988), Factorization of natural embeddings of $\ell_p^n$ into $L_r$, I, *Studia Mathematica*, **89**, 79–103.

Frazier, M. and Jawerth, B. (1990), A discrete transform and decomposition of distribution spaces, *J. Funct. Anal.*, **93**, 34–170.

Friedman, J. H. and Stuetzle, W. (1981), Projection pursuit regression, *J. Amer. Stat. Assoc.*, **76**, 817–823.

Frolov, K. K. (1976), Upper bounds on the error of quadrature formulas on classes of functions, *Dokl. Akad. Nauk SSSR*, **231**, 818–821; English translation in *Sov. Math. Dokl.*, **17**.

Frolov, K. K. (1979), Quadrature formulas on classes of functions, Candidate dissertation, Vychisl. Tsentr Acad. Nauk SSSR, Moscow.

Frolov, K. K. (1980), An upper estimate of the discrepancy in the $L_p$-metric, $2 \le p < \infty$, *Dokl. Akad. Nauk SSSR*, **252**, 805–807; English translation in *Sov. Math. Dokl.*, **21**.

Galatenko, V. V. and Livshitz, E. D. (2003), On convergence of approximate weak greedy algorithms, *East J. Approx.*, **9**, 43–49.

Galatenko, V. V. and Livshitz, E. D. (2005), Generalized approximate weak greedy algorithms, *Math. Notes*, **78**, 170–184.

Ganichev, M. and Kalton, N. J. (2003), Convergence of the Weak Dual Greedy Algorithm in $L_p$-spaces, *J. Approx. Theory*, **124**, 89–95.

Garnaev, A. and Gluskin, E. (1984), The widths of a Euclidean ball, *Dokl. Akad. Nauk USSR*, **277**, 1048–1052; English translation in *Sov. Math. Dokl.*, **30**, 200–204.

Gilbert, A. C., Muthukrishnan, S. and Strauss, M. J. (2003), Approximation of functions over redundant dictionaries using coherence, in M. Farach-Cotton, ed., *Proceedings of the 14th Annual ACM-SIAM Symposium on Discrete Algorithms* (Philadelphia, PA: SIAM).

Gine, E. and Zinn, J. (1984), Some limit theorems for empirical processes, *Ann. Prob.*, **12**, 929–989.

Gluskin, E. D. (1986), An octahedron is poorly approximated by random subspaces, *Functsional. Anal. i Prilozhen.*, **20**, 14–20.

Gogyan , S. (2005), Greedy algorithm with regard to Haar subsystems, *East J. Approx.*, **11**, 221–236.

Gogyan, S. (2009), On convergence of Weak Thresholding Greedy Algorithm in $L^1(0, 1)$, *J. Approx. Theory*, **161**, 49–64.

Gribonval, R. and Nielsen, M. (2001a), Approximate Weak Greedy Algorithms, *Adv. Comput. Math.*, **14**, 361–368.

Gribonval, R. and Nielsen, M. (2001b), Some remarks on non-linear approximation with Schauder bases, *East J. Approx.* **7**, 267–285.

Györfy, L., Kohler, M., Krzyzak, A. and Walk, H. (2002), *A Distribution-Free Theory of Nonparametric Regression* (Berlin: Springer).

Habala, P., Hájek, P. and Zizler, V. (1996), *Introduction to Banach spaces [I]* (Karlovy: Matfyzpress).

Halász, G. (1981), On Roth's method in the theory of irregularities of points distributions, *Recent Prog. Analytic Number Theory*, **2**, 79–94.

Halton, J. H. and Zaremba, S. K. (1969), The extreme and $L_2$ discrepancies of some plane sets, *Monats. für Math.*, **73**, 316–328.

Heinrich, S., Novak, E., Wasilkowski, G. and Wozniakowski, H. (2001), The inverse of the star-discrepancy depends linearly on the dimension, *Acta Arithmetica*, **96**, 279–302.

Hitczenko, P. and Kwapien, S. (1994), On Rademacher series, *Prog. Prob.*, **35**, 31–36.

Höllig, K. (1980), Diameters of classes of smooth functions, in R. DeVore and K. Scherer, eds., *Quantitative Approximation* (New York: Academic Press), pp. 163–176.

Huber, P. J. (1985), Projection pursuit, *Ann. Stat.*, **13**, 435–475.

Jones, L. (1987), On a conjecture of Huber concerning the convergence of projection pursuit regression, *Ann. Stat.*, **15**, 880–882.

Jones, L. (1992), A simple lemma on greedy approximation in Hilbert space and convergence rates for projection pursuit regression and neural network training, *Ann. Stat.*, **20**, 608–613.

Kalton, N. J., Beck, N. T. and Roberts, J. W. (1984), *An F-space Sampler*, London Math. Soc. Lecture Notes 5 (Cambridge: Cambridge University Press).

Kamont, A. and Temlyakov, V. N. (2004), Greedy approximation and the multivariate Haar system, *Studia Mathematica*, **161**(3), 199–223.

Kashin, B. S. (1975), On widths of octahedrons, *Uspekhi Matem. Nauk*, **30**, 251–252.

Kashin, B. S. (1977a), Widths of certain finite-dimensional sets and classes of smooth functions, *Izv. Akad. Nauk SSSR, Ser. Mat.*, **41**, 334–351; English translation in *Math. USSR IZV.*, **11**.

Kashin, B. S. (1977b), On the coefficients of expansion of functions from a certain class with respect to complete systems, *Siberian J. Math.*, **18**, 122–131.

Kashin, B. S. (1980), On certain properties of the space of trigonometric polynomials with the uniform norm, *Trudy Mat. Inst. Steklov*, **145**, 111–116; English translation in *Proc. Steklov Inst. Math.* (1981), Issue 1.

Kashin, B. S. (1985), On approximation properties of complete orthonormal systems, *Trudy Mat. Inst. Steklov*, **172**, 187–191; English translation in *Proc. Steklov Inst. Math.*, **3**, 207–211.

Kashin, B. S. (2002), On lower estimates for $n$-term approximation in Hilbert spaces, in B. Bojanov, ed., *Approximation Theory: A Volume Dedicated to Blagovest Sendov*, (Sofia: DARBA), pp. 241–257.

Kashin, B. S. and Saakyan, A. A. (1989), *Orthogonal Series* (Providence, RI: American Mathematical Society).

Kashin, B. S. and Temlyakov, V. N. (1994), On best $m$-term approximations and the entropy of sets in the space $L^1$, *Math. Notes*, **56**, 1137–1157.

Kashin, B. S. and Temlyakov, V. N. (1995), Estimate of approximate characteristics for classes of functions with bounded mixed derivative, *Math. Notes*, **58**, 1340–1342.

Kashin, B. S. and Temlyakov, V. N. (2003), The volume estimates and their applications, *East J. Approx.*, **9**, 469–485.

Kashin, B. S. and Temlyakov, V. N. (2007), A remark on compressed sensing, *Math. Notes*, **82**, 748–755.

Kashin, B. S. and Temlyakov, V. N. (2008), On a norm and approximate characteristics of classes of multivariate functions, *J. Math. Sci.*, **155**, 57–80.

Kerkyacharian, G. and Picard, D. (2004), Entropy, universal coding, approximation, and bases properties, *Construct. Approx.*, **20**, 1–37.

Kerkyacharian, G., Picard, D. and Temlyakov, V. N. (2006), Some inequalities for the tensor product of greedy bases and weight-greedy bases, *East J. Approx.*, **12**, 103–118.

Konyagin, S. V. and Skopina, M. A. (2001), Comparison of the $L_1$-norms of total and truncated exponential sums, *Mat. Zametki*, **69**, 699–707.

Konyagin, S. V. and Temlyakov, V. N. (1999a), A remark on greedy approximation in Banach spaces, *East J. Approx.*, **5**, 365–379.

Konyagin, S. V. and Temlyakov, V. N. (1999b), Rate of convergence of Pure Greedy Algorithm, *East J. Approx.*, **5**, 493–499.

Konyagin, S. V. and Temlyakov, V. N. (2002), Greedy approximation with regard to bases and general minimal systems, *Serdica Math. J.*, **28**, 305–328.

Konyagin, S. V. and Temlyakov, V. N. (2003a), Convergence of greedy approximation I. General systems, *Studia Mathematica*, **159**(1), 143–160.

Konyagin, S. V. and Temlyakov, V. N. (2003b), Convergence of greedy approximation II. The trigonometric system, *Studia Mathematica*, **159**(2), 161–184.

Konyagin, S. V. and Temlyakov, V. N. (2004), Some error estimates in learning theory, in D. K. Dimitrov, G. Nikolov and R. Uluchev, eds. *Approximation Theory: A Volume Dedicated to Borislav Bojanov* (Sofia: Marin Drinov Acad. Publ. House), pp. 126–144.

Konyagin, S. V. and Temlyakov, V. N. (2005), Convergence of greedy approximation for the trigonometric system, *Analysis Mathematica*, **31**, 85–115.

Konyagin, S. V. and Temlyakov, V. N. (2007), The entropy in learning theory. Error estimates, *Construct. Approx.*, **25**, 1–27.

Körner, T. W. (1996), Divergence of decreasing rearranged Fourier series, *Ann. Math.*, **144**, 167–180.

Körner, T. W. (1999), Decreasing rearranged Fourier series, *J. Fourier Anal. Appl.*, **5**, 1–19.

Korobov, N. M. (1959), On the approximate computation of multiple integrals, *Dokl. Akad. Nauk SSSR*, **124**, 1207–1210.

Kuelbs, J. and Li, W. V. (1993), Metric entropy and the small ball problem for Gaussian measures, *J. Funct. Anal.*, **116**, 133–157.

Kuipers, L. and Niederreiter, H. (1974), *Uniform Distribution of Sequences* (New York: Wiley).

Lebesgue, H. (1909), Sur les intégrales singuliéres, *Ann. Fac. Sci. Univ. Toulouse (3)*, **1**, 25–117.

Lee, W. S., Bartlett, P. L. and Williamson, R. C. (1996), Efficient agnostic learning of neural networks with bounded fan-in, *IEEE Trans. Inf. Theory*, **42**(6), 2118–2132.

Lee, W. S., Bartlett, P. and Williamson, R. (1998), The importance of convexity in learning with square loss, *IEEE Trans. Inf. Theory*, **44**, 1974–1980.

Levenshtein, V. I. (1982), Bounds on the maximal cardinality of a code with bounded modules of the inner product, *Sov. Math. Dokl.*, **25**, 526–531.

Levenshtein, V. I. (1983), Bounds for packings of metric spaces and some of their applications, *Problemy Kibernetiki*, **40**, 43–110.

Lifshits, M. A. and Tsirelson, B. S. (1986), Small deviations of Gaussian fields, *Teor. Probab. Appl.*, **31**, 557–558.

Lindenstrauss, J. and Tzafriri, L. (1977), *Classical Banach Spaces I* (Berlin: Springer-Verlag).

Liu, E. and Temlyakov, V. (2010), Orthogonal super greedy algorithm and applications in compressed sensing, IMI Preprint, http://imi.cas.sc.edu/IMI/reports/2010, **10:01**, 1–21.

Livshitz, E. D. (2003), Convergence of greedy algorithms in Banach spaces, *Math. Notes*, **73**, 342–368.

Livshitz, E. D. (2006), On the recursive greedy algorithm, *Izv. RAN. Ser. Mat.*, **70**, 95–116.

Livshitz, E. D. (2007), Optimality of the greedy algorithm for some function classes, *Mat. Sb.*, **198**, 95–114.

Livshitz, E. D. (2009), On lower estimates of rate of convergence of greedy algorithms, *Izv. RAN, Ser. Matem.*, **73**, 125–144.

Livshitz, E. D. (2010), On the optimality of Orthogonal Greedy Algorithm for $M$-coherent dictionaries, Preprint, arXiv:1003.5349v1, 1–14.

Livshitz, E. D. and Temlyakov, V. N. (2001), On the convergence of Weak Greedy Algorithms, *Trudy. Mat. Inst. Steklov*, **232**, 236–247.

Livshitz, E. D. and Temlyakov, V. N. (2003), Two lower estimates in greedy approximation, *Construct. Approx.*, **19**, 509–523.

Lugosi, G. (2002), Pattern classification and learning theory, in *Principles of Nonparametric Learning* (Viena: Springer), pp. 5–62.

Lutoborski, A. and Temlyakov, V. N. (2003), Vector greedy algorithms, *J. Complexity*, **19**, 458–473.

Maiorov, V. E. (1978), On various widths of the class $H_p^r$ in the space $L_q$, *Izv. Akad. Nauk SSSR Ser. Mat.*, **42**, 773–788; English translation in *Math. USSR-Izv.* (1979), **13**.

Mallat, S. and Zhang, Z. (1993), Matching pursuit in a time-frequency dictionary, *IEEE Trans. Signal Proc.*, **41**, 3397–3415.

Matoušek, J. (1999), *Geometric Discrepancy* (New York: Springer-Verlag).

Mendelson, S. (2003), A few notes on statistical learning theory, in *Advanced Lectures in Machine Learning, LNCS, 2600* (Berlin: Springer), pp. 1–40.

Needell, D. and Vershynin, R. (2009), Uniform uncertainty principle and signal recovery via regularized orthogonal matching pursuit, *Found. Comp. Math.*, **9**, 317–334.

Nelson, J. L. and Temlyakov, V. N. (2008), On the size of incoherent systems, Preprint http://dsp.rice.edu/cs, 1–14.

Niederreiter, H., Tichy, R. F. and Turnwald, G. (1990), An inequality for differences of distribution functions *Arch. Math.*, **54**, 166–172.

Nielsen, M. (2009), Trigonometric quasi-greedy bases for $L_p(\mathbb{T}; w)$, *Rocky Mountain J. Math.*, **39**, 1267–1278.

Nikol'skii, S. N. (1975), *Approximation of Functions of Several Variables and Embedding Theorems* (Berlin: Springer-Verlag).

Novak, E. (1988), *Deterministic and Stochastic Error Bounds in Numerical Analysis*, Lecture Notes in Mathematics 1349 (Berlin: Springer-Verlag).

Novak, E. and Wozniakowski, H. (2001), When are integration and discrepancy tractable?, *FoCM Proc., London Math. Soc. Lecture Notes Series*, **284**, 211–266.

Oswald, P. (2001), Greedy algorithms and best $m$-term approximation with respect to biorthogonal systems, *J. Fourier Anal. Appl.*, **7**, 325–341.

Pajor, A. and Tomczak-Yaegermann, N. (1986), Subspaces of small codimension of finite-dimensional Banach spaces, *Proc. Amer. Math. Soc.*, **97**, 637–642.

Petrushev, P. (1988), *Direct and Converse Theorems for Spline and Rational Approximation and Besov Spaces*, Lecture Notes in Mathematics 1302 (Berlin: Springer-Verlag), pp. 363–377.

Pisier, G. (1989), *The Volume of Convex Bodies and Banach Space Geometry* (Cambridge: Cambridge University Press).

Poggio, T. and Smale, S. (2003), The mathematics of learning: dealing with data, *Not. Amer. Math. Soc.*, **50**, 537–544.

Pollard, D. (1984), *Convergence of Stochastic Processes* (New York: Springer-Verlag).

Roth, K. F. (1954), On irregularities of distribution, *Mathematica*, **1**, 73–79.

Roth, K. F. (1976), On irregularities of distribution. II, *Commun. Pure Appl. Math.*, **29**, 749–754.

Roth, K. F. (1979), On irregularities of distribution. III, *Acta Arith.*, **35**, 373–384.

Roth, K. F. (1980), On irregularities of distribution. IV, *Acta Arith.*, **37**, 67–75.

Schmidt, E. (1906), Zur Theorie der linearen und nichtlinearen Integralgleichungen. I, *Math. Annalen*, **63**, 433–476.

Schmidt, W. M. (1972), Irregularities of distribution. VII, *Acta Arith.*, **21**, 45–50.

Schmidt, W. M. (1977), Irregularities of distribution. X, *in Number Theory and Algebra* (New York: Academic Press), pp. 311–329.

Schütt, C. (1984), Entropy numbers of diagonal operators between symmetric Banach spaces, *J. Approx. Theory*, **40**, 121–128.

Sil'nichenko, A. V. (2004), Rate of convergence of greedy algorithms, *Mat. Zametki*, **76**, 628–632.

Skriganov, M. M. (1994), Constructions of uniform distributions in terms of geometry of numbers, *Algebra Anal.*, **6**, 200–230.

Smolyak, S. A. (1960), The $\epsilon$-entropy of the classes $E_s^{\alpha k}(B)$ and $W_s^\alpha(B)$ in the metric $L_2$, *Dokl. Akad. Nauk SSSR*, **131**, 30–33.

Sobolev, S. L. (1974), *Introduction to the Theory of Cubature Formulas* (Moscow: Nauka).

Stromberg, T. and Heath Jr., R. (2003), Grassmannian frames with applications to coding and communications, *Appl. Comput. Harm. Anal.*, **14**, 257–275.

Sudakov, V. N. (1971), Gaussian random processes and measures of solid angles in Hilbert spaces, *Sov. Math. Dokl.*, **12**, 412–415.

Talagrand, M. (1994), The small ball problem for the Brownian sheet, *Ann. Prob.*, **22**, 1331–1354.

Talagrand, M. (2005), *The Generic Chaining* (Berlin: Springer).

Temlyakov, V. N. (1988a), Approximation by elements of a finite dimensional subspace of functions from various Sobolev or Nikol'skii spaces, *Matem. Zametki*, **43**, 770–786; English translation in *Math. Notes*, **43**, 444–454.

Temlyakov, V. N. (1988b), On estimates of $\epsilon$-entropy and widths of classes of functions with bounded mixed derivative or difference, *Dokl. Akad. Nauk SSSR*, **301**, 288–291; English translation in *Sov. Math. Dokl.*, **38**, 84–87.

Temlyakov, V. N. (1989a), Approximation of functions with bounded mixed derivative, *Proc. Steklov Institute*, **1**.

Temlyakov, V. N. (1989b), Estimates of the asymptotic characteristics of classes of functions with bounded mixed derivative or difference, *Trudy Matem. Inst. Steklov*, **189**, 138–168; English translation in *Proc. Steklov Inst. Math.* (1990), **4**, 161–197.

Temlyakov, V. N. (1990), On a way of obtaining lower estimates for the errors of quadrature formulas, *Matem. Sbornik*, **181**, 1403–1413; English translation in *Math. USSR Sbornik*, **71**.

Temlyakov, V. N. (1993a), *Approximation of Periodic Functions* (New York: Nova Science Publishers, Inc.).

Temlyakov, V. N. (1993b), Bilinear approximation and related questions, *Proc. Steklov Inst. Math.*, **4**, 245–265.

Temlyakov, V. N. (1995a), An inequality for trigonometric polynomials and its application for estimating the entropy numbers, *J. Complexity*, **11**, 293–307.

Temlyakov, V. N. (1995b), Some inequalities for multivariate Haar polynomials, *East J. Approx.*, **1**, 61–72.

Temlyakov, V. N. (1998a), The best $m$-term approximation and greedy algorithms, *Adv. Comp. Math.*, **8**, 249–265.

Temlyakov, V. N. (1998b), Nonlinear $m$-term approximation with regard to the multivariate Haar system, *East J. Approx.*, **4**, 87–106.

Temlyakov, V. N. (1998c), Greedy algorithm and $m$-term trigonometric approximation, *Construct. Approx.*, **14**, 569–587.

Temlyakov, V. N. (1998d), Nonlinear Kolmogorov's widths, *Matem. Zametki*, **63**, 891–902.

Temlyakov, V. N. (1998e), On two problems in the multivariate approximation, *East J. Approx.*, **4**, 505–514.

Temlyakov, V. N. (1999), Greedy algorithms and $m$-term approximation with regard to redundant dictionaries, *J. Approx. Theory*, **98**, 117–145.

Temlyakov, V. N. (2000a), Greedy algorithms with regard to multivariate systems with special structure, *Construct. Approx.*, **16**, 399–425.

Temlyakov, V. N. (2000b), Weak greedy algorithms, *Adv. Comp. Math.*, **12**, 213–227.

Temlyakov, V. N. (2001a), Lecture notes on approximation theory, University of South Carolina, Chapter I, pp. 1–20.

Temlyakov, V. N. (2001b), Greedy algorithms in Banach spaces, *Adv. Comp. Math.*, **14**, 277–292.

Temlyakov, V. N. (2002a), Universal bases and greedy algorithms for anisotropic function classes, *Construct. Approx.*, **18**, 529–550.

Temlyakov, V. N. (2002b), A criterion for convergence of Weak Greedy Algorithms, *Adv. Comput. Math.*, **17**, 269–280.

Temlyakov, V. N. (2002c), Nonlinear approximation with regard to bases, in C. K. Chui, L. Schumaker and J. Stöckler, eds., *Approximation Theory X* (Nashville, TN: Vanderbilt University Press), pp. 373–402.

Temlyakov, V. N. (2003a), Nonlinear methods of approximation, *Found. Comput. Math.*, **3**, 33–107.

Temlyakov, V. N. (2003b), Cubature formulas, discrepancy, and nonlinear approximation, *J. Complexity*, **19**, 352–391.

Temlyakov, V. N. (2005a), Greedy type algorithms in Banach spaces and applications, *Construct. Approx.*, **21**, 257–292.

Temlyakov, V. N. (2005b), Greedy algorithms with restricted depth search, *Proc. Steklov Inst. Math.*, **248**, 255–267.

Temlyakov, V. N. (2006a), Greedy approximations, in *Foundations of Computational Mathematics*, Santander 2005, London Mathematical Society Lecture Notes Series, **331** (Cambridge: Cambridge University Press), pp. 371–394.

Temlyakov, V. N. (2006b), Greedy approximations with regard to bases, in *Proceedings of the International Congress of Mathematicians*, Vol. II (Zurich: European Mathematical Society), pp. 1479–1504.

Temlyakov, V. N. (2006c), Relaxation in greedy approximation, *IMI-Preprint*, **03**, 1–26; http://imi.cas.sc.edu/IMI/reports/2006/reports/0603.pdf

Temlyakov, V. N. (2006d), Optimal estimators in learning theory, in T. Figiel and A. Kamont, eds., *Approximation and Probability*, Banach Center Publications 72 (Warsaw: Warsaw University of Technology), pp. 341–366.

Temlyakov, V. N. (2006e), On universal estimators in learning theory, *Proc. Steklov Inst. Math.*, **255**, 244–259.

Temlyakov, V. N. (2007a), Greedy expansions in Banach spaces, *Adv. Comput. Math.*, **26**, 431–449.

Temlyakov, V. N. (2007b), Greedy algorithms with prescribed coefficients, *J. Fourier Anal. Appl.*, 71–86.

Temlyakov, V. N. (2008a), Approximation in learning theory, *Construct. Approx.*, **27**, 33–74.

Temlyakov, V. N. (2008b), Greedy approximation, *Acta Numerica*, **17**, 235–409.

Temlyakov, V. N. (2008c), Relaxation in greedy approximation, *Construct. Approx.*, **28**, 1–25.

Temlyakov, V. N. and Zheltov, P. (2010), On performance of greedy algorithms, *IMI-Preprint*, **10:02**, 1–13; http://imi.cas.sc.edu/IMI/reports/2010/reports/1002.pdf

Tropp, J. A. (2004), Greed is good: algorithmic results for sparse approximation, *IEEE Trans. Inform. Theory*, **50**, 2231–2242.

Tropp, J. A. and Gilbert, A. C. (2007), Signal recovery from random measurements via orthogonal matching pursuit, *IEEE Trans. Inform. Theory*, **52**, 4655–4666.

Van de Geer, S. (2000), *Empirical Process in M-Estimaton* (New York: Cambridge University Press).

Vapnik, V. (1998), *Statistical Learning Theory* (New York: John Wiley & Sons, Inc.).

Vilenkin, I. V. (1967), Plane nets of integration, *Zhur. Vychisl. Mat. i Mat. Fis.*, **7**, 189–196; English translation in *USSR Comp. Math. Math. Phys.*, **7**, 258–267.

Wojtaszczyk, P. (1997), On unconditional polynomial bases in $L_p$ and Bergman spaces, *Construct. Approx.*, **13**, 1–15.

Wojtaszczyk, P. (2000), Greedy algorithms for general systems, *J. Approx. Theory*, **107**, 293–314.

Wojtaszczyk, P. (2002a), Greedy type bases in Banach spaces, *Construct. Funct. Theory* (Sofia: DARBA), pp. 1–20.

Wojtaszczyk, P. (2002b), Existence of best $m$-term approximation, *Functiones et Approximatio*, **XXX**, 127–133.

Wojtaszczyk, P. (2006), Greediness of the Haar system in rearrangement invariant spaces, in T. Figiel and A. Kamont, eds., *Approximation and Probability*, Banach Center Publications 72 (Warsaw: Warsaw University of Technology), pp. 385–395.

Yang, Y. and Barron, A. (1999), Information-theoretic determination of minimax rates of convergence, *Ann. Stat.*, **27**, 1564-1599.

Zygmund, A. (1959), *Trigonometric Series* (Cambridge: Cambridge University Press).

# Index